Advances in Science, Technology & Innovation

IEREK Interdisciplinary Series for Sustainable Development

Advances in Science, Technology & Innovation (ASTI) is a series of peer-reviewed books based on important emerging research that redefines the current disciplinary boundaries in science, technology and innovation (STI) in order to develop integrated concepts for sustainable development. It not only discusses the progress made towards securing more resources, allocating smarter solutions, and rebalancing the relationship between nature and people, but also provides in-depth insights from comprehensive research that addresses the **17 sustainable development goals (SDGs)** as set out by the UN for 2030.

The series draws on the best research papers from various IEREK and other international conferences to promote the creation and development of viable solutions for a **sustainable future and a positive societal** transformation with the help of integrated and innovative science-based approaches. Including interdisciplinary contributions, it presents innovative approaches and highlights how they can best support both economic and sustainable development, through better use of data, more effective institutions, and global, local and individual action, for the welfare of all societies.

The series particularly features conceptual and empirical contributions from various interrelated fields of science, technology and innovation, with an emphasis on digital transformation, that focus on providing practical solutions to **ensure food, water and energy security to achieve the SDGs.** It also presents new case studies offering concrete examples of how to resolve sustainable urbanization and environmental issues in different regions of the world.

The series is intended for professionals in research and teaching, consultancies and industry, and government and international organizations. Published in collaboration with IEREK, the Springer ASTI series will acquaint readers with essential new studies in STI for sustainable development.

ASTI series has now been accepted for Scopus (September 2020). All content published in this series will start appearing on the Scopus site in early 2021.

Baojie He · Joni Jupesta · Giuseppe T. Cirella ·
Gloria Pignatta
Editors

Urban Climate Change Adaptation

A Nature-Based Approach

A culmination of selected research papers from the International Conference on Climate Change and Environmental Sustainability (CCES) 3rd Edition—Chongqing University, China 2023

Springer

Editors
Baojie He
School of Architecture and Urban Planning
Chongqing University
Chongqing, China

Joni Jupesta
Institute for the Advanced Study
of Sustainability (UNU-IAS)
Tokyo, Japan

Giuseppe T. Cirella
Faculty of Economics
University of Gdansk
Sopot, Poland

Gloria Pignatta
School of Built Environment
University of New South Wales
Sydney, NSW, Australia

ISSN 2522-8714 ISSN 2522-8722 (electronic)
Advances in Science, Technology & Innovation
IEREK Interdisciplinary Series for Sustainable Development
ISBN 978-3-031-65087-1 ISBN 978-3-031-65088-8 (eBook)
https://doi.org/10.1007/978-3-031-65088-8

This Springer imprint is published by the registered company Springer Nature Switzerland AG
The registered company address is: Gewerbestrasse 11, 6330 Cham, Switzerland

Scientific Committee

The editors warmly thank all the Reviewers who have contributed their authority to the double-blind review process, to ensure the quality of this publication.

Preface I

When we talk about the weather and climate with citizens on the streets, many of them said that "in recent years, the summer is much hotter than before, and people have to reschedule their activities to avoid heat stress. The extreme heat, drought, flooding, and wildfire are increasingly frequent and intense as well." Climate change is an indisputable fact and has been widely recognized by the general public. Beyond the common climate-related disasters people can feel, melting glaciers, rising sea levels, and declining biodiversity are also critical climate-related disasters.

Urgent actions are needed to address climate change and its associated consequences. From the mitigation side, international agencies and national governments have already shown strong wills and efforts to decarbonize the world and to decelerate climate change. For instance, in the building and construction sector, campaigns for a net zero-carbon built environment are thriving. A series of solutions such as regenerative design, onsite renewable energy, onsite food production, upcycled building materials, and onsite wastewater treatment have been proposed and piloted. However, the world is in a shallow decarbonization stage, and an era of deep decarbonization has not yet arrived. A net zero carbon target has not been implemented in new buildings and retrofitted buildings.

From the adaptation side, society has not prepared for the normal of climate-related disasters well. For example, in 2003, Western Europe was attacked by a heatwave event, with 73,000 additional deaths. However, by 2022, when the heatwave event came back to Western Europe, approximately 60,000 lost their lives. Moreover, many nations and cities have formulated and launched heat health action plans to instruct citizens to avoid heat-related threats. However, these plans are far from sound to deliver detailed information to the citizens, and such plans have not acted as an intervention to perform. More countries have not paid attention to climate-related impacts and leave the people to expose to climate change.

Beyond climate change, society is threatened by many other challenges such as urbanization, economic growth, and population increase. The conflicts among many countries in environmental, social, and economic dimensions bring uncertainties to the efforts and actions for climate change mitigation and adaptation, and sustainable development. The Sustainable Development Goals (SDGs) chart a bright future for human beings. However, goals are just goals, they call for actions of the society in all aspects.

During the Chinese Spring Festival of 2024, Chongqing, the city I reside in, had the warmest winter, while other cities in the north part of China were under the threat of blizzard. This prominent difference leads me to think about how to create a certainly comfortable, safe, and resilient built environment in an increasingly uncertain era.

Chongqing, China

Baojie He

Preface II

Owing to anthropogenic activities that enhances greenhouse gas (GHG) emissions and modifies natural landscapes, climates at the global, regional, and local scales have been ever-changing. Climate change under most conditions is harmful to not only human being but also ecosystems. Climate change is one of the main causes of environmental deterioration. Extreme heat, drought, wildfire, and flooding are common climate-related disasters with significant environmental impacts. Economic sustainability and social sustainability are also under threats due to their dependence on physical environments. Climate-induced environmental, economic, and social consequences are likely to be irreversible with the warming intensification.

How to address climate change and secure environmental, social, and economic sustainability is a major question of the 21st century. The solutions determine human survival and prosperity, and cities are the main field of implementing solutions. Mitigation efforts and actions have been widely performed in recent years along with the collaborative and extensive calls for decarbonizing the society in many fields such as transportation, industry, building, energy generation, agriculture, and land use and forestry. In comparison, preparation for urban climate change adaptation is far from sufficient, where people have to live with different levels of climate-related disasters according to different warming scenarios. There is also an urgent need to chart a comprehensive, tangible, and implementable adaptation system, with full respects to urban contexts.

The international conference of Climate Change and Environmental Sustainability mainly covers four topics such as Climate Change Mitigation and Adaptation, Sustainable Urban-Rural Planning and Design, Decarbonisation of Built Environment, and Climate-related Governance and Challenges. These four topics can contribute to (i) measurement of climate change and assessment of environmental, economic, and social impacts, (ii) understanding of the mechanisms for effective environmental planning and sustainable resource management, (iii) advancement of strategies and techniques for climate change mitigation and adaptation, (iv) improvement of the involvement and integration of different stakeholders across international, national, state, city, and community level for environmental sustainability, and (v) demonstration of pilot cases and exemplars that can help address climate change issues.

This conference, in alignment with the most current international and national advocations and actions such as the Paris Agreement, United Nations Sustainable Development Goals, and Carbon Neutrality, is a step toward empowering decision-makers and energy stakeholders to join forces and proactively address the challenges of climate change. By bringing together international researchers who are working on topics relevant to climate change and environmental sustainability to share their latest accomplishments and research findings, voices can be amplified and actions towards a more resilient, liveable, sustainable future taken. This conference is expected to serve as an important resource to inform people and provide them with a comprehensive understanding of the possible opportunities for environmental sustainability to address climate change.

Chongqing, China

Baojie He

Acknowledgements

The Conference Chair, Co-chairs, and Editors dedicate this book to all conference attendees from the industry, governmental organizations, and academia who have been actively engaged with conference preparation, keynote speech moderation, and presentation.

Many thanks go to the valuable keynote speeches by four outstanding researchers.

Professor Dr. Regiane Relva Romano
Head Smart Campus Facens, Brizal
"Smart campus FACENS: A prototype of a smart and sustainable city"

Hesham El Said Eissa
Senior Consultant/Climate Change and Environmental Expert at DCarbon, Egypt
"Climate finance considering the global economic crisis"

Dr. Joni Jupesta
United Nations University, Japan
"The digital transformation and green transformation towards sustainable development"

Dr. Soufiane Haddout
Ibn Tofail University, Morocco
"Assessment of the properties of environmental surface sediments and heavy metal pollution in the Moroccan estuary"

Special appreciation to the moderators of the conference.

Dr. Federica Rosso
RTD-BDICEA—Sapienza University of Rome, Italy

Professor Dr. Didem Evci Kiraz
Aydın Adnan Menderes University, Türkiye

Dr. Siti Aekbal Salleh
University of Technology MARA, Malaysia

Professor Alex Opoku
University of Sharjah, UAE

Professor Waqas Ahmed Mahar
Aror University, Pakistan

Siliang Yang
Leeds Beckett University, UK

We also appreciate the support from the projects of Research on the Establishment and Application of Green and Low Carbon Habitat Environment in Chongqing East Railway Station Area (Grant No. 2022-KJ-JY-20, CSTB2022TIAD-KPX0195).

About the Conference

Since the beginning of the 21st century, humans have faced unprecedented challenges. On the one hand, human society has benefited from rapid economic development and globalization. However, environmental deterioration, climate change, conflicts and wars, and inequality have become increasingly prominent. Sustainable development is an inevitable choice for Earth and humankind's future survival, and actions to build and improve sustainability capacity are urgent.

However, since the outbreak of coronavirus in early 2020, urban production and life have been severely affected, and communications and actions towards sustainable development have significantly decelerated. Therefore, the 1st version of the Climate Change and Environmental Sustainability (abbreviated as "CCES") Conference was formally launched by Prof. Baojie He at the School of Architecture and Urban Planning, Chongqing University, China, in collaboration with International Experts for Research Enrichment and Knowledge Exchange (IEREK), Egypt in April 2021.

The 1st CCES conference was held online on November 9–11, 2021, in order to overcome the spatial constraints related to pandemic prevention and control management measures, while bringing together researchers who work on topics relevant to climate change and environmental sustainability. The CCES conference is expected to be a step toward empowering decision-makers and energy stakeholders to join forces and proactively address the challenges of climate change so that actual progress can be achieved.

With the success of the 2nd CCES conference in 2022, the conference grew into a strong and robust body with participants from more than 60 countries and nations worldwide. The CCES conference is an emerging term for enthusiasm and vitality to actively address climate change and promote environmental sustainability. The CCES conference is also a representative platform in alignment with the Paris Agreement, United Nations Sustainable Development Goals, and Carbon Neutrality by advancing a broad scope of research topics on climate change, decarbonization, and urban sustainability. The CCES conference covers four main topics: (i) *Climate Change Mitigation and Adaptation*, (ii) *Sustainable Urban-Rural Planning and Design*, (iii) *Decarbonisation of Built Environment*, and (iv) *Climate-related Governance and Challenges*.

The 3rd CCES conference was successfully held online on September 17–18, 2023, with valuable leadership and efforts from conference chairs and scientific committee members.

Conference Chair

Prof. Baojie He
School of Architecture and Urban Planning, Chongqing University, China

Conference Co-chair
Prof. Ayyoob Sharifi
The IDEC Institute, Hiroshima University, Hiroshima, Japan
Prof. Junqing Tang
School of Urban Planning and Design, Peking University, Beijing, China
Prof. Zhijian Liu
Department of Power Engineering, North China Electric Power University, Beijing, China

Overall, we expect the conference to serve as an important resource for informing people and providing them with a comprehensive understanding of the possible opportunities for environmental sustainability to address climate change.

Contents

Actions for Urban Climate Change Adaptation

Bao-Jie He, Joni Jupesta, Giuseppe T. Cirella, and Gloria Pignatta

Abstract

Climate change has significantly threatened the health, safety, and prosperity of human beings. Decarbonization movement has been started and deepened with the implementation of carbon neutral goals at different scales to decelerate climate change and alleviate associated disasters. Despite efforts being made late and being slow to mitigate, climate-related impacts cannot disappear or reduce in the short term. If mitigation goals and targets are not achieved, living with climate-related disasters will become normal for human beings in the future. Typically, cities are the main human settlements along with urbanization, but they are under severe climate-related challenges due to unstainable anthropogenic interventions. Systematically charting urban climate change adaptation pathways, actions, plans, and policies are highly needed. To support this, the 3rd International Conference on Climate Change and Environmental Sustainability was held to widely collect experience and lessons tailoring to different contexts, to promote urban climate change adaptation implementation. A total of 21 conference papers were accepted for publication. The book contains four parts including "Climate Change Impacts, Mitigation, and Adaptation", "Sustainable Planning and Design of Urban and Rural Areas", "Exploring Techniques Towards Environmental Sustainability", and "Climate-related Governance and Challenges". The contents in this book are expected to provide a reference for a comprehensive, tangible, and implementable system of urban climate change adaptation with the case studies in various cities all over the world.

Keywords

Climate actions · Mitigation and adaptation · Environmental resilience · Pathways · Built environment

1 Climate Change and Urban Sustainability

The accumulation of greenhouse gases (GHGs) in the atmosphere has indisputably led to a discernible rise in Earth's temperature. In the long term, the average temperature increase has reached 1.1–1.2 °C in many regions, and the warming is accelerating (IPCC, 2023; NOAA, 2020). The increasing temperature and its complex linkage to atmospheric circulation lead to a series of climate disasters

B.-J. He (✉)
Centre for Climate–Resilient and Low–Carbon Cities, School of Architecture and Urban Planning, Key Laboratory of New Technology for Construction of Cities in Mountain Area, Ministry of Education, Chongqing University, Chongqing 400045, China
e-mail: baojie.he@cqu.edu.cn; baojie.unsw@gmail.com

Institute for Smart City of Chongqing University in Liyang, Chongqing University, Liyang 213300, Jiangsu, China

J. Jupesta
Institute for the Advanced Study of Sustainability, United Nations University, Shibuy 150-8925, Tokyo, Japan
e-mail: jupesta@unu.edu

Center for Transdisciplinary and Sustainability Science (CTSS), Bogor Agriculture University, Bogor 16127, Indonesia

G. T. Cirella
Faculty of Economics, University of Gdansk, 81-824 Sopot, Poland
e-mail: gt.cirella@ug.edu.pl

G. Pignatta
School of Built Environment, Faculty of Arts, Design and Architecture, University of New South Wales, Kensington, NSW 2052, Australia
e-mail: g.pignatta@unsw.edu.au

High Performance Architecture Research Cluster, Faculty of Arts, Design and Architecture, University of New South Wales, Kensington, NSW 2052, Australia

City Futures Research Centre, Faculty of Arts, Design and Architecture, University of New South Wales, Kensington, NSW 2052, Australia

B. He et al. (eds.), *Urban Climate Change Adaptation*, Advances in Science, Technology & Innovation,
https://doi.org/10.1007/978-3-031-65088-8_1

such as strong precipitation, drought, and hurricanes. From a short-term perspective, the warming is marked by the sharp increase of temperature, resulting in severe extreme heat or heatwave events. The occurrence of extreme heat events usually results in the significant increase of morbidity and mortality. In 2003, an estimated 72,000 Europeans tragically lost their lives due to heat-related causes (Robine et al., 2008), while around 61,000 Europeans were reported to have succumbed to similar circumstances in 2022 (Ballester et al., 2023). This alarming death toll is accompanied by several billion euros in damages (Cirella et al., 2014; Ward et al., 2020). Extreme heat events can exacerbate drought and wildfire conditions. For example, the 2019–2020 Australian bushfire, a megafire lasting over a year, resulted in the loss of approximately 480 million animals and inflicted damage on 24.3 million hectares (Binskin et al., 2020).

With rapid urbanization sweeping the globe, cities have cemented their status as the primary human settlements. As a response to this trend, green development has emerged as a vital solution for achieving sustainable progress. However, the diverse spatiotemporal patterns observed across different cities underscore the need for customized measures to not only facilitate urbanization but also to propel green development forward (Jiao et al., 2021). Urbanization, celebrated as a campaign for human prosperity, also serves as a catalyst for numerous mega challenges and gives rise to an array of emerging threats to humanity. As cities burgeon and populations concentrate, they strain infrastructure, exacerbate resource scarcity, and intensify environmental degradation. Furthermore, rapid urban expansion often leads to socioeconomic disparities, inadequate housing, and compromised public health. Thus, while urbanization offers opportunities for growth and development, it simultaneously presents complex challenges that demand careful consideration and strategic intervention.

First, cities are main field of carbon emission, having a higher carbon emission intensity than rural areas. Cities account for more than 70% of the global CO_2 emissions due to the burning of fossil fuels. Second, the resilience of cities to climate change is decreasing along with the replacement of natural features with artificial surfaces and infrastructures. Hard surfaces cannot deal with intense precipitation well and often result in waterlogging. For instance, the Typhoon Doksuri brought a record-breaking rainfall for 140 years to Beijing, while the small proportion of green–blue infrastructures in this compact and dense metropolitan cannot absorb the extremely large volume of rainwater, thereby turning roads into rivers. Third, the continuous modifications to living environments have also caused citywide climate change, such as heat island and pollution island effects. Even though, the local climate change can interact with regional and global climate change with prominent spillover effects.

The pervasive effects of climate change, spanning global, regional, and local scales, have starkly compromised the sustainability of human settlements worldwide. The very suitability of cities for future habitation is now called into question amid escalating environmental challenges. To confront the multifaceted impacts of climate change, a dual approach comprising both mitigation and adaptation strategies is imperative. Mitigation efforts are essential to curtail GHG emissions and mitigate its adverse effects on the environment and human health. Concurrently, adaptation measures are crucial to enhancing resilience, safeguarding infrastructure, and protecting vulnerable populations from the escalating threats posed by climate-related disasters. In mitigating climate change, initiatives aimed at transitioning to renewable energy sources, improving energy efficiency, and implementing sustainable land use practices are paramount (Feigin et al., 2023). Similarly, adaptation strategies encompass a spectrum of measures ranging from infrastructure upgrades and disaster preparedness initiatives to the implementation of nature-based solutions such as green infrastructure and ecosystem restoration. By addressing both mitigation and adaptation needs, communities can effectively mitigate the severity of climate change impacts while enhancing their resilience to future challenges.

Moreover, fostering international cooperation and mobilizing resources are critical components of any comprehensive climate action plan. Collaborative efforts are essential to scaling up climate resilience measures, transferring knowledge and technology, and ensuring equitable access to resources and support. Ultimately, confronting the complexities of climate change demands a concerted and multifaceted approach that integrates mitigation, adaptation, and international cooperation to safeguard the sustainability and livability of human settlements now and in the future. The decarbonization movements have already been started and deepened at the international, national, and local scales. Many nations have given priorities to carbon neutral goals by specifying decarbonization as the national strategy and launching diverse policies, plans, and actions for actual implementation. Even though, this cannot subside plans and actions for climate change adaptation because reversing climate to the pre-industrial level is impossible, at least by 2100.

Living with climate-related disasters and threats is anticipated to become increasingly commonplace throughout the remainder of the twenty-first century and potentially beyond, should mitigation efforts fail to meet their targets. The looming specter of climate change necessitates urgent and decisive action to implement comprehensive adaptation strategies that can effectively mitigate the impacts of extreme weather events, rising sea levels, and other climate-related challenges. Of particular concern are urban areas, which serve as the primary human settlements and are particularly vulnerable to the adverse effects of climate change. To address this pressing issue, there is an urgent need for the development and implementation of robust climate change adaptation plans tailored specifically to urban systems. These plans should encompass a range of measures aimed at enhancing the resilience of cities and urban communities to climate-related hazards. This includes investing in climate-resilient infrastructure, implementing nature-based solutions such as green spaces and flood management systems, and integrating climate considerations into urban planning and development processes. Furthermore, effective adaptation strategies must be accompanied by efforts to raise awareness, build capacity, and foster collaboration among stakeholders at all levels of government, as well as within the private sector and civil society. By taking proactive steps to adapt to the realities of a changing climate, cities can mitigate the potential for widespread disruption and damage, and ensure the long-term sustainability and livability of urban environments for future generations.

2 Pathways to Urban Climate Change Adaptation

Given the multifaceted nature of climate-related disasters and their impacts, which encompass a wide range of types and severity levels alongside intricate socioeconomic dynamics, bolstering the resilience of cities to climate change demands comprehensive approaches. Typically, the solutions and strategies should be disaster-specific and impact-specific. The difference and diversity of urban contexts make it more difficult to develop "one-size-for-all" plans and actions. There is a need to broadly collect experience and lessons tailoring to different contexts, to better promote the transformation towards urban climate change adaptation implementation. Accordingly, the authors of this Introduction chapter actively supported the 3rd International Conference on Climate Change and Environmental Sustainability held in Chongqing University (China) in collaboration with the Hiroshima University (Japan), Peking University (China), and North China Electric Power University (China). This conference was aimed at bringing together researchers and practitioners who are engaged in climate change and environmental sustainability to share thoughts, findings, and actions towards a more resilient, livable, sustainable future. After critically screening abstract submissions and blindly reviewing full submissions, a total of 21 conference papers were accepted and edited into the book titled "**Urban Climate Change Adaptation**". As shown in Fig. 1, these papers are divided into four parts, including "Climate Change Impacts, Mitigation, and Adaptation" (Part 1), "Sustainable Planning and Design of Urban and Rural Areas" (Part 2), "Exploring Techniques Towards Environmental Sustainability" (Part 3), and "Climate-related Governance and Challenges" (Part 4).

The first part of ***"Climate Change Impacts, Mitigation, and Adaptation"*** consists of five chapters, covering topics on urban energy footprint management, building environmental performance, nature-based urban regeneration,

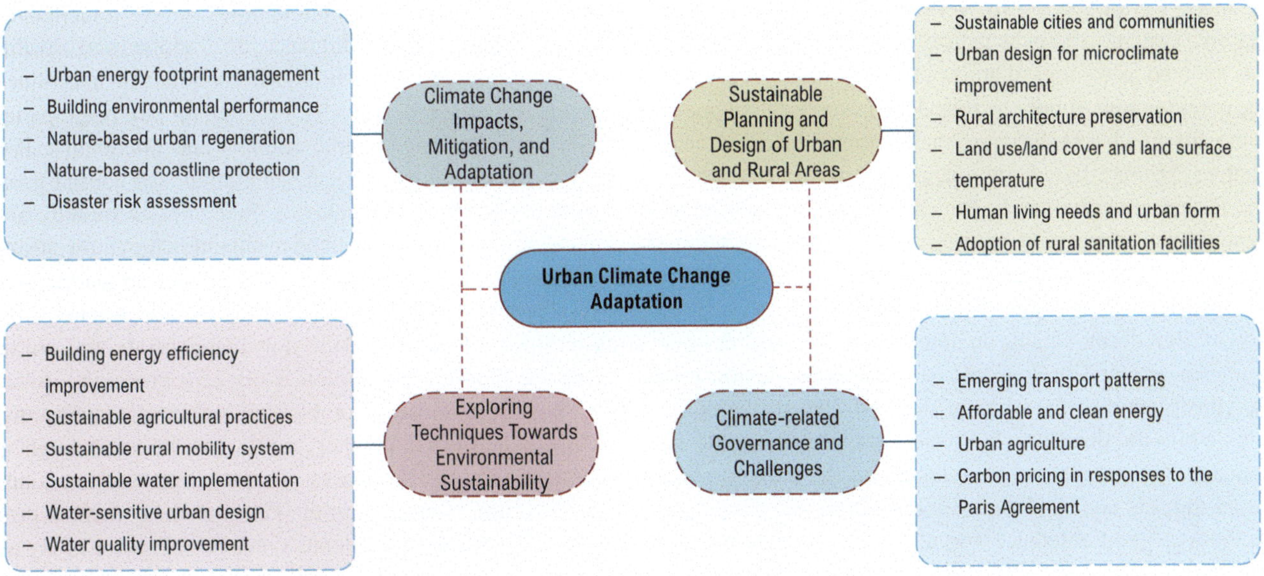

Fig. 1 Four dimensions and associated key topics under urban climate change adaptation

nature-based coastline protection, and disaster risk assessment. Chapter "Review of Urban Energy Management in India towards Sustainable Approaches" highlights a new sustainable paradigm of smart urban energy management for energy use optimization, cost reduction, and sustainability improvement. Based on the review of national documents and policies and questionnaire survey in Tamil Nadu, India, smart energy challenges in building, water management, water treatment, transport, and public service sectors were identified. Solutions to such challenges were provided in aspects of proper utilization of renewable energy resources, CO_2 emission reduction, optimization of existing energy management model, ocean thermal energy conversion, open access to the data, public awareness increase, off-grid energy production for public facilities, and integrated district-wise energy management. Chapter "Optimizing daylighting performance for educational buildings in Cairo by using parametric material system" presents the adoption of genetic algorithms to carry out parametric optimization for improving the daylighting performance of an educational building in Cairo, Egypt. The optimization of design variables such as window-to-wall ratio, window height, windowsill height, exterior shading device, and interior wall finishing materials was found to lead to a significant improvement in spatial daylight autonomy and energy efficiency, in alignment with the LEED v4.1 standard. Chapter "Vegetation as Adaptive Project: Regenerating Porta Romana Shaping a New Character" focuses on the role of nature-based solutions in structuring new design visions to regain the proximity between nature and architecture. It explores the regeneration of Porta Romana, a neglected railyard which was opened at the end of the nineteenth century, for the Olympic Village of the Winter Olympic games in 2026. In this project, buildings and public space were expected to work in synergy to set a new adaptive system, where the nature, especially for vegetation, was adopted as ecological and social infrastructure, presenting a unique opportunity to introduce new green fragments in the city to address the ongoing environmental emergency. Chapter "Nature-based approaches to protect the shoreline in urban environments: the case of Benidorm (Spain)" presents the adoption of nature-based solutions to address the coastline erosion caused by the lack of sediment supply, the increase in storms intensity and sea level rise. The city of Benidorm, Spain, an international tourist destination, was adopted as the case study area to demonstrate the construction of an artificial dune parallel to the coastline. While the dune was built with coarse sand inside for drainage, low vegetation was planted on the dune to avoid visual impacts and sand deposition. Overall, the case study provides a good reference for all other coastline protection projects all over the world. Chapter "Urban planning and climate adaptation in Peru. Analysis of the main methods of risk prevention and mitigation in cities" presents the methodological framework of climate-related risk prevention and mitigation based on the National Disaster Risk Management System in Peru. Typically, it mandatorily standardizes the assessment of disaster risks with the consideration of hazards and vulnerability, where the former calculates susceptibility and parameters, and the latter calculates exposure, fragility, and resilience in social, economic, and environmental dimensions. The mandatory documents are useful to characterize the territory and identify the vulnerable areas that cannot be mitigated and therefore cannot be urbanized.

The second part of ***"Sustainable Planning and Design of Urban and Rural Areas"*** includes six chapters covering topics of sustainable cities and communities, urban design for microclimate improvement, the preservation of rural architecture, the relationship between land use/land cover and land surface temperature, the relationship between human living needs and urban form, and the adoption and upgrade of rural sanitation facilities. Chapter "Evaluating the Implementation of Sustainable Development Goal 11: A Focus on Sustainable Cities and Communities in Ghana" demonstrates Ghana's ongoing efforts to achieve Sustainable Cities and Communities, for creating inclusive, safe, resilient, and sustainable cities and communities based on the document analysis and secondary data. The results indicate that Ghana, especially for the major cities of Accra and Kumasi, most targets under SDG 11 are unlikely to be achieved by 2030. Rather than reduction, the percentage of people living in slums saw a 3.24% increase since 2016, associated with the rapid population growth and urban–rural migration. Therefore, the government and industry should collaborate, prepare, and implement tangible actions for creating sustainable and environmentally-friendly cities and communities, even after 2030. Chapter "Exploring the Influence of Architectural Spatial Configurations on Residential Microclimates" underlines the role of urban design, by optimizing building spatial layout, in improving thermal comfort and air quality. The study presents the numerical simulation and comparison of summertime microclimates of three sites in humid subtropical Kumamoto, Japan. The results indicate the relationship between architectural configuration and wind direction have a pronounced effect on temperature and human thermal comfort. Meanwhile, the downwind residential areas, where the building orientation is aligned with wind direction, have the most favorable microclimatic conditions. Architects and urban planners must consider wind patterns when designing urban spaces. Chapter "Research on Zhe-style Dwelling Houses from the Perspective of Architecture Culture -Taking the Traditional Village in Xianju County as an Example" focuses on the revitalization of rural architecture which is the combination of local natural scenery and customs. Based on the landscape gene identification, this study extracts nine distinctive cultural elements of Zhe-style dwelling houses in China, in terms of house layout, facade composition, gable form, street scale, wind and rain corridors, roof design, building materials, overall color scheme, and decorative carvings. A good understanding of these elements is conducive to promoting their integration with local natural environments to reduce environmental damage and improve the adaptability and sustainability of residential buildings. Chapter "Exploring the Interplay between Urban Landscape and Land Surface Temperature: Case Study of South Twenty-Four Parganas" presents empirical results of how land use/land cover affects land surface temperature which is an indicator presenting the condition of urban thermal environments. The study is performed in a rapidly urbanizing Indian city of South Twenty-Four Parganas, showing that built-up features promoted urban temperature increase, while waterbody and vegetation features displayed varying degrees of cooling effects on urban temperature. The results prioritize the preservation and expansion of green spaces for sustainable urban growth and the well-being of communities. Chapter "Adapting Modern Cities to the Evolved Needs of People" explores the relation between hierarchy of needs of people living in gathered settlements from ancient ages to present along with their urban forms. The needs have been ever developing and changing as humans keep civilizing and tending to involve and implement technology. Typically, smart cities are now one of the emerging and ongoing trends of urbanism, along with the huge revolution in technology. Nevertheless, detailed description of humans needs can appear very clearly on the cities form and human settlements type such as spatial form, components, and land uses. Chapter "A Study on Participatory Attitudes and Behaviors of Rural Residents in Sanitation Facilities" focuses on attitude and willingness of village residents to improve sanitation and toilet facilities, given the high level of significance of rural sanitation facilities in curbing infectious diseases. The review results indicate that age, education, and income can significantly affect rural resident's attitudes of sanitation facilities to different degrees. Compared with sanitation facilities, rural residents are more willing to improve toilets, especially when the original toilets are not in good condition, and rural residents generally refuse to pay for sanitation improvement. Accordingly, some suggestions are given to enhance the participation in sanitation improvements.

The third part of ***Exploring Techniques Towards Environmental Sustainability*** includes six chapters covering building energy efficiency improvement strategies and performance, sustainable agricultural practices, sustainable rural mobility system, sustainable water implementation assessment, water-sensitive urban design, and water quality improvement. Chapter "Energy Performance Improvement in Lebanese Family House" aims to learn building energy efficient strategies from Cyprus's existing buildings to improve the energy performance of Lebanese terrace family house, given the similarity in climate, energy sources, and economic crises of these two countries. The impact of low-e double glazing and wall insulation were chosen as the major techniques to control the thermal transmittance of building materials for energy efficiency. The empirical results indicate that the adoption of these two techniques can improve energy efficiency by 23%. The results are expected to be a starting point to specify rules for national standards and energy codes of existing and new residential building. Chapter "Understanding Farmers Utilization of Rice and Maize Residue in the Coastal Region of Togo: A Survey-Based Study" investigates the adoption of rice and corn residues by local farmers, in the lens of sustainable agricultural practices and agricultural leftover carbon emission reduction. The survey is conducted in coastline of Togo, with the target groups of maize growers in Ave, Bas Mono, Lacs, and Yoto, and rice producers in Zio, Yoto, and Bas-Mono. The results show that maize straw, maize stalks, rice paddy straw, and rice husks are mainly burned, while rice bran is mainly used as animal feed. The study also reveals the key factors that constrain farmers' sustainable behaviors, such as time constraints like pressing schedules and labor demands, the lack of access to advanced machinery or appropriate equipment, and economic constraints like the lack of cost-effective and accessible means of residue disposal. Chapter "Planning and Designing a Sustainable Mobility System in Rural Areas" presents a systematic analysis of existing literature of sustainable mobility in rural European regions, in order to identify solutions for the sustainable transformation of transport system in marginal territories. The analysis of the sustainable mobility concept is performed in the definition (i.e., underscoring social, environmental, and economic goals and integrating them all) and criteria (e.g., low-emission, energy-efficient, and environmentally friendly). Sustainable strategies for mobility in rural areas are considered in aspects of all-inclusive European vision, climate neutral EU mobility, Italian strategy for sustainable mobility, safe transportation system, and budget friendly mobility modes. The study typically points out that user engagement is crucial for effective solution design, and existing public transport operators play an important role in addressing rural transportation challenges. Chapter "Assessing the Progress of Clean Water and Sanitation Goal (SDG 6) Implementation in Ghana" aims to examine the implementation progress of clean water and sanitation in Ghana, and estimate if the nation can achieve all targets of SDG 6 by 2030. The results show that Ghana has made credible progress, while the progress was only prominent in terms of Target 6.1 for universal and equitable access to affordable and safe drinking water. The progress in sanitation is not guaranteed, and extensive efforts are required from a wide range of stakeholders in the water and sanitation sector to deliver adequate and high-quality water and sanitation services to urban and rural communities. Overall, this study provides a good reference for many other nations to perform implementation progress assessment, identify the critical challenges, and determine the future actions for SDGs implementation. Chapter "Design Factors towards Water Retention Ability of Water-Sensitive Urban Design (WSUD) in the context of Tropical and Sub-tropical Climates: An Exploratory Literature Review" investigates what kinds of design factors can influence the water retention ability and analyzes how these factors affect retention ability in tropical and sub-tropical context, based on the literature review of English papers published in 2000–2023. This study identifies the principle of water sensitive urban design, factors in landscape design, and the differences between planting designs in rain garden landscape design and management. The factors are divided into hardscape (e.g., ponding depth, soil mixed/media, and infiltration layer) and softscape (i.e. planting). The results can improve the mitigation capacity of stormwater management in Southeast Asia, and can provide a reference for global countries. Chapter "Using Phytoremediation to Reduce Algal Blooms in Campus Green–blue Space: A Research-through-Design Study" presents a case study in Ang-Kaew University, Chiang Mai, Thailand to analyze the adoption of phytoremediation to mitigate algal blooms within campus green–blue spaces. Based on the landscape architectural approach, the research group implemented the design-planting schemes, assessed plant survivability, and tested water quality. The results suggested that planting depth is important for survival associated with the plants' capability of resisting wind and water currents. The water quality was improved in biochemical oxygen demand and phosphorus, while the nitrogen values were higher. Overall, this study presents a good example of algal bloom control through research-through-design methodology.

The fourth part of ***"Climate-related Governance and Challenges"*** includes four chapters covering emerging transport patterns, affordable and clean energy, urban agriculture, and carbon pricing in responses to the Paris Agreement. Chapter "Six Hidden Changes in Mobility during COVID-19 and Their Impact on Future Transport Technologies and Research Direction" aims to reveal key dimensions and changing patterns of transport, as well as implications and future research directions for maintaining a resilient and equitable transport system. This study identifies transport system changes in terms of transport sector employment, essential workers, personal vehicle ownership, telecommuting, ridesharing, and gender equality including ethnic minorities. In the post-pandemic era, efforts such as effective policy intervention, policy initiatives and training, and microtransit based on electrification, shared mobility, and automation are needed to adapt to the mobility change. Research priorities should be given to the integration of new technologies and urban ecosystem, travelers' willingness to ride with strangers, and on-demand automated mobility. Chapter "An Evaluation of Sustainable Development Goal Seven (Affordable and Clean Energy) in Ghana" presents a critical assessment of the implementation progress of affordable and clean energy in Ghana. The results based on document analysis and secondary data suggest only Target 7.1 which aims at ensuring all-inclusive access to affordable, reliable, and modern energy services has achieved prominent progress. However, other targets have not been implemented well. Typically, due to the high purchase and installation costs, the renewable energy technologies and systems are still facing significant challenges. The results make people reflect on the progress and efforts towards the Sustainable Development Goals. Therefore, progress monitoring and assessment should be included into SDGs implementation to identify the barriers and challenges in each stage. Chapter "Emirati Local Communities as Balanced Ecosystems: Pilot Study for Individualistic Residents-Led Urban Agriculture Within Housing Clusters" aims to understand the magnitude and aspects of individualistic residents-led urban agriculture activities through an exploratory case study in Al Ain, United Arab Emirates. Based on a series of qualitative methods such as map analysis, field observations, and semi-structured interviews in the open spaces of five housing clusters, this study points out that the urban agriculture activities are mainly driven by residents' desire to grow eatable crops. Meanwhile, the underused and undeveloped semi-public open spaces give residents opportunities to cultivate, and the residents usually cultivate the semi-public areas attached to the back and side fences of their housing plots. Overall, the individualistic residents-led urban agriculture activities are conducive to achieving the SDG 15 of Life on Land. Chapter "The Carbon Pricing Policy in Indonesia towards Paris Agreement Implementation" highlights the role of carbon pricing in accelerating the Paris Agreement goals. In the context of Indonesia, this study identifies the impact of carbon pricing policy in three sectors, such as the potential of Forestry and Land Use (FOLU) sector for carbon trade, the deployment of renewable and clean energy, and waste management (e.g., plastic waste). Accordingly, Indonesia may navigate the carbon markets by leveraging Nature-Based Solutions and clean energy initiatives, including carbon capture, utilization and storage (CCUS) and renewable energy sources. Even though, it is essential to balance the potential trade-offs like those between food and energy security in biofuel production.

3 Concluding Remarks

As we progress further into the twenty-first century, the reality of living with climate-related disasters and threats becomes increasingly apparent, exacerbated by delayed responses and actions to address climate change. The pressing need to bolster resilience against climate change is especially critical in urban areas, as they function as the primary hubs of human habitation and are disproportionately affected by environmental impacts. To effectively navigate the challenges posed by climate change, it is imperative to develop and implement comprehensive adaptation strategies tailored to the unique needs and vulnerabilities of diverse cities. Charting a path toward resilience requires a holistic approach that considers the interconnectedness of environmental, social, and economic factors. This entails not only addressing physical vulnerabilities to climate hazards such as floods, heatwaves, and storms but also tackling underlying socioeconomic disparities and governance challenges that exacerbate vulnerability. Moreover, efforts to enhance resilience must be inclusive, engaging diverse stakeholders and communities in the planning and implementation process to ensure that adaptation measures are equitable and effective.

A key aspect of building resilience involves investing in climate-resilient infrastructure, adopting nature-based solutions, and integrating climate considerations into urban planning and development policies. Additionally, enhancing adaptive capacity through education, capacity building, and knowledge sharing is essential to empower communities to anticipate, respond to, and recover from climate-related disasters. By prioritizing resilience-building efforts and taking proactive measures to address the complex challenges posed by climate change, cities can mitigate the impacts of disasters, protect vulnerable populations, and ensure a more sustainable and resilient future for all. In an effort to address the pressing challenges posed by climate change, the 3rd International Conference on Climate Change and Environmental Sustainability convened in 2023, bringing together experts, researchers, policymakers, and practitioners from around the world. The invaluable insights and findings presented during this pivotal event have been compiled into this edition of the book. With a keen focus on critical themes such as "Climate Change Impacts, Mitigation, and Adaptation," "Sustainable Planning and Design of Urban and Rural Areas," "Exploring Techniques Towards Environmental Sustainability," and "Climate-related Governance and Challenges," the contents of this book serve as a comprehensive resource for stakeholders grappling with the complexities of climate change adaptation.

By delving into these diverse topics, the book endeavors to offer a nuanced understanding of the multifaceted nature of climate change and its implications for urban and rural environments alike. Through case studies, analyses, and best practices, readers are presented with a wealth of lessons and experiences aimed at informing and guiding city-specific climate change adaptation efforts. Moreover, the book serves as a catalyst for interdisciplinary dialogue and collaboration, fostering innovative approaches and strategies to build resilience and promote sustainability in the face of an uncertain climate future.

References

Ballester, J., Quijal-Zamorano, M., Méndez Turrubiates, R. F., Pegenaute, F., Herrmann, F. R., Robine, J. M., Basagaña, X., Tonne, C., Antó, J. M., & Achebak, H. (2023). Heat-related mortality in Europe during the summer of 2022. *Nature Medicine, 29*, 1857–1866. https://doi.org/10.1038/s41591-023-02419-z

Binskin, M., Bennett, A., & Macintosh, A. (2020, October 28). Royal commission into natural disaster arrangements. Commonwealth of Australia (p. 115). Australian Capital Territory.

Cirella, G. T., Semenzin, E., Critto, A., & Marcomini, A. (2014). Natural hazard risk assessment and management methodologies review: Europe. In I. Linkov (Ed.), *Sustainable cities and military installations* (pp. 329–358). Springer Netherlands. https://doi.org/10.1007/978-94-007-7161-1_16

Feigin, S. V., Wiebers, D. O., Lueddeke, G., Morand, S., Lee, K., Knight, A., Brainin, M., Feigin, V. L., Whitfort, A., Marcum, J., Shackelford, T. K., Skerratt, L. F., & Winkler, A. S. (2023). Proposed solutions to anthropogenic climate change: A systematic literature review and a new way forward. *Heliyon, 9*(10), e20544. https://doi.org/10.1016/j.heliyon.2023.e20544

IPCC. (2023). Summary for policymakers. In [Core Writing Team, H. Lee, & J. Romero (Eds.)], *Climate change 2023: Synthesis report. contribution of working groups I, II and III to the sixth assessment report of the intergovernmental panel on climate change* (pp. 1–34). IPCC, Geneva, Switzerland. https://doi.org/10.59327/IPCC/AR6-9789291691647.001

Jiao, G., Lu, L., Chen, G., Huang, Z., Cirella, G. T., & Yang, X. (2021). Spatiotemporal characteristics and influencing factors of tourism revenue in the Yangtze River delta urban agglomeration region during 2001–2019. *Sustainability, 13*(7), e3658. https://doi.org/10.3390/su13073658

NOAA. (2020). Annual 2020 global climate report. national centers for environmental information. National Oxygen and Acetylene Association. Available at: https://www.ncei.noaa.gov/access/monitoring/monthly-report/global/202013

Robine, J. M., Cheung, S. L. K., Le Roy, S., Van Oyen, H., Griffiths, C., Michel, J. P., & Herrmann, F. R. (2008). Death toll exceeded 70,000 in Europe during the summer of 2003. *Comptes Rendus Biologies, 331*(2), 171–178. https://doi.org/10.1016/j.crvi.2007.12.001

Ward, P. J., Blauhut, V., Bloemendaal, N., Daniell, J. E., de Ruiter, M. C., Duncan, M. J., Emberson, R., Jenkins, S. F., Kirschbaum, D., Kunz, M., Mohr, S., Muis, S., Riddell, G. A., Schäfer, A., Stanley, T., Veldkamp, T. I. E., & Winsemius, H. C. (2020). Review article: Natural hazard risk assessments at the global scale. *Natural Hazards and Earth System Sciences, 20*(4), 1069–1096. https://doi.org/10.5194/nhess-20-1069-2020

Urban Environmental Sustainability

Nature-Based Approaches to Protect the Shoreline in Urban Environments: The Case of Benidorm (Spain)

Ignacio Toledo, José Ignacio Pagán, Isabel López, and Luis Aragonés

Abstract

Erosion is becoming one of the biggest problems affecting coastlines around the world. The lack of sediment supply, the increase in storms intensity and sea level rise are behind the continuous damage that affects the coast, especially in urban areas. In this study, a review of solutions has been carried out in various parts of the world to mitigate the unwanted effects of erosion in urban environments, with special emphasis on the city of Benidorm (Spain). Benidorm is an international tourist destination, whose shoreline is no exception to the effects of climate change, with a greater storms intensity that causes continuous flooding, both on the beach berm and on the promenade. For this reason, the construction of an artificial dune parallel to the coastline is proposed in order to contain the strength of the waves and the storm surge. The dune will be made up of a coarser sand size inside, favoring the drainage of the dune and being able to mix with the sand from the beach once it is destroyed by the strong waves, having to be rebuilt after the passage of the storm. Low vegetation will be planted on this dune to prevent part of the sand from ending up deposited on the promenade, also minimizing the visual impact. The solution proposed here to contain the flood level can be exported to other parts of the world, forming part of the catalog of nature-based solutions that the coastal engineer must have for proper management of the shoreline.

Keywords

Sand dunes · Vegetated dunes · Natural-based solutions · Storms · Spain

1 Introduction and Literature Review

The growing problem of coastal erosion around the world may be due to factors related to anthropogenic pressure or those related to global climate change (Pagán et al., 2016; Toimil et al., 2020). Within this last group, we can find changes in wave direction, in intensity and frequency of maritime storms, which are an aspect that influences the coastline evolution (Toledo et al., 2022; Tsimplis et al., 2005). In addition, the increase in sea temperature is directly correlated with the potential of cyclones (Mei et al., 2015; Vecchi & Soden, 2007). Similarly, the deepening of the storms and the increase in wind speed result in a rise in storm surge and, therefore, in the risk of flooding in coastal areas (Jisan et al., 2018; Vousdoukas et al., 2016).

Two thirds of the largest cities are located on the coast (Crooks & Turner, 1999), where almost 70% of the world's population lives (Brown et al., 2012), so these changes are more important when they occur in populated areas. Public administrations are forced to provide coastal cities with elements of protection against erosion. Until a few decades ago, 'hard' defense solutions have been the most used in coastal protection (Edward et al., 2006; Schoonees et al., 2019). However, these engineered solutions are becoming less and less ecologically and economically sustainable. Therefore, it is necessary to broaden the range of coastal protection actions to other 'soft' or more natural solutions (Morris et al., 2018).

Nature-based Solutions (NbS) represent actions aimed at mitigation and adaptation to climate change that have an ecosystem approach based on the scientific understanding of the carbon cycle and the sustainability of natural resources (WWF, 2021). In coastal areas, multiple

I. Toledo (✉) · J. I. Pagán · I. López · L. Aragonés
University of Alicante, Carretera Sant Vicent del Raspeig s/n, 03690 Alicante, Spain
e-mail: nacho.toledo@ua.es

J. I. Pagán
e-mail: jipagan@ua.es

I. López
e-mail: lopez.ubeda@ua.es
L. Aragonés e-mail: laragones@ua.es

B. He et al. (eds.), *Urban Climate Change Adaptation*, Advances in Science, Technology & Innovation,
https://doi.org/10.1007/978-3-031-65088-8_2

alternatives have been proposed to be developed in urban environments, from protection to management and restoration of green infrastructures (Kabisch et al., 2017). One example is the innovative sand breakwater built to protect the port of Lekki in Nigeria (van der Spek et al., 2020). This solution proposes to counteract the retreat of the coastline downstream of the port by feeding with sand that comes off the breakwater. Other solutions propose using plant species, such as *Posidonia Oceanica*, to combat erosion, benefiting of the wave attenuation property (Maiolo et al., 2020). On eroded muddy coasts, mangrove restoration or bamboo fencing is also a sustainable option (Dao et al., 2020; Verhagen, 2019). Other times, a complete project that combines the restoration of the riparian vegetation is more appropriate, as well as the construction of an artificial dune, stabilized with native vegetation and a breakwater (Rivillas-Ospina et al., 2020).

The beaches of the Spanish city of Benidorm, an international benchmark in the tourism sector, are not immune to the consequences of climate change. The storms that have occurred in recent years have endangered the infrastructures located on the seafront. Despite this, no solutions have yet been proposed to counteract these effects. The objective of this research is to define which is the most appropriate Nature-based Solution to problems related to erosion and the continuous flooding that occurs, both on the surface of the beach and on the promenade.

2 Methodology

2.1 Study Area

Benidorm city (Spain) is located on the Mediterranean coast, specifically in the southeast sector of the Iberian Peninsula (Fig. 2a), within the Province of Alicante (Fig. 2b). This city symbolizes a highly relevant tourist destination both for the Valencian Community and for the rest of Spain (Femenia-Serra & Ivars-Baidal, 2021). In the 60s Benidorm city underwent a significant urban transformation. It went from being a small fishing town to a kind of resort produced during the mass tourism boom in the Mediterranean region (Nolasco-Cirugeda et al., 2020). The urban expansion developed during the following years caused important changes in land use (Fig. 1).

Part of the city's tourist success is due to its two main beaches: (i) Poniente beach, with a length of 3 km, an average width of 68 m and an average size (D_{50}) of 0.28 mm and (ii) Levante beach, with a length of 2.3 km, an average width of 54 m and a D_{50} of 0.25 mm (Toledo et al., 2022). The orientation of both beaches to the south and the location of Sierra Helada massif to the east mean that the storm impacts from the east (the most frequent direction in the Spanish Mediterranean) is less than in other parts of eastern Spain (Amores et al., 2020). In addition to Sierra Helada, waves in the area are conditioned by Cape Gamell to the west, and by Benidorm Island to the south (Fig. 2c). *P. oceanica* abounds on the seabed, a marine plant species that forms extensive meadows on sandy bottoms.

Benidorm has a typical Mediterranean climate. This climate stands out for a regime of mild temperatures in winter (5 °C on average for minimum temperatures) and high temperatures in summer (30 °C on average for maximum temperatures), and annual precipitation in the range of 400–450 mm. Despite the low rainfall, it tends to be concentrated on a few days a year, especially in autumn. During this season, one or several meteorological elements appear simultaneously, which can cause important damage to the infrastructures located on the coastline (Bevacqua et al., 2017).

As in other regions of the Mediterranean Sea, Benidorm area is in a microtidal zone, that is, the oscillations produced by variations in atmospheric pressure are greater than the tides themselves. Storm Surges can reach values

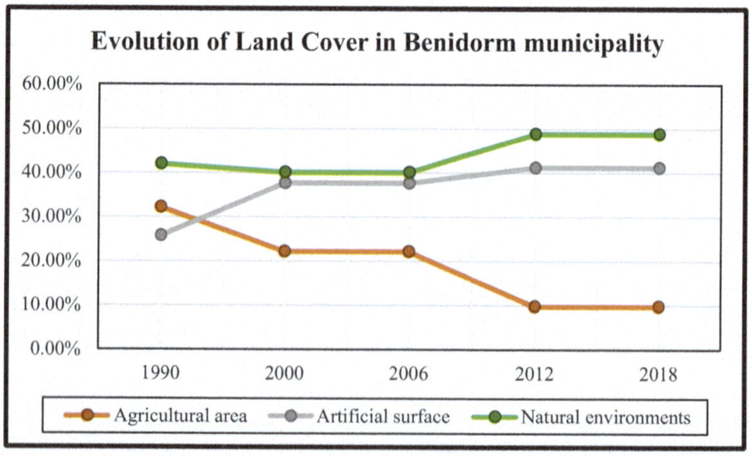

Fig. 1 Temporal evolution of CORINE Land Cover (CLC) in Benidorm municipality. *Source* Author

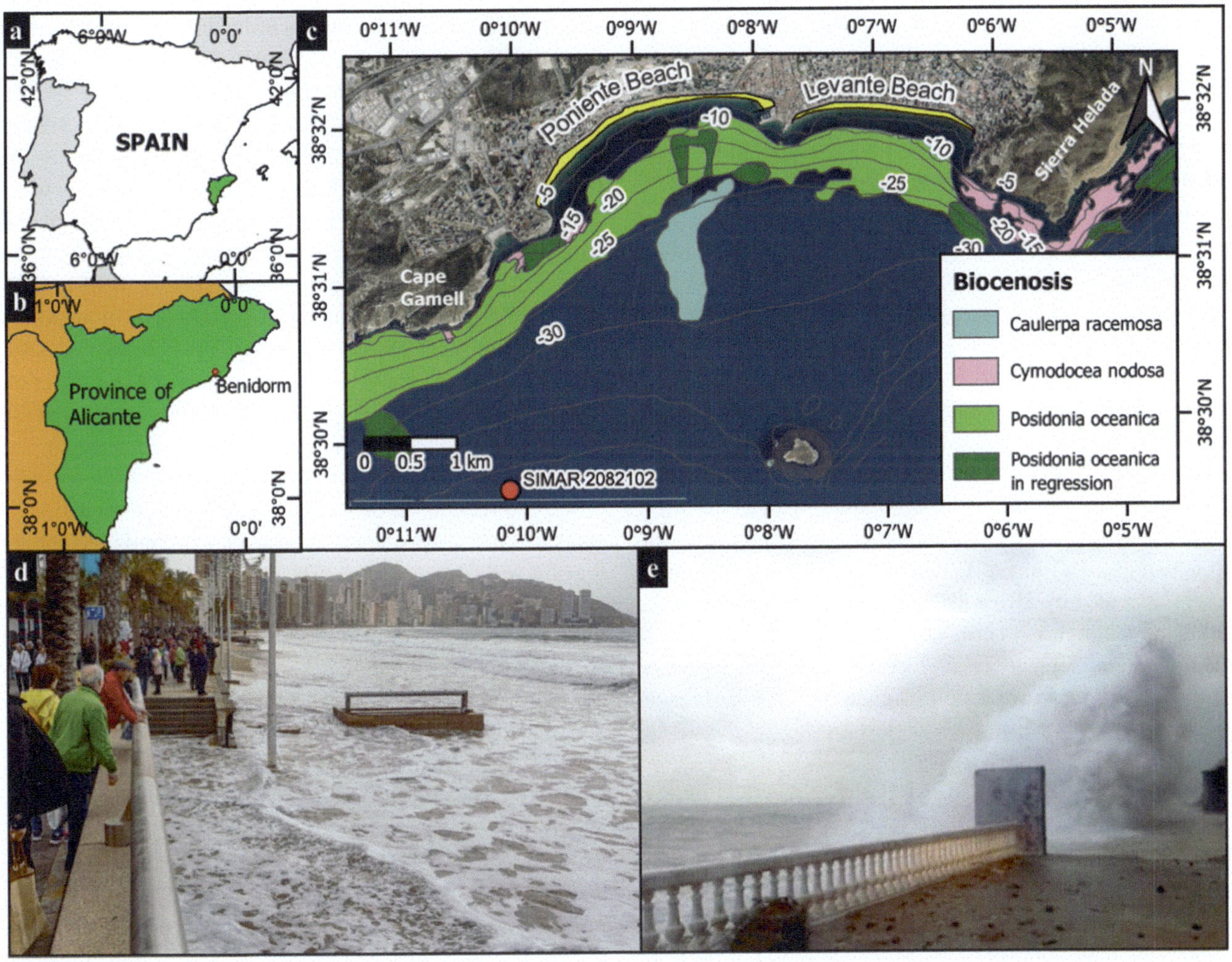

Fig. 2 **a** Study area located in Spain and **b** in the province of Alicante, **c** detail of study area, with SIMAR node and location of Benidorm beaches, **d**, **e** State of Levante Beach and Poniente Beach during the January 2020 storm in which the waves hit the promenade. *Source Diario Información* and *El Periòdic*

greater than 50 cm, while Astronomical Tides do not usually exceed 30 cm (Ecolevante, 2006). Despite this, the Benidorm promenade is subject to constant coastal flooding, where sea water floods the entire beach surface and waves hit the wall of the promenade, causing overflows (Fig. 2d, e).

2.2 Study of Waves and Flood Level Calculation

Wave data (height, period and direction) were provided by Puertos del Estado, based on SIMAR series. SIMAR dataset is made up of simulations carried out by numerical modeling of the atmosphere and waves that cover the entire Spanish coast. Wave fields have been generated by WAM and WaveWatch models, fed by the wind fields of the model provided by the Spanish Meteorological Agency (Puertos del Estado, 2020). Additionally, SWAN model is used to take into account the transformations suffered by the waves when approaching the coast. In general, these data are collected during the period 1958–2022 with an hourly frequency and a spatial resolution lower than 3 km. For this research, the database of Node 2082102, located about 5 km south of the study area, was used (Fig. 2c).

These data were processed by CAROL v1.0, a statistical analysis software developed by the IH-Cantabria. This software characterizes the oceanographic variables defined from time series, adjusting them to the theoretical distribution functions used to estimate the mean and extreme wave regime. Results are provided through fit graphs, wind roses, histograms and tables with statistics. All this allows obtaining the wave height $H_{s,12}$ (wave height with a 0.137% probability of not being exceeded), the mean and peak periods, and their corresponding directions and probabilities of occurrence.

An analysis of maritime storms was also carried out from the SIMAR database. In this study, it was considered 'storm' when a significant wave height corresponding to the 95th percentile was exceeded, that is, 1.15 m for Levante Beach during a minimum period of 6 consecutive hours (Harley et al., 2013; Valiente et al., 2019). The end of the storm is determined when that wave height is not exceeded for at least 24 h in a row. In this study, the average duration and maximum height of the storm were measured.

Next, the flood level is estimated as the maximum level that sea water reaches under the action of the most unfavourable storm produced in the study area since records are available. The calculation of the flood level was carried out as the combined action of the Storm Surge (SS), the Astronomical Tide (AT), and the run-up ($R_{u,2\%}$). To calculate the run-up, the formulation proposed by (Van der Meer et al., 2018) was used. For a maximum value approach, run-up is expressed as in Eq. 1: where $R_{u2\%}$ is the wave run-up height exceeded by 2% of the incoming waves (m), γ_b is the influence factor for a berm (−), γ_f is the influence factor for roughness elements on a slope (−), γ_β is the influence factor for oblique wave attack (−) and $\xi_{m-1,0}$ is the breaker parameter (-).

$$\frac{R_{u2\%}}{H_{m0}} = 1,00 \cdot \gamma_f \cdot \gamma_\beta \cdot \left(4 - \frac{1,5}{\sqrt{\gamma_b \cdot \xi_{m-1,0}}}\right) \quad (1)$$

The Astronomical Tide and Storm Surge data have been obtained from the Gandía Port Tide Gauge (0° 9′ 5″ W, 38° 59′ 43″ N), which is part of Puertos del Estado network. This data point is the closest, being 50 km north of the study area. These are hourly data collected during the period 2007–2022. The Storm Surge, an unpredictable factor influenced by atmospheric pressure and wind, is calculated as the residual or variance between the total sea level (initial data) and the Astronomical Tide, a deterministic factor resolved through harmonic analysis (Tapia et al., 2016).

3 Results and Discussion

Coastal territories are subject to profound changes that occur over time, and Benidorm is no exception. The city has been subjected to important anthropic pressures, the most important being the urbanization process carried out during the second half of the twentieth century. This situation has caused a decrease in the contribution of sediments to the beaches, eroding them (Aragonés et al., 2015; Toledo et al., 2022). Beaches are elements of coastal defense that protect all infrastructures (buildings, urban roads…) against storms. The increase in its duration and intensity will accelerate the erosion of the beaches.

Sierra Helada is a geographical element that protects the beaches from deterioration caused by storms from the east, but does not protect them from storms from the south (Tros De Ilarduya Fernández, 2013). In fact, although $H_{s,12}$ has shown a significant increase for almost all the incoming directions, it is south direction that presents a greater increase in the frequency and intensity of waves during the last 2 periods analyzed (Fig. 3a). For example, the wave height increased 0.62 m for the south (S) direction and 0.40 m for the SW and WSW directions, compared to the previous period. Waves coming from the southwest quadrant also increased in frequency relative to incoming waves compared to the 1991–2000 period. This behavior has also been detected in other parts of the western Mediterranean (Amarouche & Akpinar, 2021). Furthermore, an increase in the input frequency of 12% and 3% was detected for the SSW and SW directions, respectively, while 14% was lost in the S direction. In these cases, a linear regression analysis is of great help when it is necessary to know the behavior of a variable over time, especially under the influence of climate change. In Benidorm, a growth trend is observed, with a 15% increase in wave height in the 50-year period studied. Regarding the study of storms, the average storm duration remains stable between 15 and 20 h, with a small punctual decrease between 2003 and 2007 (Fig. 3b), even the long-term trend is slightly negative. However, this stability is not perceived in the maximum storm wave height, where there was a strong increase in it in the last 15 years, with a maximum of 3.31 m in 2018, representing 65% more than the average of the first 30 years studied (Fig. 3c).

The evolution of the main wave parameters at Levante Beach was also analyzed (Table 1). The significant wave height in the last period stood out clearly, with an increase of 14% compared to the period 2001–2010, as has already been observed previously. As far as the periods are concerned, these have not undergone important changes, with some variations in the last two periods studied. However, the mean wave direction (MWD) stood out above the rest, with a significant clockwise rotation of 4.9° between the period 1981–1990 and 2011–2020.

In these cases, it will be the extreme regime necessary to define the design parameters when proposing solutions to contain the waves and protect the frontline. The most unfavorable waves in the study area ($H_{m,0}$=2.83 m, $T_{m,0}$=7.53 s and SWL=63 cm) caused a flood level of 1.71 m at Levante Beach. This water level covers the entire beach berm during the most unfavorable storm in the historical series (1958–2022). However, at specific times a maximum flood level of 2.49 m can occur (Fig. 4). The promenade, given the low crest level (2.3–2.4 m), is prone to flooding, causing material damage to road infrastructure and businesses located along the beach front.

In order to protect infrastructures located on the beach front from possible floods of maritime origin that may occur in Levante Beach, the construction of a dune is

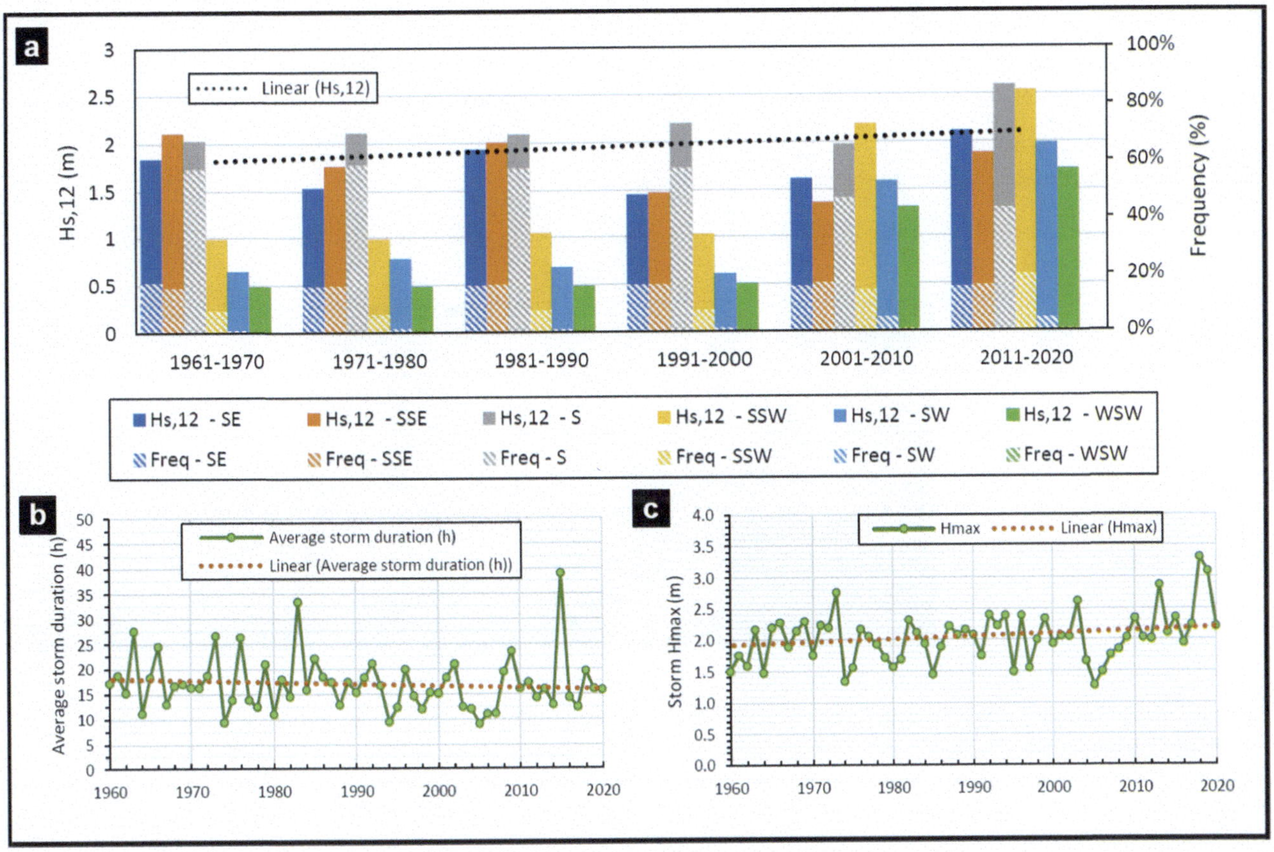

Fig. 3 **a** Wave height ($H_{s,12}$) and frequency for each direction and period, **b** evolution of the storm duration average, **c** evolution of the storm H_{max}. *Source* Author

Table 1 Evolution of wave parameters in Benidorm: significant wave height ($H_{m,0}$), mean period (T_m), peak period (T_p), mean wave direction (MWD)

	$H_{m,0}$ (m)	T_m (s)	T_p (s)	MWD (°)
1961 1970	0.451	3.433	4.979	172.66
1971–1980	0.451	3.475	5.098	173.08
1981–1990	0.436	3.433	5.013	172.77
1991–2000	0.447	3.424	5.018	172.94
2001–2010	0.472	3.213	4.951	176.27
2011–2020	0.537	3.266	5.247	177.66

proposed on the backshore. The sea defense dune must be designed in such a way that overflow is limited. The flood level for this new case has been calculated according to (Van der Meer et al., 2018), considering that it is a vegetated sand dam. The most unfavorable waves generate a flood level of 3 m at its maximum point (Fig. 5a). For this reason, a dune with a maximum height of 3.2 m is proposed (Fig. 5b), which can be reduced in those sections where the height of the promenade is sufficient to contain the waves and thus minimize the visual impact it may generate (Fig. 5c). A front slope of 1:4 was applied, since gentle slopes require a lower volume of sand (lower cost), stability

Fig. 4 Flooding level of the waves in the current situation of the beach. *Source* Author

Fig. 5 **a** Flooding level of the waves in the future situation of the dune, **b** urban dune at low points of the promenade, **c** urban dune where the level of the promenade is sufficient to contain the maritime flooding. *Source* Author

problems will not appear and will supply a greater ecological benefit (Verhagen, 2019).

Unlike the usual structure of breakwaters, the core of our breakwater will be made of a thicker and more permeable material than that of the armor layer, with the aim of facilitating the outflow of infiltrated water, as well as being a reservoir for the beach itself. Therefore, the dune will be composed of two layers of different size of sediment (Fig. 5a). The core will be made up of coarse sand with a D_{50} of 1.78 mm (washed sand from a quarry and free of fines), while the armor layer will be covered with the size of the current sand, which has been estimated to be 0.27 mm (Toledo et al., 2022). The material costs lead us to obtain a volume of 2.55 m^3/m of fine sand and 10.42 m^3/m of coarse sand.

When the sea level floods the entire beach surface, the frontal slope of the dune will begin to erode, thus adding sand to the beach, thus recovering part of the volume lost during strong waves. Once the dune is destroyed, it must be rebuilt again. This dune must be vegetated with autochthonous shrub planting to prevent sand mobility when the wind blows in the direction of the land (wind from the south). In addition, it will be necessary to install an irrigation network attached to the promenade in order to maintain the vegetation in a healthy state.

The construction of a sand dune is a recurring solution to face the problems of beach erosion and flooding on the coast (Rivillas-Ospina et al., 2020; van der Spek et al., 2020). However, the design of the vegetated dune in urban environments with high risk of coastal flooding is a relatively recent alternative. The property of the dune vegetation to dissipate wave energy will be useful to decrease its speed, reducing its erosive capacity (Maiolo et al., 2020). The design of the urban dune must be in accordance with the environment in which it is located, but also that it be compatible with the tourist and leisure activities that take place there. It will be essential that the dune, in addition to having a sufficient height to contain the waves, allows a direct view of the beach from the promenade, given the importance of this element for city tourism (Fig. 5). The use of a coarse sand size would facilitate drainage, but would also avoid having to dredge the seabed to get beach-size sand, with the environmental problems that this entails (Dan Gavriletea, 2017).

Climate change will cause flooding on the beach berm and the damage caused to the coastline to occur more frequently in the coming years (Olcina, 2020). This type of damage will jeopardize the proper functioning of the city, reducing its capacity when facing these climatic risks. It is essential that the Public Administrations design action plans to limit this damage, and the chosen solutions must be based on nature (Castelle et al., 2019; Kabisch et al., 2017). In the future, it may be interesting to establish a series of indicators (KPIs) that quantify the achievement of the set objectives (Mosca et al., 2023). A valid indicator could be the % of dune surface that is protected by vegetation above a certain threshold or the % of times that a coastal flood is avoided thanks to the construction of the dune. The selection of one indicator or another will depend on the type of the natural solution implemented and the objectives that we initially set.

The solution presented here, despite the fact that it has been specified for the urban environment of Benidorm (Spain), can be fully extrapolated to any part of the world, especially in those cities where tourism is the main activity, so the adequate maintenance of the beaches and the coastline is vital.

4 Conclusion

Anthropic activities in the territory and extreme weather events promoted by climate change put the proper functioning of cities in coastal areas at risk. In this study, the danger of coastal flooding in Benidorm city (Spain) has been analyzed, and solutions have been proposed to minimize it. The design of a vegetated urban dune on the back of the beach is a solution to minimize the overflow that occur during the most severe storms, which have significantly increased their wave height in recent years (setting a maximum of 3.31 m in 2018).

The proposed solutions that the competent bodies have to take must be based on nature so that they are sustainable, both economically and environmentally. The proposed case of Benidorm (Spain), together with the applied methodology, is fully exportable to any part of the world, becoming part of the catalog of alternatives that the coastal engineer must apply to protect coastal cities.

Acknowledgements Authors thank the European Commission through the project "*Smart Control of the Climate Resilience in European Coastal Cities*" (Grant agreement ID: 101003534) for funding this research.

References

Amarouche, K., & Akpinar, A. (2021). Increasing trend on stormwave intensity in thewestern mediterranean. *Climate, 9*(1), 1–17. https://doi.org/10.3390/cli9010011

Amores, A., Marcos, M., Carrió, D. S., & Gomez-Pujol, L. (2020). Coastal impacts of Storm Gloria (January 2020) over the north-western Mediterranean. *Natural Hazards and Earth System Sciences, 20*(7), 1955–1968. https://doi.org/10.5194/nhess-20-1955-2020

Aragonés, L., García-Barba, J., García-Bleda, E., López, I., & Serra, J. C. (2015). Beach nourishment impact on Posidonia oceanica: Case study of Poniente Beach (Benidorm, Spain). *Ocean Engineering, 107*, 1–12. https://doi.org/10.1016/j.oceaneng.2015.07.005

Bevacqua, E., Maraun, D., Hobæk Haff, I., Widmann, M., & Vrac, M. (2017). Multivariate statistical modelling of compound events via pair-copula constructions: Analysis of floods in Ravenna (Italy). *Hydrology and Earth System Sciences, 21*(6), 2701–2723. https://doi.org/10.5194/hess-21-2701-2017

Brown, K., Tompkins, E. L., & Adger, W. N. (2012). Making waves: Integrating coastal conservation and development. In: *Making waves: Integrating coastal conservation and development*. https://doi.org/10.4324/9781849771887

Castelle, B., Laporte-Fauret, Q., Marieu, V., Michalet, R., Rosebery, D., Bujan, S., Lubac, B., Bernard, J. B., Valance, A., Dupont, P., El Moctar, A. O., & Narteau, C. (2019). Nature-based solution along high-energy eroding sandy coasts: Preliminary tests on the reinstatement of natural dynamics in reprofiled coastal dunes. *Water (Switzerland), 11*(12). https://doi.org/10.3390/w11122518

Crooks, S., & Turner, R. K. (1999). Integrated coastal management: sustaining estuarine natural resources. In *Advances in ecological research* (Vol. 29, Issue C, pp. 241–289). Academic Press Inc. https://doi.org/10.1016/S0065-2504(08)60195-6

Dan Gavriletea, M. (2017). Environmental impacts of sand exploitation. Analysis of sand market. In *Sustainability (Switzerland)* (Vol. 9, Issue 7, p. 1118). Multidisciplinary Digital Publishing Institute. https://doi.org/10.3390/su9071118

Dao, H. T., Hofland, B., Stive, M. J. F., & Mai, T. (2020). Experimental assessment of the flow resistance of coastal wooden fences. *Water (switzerland), 12*(7), 1910. https://doi.org/10.3390/w12071910

De Ilarduya, T., & Fernández, M. (2013). West sea storms and Atlantic depressions in the province of Alicante: The case of Benidorm. *Estudios Geográficos, 74*(274), 287–310. https://doi.org/10.3989/estgeogr.201310

Ecolevante. (2006). Estudio ecocartográfico del litoral de las provincias de Alicante y Valencia. *General Service of Coasts of the State, Madrid*

Edward, J. K., Terazaki, M., & Yamaguchi, M. (2006). The impact of tsunami in coastal areas: Coastal protection and disaster prevention measures-Experiences from Japanese coasts. *Coastal Marine Science, 30*(2), 414–424. https://doi.org/10.3/JQUERY-UI.JS

Puertos del Estado. (2020). *Conjunto de datos SIMAR*. https://bancodatos.puertos.es/BD/informes/INT_8.pdf

Femenia-Serra, F., & Ivars-Baidal, J. A. (2021). Do smart tourism destinations really work? The case of Benidorm. *Asia Pacific Journal of Tourism Research, 26*(4), 365–384. https://doi.org/10.1080/10941665.2018.1561478

Harley, M. D., Andriolo, U., Armaroli, C., & Ciavola, P. (2013). Shoreline rotation and response to nourishment of a gravel embayed beach using a low-cost video monitoring technique: San Michele-Sassi Neri, Central Italy. *Journal of Coastal Conservation, 18*(5), 551–565. https://doi.org/10.1007/s11852-013-0292-x

Jisan, M. A., Bao, S., & Pietrafesa, L. J. (2018). Ensemble projection of the sea level rise impact on storm surge and inundation at the coast of Bangladesh. *Natural Hazards and Earth System Sciences, 18*(1), 351–364. https://doi.org/10.5194/nhess-18-351-2018

Kabisch, N., Korn, H., Stadler, J., & Bonn, A. (2017). *Nature-based solutions to climate change adaptation in urban areas—linkages between science, policy and practice* (pp. 1–11). https://doi.org/10.1007/978-3-319-56091-5_1

Maiolo, M., Mel, R. A., & Sinopoli, S. (2020). A stepwise approach to beach restoration at calabaia beach. *Water (Switzerland), 12*(10). https://doi.org/10.3390/w12102677

Van der Meer, J. W., Allsop, N. W. H., Bruce, T., De Rouck, J., Kortenhaus, A., Pullen, T., Schüttrumpf, H., Troch, P., & Zanuttigh, B. (2018). *Manual on wave overtopping of sea defences and related structures. An overtopping manual largely based on European research, but for worldwide application*. www.overtopping-manual.com

Mei, W., Xie, S. P., Primeau, F., McWilliams, J. C., & Pasquero, C. (2015). Northwestern Pacific typhoon intensity controlled by changes in ocean temperatures. *Science Advances, 1*(4). https://doi.org/10.1126/sciadv.1500014

Morris, R. L., Konlechner, T. M., Ghisalberti, M., & Swearer, S. E. (2018). From grey to green: Efficacy of eco-engineering solutions for nature-based coastal defence. In *Global change biology* (Vol. 24, Issue 5, pp. 1827–1842). Blackwell Publishing Ltd. https://doi.org/10.1111/gcb.14063

Mosca, F., Canepa, M., & Perini, K. (2023). Strategies for adaptation to and mitigation of climate change: Key performance indicators to assess nature-based solutions performances. In *Urban climate* (Vol. 49). https://doi.org/10.1016/j.uclim.2023.101580

Nolasco-Cirugeda, A., Martí, P., & Ponce, G. (2020). Keeping mass tourism destinations sustainable via urban design: The case of

Benidorm. *Sustainable Development, 28*(5), 1289–1303. https://doi.org/10.1002/sd.2084

Olcina, J. (2020). Clima, cambio climático y riesgos climáticos en el litoral mediterráneo. Oportunidades para la geografía. *Documents d'Anàlisi Geogràfica, 66*(1), 159. https://doi.org/10.5565/rev/dag.629

Pagán, J. I., Aragonés, L., Tenza-Abril, A. J., & Pallarés, P. (2016). The influence of anthropic actions on the evolution of an urban beach: Case study of Marineta Cassiana beach, Spain. *Science of the Total Environment, 559*, 242–255. https://doi.org/10.1016/j.scitotenv.2016.03.134

Rivillas-Ospina, G., Maza-Chamorro, M. A., Restrepo, S., Lithgow, D., Silva, R., Sisa, A., Vargas, A., Sarmiento, J. P., Caes, J., Bolivar, M., Del Rio, R., Campo, E., Casas, D., & Rudas, D. (2020). Alternatives for recovering the ecosystem services and resilience of the Salamanca Island Natural Park, Colombia. *Water (Switzerland), 12*(5), 1513. https://doi.org/10.3390/w12051513

Schoonees, T., Gijón Mancheño, A., Scheres, B., Bouma, T. J., Silva, R., Schlurmann, T., & Schüttrumpf, H. (2019). Hard structures for coastal protection, towards greener designs. *Estuaries and Coasts, 42*(7), 1709–1729. https://doi.org/10.1007/s12237-019-00551-z

Tapia, A., Lopez, R., Gili, J., Palau, V., & Pros Llavador, F. (2016). Nivelación hidrostática entre dos mareógrafos situados en el Puerto de Barcelona. *Congreso Internacional de Geomática y Ciencias de La Tierra*, 13–20. https://upcommons.upc.edu/handle/2117/101530

Toimil, A., Camus, P., Losada, I. J., Le Cozannet, G., Nicholls, R. J., Idier, D., & Maspataud, A. (2020). Climate change-driven coastal erosion modelling in temperate sandy beaches: Methods and uncertainty treatment. In *Earth-science reviews* (Vol. 202). *Earth-Science Reviews*. https://doi.org/10.1016/j.earscirev.2020.103110

Toledo, I., Pagán, J. I., López, I., & Aragonés, L. (2022). Causes of the different behaviour against erosion: Study case of the Benidorm Beaches (1956–2021). *Marine Georesources and Geotechnology, 41*(6), 648–661. https://doi.org/10.1080/1064119x.2022.2084003

Tsimplis, M. N., Woolf, D. K., Osborn, T. J., Wakelin, S., Wolf, J., Flather, R., Shaw, A. G. P., Woodworth, P., Challenor, P., Blackman, D., Pert, F., Yan, Z., & Jevrejeva, S. (2005). Towards a vulnerability assessment of the UK and northern European coasts: The role of regional climate variability. *Philosophical Transactions of the Royal Society a: Mathematical, Physical and Engineering Sciences, 363*(1831), 1329–1358. https://doi.org/10.1098/rsta.2005.1571

Valiente, N. G., McCarroll, R. J., Masselink, G., Scott, T., & Wiggins, M. (2019). Multi-annual embayment sediment dynamics involving headland bypassing and sediment exchange across the depth of closure. *Geomorphology, 343*, 48–64. https://doi.org/10.1016/j.geomorph.2019.06.020

van der Spek, B. J., Bijl, E., van de Sande, B., Poortman, S., Heijboer, D., & Bliek, B. (2020). Sandbar breakwater: An innovative nature-based port solution. *Water (switzerland), 12*(5), 1446. https://doi.org/10.3390/w12051446

Vecchi, G. A., & Soden, B. J. (2007). Effect of remote sea surface temperature change on tropical cyclone potential intensity. *Nature, 450*(7172), 1066–1070. https://doi.org/10.1038/nature06423

Verhagen, H. J. (2019). Financial benefits of mangroves for surge prone high-value areas. *Water (Switzerland), 11*(11). https://doi.org/10.3390/w11112374

Vousdoukas, M. I., Voukouvalas, E., Annunziato, A., Giardino, A., & Feyen, L. (2016). Projections of extreme storm surge levels along Europe. *Climate Dynamics, 47*(9–10), 3171–3190. https://doi.org/10.1007/s00382-016-3019-5

WWF. (2021). Soluciones basadas en la Naturaleza: Soluciones urbanas basadas en la naturaleza. In *Ciudades que lideran el camino 2021*. Comité Científico de Cambio Climático; Ministerio de Ciencia, Tecnología, Conocimiento e Innovación. https://doi.org/10.5281/ZENODO.5736938

Review of Urban Energy Management in India Towards Sustainable Approaches

Kiruthiga Kandasamy and Rahul P. Menon

Abstract

The utmost of every urban space experiencing rapid urbanization due to unprecedented population growth leads to a serious impact on energy consumption, eventually resulting in carbon emissions. Containing and supporting over half the global population within cities has become a challenge, which will surge soon. Cities deal with several intricately woven problems in numerous categories, and it is necessary to address these issues by combining energy management standards for accomplishing sustainable urban development. In the process of urban transformation, the existing cities are initiated into the smart cities transformation with smart energy management becoming an integral component in which city energy footprint management becomes the most challenging goal. Renewable energy sources are considered a feasible substitute for conventional energy sources in contemporary society. Thus, this study comprehensively reviews the various renewable energy sources for different urban sectors in Tamil Nadu, India, and the crisis faced by the urban centers due to rapid urbanization. With proper urban energy management, we can devise appropriate sustainable solutions and guidelines for the future of urban wellbeing.

Keywords

Urban energy · Energy management · Renewable energy · Sustainability · Tamil Nadu

K. Kandasamy (✉)
Department of Architecture and Planning, National Institute of Technology Calicut, Kozhikode, India
e-mail: kiruthigak@nitc.ac.in

R. P. Menon
School of Planning Architecture and Design,
Hindustan Institute of Technology and Science, Chennai, India

1 Introduction

Increasing urbanization and high demand for energy is a problem that is acknowledged on a global scale. According to UN-Habitat, 75% of extensive primary energy is consumed by the Urban environment in which 50–60% of the energy is emitted as greenhouse gases. The percentage may rise up to 80% as indirect emissions if we include urban inhabitants (UN-Habitat, 2016). As per the findings of the World Resource Institute Report 2017, India's contribution to the overall global carbon emissions is approximately 6.65%, placing it in the fourth position after China (26.83%), the USA (14.36%), and the EU (9.66%) (Kumar, 2020). Majority of the Indian cities are confronting a series of issues pertaining to Energy Management. Since the non-renewable energy sources endure growing strain to supply the energy demands of cities. Further, the environment has been negatively impacted by the scarcity of energy resources, and the high levels of greenhouse gas emissions by urban energy use, which made climate change the way worse.

According to the International Energy Agency (IEA) (IEA, 2019, 2021), in 2019 and 2021, it is projected that the urban populace of India will experience an increase of 270 million individuals from 2020 to 2040. India's energy consumption has experienced a significant increase of almost 100% since the year 2000, indicating substantial potential for further rapid expansion. India has experienced an annual electricity supply deficit since 2009, as reported by the Overview provided by the Ministry of Power under the Government of India in 2022 (India Energy Outlook, 2021). The contemporary world is predominantly powered by a set of primary energy sources, namely fossil fuels, biofuels, nuclear fuels, wind, and solar radiation, as reported by the International Energy Agency (IEA, 2021) and corroborated by data and statistics (India Energy Outlook, 2021). According to the International Energy Agency's analysis of the India Energy Outlook 2021, thermal power plants,

B. He et al. (eds.), *Urban Climate Change Adaptation*, Advances in Science, Technology & Innovation,
https://doi.org/10.1007/978-3-031-65088-8_3

which use coal as a primary source of energy, produce about 70% of India's electricity, which results in significant carbon dioxide emissions (IEA, 2019).

According to the projections made by the International Energy Agency (IEA), there is a forecasted increase of 145 TWh (Terawatt-hour) in global solar energy production by the year 2031 (IEA, 2019). The proportion of renewable power generation has reached a record high of 30% in 2021, owing to an upsurge in renewable energy sources (Raghuwanshi & Arya, 2019). It is anticipated that this figure will escalate to approximately 50% by 2030 and 80% by 2050, as per the International Energy Agency (2021) (IEA, 2019). Renewable energy sources are considered a feasible substitute for conventional energy sources in contemporary society due to their environmentally friendly nature, noiseless operation, and inexhaustible availability. This study reviews the current urban energy situation in India and takes Tamil Nadu as a case study to overview energy management through sustainable approaches.

2 Literature Review

2.1 Urban Energy Management

The methods used to extract, distribute, and use energy are major contributors to environmental degradation among developing and OECD nations. Therefore, there is an especially pressing need to move away from the existing demand for energy that comes from limited fossil fuels and towards cleaner, low-carbon, renewable energy sources. The demand for cleaner, greener energy is essential to solving many of the present international environmental concerns, such as desertification, environmental degradation, and loss of biodiversity. Reducing the use of fossil fuels must be combined with (a) investigating alternative sources of energy, (b) connecting energy to present-day environmental problems (such as global warming or the effects of climate change), (c) linking the use of energy to environmental management, and (d) modifying lifestyles and stepping up community involvement (Modrzyński & Karaszewski, 2022). Urban energy management is a critical component of creating sustainable, resilient, and livable cities. It addresses environmental concerns, supports economic growth, enhances infrastructure planning, and improves the overall well-being of urban populations. As cities continue to grow and face evolving challenges, effective energy management becomes increasingly vital for their success and sustainability (Yenneti et al., 2019). Thus, Urban energy management involves the planning, optimization, and efficient use of energy resources within urban areas. It encompasses a variety of strategies and technologies aimed at meeting the energy needs of a growing urban population while simultaneously addressing environmental, economic, and social considerations.

2.2 Smart Energy Management (SEM)

In today's energy-intensive world, smart energy management (SEM) has emerged as a critical strategy for optimizing energy consumption, minimizing costs, and enhancing sustainability. Leveraging the power of information and communication technologies (ICT), SEM systems are transforming the way we manage energy across residential, commercial, and industrial sectors (Calvillo et al., 2016; Pappas & Chalvatzis, 2017). The adoption of SEM is driven by a compelling set of imperatives such as (i) Economic Benefits: SEM systems effectively identify and eliminate energy wastage, leading to substantial savings on energy bills. These cost savings can be reinvested in business growth, infrastructure improvements, or consumer rebates, (ii) Environmental Stewardship: SEM contributes to minimizing greenhouse gas emissions and alleviating other environmental impacts associated with excessive energy consumption. As the world transitions towards a low-carbon economy, SEM plays a crucial role in achieving sustainability goals, and (iii) Enhanced Comfort and Convenience: SEM systems automate numerous energy-related tasks, such as adjusting heating and cooling settings, promoting greater comfort and convenience in homes and businesses. By tailoring energy usage to occupancy patterns and occupant preferences, SEM enhances user satisfaction (Yenneti, 2016).

SEM is defined as "*Smart energy management is a component of smart city development aiming at a site-specific continuous transition towards sustainability, self-sufficiency, and resilience of energy systems while ensuring accessibility, affordability, and adequacy of energy services, through optimized integration of energy conservation, energy efficiency, and local renewable energy sources. It is characterized by a combination of technologies with information and communication technologies that enables integration of multiple domains and enforces collaboration of multiple stakeholders while ensuring sustainability of its measures.*" (Mosannenzadeh et al., 2017).

SEM finds diverse applications across various domains, offering tailored solutions to specific energy challenges. In residential Homes, SEM systems empower homeowners to reduce energy costs, augment the comfort of their living spaces, and minimize their environmental impact. By providing real-time energy consumption data and automating energy-related tasks, SEM systems encourage homeowners to adopt energy-conscious behaviors.

In commercial buildings, SEM systems assist businesses in minimizing energy expenses, bolstering sustainability efforts, and enhancing occupant satisfaction. By optimizing energy usage across various building systems, SEM systems contribute to cost savings, reduced carbon emissions, and improved indoor environmental quality. In industrial facilities, SEM systems enable industrial facilities to optimize energy consumption, reduce costs effectively, and enhance operational efficiency. By monitoring and controlling energy-intensive processes, SEM systems minimize energy wastage and improve overall production efficiency (Ministry of New and Renewable Energy, 2023a; Mosannenzadeh et al., 2017).

Creating a path for a future of sustainable energy as SEM technology develops, the possible advantages would be (i) Further reductions in energy costs: The growing sophistication of SEM systems promises to uncover additional energy-saving opportunities, leading to even greater cost savings for consumers and businesses, (ii) Improved grid stability: SEM systems contribute to grid stability by managing demand and reducing reliance on peak power plants, enhancing overall grid reliability and resilience, and (iii) Greater adoption of renewable energy: SEM systems facilitate the seamless integration of renewable energy sources into the grid, promoting the reduction of greenhouse gas emissions and improving air quality. Smart energy management plays a pivotal role in addressing global energy challenges, fostering sustainable practices, and ensuring a secure and efficient energy future (Yenneti et al., 2019).

3 Review Methodology

The methodology for this review paper consists of three levels: (i) Review of existing renewable energy sources in India, (ii) Quantitative analysis, and (iii) Qualitative analysis. The study starts with the existing literature review of urban energy management in the global scenario. The initial level involves the review of energy demand and the contribution of renewable energy resources in India, in which it also discusses the existing policies and guidelines. The second level takes a case study of Tamil Nadu and includes a precise review based on quantitative analysis with the available secondary data (recent reports from the Ministry of India and existing research studies). The qualitative analysis includes the survey results conducted among the group of research scholars from various institutions (Centrally Funded research institutions in Tamil Nadu), working in the energy and sustainability field in Tamil Nadu. The pilot survey was conducted through Google form and 34 respondents were collected. Based on the results, the challenges of using energy in the city infrastructure were discussed.

4 Review of Existing Renewable Energy Resources in India

4.1 Contribution of Renewable Resources to the Energy Demand in India

Several energy management trends are fueling the demand for cleaner, low-carbon, and renewable energy sources. Numerous environmental challenges on a global scale, such as climate change, biodiversity loss, and desertification, necessitate a concerted effort toward the adoption of cleaner and more sustainable energy sources (Akadiri & Adebayo, 2022; Sahoo & Sahoo, 2022). This approach is expected to catalyze action in other areas, such as governance, education, and technological advancements. The Government of India (GoI) has been formulating policies and initiatives to facilitate the integration of smart energy management into urban planning (Ministry of New & Renewable Energy, 2023a). The Indian government has implemented a dual strategy that concentrates on supply and demand to satisfy its populace's energy requirements and mitigate carbon emissions. The power generation sector is currently emphasizing the increased utilization of sustainable energy sources, specifically solar and wind power.

4.2 Ministry of New and Renewable Energy Sources

India is now committed to achieving about 50% of its cumulative installed electric power capacity from non-fossil fuel-based energy resources by 2030 (Akadiri & Adebayo, 2022; Kang et al., 2019). As of December 31st, 2022, the country had installed a cumulative renewable energy capacity of 167.75 GW. Additionally, there are ongoing projects with a total capacity of 78.75 GW at different stages of implementation and 32.60 GW at various stages of bidding. According to the REN21 Renewables 2022 Global Status Report, India ranks fourth globally in terms of installed capacity for renewable energy, wind power, and solar power (Ministry of New & Renewable Energy, 2023a; Ministry of New and Renewable Energy, 2023b). Figure 1 shows the major ongoing schemes under the Ministry of New and Renewable Energy Sources in India.

The installed capacity of renewable energy, which encompasses large hydro, has experienced a notable increase from 76.37 GW in March 2014 to 167.75 GW in December 2022, representing a growth rate of approximately 2.20 times. The nation's solar power capacity has experienced a significant increase, rising from 2.63 GW in March 2014 to 63.30 GW in December 2022, indicating a

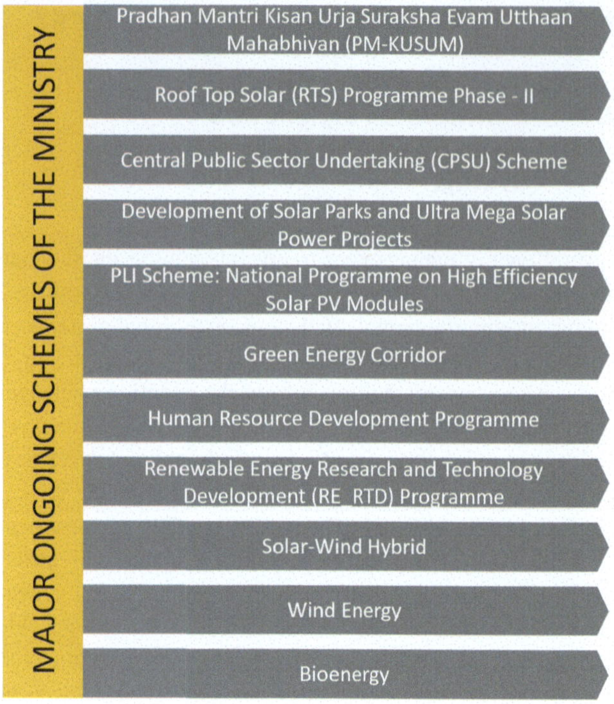

Fig. 1 Major Schemes under the Ministry of New and Renewable Energy Sources (2023a) in India. (Kumar & Majid, 2020)

growth of 24.07 times. The growth of the wind sector led to a robust environment, project operating skills, and an annual production base of over 12,000 MW (Ministry of New & Renewable Energy, 2023a).

4.3 Existing Key Policies for Energy Management in India

To help mainstream Smart Energy Management (SEM) in urban planning, the Government of India (GoI) started implementing policies and programmes. Figure 2 shows the key policies for energy management in India. The Indian government has taken a two-pronged strategy, i.e., concentrating on supply and demand, in order to fulfill the energy requirements of inhabitants and decrease carbon emissions. For the generation side, increased usage of renewable energy is emphasized, particularly solar and wind energy. Whereas, for the demand side, Initiatives are being taken to increase energy efficiency by using a number of novel policy approaches (Yenneti et al., 2019).

4.4 Case Study of Tamil Nadu

Tamil Nadu, which has a land size of 130,058 square kilometers, is located in the southmost region of India. Tamil Nadu established a separate division called the Tamil Nadu Energy Development Agency (TEDA) in 1985 to advance non-conventional energy development and mitigate the consequences of climate change by developing supportive policies (Tamil Nadu Energy Development Agency, 2019).

The State government of Tamil Nadu has ambitious plans for the energy industry and estimates that the state will need 20,000 megawatts (MW) of power in the future years, which is a new high for power consumption. With the goal of meeting the rising demand for electricity, the State of Tamil Nadu wants to double the existing capacity for power generation in the State by adding 33,000 MW by 2030. However, renewable energy sources will be used to generate an additional half of the power generation capacity. Even though Tamil Nadu's grid currently receives a respectable 20.88% of its energy from renewable sources, it is anticipated that by 2030, 50% of the grid's electricity will be green (Ministry of New and Renewable Energy, 2023b), refer to Fig. 3.

5 Results and Discussions

5.1 Renewable Energy Sources in Tamil Nadu

Tamil Nadu has the most solar energy accessible and holds the fourth position among the other states in India. The Kamuthi Solar Power Project, which covers an area of 2500 acres and includes 2.5 million solar modules with a capacity of 648 MW, is one of the largest single-location solar power plants (Tamil Nadu Energy Development Agency, 2019). The Tamil Nadu government has created a 2019 Solar Energy Policy with the goal of producing roughly 9000 MW by 2022 (Ministry of New and Renewable Energy, 2023b).

According to the Ministry of New and Renewable Energy (MNRE) (Ministry of New and Renewable Energy, 2023b), the total cumulative solar capacity installed in Tamil Nadu is about 6412.36 MW as of 31st December 2022. The estimated potential of solar energy in Tamil Nadu is about 17.67 GWp, assessed by the National Institute of Solar Energy (NISE). In collaboration with EverRenew, an NTC group EPC firm, Amplus Solar, a Gurugram-based supplier of distributed clean energy solutions, will construct a 200 MWp open-access wind-solar power plant in Tamil Nadu (Ministry of New and Renewable Energy, 2023b).

Tamil Nadu must take advantage of the fact that it is one of the few places in the world with a comprehensive value chain for wind manufacturing in order to increase its participation in the global supply chain, which keeps on generating trade surpluses (Sangroya & Nayak, 2015). Also, Tamil Nadu is one of the eight windy states with Wind power potential. 9.6 GW of the total wind installations in India are located in Tamil Nadu, making up over a quarter of the country's total capacity. The installed capacity of wind

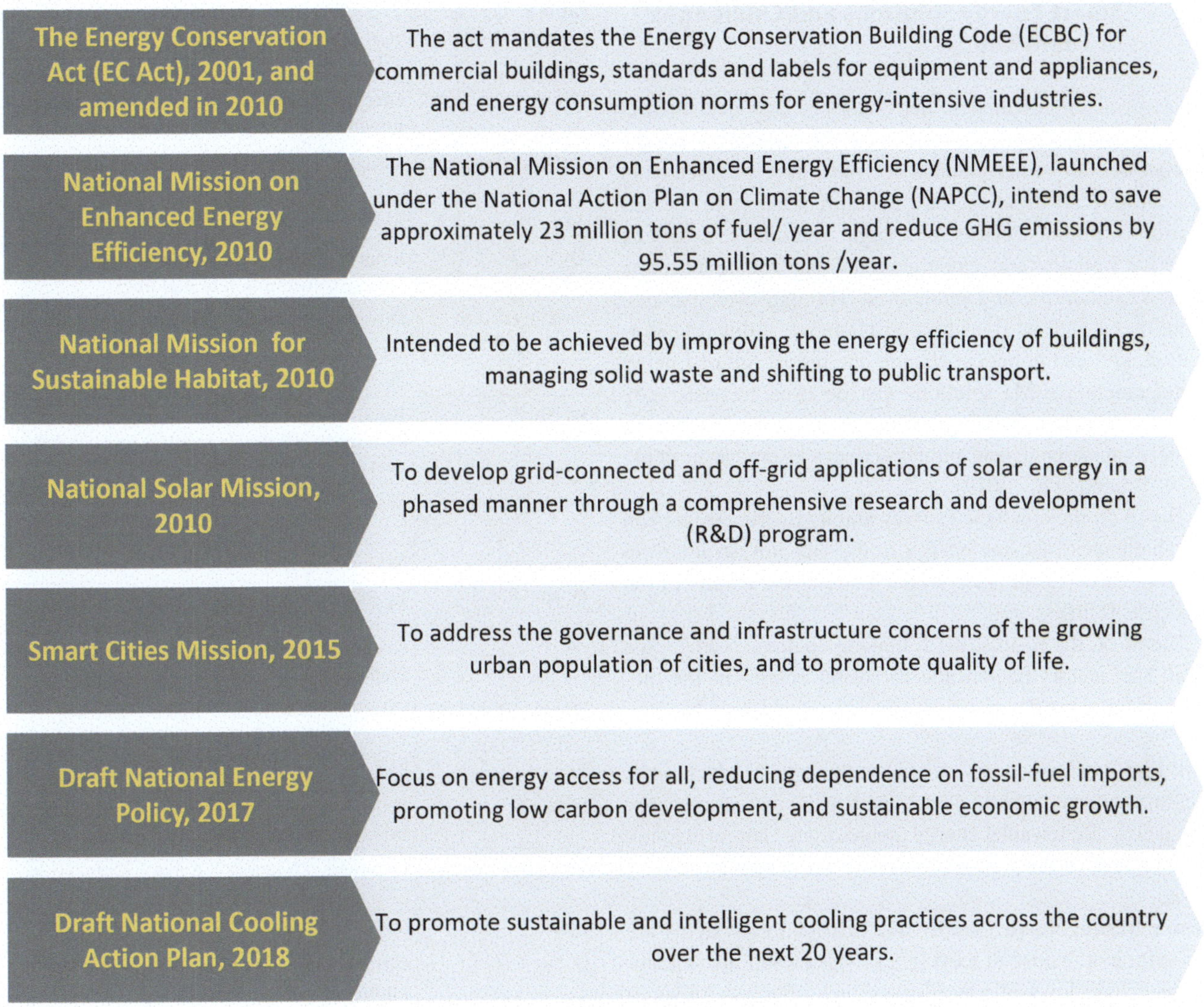

Fig. 2 Key policies for energy management in India (Modrzyński & Karaszewski, 2022)

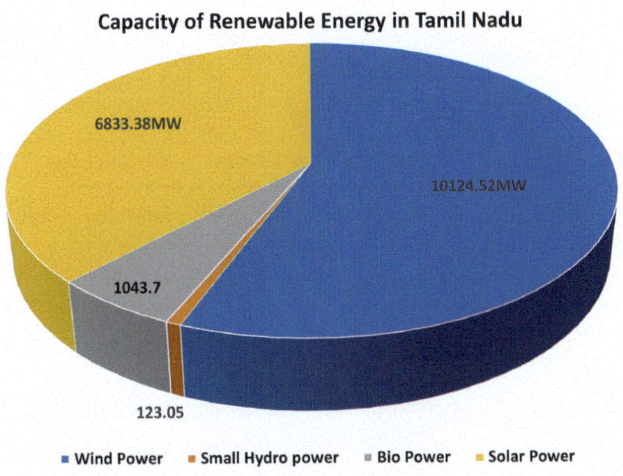

Fig. 3 The total installed capacity of Renewable Power in Tamil Nadu as of 31st May 2023 (Yenneti et al., 2019)

power in Tamil Nadu is 99.63GW. Tamil Nadu has been identified as a potential zone along with Gujarat for the adoption of offshore wind energy to produce 70GW in the initial assessment. Further, the National Institute of Wind Energy (NIWE) identified a suitable site at Dhanushkodi, Tamil Nadu, to strengthen and develop offshore wind turbine research (Tamil Nadu Solar Policy, 2019).

The capacity produced by biomass-based cogeneration fuel plants, biomass gasification plants, bio methanation plants, and vegetable waste-based power plants is added up to produce the entire amount of biomass energy. Collectively, all these facilities aid in the state's generation of biomass energy. Through the Tamil Nadu Biomass Power Producers Association, the Tamil Nadu Energy Development Agency (TEDA) has actively promoted production on a big scale in selected areas having significant potential for crops (Tamil Nadu Solar Policy, 2019).

5.2 Smart Energy Solutions and Challenges in Tamil Nadu

Smart grids, smart meters, net metering, city-level district cooling systems, and the utilization of renewable energy sources are all either planned for the future or already in use. Despite all these sources, there are barriers to the utilization of public services on a large scale due to a lack of effective policies and guidelines and inadequate business models, standards, and guidelines. Through effective distribution as well as transmission planning, utility transformation, and technological change, the use of Smart Energy Management (SEM) solutions can be used to assist city infrastructure like buildings, water management, transport, waste management, and public services (Elavarasan et al., 2020; Yenneti et al., 2019).

Based on the conducted survey among the researchers on the challenges of using energy in the city infrastructure in the context of Tamil Nadu. Table 1 shows the results of the survey as follows:

Based on the results, the following challenges that exist in all the sectors responsible for using energy in the city infrastructure are discussed.

Building. Lack of energy storage with respect to heating & cooling systems; Insufficient expertise to increase energy efficiency; Renewable energy sources are not efficiently applied; lack of innovative projects for implementation.

Water Management. Supply and distribution systems are not sufficient; Reuse & recycling of wastewater are not significant; high level of water losses; lack of access to smart technologies and real-time data collection; decentralized treatment systems are not sufficient.

Water Treatment. Lack of awareness among the inhabitants; and a lack of manpower sources; Smart management is not effectively applied in waste segregation and waste management. Waste-to-energy conversion is not utilized properly; No accountability and transparency in waste management; not much into intelligent recycling solutions.

Transport. Shortfall of low-carbon mobility development; lack of energy-efficient infrastructure; Low numbers in the Consumption of eco-friendly vehicles; Not opting the public transportation efficiently; No strict fuel efficiency norms; not integrating ICT solutions in transportation.

Public Services. No integration of smart systems; No option for renewable energy solutions; No coordination at the interdepartmental level; limited access to renewable energy resources; and no efficient policy and regulatory framework.

Table 1 Questionnaire survey results from the 34 respondents with respect to the cities infrastructure in Tamil Nadu

Q. No: 1	**Buildings are designed and constructed with energy efficient features, supported with the integration of renewable energy sources**			
Strongly disagree	Disagree	Neutral	Agree	Strongly agree
11%	65%	12%	6%	6%
Q. No: 2	**Smart technologies are aided to optimize energy consumption in buildings**			
Strongly disagree	Disagree	Neutral	Agree	Strongly agree
15%	50%	13%	12%	10%
Q. No: 3	**Buildings are opted with management systems to effectively monitor and control energy usage**			
Strongly disagree	Disagree	Neutral	Agree	Strongly agree
16%	54%	10%	12%	8%
Q. No: 4	**In cities, water pumping and distribution systems are designed to optimize energy consumption and maintains water pumping facilities**			
Strongly disagree	Disagree	Neutral	Agree	Strongly agree
13%	52%	12%	12%	11%
Q. No: 5	**Cities incorporate renewable energy sources in water treatment and distribution**			
Strongly disagree	Disagree	Neutral	Agree	Strongly agree
13%	60%	11%	8%	8%
Q. No: 6	**Cities promote the usage of energy-efficient pumps and smart technologies in water management facilities**			
Strongly disagree	Disagree	Neutral	Agree	Strongly agree
15%	55%	10%	12%	8%
Q. No: 7	**The city periodically updates and accesses water treatment processes**			
Strongly disagree	Disagree	Neutral	Agree	Strongly agree
16%	56%	8%	12%	8%
Q. No: 8	**Urban Infrastructures strongly integrate renewable energy sources in water treatment facilities like on-site renewable energy systems**			
Strongly disagree	Disagree	Neutral	Agree	Strongly agree
18%	60%	6%	8%	8%
Q. No: 9	**Cities promote the use of sustainable transportation modes like electric vehicles, cycling, walking, etc**			
Strongly disagree	Disagree	Neutral	Agree	Strongly agree
6%	56%	12%	18%	8%

(continued)

Table 1 (continued)

Q. No: 10	**Urban infrastructures provide alternative transportation mode for bike lanes and pedestrian walkways**			
Strongly disagree	Disagree	Neutral	Agree	Strongly agree
10%	52%	10%	16%	12%
Q. No: 11	**Urban infrastructures manage traffic flow to minimize energy use and fuel consumption**			
Strongly disagree	Disagree	Neutral	Agree	Strongly agree
12%	56%	12%	10%	10%
Q. No: 12	**Utilization of smart technologies to optimize energy use in public spaces**			
Strongly disagree	Disagree	Neutral	Agree	Strongly agree
36%	36%	10%	10%	8%
Q. No: 13	**Cities encourage energy conservation initiatives to reduce energy consumption**			
Strongly disagree	Disagree	Neutral	Agree	Strongly agree
5%	42%	10%	26%	17%
Q. No: 14	**Government engages the public about the energy implications of energy management and its benefits**			
Strongly disagree	Disagree	Neutral	Agree	Strongly agree
9%	48%	7%	20%	16%
Q. No: 15	**Government provides incentives and policies to encourage the use of renewable energy in energy management**			
Strongly disagree	Disagree	Neutral	Agree	Strongly agree
6%	44%	10%	27%	13%

6 Conclusion and Recommendations

This paper reviewed the existing scenario of India in energy management and took a case study of Tamil Nadu in detail to the challenges. The scope of urban energy management is emerging into a new paradigm and developing countries are still struggling little for proper management in terms of resources. In the case of India, the supply of renewable energy resources is far better compared to the previous years. From the review of the current scenario of urban energy management in India, the following recommendations are suggested to improve energy efficiency and its management in a holistic manner.

Proper Utilization of Renewable Energy Resources. India has good renewable energy resources compared to other developing countries. Increasing the existing capacity of production will be significant enough to tackle the energy demand. Not only on a large scale, even individual residential buildings can be able to produce energy from solar cells, and the government is taking good initiatives. With proper utilization of renewable energy on large-scale plants and small-scale projects, we can achieve a sustainable energy supply throughout the country.

Reduce CO_2 Emission. The emission of CO_2 is quite inevitable in the urban environment, and it is experienced in the rural sector also. Most of the emission comes from the vehicles and the building sectors. A few cities in India, have taken the initiative to control the CO_2 emission by reducing the usage of old vehicles. Though electric vehicles may not actually mean a sustainable solution (unless the energy is obtained from renewable resources), they somehow reduce air pollution. With this, we can improve the air quality and create a healthier environment for all inhabitants.

Optimization of Existing Energy Management Model. The current energy management models can be optimized to bring more efficiency with respect to spatial relationships. The process of shaping the urban infrastructure would be taken with urban building energy modeling (UBEM) to get immediate insight into the carbon and energy trade-offs for the urban environment. Based on the iteration and various planning permutations, the future energy prediction for urban planning can be made.

Ocean Thermal Energy Conversion (OTEC). OTEC and desalination synergy can offer significant benefits, especially in regions like Tamil Nadu where there is a high demand for fresh water and the potential for utilizing ocean energy resources. OTEC harnesses the temperature difference between warm surface water and cold deep water to generate electricity. By integrating OTEC systems, Tamil Nadu can produce renewable energy, contributing to its energy needs in an environmentally sustainable manner.

Open Access to the Data. Large-scale data at the urban level is still not accessible even for research and development in the area of urban energy management. An inadequate data-sharing platform leads to delays in the research process. Setting up an urban lab state-wise and providing open access to large-scale data would provide a platform for more innovative ideas and solutions to the existing problems.

Public Awareness. Climate change and its associated effects need to be brought to the public notice. The young generation should be nurtured about energy efficiency and its significance to the environment. Sustainability should be brought into practice in all sectors, for instance, young people can avoid gasoline vehicles and can practice bicycles. Like this, we should educate the young generation to practice and respect the environment, instead of adopting the technologies in a recreational way.

Off-grid Energy Production for Public Facilities. Public spaces, facilities and infrastructures could use off-the-grid power management where the electricity is generated via PV panels & stored in the battery to be used when needed. This would decrease the dependency on a centralized power supply.

Integrated District-Wise Energy Management. For new and developing cities the energy generation, supply a management could be done district-wise, where a centralized plant relies on renewable/low-emission sources. This method could better optimize energy management within the city districts and help monitor the usage efficiently. During the crisis, this method could help in connecting each district and sharing power as the grids are interconnected.

References

Akadiri, S. S., & Adebayo, T. S. (2022). Asymmetric nexus among financial globalization, non-renewable energy, renewable energy use, economic growth, and carbon emissions: Impact on environmental sustainability targets in India. *Environmental Science and Pollution Research, 29*(11), 16311–16323.

Calvillo, C. F., Sánchez-Miralles, A., & Villar, J. (2016). Energy management and planning in smart cities. *Renewable and Sustainable Energy Reviews, 55*, 273–287.

Elavarasan, R. M., Shafiullah, G. M., Padmanaban, S., Kumar, N. M., Annam, A., Vetrichelvan, A. M., & Holm-Nielsen, J. B. (2020). A comprehensive review on renewable energy development, challenges, and policies of leading Indian states with an international perspective. *IEEE Access, 8*, 74432–74457.

IEA. (2019). World Energy Outlook 2019, Renewables. Accessed online: https://www.iea.org/reports/world-energy-outlook-2019/renewables

IEA. (2021). Renewable electricity growth is accelerating faster than ever worldwide, supporting the emergence of the new global energy economy. Accessed online: https://www.iea.org/news/renewable-electricity-growth-is-accelerating-faster-than-ever-worldwide-supporting-the-emergence-of-the-new-global-energy-economy

India Energy Outlook. (2021). Accessed online: https://iea.blob.core.windows.net/assets/1de6d91e-e23f-4e02-b1fb-51fdd6283b22/India_Energy_Outlook_2021.pdf

Kang, S. H., Islam, F., & Tiwari, A. K. (2019). The dynamic relationships among CO_2 emissions, renewable and non-renewable energy sources, and economic growth in India: Evidence from time-varying Bayesian VAR model. *Structural Change and Economic Dynamics, 50*, 90–101.

Kumar. J, C.R., & Majid, M. A. (2020). Renewable energy for sustainable development in India: current status, future prospects, challenges, employment, and investment opportunities. *Energy, Sustainability and Society, 10*(1), 1–36.

Ministry of New and Renewable Energy. (2023). Annual Report 2022–23. Accessed online: https://mnre.gov.in/img/documents/uploads/file_s-1683099362708.pdf

Ministry of New and Renewable Energy (2023). State-wise installed capacity of Renewable Power as on 31.05.2023. Accessed online: https://mnre.gov.in/img/documents/uploads/file_s-1686304505453.pdf

Modrzyński, P., & Karaszewski, R. (2022). Urban energy management—A systematic literature review. *Energies, 15*(21), 7848.

Mosannenzadeh, F., Bisello, A., Vaccaro, R., D'Alonzo, V., Hunter, G. W., & Vettorato, D. (2017). Smart energy city development: A story told by urban planners. *Cities, 64*, 54–65.

Pappas, D., & Chalvatzis, K. J. (2017). Energy and industrial growth in India: The next emissions superpower? *Energy Procedia, 105*, 3656–3662.

Raghuwanshi, S. S., & Arya, R. (2019). Renewable energy potential in India and future agenda of research. *International Journal of Sustainable Engineering, 12*(5), 291–302.

Sahoo, M., & Sahoo, J. (2022). Effects of renewable and non-renewable energy consumption on CO_2 emissions in India: Empirical evidence from disaggregated data analysis. *Journal of Public Affairs, 22*(1), e2307.

Sangroya, D., & Nayak, J. K. (2015). Development of wind energy in India. *International Journal of Renewable Energy Research, 5*(1), 1–13.

Tamil Nadu Solar Policy. (2019). Tamil Nadu Energy Development Agency, June 2019. Accessed Online: http://teda.in/wp-content/uploads/2019/02/SOLARPOLICY2019

Tamil Nadu Energy Development Agency. (2019). Accessed Online: http://teda.in/

UN-Habitat, 2016. Urban energy. Accessed online: http://www.unhabitat.org

Yenneti, K., Rahiman, R., Panda, A., & Pignatta, G. (2019). Smart energy management policy in India—A review. *Energies, 12*(17), 3214.

Yenneti, K. (2016). The grid-connected solar energy in India: Structures and challenges. *Energy Strategy Reviews, 11*, 41–51.

Optimizing Daylighting Performance for Educational Buildings in Cairo by Using Parametric Material System

Hoda Abd Elmonem Abd Allah Oraby,
Hossam El-Deen Hassan Elborombaly,
and Motaz Abd El Fattah Mohamed

Abstract

The research aims to conduct an empirical study on parametric material systems using the Grasshopper simulation program to enhance natural lighting in a classroom at El-Obour School in Cairo, Egypt. By manipulating variables such as window-to-wall ratio, window height, windowsill height, exterior shading device parameters, and interior wall finishing materials, the study evaluates various parametric systems. Emphasizing a low-cost strategy for advancing contemporary teaching methods, achieving cost savings, and reducing pollution, the focus is on energy efficiency in schools. Despite schools being significant consumers of energy and resources, recent advancements in materials and technology have introduced innovative building techniques, rejuvenating centuries-old practices. The study envisions the potential impact of these materials and techniques on future structures and environments, aiming to enhance sustainability and efficiency across products, structural engineering, and support systems. Utilizing parametric systems, the research seeks to elevate the effectiveness of architecture while minimizing its environmental footprint. This exploration aligns with an environmentally conscious design approach inspired by the natural evolution of smart system material properties, emphasizing adaptability to environmental changes. The combination of smart, kinetic material systems with architecture aims to explore their capacity for delivering increased comfort and significantly more energy-efficient structures.

Keywords

Daylighting · Parametric design · Algorithmic design · Energy efficiency · Computer-aided design

1 Introduction

The sector of Egyptian educational buildings uses a lot of energy as shown in Figs. 1 and 2, The market for educational buildings is expanding continuously.[1] The construction of school buildings in Egypt is given top priority by society despite the wide range of other urgent issues. Especially after raising the building ratio for public schools in new cities from 30 to 35% to improve the schools' capacity.[2] An intelligent choice could reduce these energy requirements of envelope materials. The suggested method was developed using Ladybug and Honeybee's daylight and daylighting simulation as the primary driving a design tool development. It will demonstrate the advantages of applying parametric design tools in architecture. Procedure for getting a particular thermal quality and daylight followed by an explanation of the rationale behind using a combined valuation tool a closed-loop procedure between simulation and design that optimizes the architectural design in accordance with the needs for thermal and daylight requirements is developed in conjunction with a parametric modelling tool.

The electricity sector in Egypt is currently experiencing a new crisis, prompting the government to adopt a policy of load shedding, reaching more than 10 h in some areas daily.

The Egyptian government attributes the return to load shedding to the severe heatwave currently affecting Egypt, leading to a significant increase in electricity consumption.

H. A. E. A. A. Oraby (✉) · H. E.-D. H. Elborombaly · M. A. E. F. Mohamed
Architectural Department, Faculty of Engineering, Ain Shams University, Cairo, Egypt
e-mail: hodaoraby@outlook.com

M. A. E. F. Mohamed
e-mail: moatazafatah@eng.asu.edu.eg

[1] Ministry of Electricity and Energy (2015).

[2] New Urban Communities Authority—http://www.newcities.gov.eg-2023.

B. He et al. (eds.), *Urban Climate Change Adaptation*, Advances in Science, Technology & Innovation,
https://doi.org/10.1007/978-3-031-65088-8_4

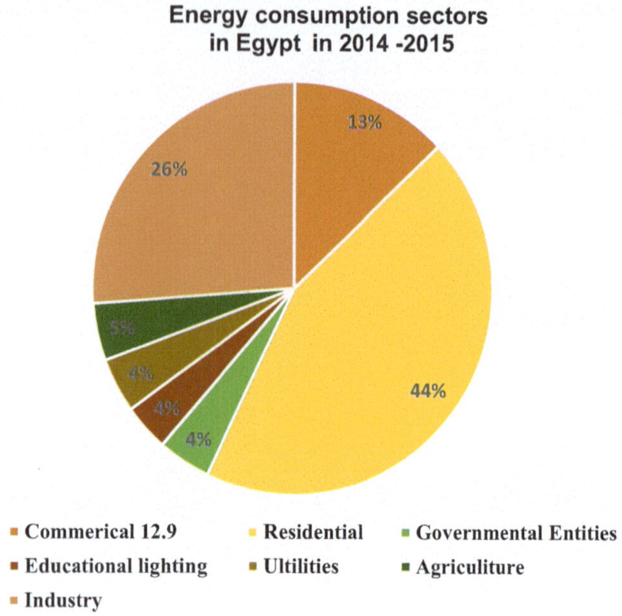

Fig. 1 Energy consumption sectors 2015. Ref: Annual report of the Egyptian Electricity holding company, ministry of Energy Cairo, Egypt Electricity consumption by sector 2022—www.statista.com

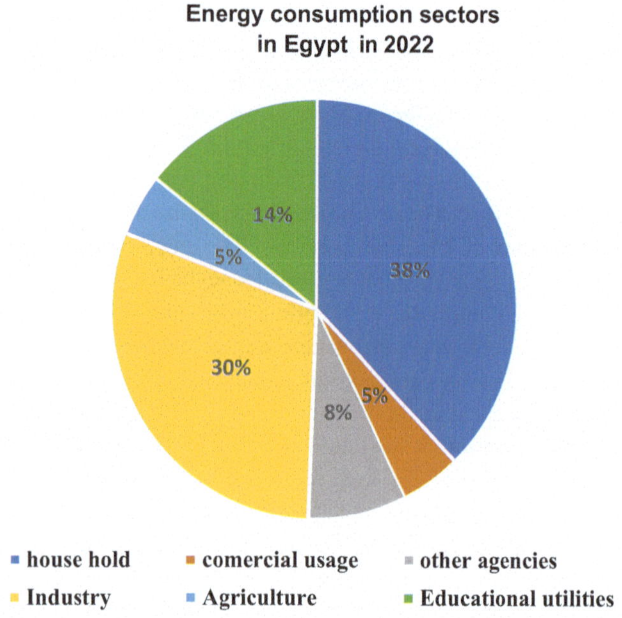

Fig. 2 Energy consumption sectors 2022. Ref: Annual report of the Egyptian Electricity holding company, ministry of Energy Cairo, Egypt Electricity consumption by sector 2022—www.statista.com

The increased electricity consumption in Egypt has resulted in a rise in the volume of gas used for power generation, putting intense pressure on its networks, and causing a decrease in gas pressure in the networks connecting to power stations.

Data from the British Energy Institute shows that the total electricity generation in Egypt declined to 200.8 terawatt-hours in the past year, compared to 209.7 terawatt-hours in 2021, representing an annual decrease of 4.2%.

Despite its decreased share last year, gas remains the largest source of electricity generation in Egypt at over 79%, but this is lower than the previous year, which was 82.9%. Electricity generated from natural gas decreased to 159.3 terawatt-hours in the past year, compared to 174 terawatt-hours in 2021, marking an 8.5% decline. Therefore, optimal utilization of natural lighting was deemed necessary as a solution to coincide with the economic crisis in Egypt.[3]

Since the nineteenth century, western nations, particularly those in Europe, have been interested in the study of daylight conditions in educational structures. Providing a view of the outside or even using it could reduce students' performance without creating a pleasant and healthy environment, even though it has been argued that natural light, which is more stable and controllable, is preferable to artificial light. There is currently a consensus that including natural light when designing spaces is crucial. By illustrating the various parametric approaches and utilizing technological solutions, this study looks at how the research community has gone about understanding and resolving the intricate connections between the local climate, user needs, and design limitations in educational facilities. Considering the Fig. 2 suggested. The reported case studies, which are based on fully comprehending the complex requirements of a daylit educational environment, when discussing experimental evaluation or simulations, emphasized the significance of a comprehensive investigation of the subject. To fully understand the complex requirements of a classroom with natural light and illustrate the significance of an in-depth approach to the subject.

2 Literature Review

The literature review provides a detailed overview of the role of optimization, parametric modeling, and simulation-based methods in building design. Here's a breakdown of the key points[4]:

2.1 Optimization Definition and Scope

- Optimization involves determining the minimum or maximum value of a function.[5]

[3] www.energyinst.org.

[4] Makhairas et al. (2014).

[5] Nguyen et al. (2014).

- Building design optimization spans various sciences, including mathematics, economics, environmental science, and computer science.[6]

2.2 Parametric Simulation Method

- Building performance is developed using the parametric simulation method.
- This involves altering one variable while keeping others constant to observe its impact on design goals.[7]

2.3 Benefits of Parametric Modeling

- Parametric modeling allows for transformations leading to various arrangements of geometric components.[8]
- Automation and software availability have facilitated the pursuit of parametric analysis.

2.4 Parametric Modeling Tools

- Tools like Grasshopper, Dynamo, and Generative Components are employed for parametric modeling.
- Plugins like Honeybee and Mr. Comfy enhance the capabilities of Grasshopper.[9]

2.5 Evolutionary Algorithms for Optimization

- Genetic algorithms and evolutionary algorithms are used for searching combinations of independent parameters within the solution space of parametric models.[10]

2.6 Challenges in Simulation-Based Optimization

- Complex nature of building simulation outputs, computational costs, and multi-objective design issues pose challenges.
- Multi-objective optimization, particularly using evolutionary algorithms, has gained popularity.-[11]

2.7 Integration of Optimization in Architecture

- Automated optimization methods, like genetic algorithms, are increasingly utilized in architecture for finding design alternatives meeting specific criteria.[12]

2.8 Case Studies and Applications

- Integration of Building Performance Simulation (BPS) with parametric modeling is employed for optimizing building performance.
- Case studies involving shading louvres for daylighting in a classroom and the use of Paragren tool for improving environmental performance are highlighted.[13]

2.9 Passive Performance Optimization Framework (PPOF)

- PPOF combines parametric modeling, BPS, and GAs in a closed-loop cycle for optimal daylighting and energy use.

2.10 Parameters for Optimization

- Independent parameters for optimization include building geometry, window-to-wall ratio (WWR), shading devices, orientation, and location.

2.11 Tools and Plugins

- Ladybird and Honeybee oversee BPS, Octopus serves as a GΛs operator, and Grasshopper oversees parametric modeling.[14]

The literature review provides a comprehensive exploration of optimization methods in building design, with a focus on the integration of parametric modeling, simulation-based techniques, and evolutionary algorithms for improved performance and sustainability. The inclusion of specific case studies and tools enhances the practical understanding of these concepts. Here is a breakdown of the case studies that the research focus on it:

---

[6] Nguyen et al. (2014).
[7] Lin and Gerber (2014).
[8] Lin and Gerber (2014).
[9] Turrin et al. (2011).
[10] Turrin et al. (2011).
[11] Konis et al. (2016).

[12] Østerga° rd et al. (2016).
[13] Turrin et al. (2011).
[14] Evins (2013).

2.11.1 El Sheikh and Gerber[15]

- Investigated the design of an intelligent building skin and its impact on daylighting performance using a parametric approach.
- Utilized Grasshopper, Rhino, Diva, and Galapagos as parametric tools to explore semi-autonomous and fully autonomous configurations of an exterior building envelope louver system.
- Found that daylighting performance improved in interior spaces when considering occupants' preferences and building requirements.
- Concluded that the automation of independent tilt angles can enhance the Building Intelligent Skin system, achieving twice the performance of a conventional system in terms of daylight harvesting.

2.11.2 Sjarifudin and Justina[16]

- Investigated automated louvers using a parametric camshaft design method.
- Simulated a workspace in Jakarta with varying shading angles to provide sufficient daylight inside the workspace.
- Concluded that occupants' visual comfort can be achieved by adjusting shading opening angles between 15 and 75°.

2.11.3 Jaepil, Taekyoung, Euisoon, and Gensong[17]

- Used parametric louvers to optimize thermal performance in buildings.
- Analyzed the best adjustments for louvers, considering the distance from glazing, louver's depth, and louver's tilt angle.
- Conducted simulations using Rhinoceros, Grasshopper, Ecotect, and Galapagos.

2.11.4 Maria and Aris[18]

- Investigated dynamic and automated shading systems using sensor points and controlled algorithms responsive to sun movement and weather conditions.
- Concluded that popular automated systems and motorized Venetian blinds can achieve energy savings of 20% and 50% for cooling and lighting, respectively, compared to conventional louvers.

[15] El-Sheikh and Gerber (2011).

[16] Sjarifudin and Justina (2014).

[17] Choi et al. (2014).

[18] Konstantoglou and Tsang (2016).

2.11.5 A Previous Study[19]

On, a parametric approach was employed to investigate louvers design and its impact on daylighting:

a. **Simulation Process**
   - A parallel simulation was conducted, linking Grasshopper with Radiance through an algorithm known as Diva.
   - Radiance and EnergyPlus interface engines were used in the simulation process.

b. **Findings**
   - Increasing louvers count from 3 to 10 led to a decrease in spatial Daylight Autonomy (sDA) from 40 to 30% at 30% WWR with a 1.5 depth ratio.
   - With the same depth ratio, sDA decreased from 76 to 61% at 50% WWR.
   - Increasing the depth ratio from 0.75 to 1.5 at 20% WWR resulted in a 40% decrease in sDA, diminishing as WWR increased until sDA reached 100% at 60% WWR.
   - Louvers count had no significant influence on sDA concerning louvers depth ratio, which was found to have a more substantial impact on sDA.

c. **Conclusion**
   - The study's utility lies in exploring all parameters in all screens, facilitating effective evaluation.
   - Accurate classification of parameters was achieved to identify optimum solutions.
   - However, studying many parameters together may not yield accurate overall performance results, suggesting the importance of individual parameter studies while keeping others fixed.

2.11.6 Pil and Steffen's Study[20]

Focuses on employing a parametric method in Grasshopper to generate diverse fenestration systems for a room that meets specific performance criteria related to indoor climate, daylighting, and energy conservation. The key points from the study are as follows:

a. **Parametric Method in Grasshopper:** The research utilizes a parametric approach within Grasshopper, a visual programming language and environment. The goal is to generate various fenestration systems that fulfill performance criteria, addressing indoor climate, daylighting, and energy-saving aspects.

b. **ICE Bear Plugin**: A new plugin in Grasshopper, called ICE bear, has been introduced.
   ICE bear can simulate the annual indoor climate, daylight, and energy performance of complex geometries. It utilizes hourly weather data for accurate simulations.

[19] Wagdy and Fathy (2015).

[20] Lauridsen and Petersen (2014).

c. **Integration with Galapagos Optimization Algorithm:** ICE Bear is used in conjunction with the Galapagos optimization algorithm. The Galapagos algorithm is employed to explore and find design solutions based on the designer's conceptual ideas.
d. **Performance Criteria**: The objective is to achieve specific performance criteria for the designed fenestration systems. The criteria include reaching a maximum energy use for building operation (25 kWh/m^2/year), meeting thermal class II requirements (not exceeding 100 h above 26 °C), and achieving a minimum daylight autonomy of 60% at a specific point in the middle of the room.
e. **Optimization Goal:** The primary aim of the method is to find solutions that maximize energy efficiency while adhering to indoor environmental requirements.
f. **Conclusion:** The study concludes that the parametric method, combined with the ICE Bear plugin and Galapagos optimization algorithm, can generate solutions that align with the designer's conceptual ideas and successfully meet the specified indoor environmental requirements.

3 The Problem Definition

The research problem of this study is divided into two sections as shown in Fig. 3:

- The main section is the Lack of daylighting in educational spaces in schools in the preliminary stage.

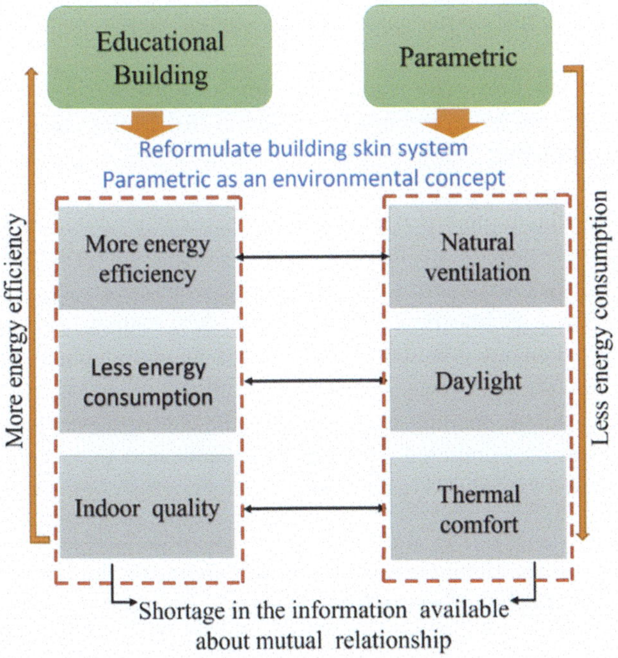

Fig. 3 The connection between parametric design and educational buildings. *Source* Authors

- The second section is to find out what kind of relationship there is between intelligent parametric materials and educational spaces.
- Non-observance of design standards and foundations in the design of most educational buildings in Cairo, especially schools Elementary—the place of study.
- Ignoring the role of sustainable design and the concept of sustainability as one of the trends affecting the quality of the educational process and the preservation on the environment.
- Identifying the applicability of intelligent parametric systems in less energy consumption and more energy efficiency in an educational space is the sub-objective of the study.

Implementing new construction methods and materials, as well as new material and technological advancements. Some of these methods even put a modern spin on age-old customs. Future structures and spaces may benefit from methods like kinetic facades and parametric brick facades.

4 The Research Objectives

The research objectives outlined focus on addressing daylighting challenges in educational buildings through the utilization of smart material techniques suitable for the Greater Cairo climate. Here's a breakdown of each specific objective.

4.1 Analyzing Parametric Material Systems

- Goal: Evaluate the effectiveness of parametric material systems in addressing environmental challenges within educational buildings.
- Approach: Conduct a comprehensive analysis of different parametric material systems, considering their adaptability and responsiveness to environmental factors.

4.2 Reformulating Building Skin Systems

- Goal: Improve energy efficiency and reduce energy consumption by rethinking the design of building skin systems.
- Approach: Explore innovative approaches to reformulate building skins, considering factors such as insulation, reflectivity, and adaptability to climatic conditions.

4.3 Identifying Environmental Considerations for Facades

- Goal: Determine specific environmental considerations relevant to the design of educational building facades.

Table 1 The optimization process involved changing the classroom parameters

Parameters	Attributes	No of value
WWR	10–50% With an increment of 10%	5
Glass material	variety compositions of thickness and materials	4
Wall construction	variety compositions of thickness and materials	5
Shading depth	70 cm With an increment of 25% of the window height	6
Shading Angle	2–10 with an increment of 2	4
Shading count	35–75% with an increment of 15	10

- Approach: Investigate how building facades can contribute to or mitigate environmental challenges, such as optimizing natural light, controlling heat gain, and managing airflow.

4.4 Enhancing User Comfort Within Educational Buildings

- Goal: Increase the likelihood of providing consistent comfort for occupants in educational buildings.
- Approach: Explore design strategies that prioritize user comfort, considering factors like temperature regulation, daylighting quality, and overall indoor environmental quality.

The significance of incorporating intelligent material systems is highlighted, emphasizing their potential to offer innovative solutions to long-standing challenges. Smart materials with responsive and adaptive properties are seen as instrumental in optimizing energy usage, creating environmentally friendly structures, and improving user comfort. The aim is to contribute to the overall efficiency and sustainability of educational facilities, aligning with contemporary environmental and energy conservation goals.

5 Methodology

Various research methods, tools, and techniques were employed to achieve the objectives and purposes of the study, including (deductive, inductive, historical… external) to construct a comprehensive analytical framework based on the selected theoretical background.

- Analytical deduction through literature review for environmental modeling of educational buildings.
- Deductive analysis of the interplay concept for modeling information in educational building.
- Inductive analysis of the precedents of strategies for modeling information in educational buildings.
- Framework creation: for a non-homogeneous executable interplay design algorithm management system.
- Application of a suitable case study to validate the proposed framework.
- The optimization in this paper's framework would automatically, evaluated on a single canvas acting as a model or simulation. The canvas of the Grasshopper plug-in contains most of this parametric framework's data as shown in Table 1.
- The research methodology of this study is divided into three sections:

A **Definition of reference case & parametric techniques of the study**.

The framework is initially created with the case study zones (one classroom) and all necessary parameters Are:

- Width: 6.20 m
- Height: 2.90 m
- Length: 8.45 m

B. **Analyzing the application of the parametric technique by grasshopper 3d Rhino (6), Honeybee and ladybug**

- A classroom 's performance is influenced by several variables, which are listed in Table 1. Classrooms perform differently in terms of energy and daylighting.
- These architectural limitations influence natural lighting and energy which are mentioned in the determinants of architectural form. such as WWR, glazing
- material, shading devices, and wall construction material, are combined. There are many indicators that can be used to evaluate performance.

C. **Simulations then evaluate and compare the results with the reference case**

- A variety of software and simulation engines are used in this process. For instance, Grasshopper will oversee

Fig. 4 Workflow for parametric optimization in Grasshopper
Source: The Authors

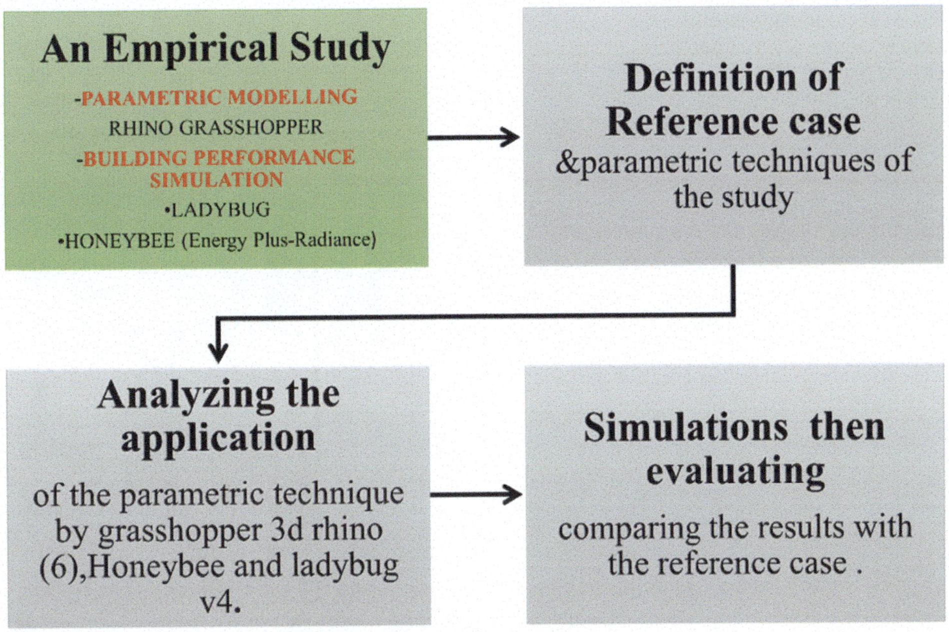

parametric modelling, and Ladybug and Honeybee will take the lead in simulating and visualizing building performance. By using honeybee radiance, the daylighting analysis will be controlled. The multi-objective

- ladybug and honeybee plug-in will be used to optimize strategies following simulation and parametric modelling.
- Metric known as "SDA300/50%" based on LEED IEQ credit daylighting.

Measures the percentage of floor space that is illuminated by 300 lux for at least 50% of the occupied hours between 8 and 2 throughout the year.[21]

- The simulation settings for daylighting and energy will then be determined. In response, honeybee plugin HB tools can run numerous simulations with various pairings even as the two indicators are optimized. The honeybee canvas depicts the front of the classroom, and the objective function will be used to determine the best solutions. Setting up the entire workflow for parametric optimization is covered in detail in this section. Site selection, weather and climate information, geometry
- creation, which includes creating masses, and window installation are the five sequential steps that make up the workflow. Optimization, energy modelling, daylight modelling, and the addition of awnings and setting materials. in the manner as demonstrated in Fig. 4.

The Research Tools

- **ConText analysis.**
- **Software program.**
- **Rhinoceros 3D version [6] program with Grasshopper program plugin.**
- The Rhinoceros 3D -computer-aided design (CAD)- application uses the Grasshopper visual programming language and environment. David Rutten created it. in the year 2007. Programs are built by dragging components onto the drawing board and connecting their outputs to the inputs of succeeding components. It creates the possibility that the Record History feature in Rhino could evolve into a comprehensive tool that aids in formulating policies based on the history of producing the project's object. As a result, Grasshopper has produced results in less time and with a significantly higher level of geometric control than can be achieved using traditional techniques.[22]
- **ladybug &honeybee version [4] plugin.**
- **Ladybug**: The Ladybug Tools software suite supports environmental design and instruction. As one of the market's most comprehensive environmental design applications, Ladybug Tools connects 3D Computer-Aided Design (CAD) interfaces to a variety of trusted simulation engines.[23]
- **Honeybee version [4]** Honeybee supports in-depth natural light and thermodynamics, modelling, which

[21] Www.usgbc.org/leed LEED v4.1 BUILDING DESIGN AND CONSTRUCTION April 2023, page, (258).

[22] www.rhino3d.com/6/new/grasshopper. Up dated 2023.

[23] www. ladybug. Tools Up dated 2023.

Fig. 5 Plans for the school's main entrances. *Source* Cairo's General Authority for Educational Buildings

are frequently crucial in the middle and later stages of design. It specifically creates, runs, and displays the results of the radiation and daylight simulations using Radiance, as well as energy models using Energy Plus/Open Studio. By integrating these engines into the Grasshopper/Rhino CAD environment, it accomplishes this. It also serves as an object-oriented Software Development Kit (SDK) for these engines.[24]

5.1 Site[25]

"El Ressala Language School" is an elementary school located in El Obour City, which is situated northeast of Cairo, Egypt. The school was constructed on a piece of land estimated to be 7,500 square meters in size. The school buildings occupy 30% of this total area.

As shown in Figs. 5 and 7, Here are some additional details about the school and its location:

- Location: El Obour City, which is 71 km northeast of the Pyramids district and 41 km from Cairo's central business district, as shown in Fig. 6.
- Project Plot: The school is in Plot 8 within El Obour City.
- Total Number of Students: The school serves a total of 700 students.
- Date of Creation: The school was established in the year 2007 A.D.

5.2 Climate and Weather Data[26]

- **Köppen Climate Classification (Group B)**: Egypt's climate falls under Group B in the Köppen Classification system. Group B represents arid, or desert climates characterized by extremely hot and dry conditions. This classification is primarily due to Egypt's geographical location within the mid-latitude global desert zone.[27]
- **Exception in the North**: The only exception to this arid desert climate in Egypt is the northern region along the Mediterranean Sea. This coastal area is influenced.
- by the Mediterranean Sea and is therefore categorized as having a different climate. It is described as hot and muggy, which is typical of Mediterranean climates.
- **El Obour's Climate**: El Obour, located within Egypt, experiences weather conditions that closely resemble the

[24] www.ladybug.tools/honeybee. Up dated 2023.

[25] Peel et al. (2007).

[26] Attia (2012).

[27] Climate-Data.org, climate: Cairo international Airport (2012) .

Fig. 6 The location of case study. *Source* www.google.com/maps/place/Resala+Language+School

Fig. 7 Bubble diagram function by Authors. Ref: Cairo's General Authority for Educational Buildings

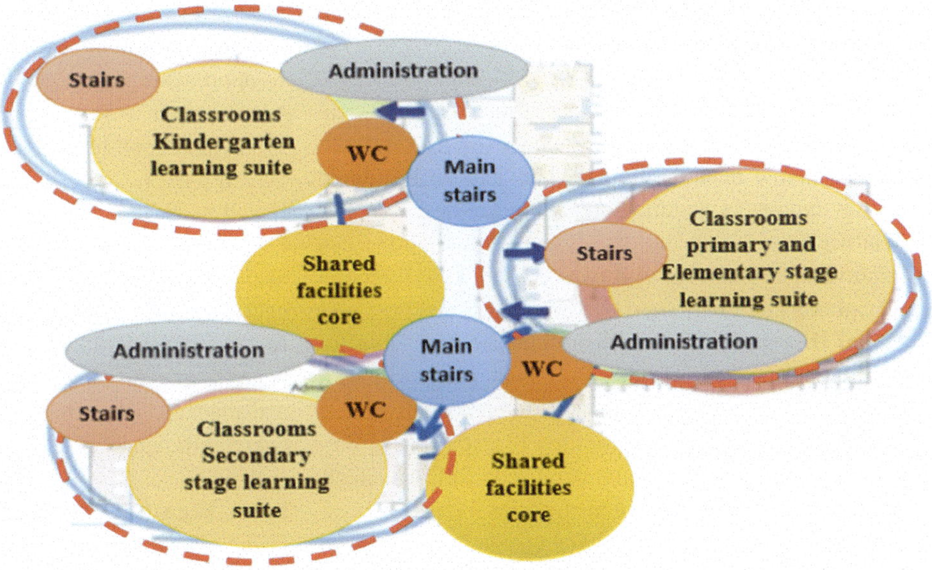

desert climate. As a result, it has extremely hot and dry weather. This type of climate is characterized by high temperatures, limited rainfall, and low humidity.

- **Daylight and Heat Gain**: Desert climates, like the one in El Obour, are known for their even distribution of daylight and significant heat gain. This is because these climates often have clear skies and receive intense direct solar radiation. As a result, buildings and outdoor spaces in desert areas can experience high levels of solar exposure and heat accumulation.

Understanding the Köppen Climate Classification and the specific climate characteristics of regions like Egypt and El Obour is essential for various purposes, including urban

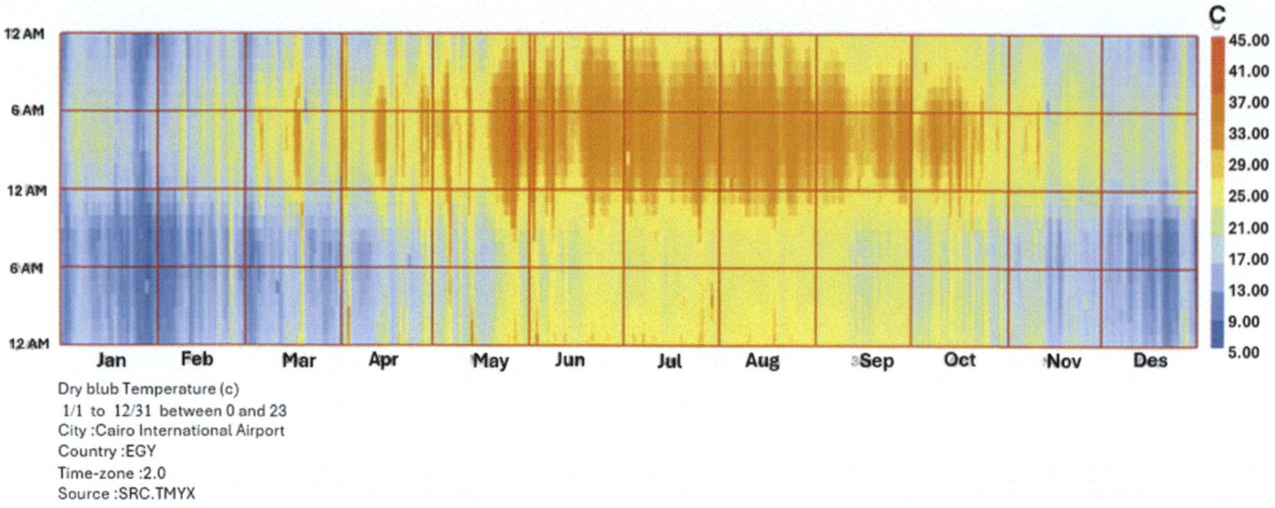

Fig. 8 Cairo International Airport's dry bulb temperature. *Source* The Authors

planning, agriculture, and architectural design, as it helps in anticipating and mitigating the challenges posed by the local climate conditions.

The clear sky conditions in a desert climate, such as the one in Cairo, can indeed offer opportunities for efficient daylighting in buildings. However, these conditions also present challenges related to visual discomfort and excessive heat gain, as mentioned. Here's a more detailed explanation: [28]

- **Efficient Daylighting**: Clear skies, especially in a location like Cairo with year-round sunny and clear weather, provide ample natural sunlight. This can be harnessed for daylighting in buildings, reducing the need for artificial lighting during the day. Effective daylighting can improve indoor comfort and reduce energy consumption for lighting.
- **Visual Discomfort**: While clear skies are beneficial for daylighting, they can also lead to visual discomfort, particularly if not properly managed. Intense sunlight can create glare, which is the sensation of excessive brightness that can cause eye strain and discomfort. Architects and designers need to consider strategies like
- shading devices, diffusing glazing, or well-designed window placements to mitigate glare and ensure visual comfort.
- **Excessive Heat Gain**: In addition to visual discomfort, clear skies and direct solar radiation can result in excessive heat gain in buildings. This can lead to higher indoor temperatures, especially during the hot summer months in Cairo when the average temperature can reach 27 °C to 34 °C. To address this issue, buildings need effective insulation, shading solutions, and energy-efficient cooling systems to maintain comfortable indoor temperatures.
- **Seasonal Temperature Variations**: It's important to note the significant difference in average temperatures between winter and summer in Cairo. Winters are relatively mild with average temperatures around 13 °C, while summers are hot, with average temperatures approaching 27 °C to 34 °C. Designing buildings that.
- can adapt to these temperature variations is crucial for occupant comfort and energy efficiency.

In summary, the clear sky conditions in a desert climate like Cairo offer opportunities for efficient daylighting, but they also come with challenges related to visual discomfort and excessive heat gain. Architects and designers must carefully consider building design and strategies to balance the benefits of natural daylight with the need to control glare and manage indoor temperatures for occupant comfort and energy efficiency.

5.2.1 Weather Data File

Using weather data from Cairo International Airport as a proxy for El Obour is a common approach when specific local weather data is not available. Given that Cairo International Airport is relatively close to El Obour, the data can still provide valuable insights into the local climate conditions. Here's a summary of the key information regarding the weather data as shown in Figs. 8 and 9a, b:

Data Source: The weather data used in the study was obtained from the website of the US Department of Energy.

[28] Computer-Aided Architectural Design (2021).

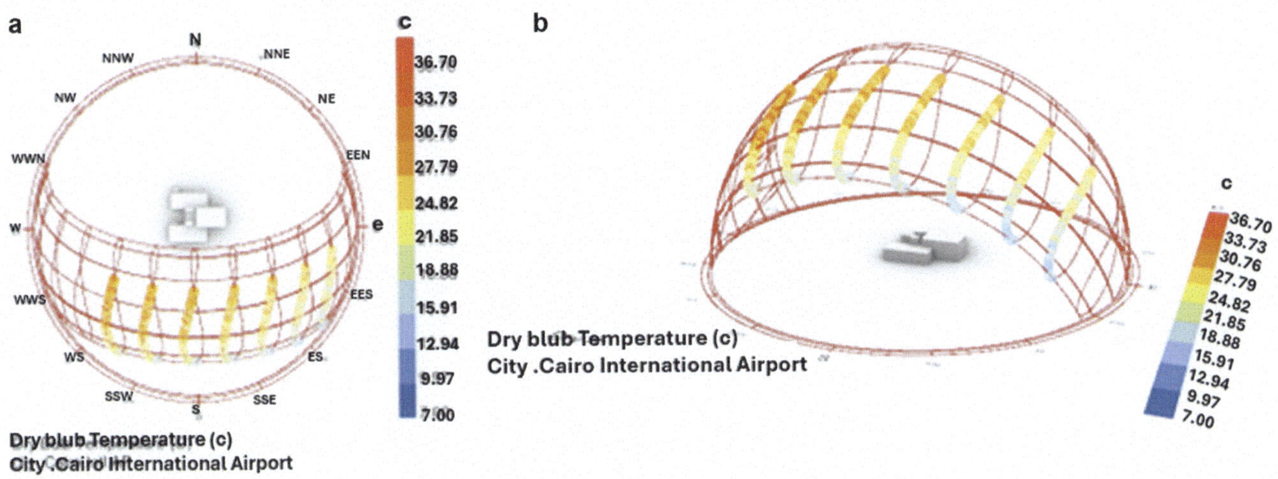

Fig. 9 **a** Diagram of the Sun's Path (By Authors) **b** Diagram of the Sun's Path (By Authors)

Proxy Location: Since there is no specific weather data file for El Obour, use weather data from Cairo International Airport, which is located 32 km away from El Obour. This is a reasonable choice, as weather conditions in nearby areas tend to be similar.

Temperature and Relative Humidity: According to the data from Cairo International Airport, the average summer temperature is approximately 27.6 degrees Celsius, and it is associated with a relative humidity of 52.3%. In contrast, the average winter temperature is 15 degrees Celsius, with a relative humidity of 56.7%. These temperature and humidity values provide a good representation of the climate in the El Obour region.

Data Accuracy: As mentioned, the most recent readings and data from El Obour city appeared to be very close to the data from Cairo International Airport. This suggests that the airport data is a reliable approximation for El Obour's climate, especially for study's purposes.[29]

5.2.2 The Massing

1. Zones for the primary school case study as shown in Figs. 10, 11 and 12 were first created in Rhino, as seen in Fig. 13, and then exported to Grasshopper. Zones had separated. classrooms, communal restrooms, multipurpose rooms, kindergarten classrooms, stairwells, and water closets are all exported.
2. All these zones were sent to the (Honeybee_Intersect Mass) component, which separates the surfaces of each zone to make sure that each adjacent surface has a matching surface. This is required to guarantee that, during an energy simulation, the conductive heat flow calculation is performed accurately across the surfaces.
3. All zone categories created by the previous component were sent to the (Honeybee_Masses2 Zones) component, which transformed each one into a Honeybee zone with all the necessary properties for an energy simulation. As can be seen in the Fig. 14, a primary school zone program was created from components of the "Honeybee_Bldg".

Programs" and "Honeybee_List Zone Programs" processes. In addition, the zone's conditional status was set to true. Zone names were also established.

5.2.3 Windows

1. Eliminating Adjacencies:
   - The first step involves eliminating adjacent surfaces, combining them into a single surface to represent interior walls. This step is illustrated in Fig. 15.
2. Generating Glazing Based on Window-to-Wall Ratio:
   - The next step involves using the "Honeybee_Glazing Based Ratio" component.
   - This component generates primary classes of windows based on the window-to-wall ratio (WWR).
   - Two parameters, WWR-East and WWR-West, are set as sliders with a step size of 0.1, ranging from 0.10 to 0.30.
   - It's mentioned that Honeybee zones were split into three groups before proceeding with this component.

Groups:

- A. Kindergarten Garden Classes: This group consists of three classes and a reception area. It appears that this group did not go through the optimization process.
- B. Primary Category: This group includes six classes and a corridor. The default WWR was set to 0.3 for this

[29] Oraby et al. (2023).

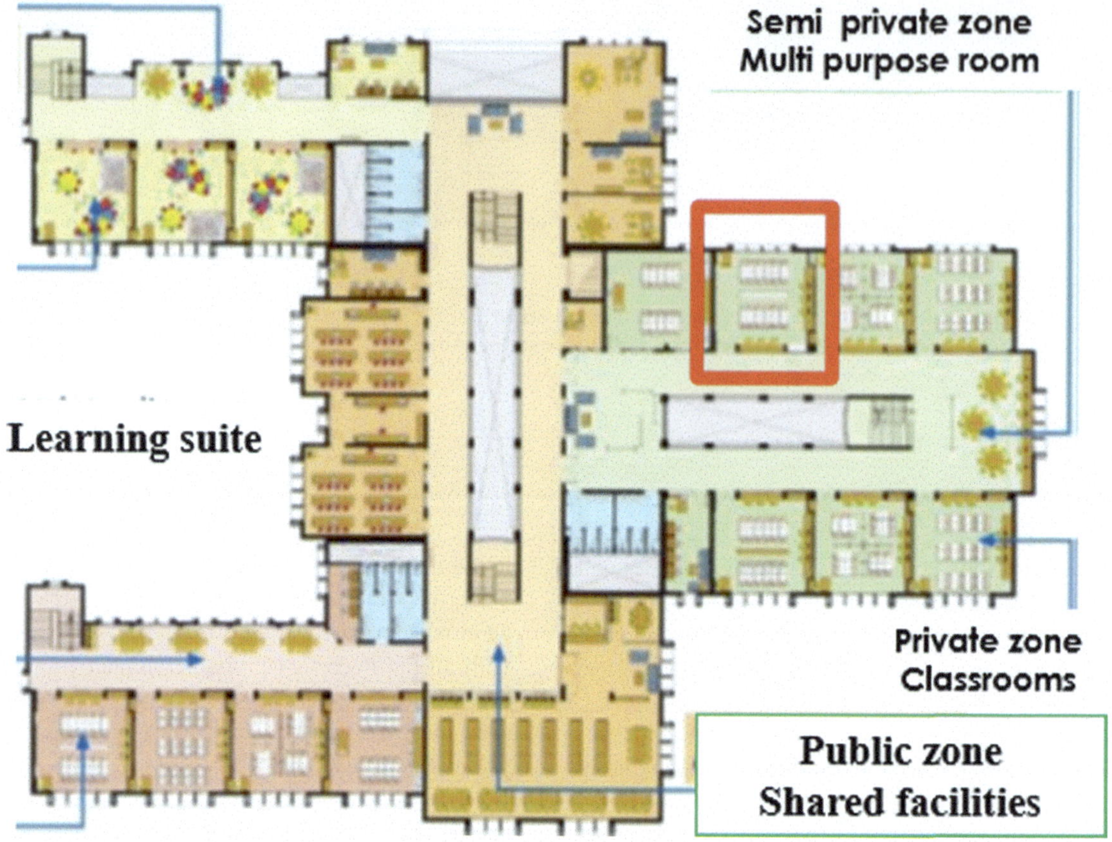

Fig. 10 Primary school El ressala's first floor plan. Ref: Egyptian Educational Buildings Authority

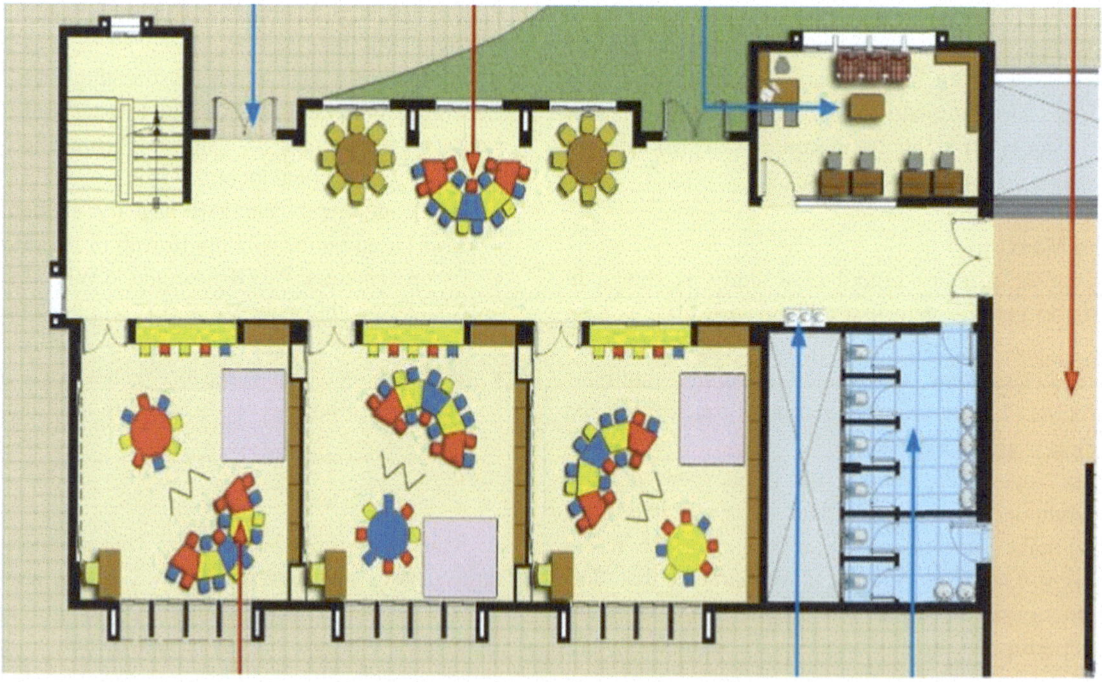

Fig. 11 Zone for kindergarten classes. Ref: Egyptian Educational Buildings Authority

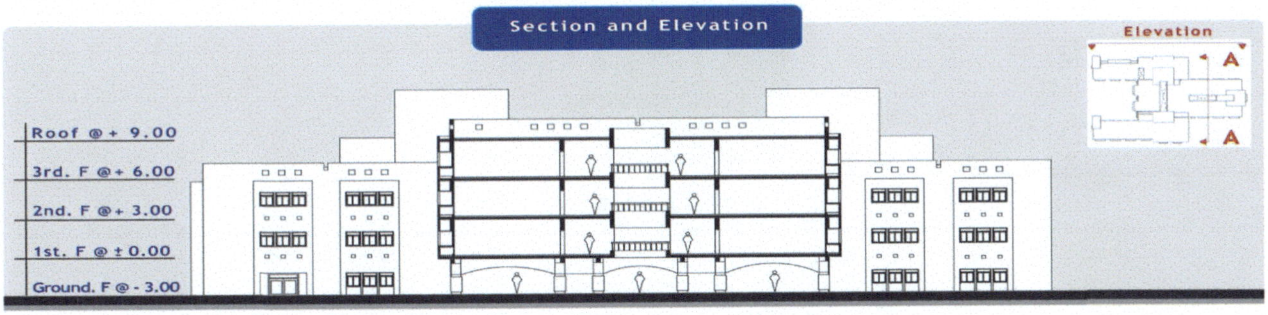

Fig. 12 The main section of El Ressala Primary School, displaying the floor levels. Ref: Egyptian Educational Buildings Authority

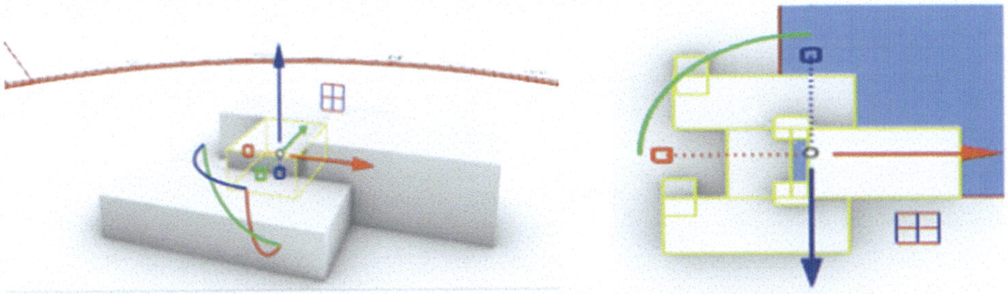

Fig. 13 Rhino was used to create a primary school's Zones paleosurfaces and the top view case study. Source The Authors

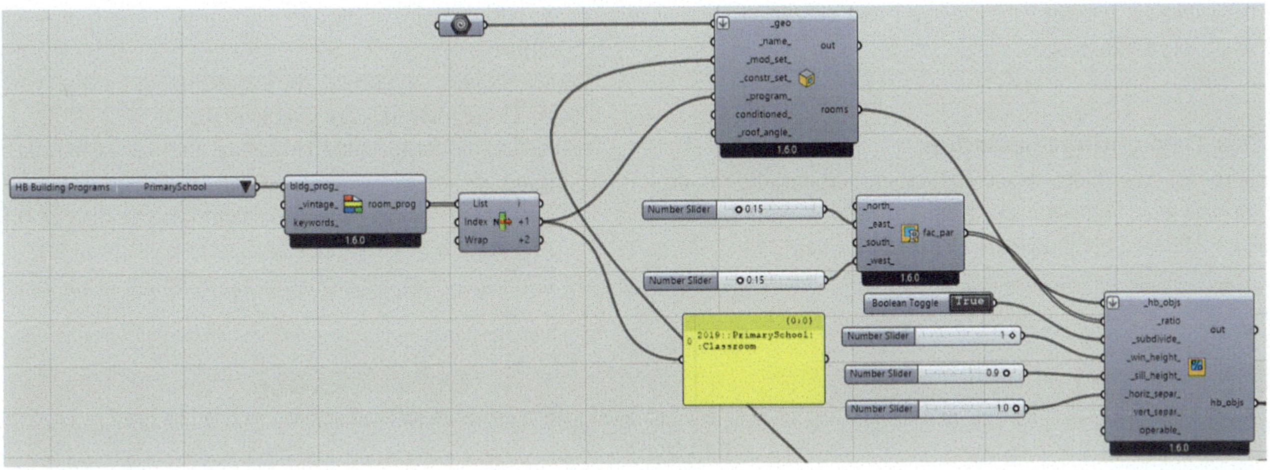

Fig. 14 Convert case study to primary school zone definition. Ref: (The Authors)

category, and it was created with the intention of focusing on the optimization framework.

- C. Preparatory School Classes: This group comprises seven classes and a corridor. Like Group A, it's mentioned that this group was not subject to the Optimization framework.

5.2.4 External Shading Devices

In the south elevation, horizontal shading structures were built to limit the amount of direct sunlight that entered zones. Three parameters were set for the (Honeybee_Energy Plus Window Shade Generator) component, as shown in Table 2.

Set to a shading angle of (75–75) with a 15-degree increment as shown in Fig. 18. A restriction was set at 0.05 m for the distance to the glass as shown in Fig. 16a, b. The (Honeybee_ EP Context Surface) component received output shading devices, which Honeybee then processed. in the final simulation Stage's context.

5.2.5 Materials

The Honeybee_Set Hb modifier set received all zones in both categories. Construction materials were placed to create the walls, floors, shading devices, and windows in this component of Fig. 17. As shown in Table 3.

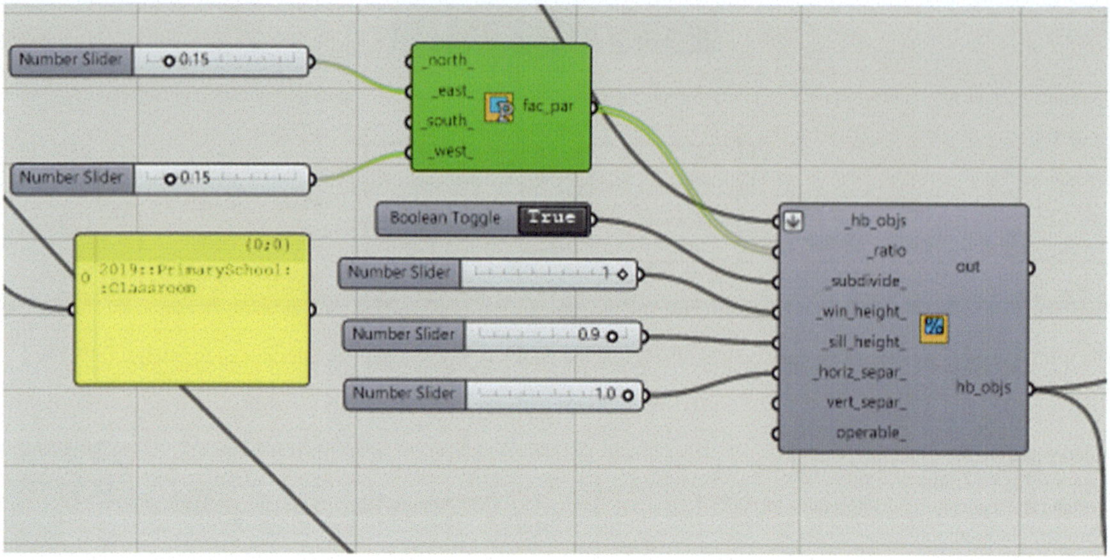

Fig. 15 Honeybee_Glazing Based Ratio developed by the researcher

Table 2 Set shading devices parameters by the researcher

The shading device depth	0.5–0.7 m with a 0.5 increment
The shading device count	2–8 with an increment of 2
The shading device distance	0.2 to 0.8 m

5.2.6 Daylighting Simulation

In Fig. 17, the "Honeybee Set Radiance Material" component was used to assign materials to different building elements. Here's a breakdown of the materials set:

A. For the walls, floors, and ceilings:
   - Off-white plaster was used for the walls.
   - Grey floor tiles were applied to the floors.
   - A white paint material was used for the room ceilings.

B. Radiance glass materials were developed for the windows. These materials are characterized by:
   - Having no reflectance (meaning they do not reflect light).
   - Being transparent with a transmittance value of 0.3 (meaning they allow 30% of light to pass through).

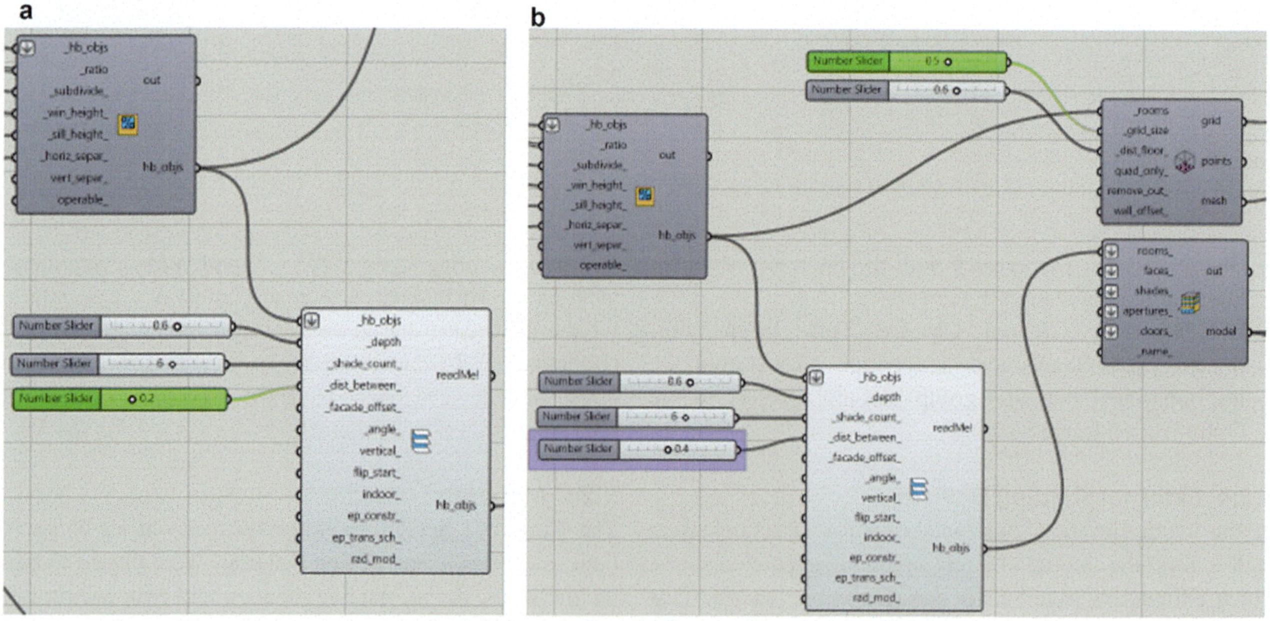

Fig. 16 **a** Window Shading Generator by Authors **b.** Window Shading Generator by Authors

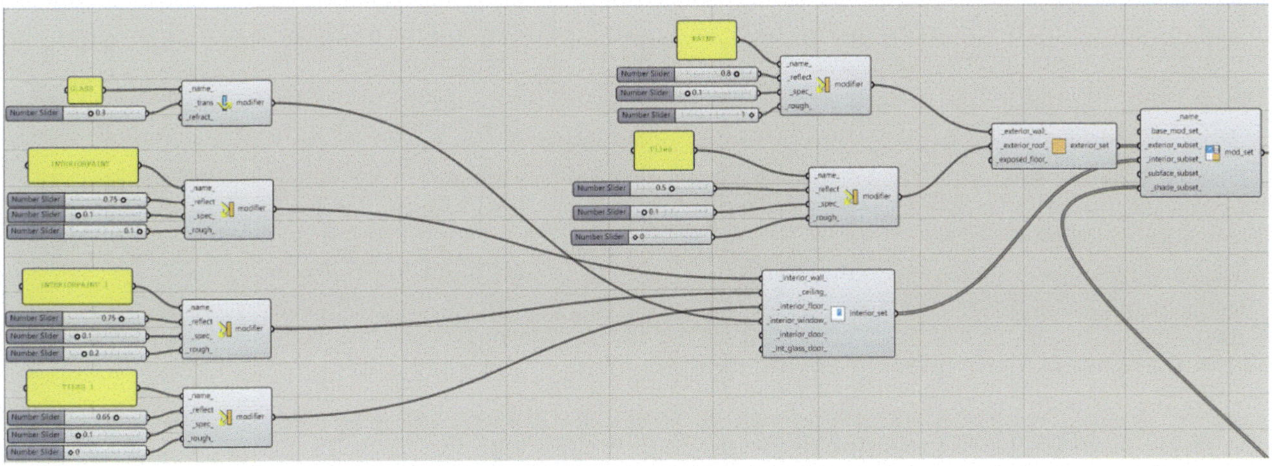

Fig. 17 Honeybee exterior modifier set (adding materials) Honeybee interior modifier set (adding materials) by Authors

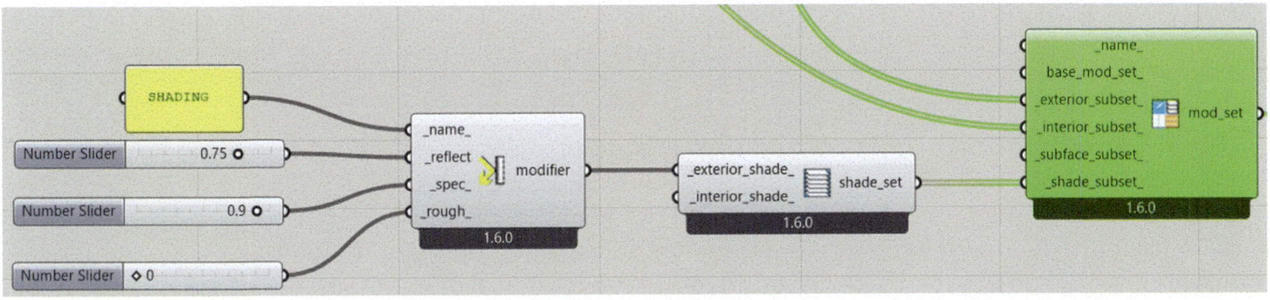

Fig. 18 Honeybee shade modifier subset by Authors

C. Shading settings were configured for the exterior shades:

- A shade modifier subset was defined.
- The reflectance of the shades was set to 0.75 (meaning they reflect 75% of incoming light).
- The specularity of the shades was set to 0.9, indicating a high degree of glossiness or shininess.

In the daylighting simulation process using the "Honeybee Annual Daylight Simulation" component, the following steps were taken:

- The recipe for the simulation was prepared to assess daylighting conditions as shown in Fig. 19a, b.
- Test points, vectors, and meshes representing test surfaces were generated using the "Honeybee Generate Test Points" component.
- The test surfaces had a grid size of 0.50 m by 0.50 m and were located at a height of 0.6 m above the finished ground level. This setup aligns with the recommendations of LEED V4 for daylighting analysis as shown in Fig. 19c, also in Figs. 19d and 20, show the results for test point 66.

Additionally, the simulation considered weather data from Cairo International Airport to accurately model the daylight conditions in the given location. This information would be valuable for evaluating the building's daylight performance and its compliance with LEED V4 standards.

The main element to produce a daylighting analysis is (Honeybee_ Run Daylight Simulation). Contextual structures, output zones, and shading tools were added to HB.

Objects. Additionally, this component's working directories, analysis recipe, and CPU count were all set.

The (Honeybee_ Read Annual Analysis Results) component was used to translate the analysis results into SDA values.

5.3 Optimization

As shown in Fig. 21 show the results for daylighting autonomy, continuous daylighting autonomy and daylighting useful illuminance for test point 66.

5.4 SDA (Special Daylighting Autonomy)

Using the Honeybee Read Annual Analysis Results component, analysis results were converted to SDA values as shown in Fig. 22.

Table 3 Material modifier set sent to honeybee plugin by Authors

Honeybee modifier set																			
Exterior modifier subset						Interior modifier subset													
Exterior wall			Exterior roof			Wall			Ceiling			Floor				Window	
Reflectance	Specularity	Roughness	Reflectance	Specularity	Roughness	Reflectance	Specularity	Roughness	Reflectance	specularity	Roughness	Reflectance	Specularity	Roughness	Transparent	Reflectance
Paint			Tiles			Interior paint			Interior paint			Tiles			Glass	
0.6	0.1	1	0.3	0.1	0	0.75	0.1	0.1	0.75	0.1	0.1	0.65	0.1	0	0.3	0

Honeybee modifier set					
Shade modifier subset					
Exterior shade subset			Interior shade subset		
Reflectance	Specularity	Roughness	Reflectance	Specularity	Roughness
0.75	0.9	0	0	0	0

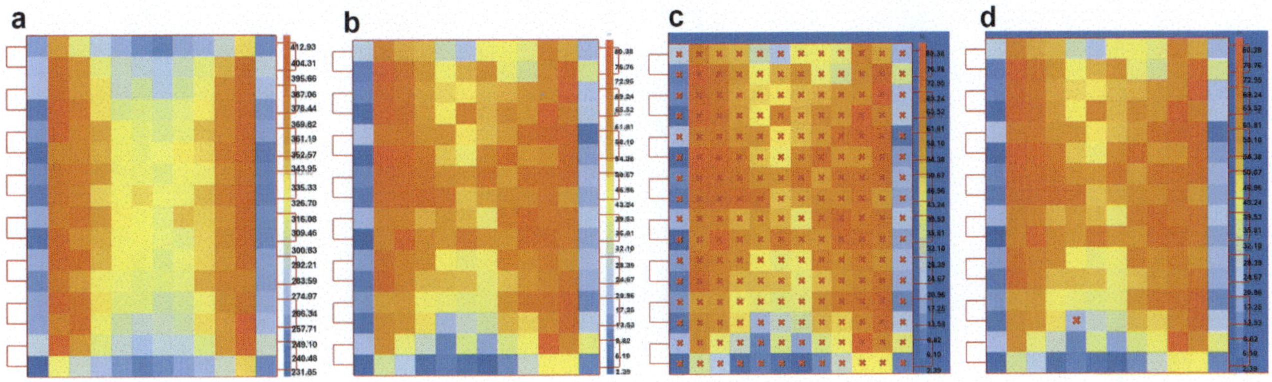

Fig. 19 **a** Daylighting energy simulation results of base case model, **b** DA results, **c** DA results grid point, **d** DA results on point (66). *Source* The authors

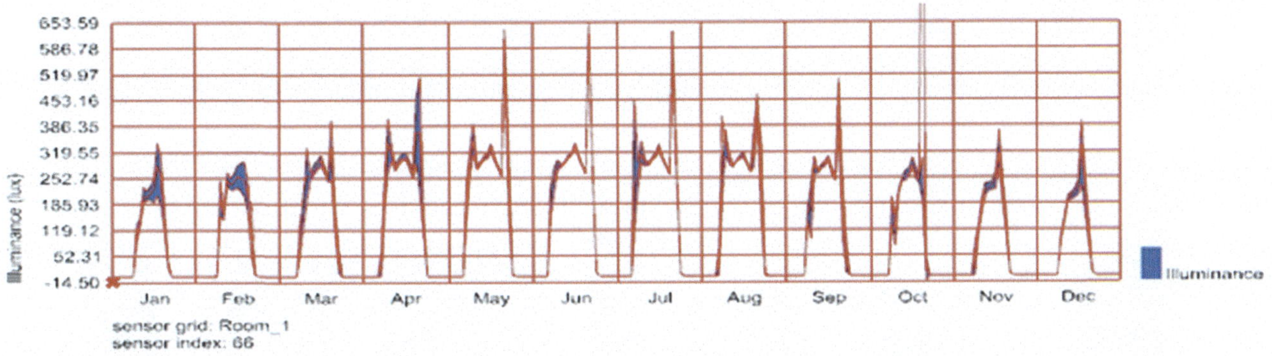

Fig. 20 Ladybug Monthly plot data in lux results on point (66). *Source* The Authors

5.5 Results and Discussion

Discussing the outcomes of a building design simulation and the factors affecting daylighting and energy performance. Here's a summary of the key points mentioned:

5.5.1 Low Fitness Function Value

The simulation of the current building design state has revealed unsuccessful daylighting results, which have led to a low fitness function value. This suggests that the building design may not be effectively utilizing natural daylight.

5.5.2 SDA (Spatial Daylight Autonomy) Results

- More than half of the simulation area fails to reach 50% in terms of SDA.
- The other half also performs poorly in terms of daylighting.
- The primary reason for these disappointing SDA values is attributed to the building parameters, specifically:
  - A small Window-to-Wall Ratio (WWR) in the southern elevation.
  - Lack of fenestrations (windows) in the eastern elevation.
  - Absence of shading structures on the southern elevation.

5.5.3 Shading Device Adjustments

- To improve SDA results, shading devices were introduced as shown in Fig. 23.
- The shading devices were set with an increment depth of 0.7 m, a shading count of 7, and a distance between them of 0.2 m. They were also angled at 45 degrees.
- Additionally, the WWR in the east was increased by 0.2%, and in the west by 0.15%.
- These adjustments resulted in more than half of the simulation area reaching 50% in SDA, indicating an improvement in daylighting performance.

5.6 Base Case Analysis

See Table 4.

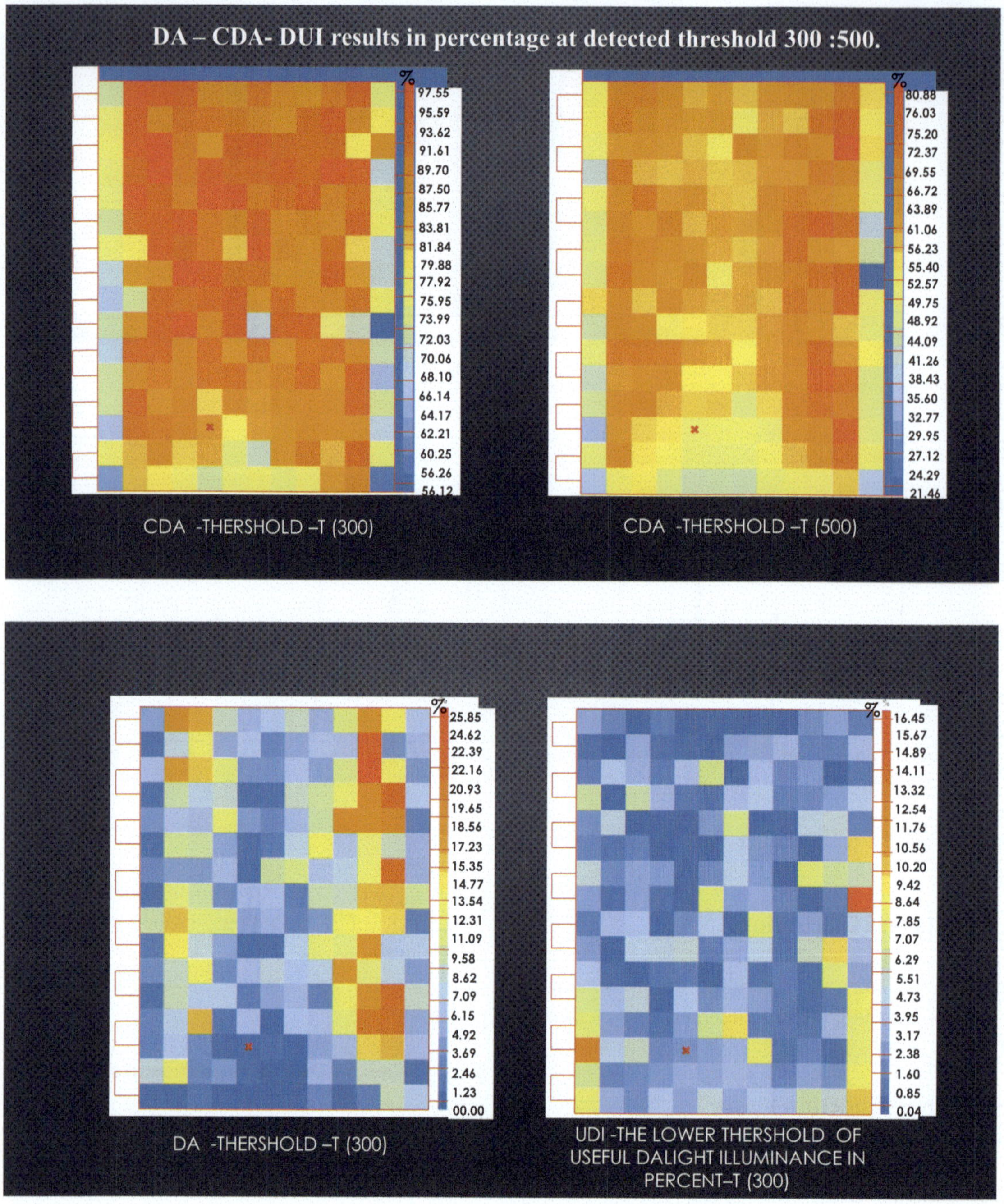

Fig. 21 DA–CDA–DUI results in percentage at detected threshold 300:500. *Source* The Author

5.7 The Absolute Optimum Solution

See Table 5.

6 Conclusion

The study aimed to utilize genetic algorithms within Grasshopper Canvas to perform parametric optimization. The specific focus was on improving daylighting in a classroom at El-Obour Elementary School, using SDA (Spatial Daylight Autonomy) 300/50% metrics for daylighting and energy efficiency, in accordance with LEED v4.1 standards set by the USGBC (United States Green Building Council).

To achieve this, the researchers defined a set of variables and parameters based on the optimal solution model and data shown in Fig. 23.

From the base case analysis as shown in Table 4 to absolute optimum solution as shown in Table 5,

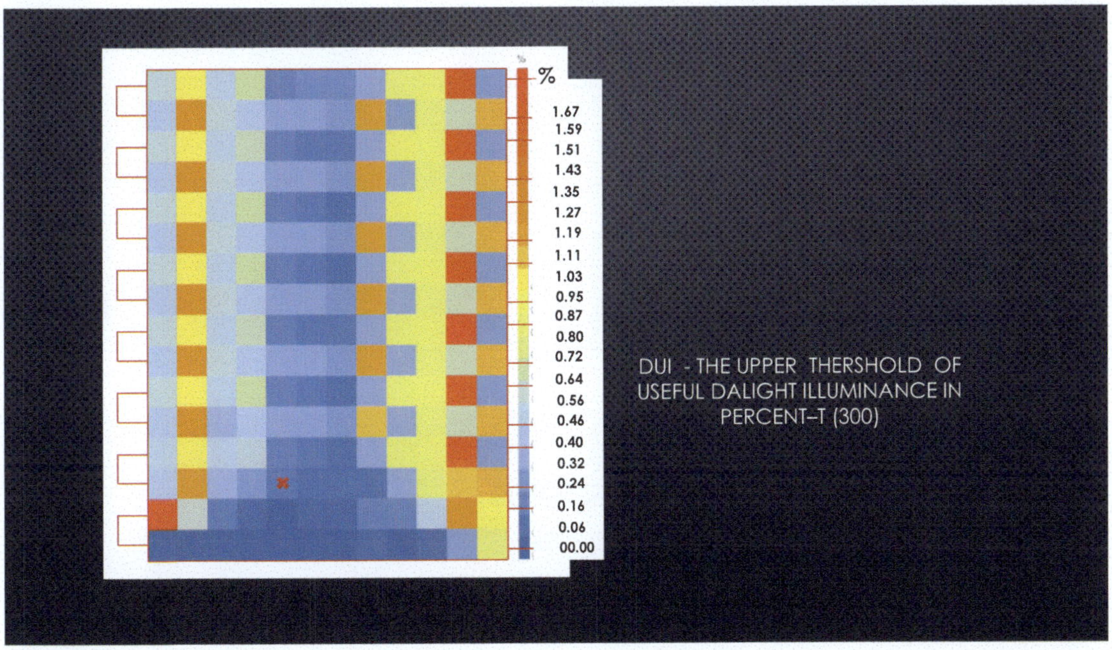

Fig. 21 (continued)

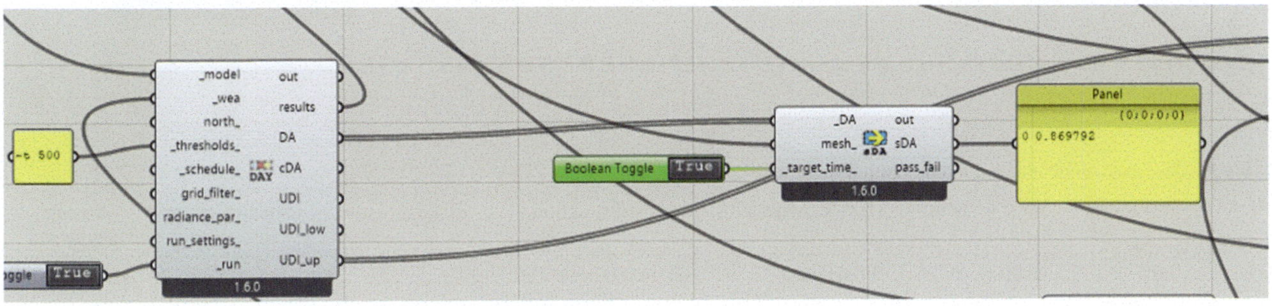

Fig. 22 SDA results special daylighting autonomy as percentage of area for each analysis grid. *Source* The Authors

Window-to-Wall Ratio (WWR) for the east-facing side at 0.2% and the south-facing side at 0.15%. Additionally, they set specific parameters for the honeybee shading modifier, such as a shading width of 0.7 m, a shading count of 7 for shading devices, and a shading angle of 45 degrees. Furthermore, the study integrated materials with the Honeybee interior modifier set.

Using genetic algorithms, the study generated nearly 300 solutions over six generations. Each generation represented an improvement over the previous one in terms of optimization. The goal was to create a honeybee symbolizing the ideal result for each analysis grid, the special daylighting autonomy is expressed as a percentage of the area.

After examining all the solutions represented by these points, the researchers identified the best solution, which scored 86.97% in terms of the objective metric. This solution demonstrated an SDA improvement of nearly 145% compared to the base case design.

In summary, the study demonstrated how genetic algorithms can be employed to optimize daylighting conditions in a classroom, resulting in a significant improvement in spatial daylight autonomy and energy efficiency, aligned with LEED v4.1 standards.

7 Recommendation

Using computational design in architecture can significantly enhance the design process, improve efficiency, and lead to innovative and sustainable solutions. Here are some

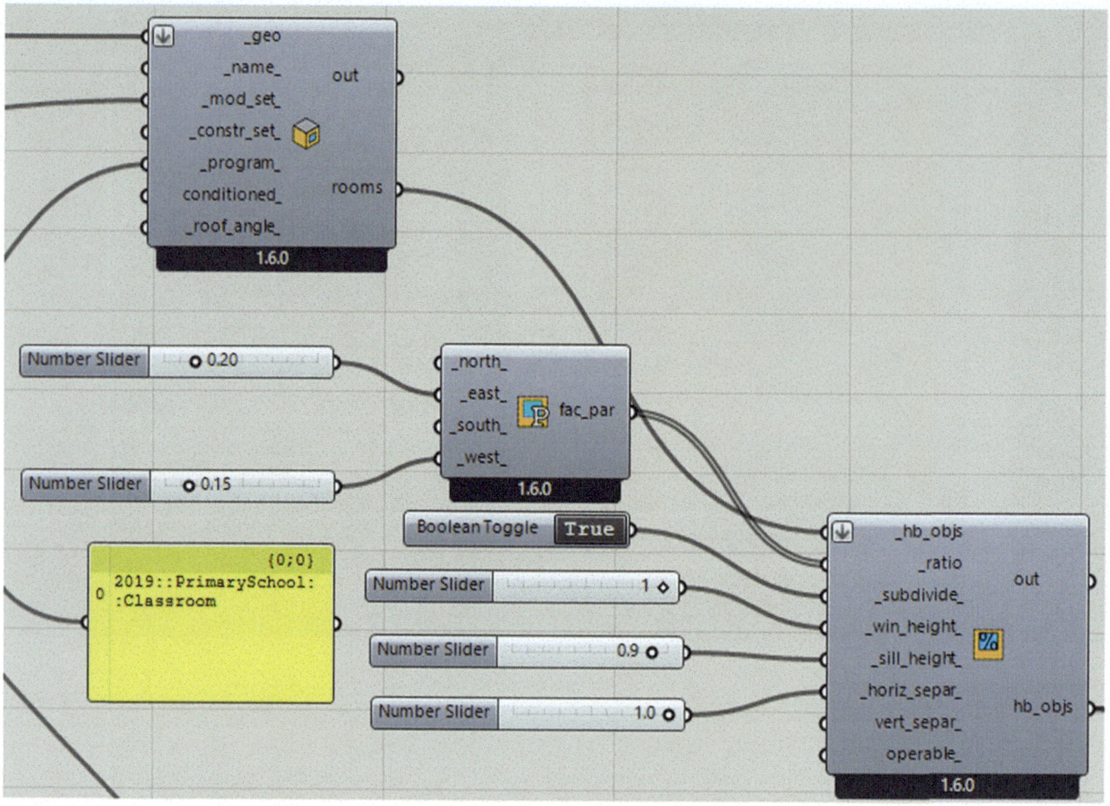

Fig. 23 Set the WWR ratio and shading device parameters for optimization. Ref: The Authors

Table 4 Base case model and parameters Ref: The Authors

SDA	WWR east	WWR south	GLASS material	Wall construction	Shading width	Shading count	Shading angle
35	0.6	0.20	Single clear	0.2 m red beak wall	–	–	–

Table 5 The optimum solution model and parameters. Ref: The Authors

SDA	WWR east	WWR south	GLASS material	Wall construction	Shading width	Shading count	Shading angle
86.97	0.20	0.15	Single clear	0.2 m red beak wall	0.70	7	45

recommendations for leveraging computational design in architecture:

1. **Parametric Modeling and Generative Design**: Utilize parametric modeling tools to create flexible and dynamic 3D models. Generative design algorithms can automatically generate and evaluate numerous design options based on specified constraints, helping architects explore a wide range of possibilities quickly.
2. **Performance-Based Design**: Integrate computational simulations and analysis tools to assess the performance of the design in real-time. This includes energy analysis, daylight studies, thermal comfort simulations, and structural analysis, which can lead to optimized and energy-efficient designs.
3. **Digital Fabrication and Building Information Modeling (BIM)**: Combine computational design with digital fabrication technologies, such as 3D printing and CNC machining, to produce complex and customized building components efficiently. BIM allows for seamless collaboration and data exchange among various project stakeholders.

4. **Responsive Architecture**: Employ sensors and interactive elements in buildings to create responsive environments. Computational systems can enable adaptive facades, kinetic structures, and other dynamic elements that respond to environmental conditions and user interactions.
5. **Data-Driven Design Decisions**: Utilize data analysis and machine learning algorithms to derive insights from previous projects, user behavior, and site conditions. This data-driven approach can inform design decisions and lead to more informed and user-centric solutions.
6. **Optimization and Efficiency**: Employ optimization algorithms to find the best design solutions based on specific criteria, such as cost, material efficiency, or environmental impact. This can lead to more sustainable and resource-efficient buildings.
7. **Virtual Reality (VR) and Augmented Reality (AR)**: Use VR and AR technologies to create immersive experiences for clients and stakeholders, allowing them to visualize and experience the design in a more interactive and realistic manner.
8. **Collaborative Design Platforms**: Utilize cloud-based collaboration platforms to facilitate teamwork and streamline communication among architects, engineers, and other project stakeholders. This promotes efficient coordination and reduces errors during the design process.
9. **Digital Prototyping and Simulation**: Create digital prototypes and conduct simulations for complex building systems and assemblies. This enables architects to identify potential issues early in the design phase and make informed design decisions.
10. **Education and Training**: Invest in educating architects and design professionals in computational design tools and techniques. Encourage continuous learning and exploration of new technologies to stay at the forefront of architectural innovation.

By incorporating computational design methods and technologies into the architectural workflow, architects can push the boundaries of creativity, sustainability, and efficiency, resulting in more innovative and user-centered architectural solutions.

8 Declaration

- The authors declare that they have no competing interests
- .This research received no funding.

References

Attia, S. (2012). A tool for design decision making: Zero energy residential buildings in hot humid climates, Presses univ de Louvain (2012)

Choi, J., et al. (2014). Parametric louver design system based on direct solar radiation control performance. *Journal of Asian Architecture and Building Engineering, 13*(1), 57–62.

Climate-Data.org, climate: Cairo international Airport. (2012). https://ar.climate-data.org/location/3677. Accessed November 18, 2017.

Computer-Aided Architectural Design. (2021). Design Imperatives: The Future is Now 19th International Conference. David Gerber, CAAD Futures, Los Angeles, CA, USA, July 16–18

El-Sheikh, M., & Gerber, D. (2011). *Building skin intelligence*. In: Proceedings of ACADIA

Evins, R. (2013). A review of computational optimization methods applied to sustainable building design. *Renewable and Sustainable Energy Reviews, 22*, 230–245. https://doi.org/10.1016/j.rser.2013.02.004

Konis, K., Gamas, A., & Kensek, K. (2016). Passive performance and building form: an optimization framework for early-stage design. *Solar Energy Support Programme, 125*, 161–179. https://doi.org/10.1016/j.solener.2015.12.020

Konstantoglou, M., & Tsangrassoulis, A. (2016). Dynamic operation of daylighting and shading systems: A literature review. *Renewable and Sustainable Energy Reviews, 60*, 268–283.

Lauridsen, P. K. B., & Petersen, S. (2014). Integrating indoor climate, daylight and energy simulations in parametric models and performance-based design. In *Proceedings of the Third International Workshop on Design in Civil and Environmental Engineering*

Lin, S. H. E., & Gerber, D. J. (2014). Designing-in performance: A framework for evolutionary energy performance feedback in early-stage design. *Automation in Construction, 38*, 59–73. https://doi.org/10.1016/j.autcon.2013.10.007.5

Michelle, B., & Gemilang, M. (2022). Bibliometric analysis of generative design. Algorithmic Design, and Parametric Design in Architecture 2022 JARINA

Ministry of Electricity and Energy (2015). Egyptian Electricity Holding Company Annual Report 2014/2015, Cairo. http://www.moee.gov.eg/english_new/EEHC_Rep/2014-2015en.pdf

Makhairas, V., Tsang Rassoulis, A., & Axarli, K. (2014). Algorithms for optimization of building design: A review. *Renewable and Sustainable Energy Reviews*, **31** (2014) 101–112, https://doi.org/10.1016/j.rser.2013.11.036.

New Urban Communities Authority—http://www.newcities.gov.eg

Nguyen, A.-T., Reiter, S., & Rigo, P. (2014). A review on simulation-based optimization methods applied to building performance analysis. *Applied Energy, 113*, 1043–1058. https://doi.org/10.1016/j.apenergy.2013.08.0613

Østerga° rd, T., Jensen, R. L., & Maagaard, S. E. (2016). Building simulations supporting decision making in early design—A review. *Renewable and Sustainable Energy Reviews, 61* 187–201, https://doi.org/10.1016/j.rser.2016.03.045.

Peel, M., Finlayson, B. C. L., & McMahon, T. A. (2007). Updated world map of the Koppen-Geiger climate classification. Hydrology and Earth System Sciences, pp. 439–473.

Sjarifudin, F. U. & Justina, L. (2014). Daylight adaptive shading using parametric camshaft mechanism for SOHO in Jakarta. In *EPJ Web of Conferences*. EDP Sciences.

Turrin, M., Von Buelow, P., & Stouffs, R. (2011). Design explorations of performance driven geometry in architectural design use. Parametric modeling and genetic algorithms. *Advanced Engineering Informatics, 25*, 656–675. https://doi.org/10.1016/j.aei.2011.07.009

Oraby, H., & Elborombaly, M. (2023) Abdelfattah, Optimizing daylighting performance for educational buildings by using parametric material system, 2023, Ain Shams University, www.researchsquare.com/article/rs-3079465/v1, https://doi.org/10.21203/rs.3.rs-3079465/v1 licensed under a CC BY 4.0 License.

Wagdy, A., & Fathy, F. (2015). A parametric approach for achieving optimum daylighting performance through solar screens in desert climates. *Journal of Building Engineering, 3*, 155–170.

References Courses

Daylighting study in Ladybug Tools V1.4 - Arabic course -Mona risk-www.udemy.com 2023.

Grasshopper Arabic Course—Omar Elgendy- JEEZAR Academy -2019

Exploring the Influence of Architectural Spatial Configurations on Residential Microclimates

Jiahao Liu, Tomoyuki Tanaka, Lei Tan, and Yuan Chen

Abstract

In urban residential areas, the the architectural spatial configuration is instrumental in modulating local micro-climatic parameters, which in turn exert a profound effect on the inhabitants' comfort and overall quality of life. This paper is concentrated on the architectural and urban planning intricacies encountered in Kumamoto, Japan, particularly scrutinizing the impact of these spatial arrangements on microclimatic variances. This study bridges the existing knowledge gap by demonstrating a clear link between particular spatial arrangements and their effects on microclimates. It underscores the role of urban design in enhancing thermal comfort and air purity. Through the use of both computerized simulations and research site observational data, the research evaluates how different architectural layouts influence wind flow and the dispersion of temperature. The results demonstrate that specific architectural designs markedly improve air circulation and lower ambient temperatures, thereby presenting notable advancements compared to conventional urban planning approaches. These findings emphasize the importance of building spatial layout, while also providing crucial insights into climate-adaptive urban design.

J. Liu (✉) · L. Tan
Graduate School of Science and Technology/Doctoral Course Education Program for Architecture and Environmental Planning, Kumamoto University, Kumamoto, Japan
e-mail: liujiahao983@gmail.com

L. Tan
e-mail: tanlei0708@gmail.com

T. Tanaka
Faculty of Advanced Science and Technology, Kumamoto University, Kumamoto, Japan
e-mail: tanaka@arch.kumamoto-u.ac.jp

Y. Chen
Faculty of Advanced Science and Technology/Division of Social Infrastructure and Environment, Kumamoto University, Kumamoto, Japan
e-mail: chenyuan002@yeah.net

Keywords

Microclimates · Living conditions · Architectural spatial configurations · Residential blocks · Urban design

1 Introduction

Climate change and its accompanying impacts pose significant challenges to urban environments worldwide. In the context of increasing urbanization and climate change, understanding the intricate dynamics of urban microclimates has become more crucial than ever. One of the key factors shaping these microclimates is the spatial configuration of architectural structures in urban residential areas. This study aims to explore this relationship in detail, focusing on the residential areas of Kumamoto City, Japan.

Residential spatial configurations, generally speaking, can be categorized into several types based on factors such as building density, orientation, height variation, and the presence of vegetation. For example, configurations could range from low-rise, low-density neighborhoods with substantial vegetation and open spaces, to high-rise, high-density urban blocks with minimal green spaces (Liu et al., 2022). The type and design of these configurations can significantly influence the local microclimate, affecting variables such as temperature, humidity, wind speed and direction, and solar access (Bourbia & Boucheriba, 2010).

Previous research has established that urban microclimates are affected by built environment configurations in several ways (Jiang et al., 2020). This is due to factors such as reduced vegetation, increased heat absorption by built structures, and decreased wind speed due to obstruction by buildings. On the other hand, configurations with substantial green spaces can help moderate temperatures, improve air quality, and enhance wind flow, creating a more favorable microclimate (Rui et al., 2018).

B. He et al. (eds.), *Urban Climate Change Adaptation*, Advances in Science, Technology & Innovation,
https://doi.org/10.1007/978-3-031-65088-8_5

Our study start in this context in Kumamoto. With its diverse architectural styles and distinct seasonal climate variations, it offers a rich environment to examine the influence of architectural spatial configurations on residential microclimates. This study uses ENVI-met to model and analyze the microclimates of three different residential areas within the city, each exhibiting unique spatial configurations.

The primary aim of this research is to delve into the architectural spatial configurations of residential areas in Kumamoto City, with an emphasis on their impact on local microclimates. This study seeks to bridge the significant research gap in the understanding of microclimate adaptability in these urban settings. And the results could inform sustainable urban planning and architectural design strategies, potentially contributing to improved human comfort, energy efficiency, and climate change mitigation efforts.

2 Methods

2.1 Study Area

This study is conducted in Kumamoto City, which is located on the island of Kyushu in southern Japan (Fig. 1a), serves as the study area for this research. Geographically, it lies between latitudes 32.47° N and longitudes 130.36° E. (Fig. 1b) With a land area of approximately 389.53 km^2, it represents a significant urban entity in the region. Kumamoto City experiences a humid subtropical climate, characterized by mild winters and hot summers. Annual average temperatures range from 6 °C in January to around 31 °C in August, with precipitation peaking in the summer months due to the East Asian monsoon.

Kumamoto City is characterized by a humid subtropical climate, which involves mild winters and hot summers. Average annual temperatures oscillate from a low of 6 °C in January to a high of around 31 °C in August. In addition, the city experiences peak precipitation during the summer months due to the East Asian monsoon (Japan Meteorological Agency). The period of May 1st to June 30th, 2023, was specifically chosen for this study, despite not being a typical summer season. This timeframe was selected due to the transitional climatic conditions it presents, offering a diverse range of temperatures and humidity levels that are crucial for assessing microclimatic variations. During this period, Kumamoto City recorded temperatures ranging from a high of 34 °C to a low of 17 °C, providing a comprehensive spectrum of environmental conditions for analysis. The selection of Kumamoto City as the study area was strategic; its unique climate patterns offer an ideal context for examining the influence of architectural spatial configurations on residential microclimates. Additionally, understanding the interplay between these spatial configurations and local microclimates will not only guide climate-responsive architectural and urban planning strategies in Kumamoto City, but also in other regions with similar climatic conditions.

For the purpose of this research, three distinct residential areas within Kumamoto City were selected based on their varying spatial layouts. The residential areas under study exhibit three types of architectural layouts: open irregular layouts (Spot A) (Table 1; Figs. 2 and 3), enclosed layouts (Spot B) (Table 1; Figs. 5 and 6), and open parallel layouts (Spot C) (Table 1; Figs. 8 and 9). Spot A, locateded in Kumamoto City's Chuo Ward, specifically Kurokami 5-chome, boasts an open and irregular layout, marking a unique urban design (Table 1; Fig. 4). Its architectural identity is shaped by non-linear and asymmetrical structures,

Table 1 Classification of Research Sites and their models in ENVI-met

Spot A (Open irregular layouts)	Figure 2	Figure 3	Figure 4
Spot B (Enclosed layouts)	Figure 5	Figure 6	Figure 7
Spot C (Open parallel layouts)	Figure 8	Figure 9	Figure 10

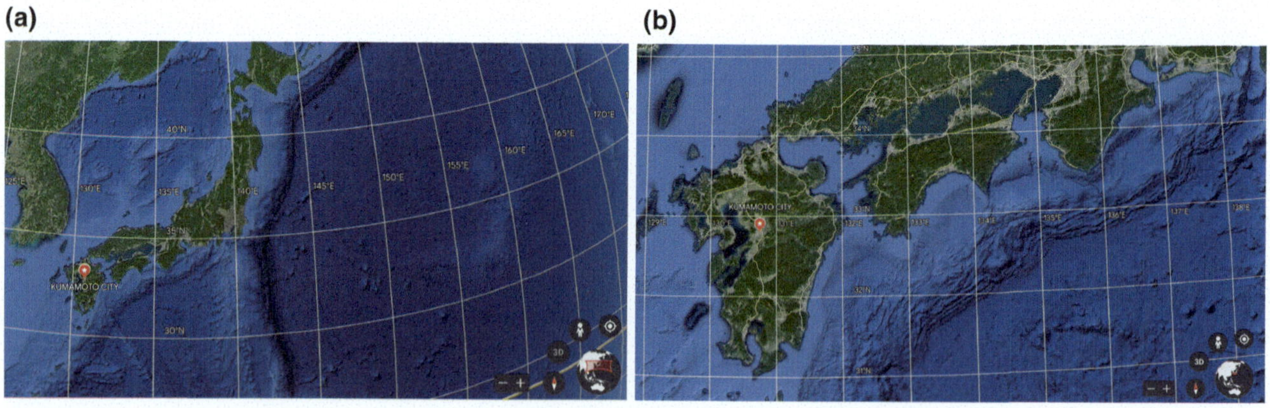

Fig. 1 **a** Geographical location of Japan(Google Earth). **b** Geographical location of Kumamoto (Google Earth)

Fig. 2 Spot A vertical view (Google Earth)

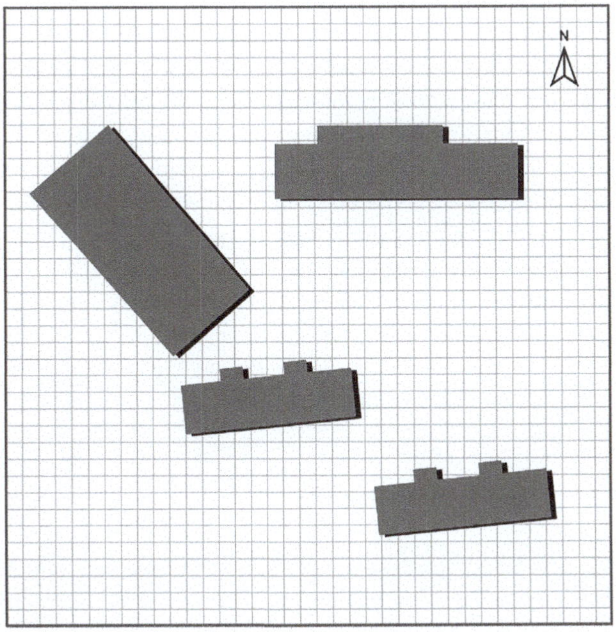

Fig. 3 Spot A plane Fig. 5 (Envi-met)

offering a varied and vibrant cityscape. In contrast, Spot B, located in Kurokami 2-chome of the same ward, adopts an enclosed layout. This design, sharply differing from Spot A's openness, features buildings that collectively create a well-defined and unified boundary, fostering a sense of community and enclosed space (Table 1; Fig. 7). Meanwhile, Spot C in Watoka 1-chome, Chuo Ward, is recognized for its straightforward, parallel layouts. This area's architectural essence, defined by linear and orderly arrangements, fosters a sense of uniformity and structured urban planning (Table 1; Fig. 10). This particular design aids in facilitating smooth wind currents and better sunlight penetration, thereby potentially improving the area's microclimate.

2.2 Research Method

Our research methodology involves constructing a detailed microclimate model using the software ENVI-met. ENVI-met provides a comprehensive three-dimensional analysis

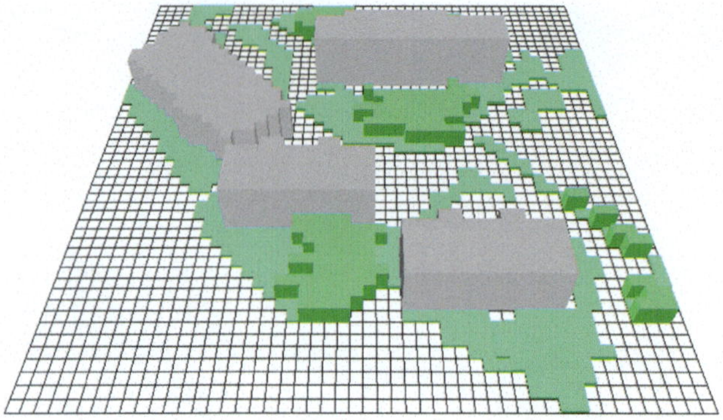

Fig. 4 Spot A model (Envi-met)

Fig. 5 Spot B vertical view (Google Earth)

that allows users to model and examine complex environmental processes in detail. With ENVI-met, researchers can simulate how different factors, such as buildings, vegetation, and weather conditions, can impact the local microclimate over a specified period (Paramita et al., 2018).

The architectural layout of urban environments plays a pivotal role in shaping their microclimates, particularly affecting temperature, humidity, and consequently, human comfort. The arrangement of buildings, their density, height, and orientation, directly influences microclimatic conditions by altering wind patterns, sunlight exposure, and heat retention. In addressing the impact of architectural layouts on residential microclimates, our study focused on two key indicators: temperature and humidity. These were selected as they are primary determinants of human thermal comfort. The temperature is essential for assessing thermal comfort. We analyzed how building density, height, and orientation affect temperature levels in residential areas. And humidity is in conjunction with temperature, significantly impacts human comfort.

These three distinct residential areas were modeled in ENVI-met software. The models incorporate site-specific data such as building dimensions, material properties, orientations, and local vegetation, providing an accurate reflection of the real-world conditions of the selected areas. Local climatic conditions, consistent with Kumamoto City's seasonal variations, were also factored into the models. This methodology is in line with prior studies that underscore the profound influence of architectural design and spatial arrangement on the microclimatic features within residential zones (Wang et al., 2019).

Kumamoto City experiences a humid subtropical climate, characterized by mild winters and hot summers. The period of study, May to June 2023, recorded daily maximum temperatures peaking at approximately 3 °C around 16:00, and daily minimum temperatures of around 17 °C at 06:00. These temperature patterns and the overall climate context make Kumamoto City an interesting site for studying the influence of architectural spatial configurations on residential microclimates. In selecting the time frame of May and June for our study, we aimed to capture the immediate effects of architectural spatial configurations on microclimates during a transitional seasonal period. This approach provides valuable insights into short-term microclimatic responses, although we recognize it may not fully represent long-term climatic trends. This limitation was considered in the context of our study's objectives.

To ensure that our simulations accurately reflect the real-world conditions of Kumamoto City, we configured the ENVI-met model based on a number of specific parameters. Included in our model parameters is a grid size of 2 m, with dimensional parameters set at 50, 50, and 25 for length, width, and height respectively. The rationale behind this particular size selection lies in its ability to accurately encapsulate the diversity inherent in the architectural spatial configurations across the selected residential areas, while also maintaining computational practicability. The simulation spanned a 24-h period commencing on May 1st, 2023, a time frame deliberately chosen to ensure a representative analysis of the microclimatic conditions typifying the late spring season in Kumamoto City (Table 2).

2.3 Simulation Results

The simulation results derived from ENVI-met are anticipated to provide a comprehensive depiction of the microclimatic conditions within each of the three residential areas. These results will include a range of climatic parameters

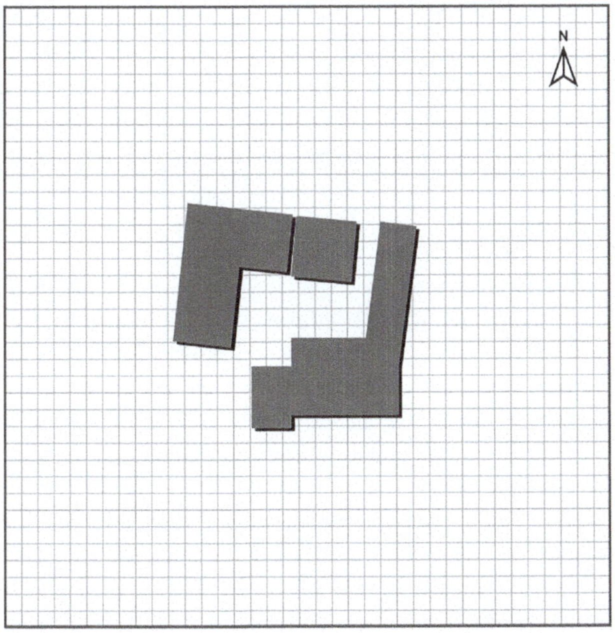

Fig. 6 Spot B plane (Envi-met)

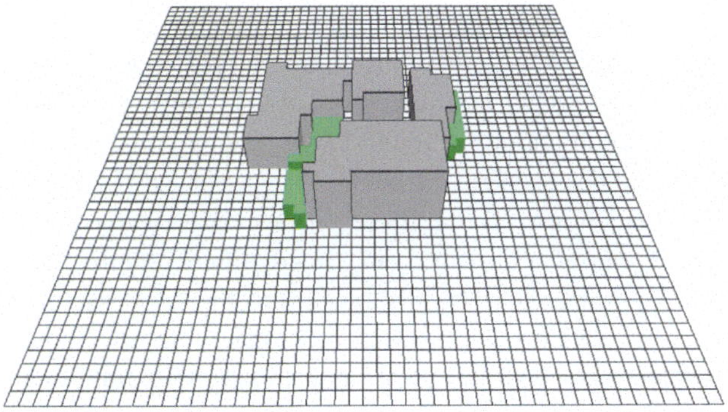

Fig. 7 Spot B model (Envi-met)

such as air temperature, relative humidity, wind speed and direction, and possibly more. The data will be systematically analyzed to understand the potential correlations and variations linked with the differing architectural spatial configurations of the residential areas.

In our simulations, we meticulously analyzed the impact of building density, height, shape, and orientation on microclimates. For each residential area, we varied these parameters to observe their effects. Building density was manipulated by altering the spacing between structures, height by varying the number of floors, shape by changing the buildings' geometric configurations, and orientation by adjusting their alignment relative to the sun and prevailing winds.

Our findings reveal a clear correlation between building density and microclimate comfort. In areas like Spot B, where building density was high, the reduced ventilation led to lower comfort levels due to increased temperatures

Fig. 8 Spot C vertical view (Google Earth)

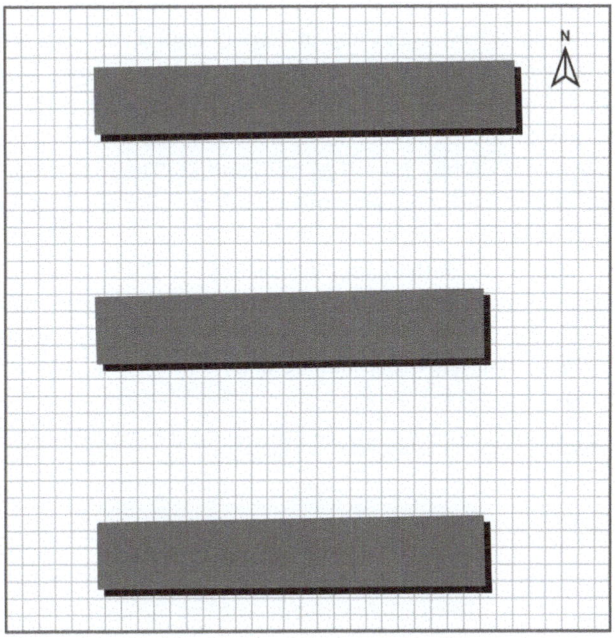

Fig. 9 Spot C plane (Envi-met)

and stagnant air. "The PMV (Predicted Mean Vote Model) in the context of ENVI-met is a key indicator for assessing thermal comfort, reflecting the average thermal sensation of people based on environmental factors such as temperature, humidity, and wind speed." (Fanger, 1982). The PMV results revealed interesting patterns across the different architectural spatial configurations. In the residential areas characterized by a favorable wind direction,

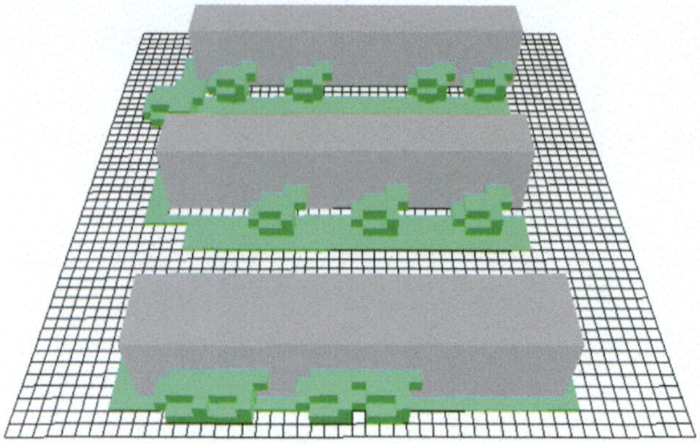

Fig. 10 Spot C model (Envi-met)

where buildings were aligned to maximize the airflow, the PMV values indicated a higher level of thermal comfort. At locations like Spot B, where building density was high, we observed a marked decrease in comfort due to poor ventilation and increased temperatures. Conversely, areas like Spot A & C with lower building densities exhibited better air circulation, enhancing comfort levels. Taller buildings significantly affected local microclimates, trapping heat and hindering airflow, especially in densely built areas. This led to higher temperatures and reduced comfort levels (Fig. 11). On the other hand, in areas where buildings were perpendicular to the prevailing wind direction, the PMV values indicated a slightly lower level of comfort. The reduced air movement and higher air temperatures in these areas had a slight negative impact on the thermal sensation experienced by occupants. Furthermore, the analysis of PMV results in enclosed residential layouts revealed a slightly lower level of comfort compared to the areas with a favorable wind direction. The enclosed layouts tended to have higher air temperatures and relatively lower air circulation, resulting in a less favorable thermal environment. Although the presence of shading and wind protection in these areas provided some benefits, the increased heat retention within the enclosed spaces offset these effects to some extent.

3 Results

In our simulations, the influence of architectural spatial configurations on the residential microclimates of the selected areas in Kumamoto City was strikingly apparent. Three distinctive patterns emerged with respect to downwind, perpendicular-to-wind, and enclosed residential configurations. The simulations using the ENVI-met model provided comprehensive temperature field data for each of the selected residential areas, and specific characteristics emerged in relation to the architectural spatial configurations.

The PMV values in our study indicate the level of thermal comfort experienced in different architectural spaces. A smaller fluctuation in PMV values, especially those approaching zero, denotes a higher level of comfort. For instance, at SPOT A, the most comfortable conditions were observed between 9:00 and 10:00 AM. Conversely, during the afternoon period of 16:00–17:00, among the three locations studied, residents at SPOT C experienced the highest level of comfort. The collected data clearly indicates that Spot A enjoys a slightly better microclimate comfort level than Spots B and C. This advantage is likely due to its unique open and varied layout, which promotes more effective air movement and less heat build-up. An interesting point in our study is the PMV measurement at 1 PM, showing that Spot A offers notably better thermal comfort compared to Spots

Table 2 ENVI-met Simulation Settings

Model domain	Size of grid cells (Δx, Δy, Δz)	(2, 2, 2)
	Number of grid cells	(50, 50, 25)
Time frame	Start simulation at day	(05. 01. 2023)
	Start simulation at time	06:00:00
	Full-scale simulation time (h)	24
Building material (Cement and concrete)	Wall material	C5 Concrete Wall (Cast dense)
		C4 Concrete Wall (Filled block)
	Roof material	R1 Tile
		IR Iron (Single layer)
Temperature (°C)	Minimum	9.9
	Maximum	23
Relative humidity (%)	Minimum	37
	Maximum	91

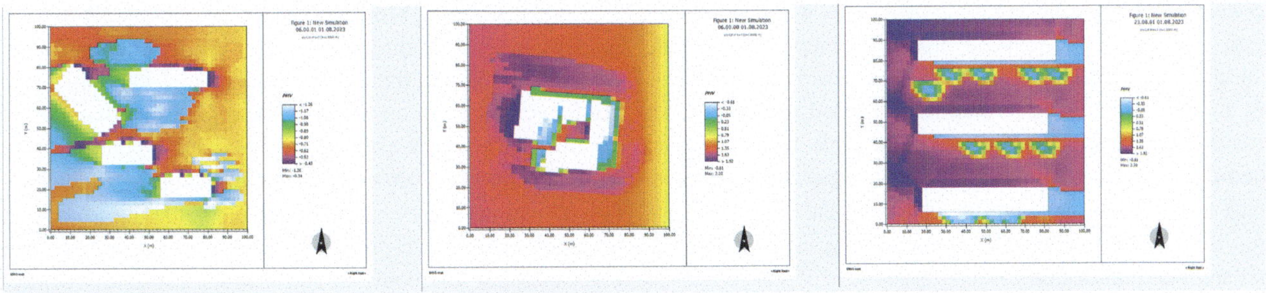

Fig. 11 Visualization of PMV values in different spatial configurations (ENVI-MET)

Table 3 Hourly PMV Values in one day

Time/spots	Spot A	Spot B	Spot C
9:00–10:00	– 1.85–1.93	– 2.01–2.95	– 1.82–1.98
10:00–11:00	– 2.10–2.49	– 2.34–2.81	– 2.62–2.37
11:00–12:00	– 2.09–2.56	– 3.15–3.01	– 2.91–2.92
12:00–13:00	– 2.42–2.12	– 3.91–2.92	– 3.58–3.21
13:00–14:00	– 3.36–2.67	– 4.11–3.47	– 3.68–3.08
14:00–15:00	– 3.31–3.01	– 3.90–3.65	– 3.54–3.02
15:00–16:00	– 2.85–2.73	– 3.62–2.91	– 2.90–2.86
16:00–17:00	– 2.74–2.31	– 3.31–2.32	– 2.69–2.40
17:00–18:00	– 1.96–2.30	– 1.90–2.31	– 2.11–1.98
18:00–19:00	– 1.90–1.91	– 1.95–2.09	– 1.89–1.63

B and C. This significant difference underscores the success of Spot A's design in creating a more pleasant microclimate, especially during the sunniest parts of the day. Our thorough analysis of the PMV values, covering a 10-h period from 9 a.m. to 19 p.m. (Table 3), reveals a clear pattern of comfort levels across the three spots. Particularly during the afternoon, when the sun is strongest, Spot A consistently shows higher PMV values, indicating a more comfortable thermal environment. This is largely due to its layout, which effectively balances sunlight exposure and air flow.

When these PMV results are juxtaposed with prior research, it becomes clear that urban configurations featuring open and less uniform designs are more conducive to thermal comfort. This observation is consistent with the conclusions drawn in the research conducted by Stark da Silva et al. (2023). In the downwind residential areas, where buildings were aligned with the wind direction, a certain pattern was observed. Detailed examination of the temperature distribution showed a slightly lower average temperature compared to the other areas. This led to an increased level of human comfort within these residential areas. In the residential areas where buildings were arranged perpendicular to the wind direction, different temperature behaviors were noted. The average temperature in these areas was found to be slightly higher compared to the downwind areas, resulting in a lower level of human comfort. In the enclosed residential areas, where buildings form a more or less closed configuration, distinctive temperature features were noticed. The temperature in these areas was found to be marginally higher than in the downwind residential areas, but slightly lower than the areas with buildings arranged perpendicular to the wind direction. This placed the human comfort level in these areas between the other two types.

In our study, we also carefully considered the impact of varying spatial arrangements, particularly focusing on the implications of dispersed and open layout configurations. For instance, our research uncovered that location 'A', characterized by its open and dispersed building arrangement, exhibited lower ambient temperatures and relative humidity levels. This distinctive feature contributed to a more agreeable microclimate, enhancing comfort for the residents. Conversely, in locations 'B' and 'C', which are categorized by their relatively enclosed configurations and smaller open spaces, we observed a consistent rise in environmental temperatures and relative humidity. Despite the potential benefits offered by such layouts, such as increased shade and wind protection, the amplified heat retention seemed to counterbalance these advantages, leading to a less comfortable microclimate. Our findings also suggest that residential areas with larger open spaces, even if they do not exhibit the lowest temperature and humidity values, provide an enhanced level of thermal comfort. This can be attributed to the improved air circulation and the psychological benefits associated with the perception of open spaces, both of which seem to amplify the overall sense of well-being for the inhabitants.

4 Discussion

The results of our simulations provide a compelling insight into the influence of architectural spatial configurations on the microclimates of residential areas in Kumamoto City. The findings of this study underscore the substantial influence of architectural spatial configurations on microclimatic conditions within residential areas. Notably, the interplay between architectural configuration and wind direction seems to have a pronounced effect on temperature fields and, in turn, human comfort levels. The downwind residential areas, characterized by buildings aligned with the wind direction, demonstrated the most favorable microclimatic conditions. The slightly lower average temperatures in these areas enhanced human comfort, affirming the potential benefits of considering wind direction during urban planning and architectural design.

These findings contribute to a more nuanced understanding of the role of architectural spatial configuration in shaping microclimatic conditions. However, it is worth noting that while this study focused on temperature as a measure of microclimate and human comfort, other factors such as humidity, sunlight exposure, and noise levels also contribute to the overall microclimatic conditions and human comfort (Ali et al., 2018). Therefore, future research could adopt a more holistic approach, considering these other variables in conjunction with architectural configuration. While this study provides valuable insights, it is not without limitations. The study focused on a single city and used only three residential areas for analysis. Future studies could benefit from considering a broader range of cities and architectural configurations.

We acknowledge that this limited duration may not capture the long-term trends and seasonal variations in microclimate changes. To address this, future research could extend the timeframe, potentially spanning multiple seasons or years, to observe and analyze long-term trends in microclimatic conditions. Such extended studies would complement our findings by providing a broader perspective on how architectural designs impact microclimates over different timescales. Although our study focuses on short-term observations, the findings lay the groundwork for understanding how architectural layouts can influence immediate environmental conditions, serving as a stepping stone for more extensive research on long-term microclimatic trends.

In our study, while focusing on the impact of architectural spatial configurations, we acknowledge that other factors such as vegetation, land forms, and the characteristics of surrounding buildings also play a significant role in shaping microclimates. Vegetation, for instance, can affect local temperatures and humidity levels. Although these factors were not the primary focus of our study, their potential impact on the results is recognized and merits consideration in future research.

5 Conclusion

This research integrates and evaluates how different urban spatial arrangements influence microclimate comfort, with a particular focus on three distinct locations within a residential area. Motivated by the urgent necessity to improve urban livability amidst swift urbanization and environmental changes, this study fills a crucial void in current academic discourse on the tangible effects of urban design on microclimates, especially in residential contexts. Our approach included an in-depth examination of PMV indices at various times, capturing the fluctuating characteristics of urban microclimates. Notably, the study reveals that open and less conventional layouts, especially evident in Spot A, offer enhanced thermal comfort compared to denser constructions. This finding underscores the importance of innovative urban planning in elevating the comfort of inhabitants and the overall ecological state. The research recognizes certain limitations, such as its concentration on particular locales and the inherent limitations of simulated data. Prospective research paths could involve delving into a broader range of urban environments and integrating a more varied set of climate data, aiming to corroborate and expand upon these insights.

To sum up, this study reiterates the importance of sustainable and considerate urban planning and design, with architectural spatial configuration playing a pivotal role in enhancing residential microclimates and promoting human comfort (Baker & Steemers, 2000). The influence of architectural configurations on wind patterns was clearly evident. Areas with high-rise buildings obstructed wind flow, leading to less wind movement at the ground level. In contrast, more open spaces facilitated greater wind movement. Thus, architects and urban planners must consider wind patterns when designing urban spaces, as this can significantly affect the comfort and well-being of residents (Baker & Steemers, 2019).

References

Ali, S., & Li, B. (2018). Evaluating the impact of the morphological transformation of urban sites on the urban thermal microenvironment. *Buildings, 8*(12), 182. https://doi.org/10.3390/buildings8120182

Baker, N., & Steemers, K. A. (2000). *Energy and environment in architecture: A technical design guide*. E. & F.N. Spon.

Baker, N., & Steemers, K. (2019). *Healthy Homes: Designing with light and air for sustainability and Wellbeing*. RIBA Publishing.

Bourbia, F., & Boucheriba, F. (2010). Impact of street design on urban microclimate for semi arid climate (Constantine). *Renewable Energy, 35*(2), 343–347. https://doi.org/10.1016/j.renene.2009.07.017

Fanger, P. O. (1986). Thermal environment—Human requirements. *The Environmentalist, 6*(4), 275–278. https://doi.org/10.1007/BF02238059

Jiang, Y., Wu, C., & Teng, M. (2020). Impact of residential building layouts on microclimate in a high temperature and high humidity region. *Sustainability, 12*(3), 1046. https://doi.org/10.3390/su12031046

Liu, Y., Long, Y.-D., Wang, B.-H., & She, X. (2022). Construction of multiobjective planning decision-making model of ecological building spatial layout under the background of rural revitalization. *Journal of Environmental and Public Health, 2022*, 1–9. https://doi.org/10.1155/2022/7021770

Paramita, B., Fukuda, H., Perdana Khidmat, R., & Matzarakis, A. (2018). Building configuration of low-cost apartments in Bandung—Its contribution to the microclimate and outdoor thermal comfort. *Buildings, 8*(9), 123. https://doi.org/10.3390/buildings8090123

Rui, L., Buccolieri, R., Gao, Z., Ding, W., & Shen, J. (2018). The impact of green space layouts on microclimate and air quality in residential districts of Nanjing. *China. Forests, 9*(4), 224. https://doi.org/10.3390/f9040224

Stark da Silva, P. W., Duarte, D., & Pauleit, S. (2023). The role of the design of public squares and vegetation composition on human thermal comfort in different seasons a quantitative assessment. *Land, 12*(2), 427. https://doi.org/10.3390/land12020427

Wang, Y., Zhou, D., Wang, Y., Fang, Y., Yuan, Y., & Lv, L. (2019). Comparative study of urban residential design and microclimate characteristics based on Envi-Met Simulation. *Indoor and Built Environment, 28*(9), 1200–1216. https://doi.org/10.1177/1420326x19860884

Sustainable Urban Design and Planning

Design Factors Towards Water Retention Ability of Water-Sensitive Urban Design (WSUD) in Tropical and Subtropical Climates: An Exploratory Literature Review

Chulalux Wanitchayapaisit, Damrongsak Rinchumphu, and Pongsakorn Suppakittpaisarn

Abstract

Water Sensitive Urban Design (WSUD) is a comprehensive approach to managing the quality and quantity of rainwater within urban settings and addressing the challenges posed by changes in natural infiltration and precipitation processes. Green Stormwater Infrastructure (GSI) is a key component of WSUD and is extensively utilized in Europe and the United States. GSI can take many forms. One example of GSI is the rain garden, which is frequently integrated into WSUD practices across various regions and communities. The research question for this study is to find guidance for rain gardens in Southeast Asia. The objective is to identify the design factors that affect water retention ability in rain gardens and to determine the specific measurements that can be applied in Southeast Asia and across the world In similar climates. An exploratory literature review method delves into current examples of design factors influencing the water retention ability of WSUD in the tropical and sub-tropical contexts of Southeast Asia. This research found limited studies focussing on WSUD in tropical and sub-tropical climates, limited discussion on local materials, and how planting design may impact WSUD performance. The research gap highlights the need for standardized measurement units and further exploration into the specific design factors influencing the performance of plants in the context of WSUD in tropical and subtropical climates. Moreover, future studies should explore additional aspects, such as investigating input values for modeling, to improve the capability of calculating water retention ability based on relevant plant factors. Furthermore, it could involve examining the water-absorbing performance of different tree species. Designers must work with researchers to create case studies and compile evidence for effective WSUD in the near future.

Keywords

Water-Sensitive Urban Design (WSUD) · Green Stormwater Infrastructure (GSI) · Rain garden design · Absorption ability · Southeast Asia

1 Introduction

1.1 Background Information

The current, rapid, and widespread urbanization has resulted in changes in the water cycle. More impervious surfaces inhibit rainwater from their typical infiltration and precipitation processes. In addition, densely populated urban and agricultural communities have created unique pollution in urban areas that affect water sources both within the source area and the natural ecosystems downstream (Baird et al., 2020; Li et al., 2022). In addition, cities and towns sometimes face water restrictions in response to drought and water scarcity (Coutts et al., 2013). These changes challenge the conventional urban drainage infrastructure, which uses hard, impervious surfaces to transport stormwater away from the city as soon as possible (Norton et al., 2018).

In various cities worldwide, particularly in Europe and the U.S., a sustainable utility known as Water Sensitive

C. Wanitchayapaisit · P. Suppakittpaisarn (✉)
Landscape Design and Environmental Management Studio, Faculty of Agriculture, Chiang Mai University, Chiang Mai, Thailand
e-mail: Pongsakorn.sup@cmu.ac.th

C. Wanitchayapaisit
e-mail: Chulalux.w@cmu.ac.th

C. Wanitchayapaisit · D. Rinchumphu
Faculty of Engineering, Chiang Mai University, Chiang Mai, Thailand
e-mail: Damrongsak.r@cmu.ac.th

B. He et al. (eds.), *Urban Climate Change Adaptation*, Advances in Science, Technology & Innovation,
https://doi.org/10.1007/978-3-031-65088-8_6

Urban Design (WSUD) is being implemented. WSUD is a comprehensive concept that aims to regulate the hydrological cycle and enhance water quality by integrating local facilities. Green Stormwater Infrastructure (GSI) is incorporated into WSUD to manage the quality and quantity of rainwater (Ferreira et al., 2021) and reduce the negative impacts of urban stormwater runoff. The concept of GSI is to manage rainwater and retain it within the urban landscape through stormwater harvesting and reuse, which can help reduce urban temperatures and enhance evapotranspiration and surface cooling when integrated with vegetation (Coutts et al., 2013).

WSUD is a comprehensive approach to managing water in urban areas. It encompasses various sustainable water management strategies, including GSI. Within GSI, multiple system types exist, such as rain gardens, bio-infiltration systems, bio-retention systems, and bio-swales. Although there is some overlap in the definitions of these terms, GSI is an umbrella term that encompasses a wide range of stormwater management practices. These practices use infiltration, retention, evapotranspiration, and detention to manage stormwater effectively. Specifically, a rain garden illustrates GSI due to its structured bioretention systems incorporating integrated infiltration, storage, and evapotranspiration processes within an exposed basin. This design closely adheres to the fundamental principles of bioretention systems (Reis, 2019). Therefore, the functional framework of rain gardens portrays them as integral components of GSI, designed in alignment with structural patterns and WSUD principles to facilitate efficient urban rainwater drainage.

1.2 Principle of Water-Sensitive Urban Design

Water management in urban areas has become a worldwide concern, focusing on sustainability and resilience. The Water Sensitive Urban Design (WSUD) is a guiding principle that has been implemented globally to achieve these goals. The following highlights the fundamental principles of WSUD and how it is being implemented worldwide. WSUD principles utilize the function of GSI to control the hydrological cycle and enhance water quality by implementing local facilities such as rain gardens, green spaces, parking zones, open areas of building blocks, and street infrastructure (Ristianti et al., 2022). Examples of local facilities are shown in Fig. 1.

The principle of WSUD is collecting, retaining, and infiltrating surface runoff to promote groundwater recharge and replenish soil moisture. This design strategy emphasizes stormwater retention and contributes to preserving culturally significant organic layers below ground (Venvik & Boogaard, 2020). Guidelines for designing rain gardens, a key component of WSUD, have been widely adopted worldwide under various names, providing paradigms for each principle, and their origins are as follows.

Water sensitive urban design (WSUD) in Australia takes a comprehensive approach to managing water within urban areas. It recognizes the importance of considering the entire water cycle, including wastewater, water supply, and stormwater management. It aims to create more sustainable and resilient cities that effectively manage water resources while promoting environmental conservation (Radcliffe, 2019).

Low Impact Development (LID) in the USA: The Low Impact Development Stormwater Management Planning and Design Guide was adopted in 1999 (Eckart et al., 2018; Valley, 2010). LID aims to maintain natural or predevelopment hydrologic conditions, minimizing runoff volume from neighbourhoods, subdivisions, or individual lots. Urban tree canopy interception, evaporation, rainwater harvesting, and engineered infiltration are recommended strategies (Eckart et al., 2018).

Sustainable Urban Drainage Systems (SuDS) in the UK are integral to WSUD. This principle aims to mitigate flooding, efficiently manage drainage, and enhance water quality through effective pollutant filtration and natural treatment processes. Moreover, SuDS contributes to the preservation of biodiversity by creating habitats for wildlife and fostering ecological equilibrium within urban landscapes (Ferrans et al., 2022).

Low Impact Urban Design and Development (LIUDD) in New Zealand is a sustainable living concept. Urban sustainability and health are achieved through effectively managing stormwater, waste, energy, transport, and ecosystem services (van Roon et al., 2008). Greening cities by planting ecologically with local species is also a vital part of the overall well-being of ecosystems and citizens (Ignatieva et al., 2008).

Sponge Cities in China exemplify an innovative approach to urban water management. The concept incorporates sustainable drainage systems, green infrastructure, and water-sensitive design principles that promote the sustainable utilization of water resources (Liu et al., 2022; Tu & Tian, 2015).

Active, Beautiful, Clean Waters (ABC Waters) design features in Singapore: The natural systems consist of plants and soil that detain and treat rainwater runoff, which improves the water quality of downstream waterways or water bodies. At the same time, using natural biological features encourages ecological habitats for plants and animals and enhances the aesthetics of the surroundings (Neo et al., 2022).

The components of drivers can vary in different contexts or situations. It is crucial to consider the suitability of the area, such as water sources, existing infrastructure, seasonal

Fig. 1 Examples of local facilities: Rain garden of a building block in Chicago Art Institute (top left), Bioswale along the street in Portland 10,000 Rain Garden Project (top right), Green spaces in Angkaew Phytoremediation Chiangmai Thailand (Bottom). (Photographs by author)

weather patterns, local water cycle dynamics, and societal expectations (Matthews et al., 2015; Zuniga-Teran et al., 2020).

Most studies in the literature have been conducted in temperate countries such as Australia, China, the UK and the USA. Principles from countries with climates similar to Southeast Asia are only those from Singapore. Singapore represents a group of tropical climates that falls within the Equatorial rainforest (Af) and Equatorial monsoon (Am) climate zones. The Equatorial rainforest and equatorial monsoon climates, both hot and humid climates, differed by the consistency of the humidity. While the Equatorial rainforest climate is characterized by consistent rainfall throughout the year, the Equatorial monsoon climate's humidity levels are influenced by monsoonal winds (Peel et al., 2007). The design principles for rain gardens that are most suitable for Southeast Asia may need to be based on climate characteristics specific to the region. Applying data from countries with similar climates, such as Singapore, can provide valuable insights and guidance for designing practical rain gardens tailored to the unique climatic conditions of Southeast Asia.

1.3 Research Question

The practical application of Green Stormwater Infrastructure (GSI) principles varies depending on specific geographical areas and constraints (Suppakittpaisarn et al., 2020; Wanitchayapaisit et al., 2022). Therefore, context-specific guidelines need to be developed to implement GSI effectively. However, there are insufficient examples and principles of Water Sensitive Urban Design (WSUD), resulting in the lack of design guides for GSI and rain gardens in Southeast Asia. This has resulted in the absence of standard measures to assess the performance of different designs for WSUD in Southeast Asia.

Moreover, more global information about planting design to impact WSUD performance positively is needed. Recommendations across guidelines suggest using native plants to increase root mass and survivability. However, no calculations were proposed to confirm these recommendations (Emanuel et al., 2010b; Franti & Rodie, 2007). For example, in current calculation models, an open lawn and a field of complex native ecosystems in a rain garden are considered the same.

Therefore, the research question for this study is to find guidance for rain gardens in Southeast Asia. The objective is to identify the design factors that affect water retention ability in rain gardens and to determine specific measurements that can be applied in Southeast Asia and similar climates worldwide.

2 Methodology

To explore rain garden design factors that affect water retention ability in Southeast Asia and worldwide, we have set specific research objectives. Our aim is to conduct an exploratory literature review to categorize relevant information and gain a preliminary understanding of the subject matter (Swedberg, 2020). This can help create a conceptual framework or proposal for future research. In the subsequent sections, we will explain our approach to conducting the literature search, establishing inclusion criteria, choosing, extracting, evaluating the studies, and synthesizing collected information. We hope to identify specific measurements that can be applied in similar climates, and thus, provide guidance for rain garden design in Southeast Asia and around the world (Suppakittpaisarn et al., 2017).

2.1 Literature Search

The investigation methodology systematically explored thematic relevance across article titles, abstracts, and indexes using diverse keyword permutations.

The first category contained keywords related to the WSUD principle, including: "Water sensitive urban design", "Green stormwater Infrastructure", "Rain Garden Design", "Bioretention", "Water management principle", "Guide", "Design criteria", "Design factor", and "Planting design".

The second category contained keywords related to performance: "Water retention ability", "Absorb", "Layers", "Construction", and "Experiment".

The third category contained keywords related to the area, including: "Thailand" and "Tropical". Each document was scrutinized by two independent researchers to ensure comprehensive coverage. We combined terms from these three classifications using the connector 'AND'. For instance, in our search for a relationship between Rain gardens and Layers and Tropical, our search terms were "'Rain gardens' AND 'Layers' AND 'Tropical'". All possible combinations of WSUD principle, performance, and area keywords were used.

We discovered that only a few studies have explored the utilization of WSUD in tropical and sub-tropical regions, such as rain gardens, guides, and tropical areas. Therefore, we expanded our search parameters by integrating the "WSUD principle," "performance," and "tropical" vocabulary related to the principle, performance, and area, respectively. We also incorporated "Ability," "Procedure," "Pattern," and "Southeast Asia" to broaden our search. Additionally, we included common greenspace features in the design. We added a fourth term, such as "native plants," "wetlands," and "biodiversity," in the same manner as the first search. We match all possible combinations again.

After following the method above to match the search terms, the review can be divided into three sections to address the research questions and objectives on different topics, as follows:

Section 1: Principle of water sensitive urban design.

Section 2: The factors in landscape design that impact water retention ability.

Section 3: The role of plants and design factors.

2.2 Criteria

Inclusion: We searched for English articles published in journals, scholarly literature, governmental publications, online databases, and expert recommendations, as well as through the snowball literature review method. This paper provides a comprehensive analysis of research publications from January 2000 to January 2023, particularly emphasizing design principles concerning materials, layers, size, and planting. The design principle was to measure the water retention ability of WSUD, which will be discussed in the following section.
Exclusion: Regarding the excluded content, several selection criteria were applied, including articles not written in English, design papers, single case studies, and documents dated before 2000.

2.3 Data Analysis and Synthesis

In terms of the analysis process, the researchers were actively involved. Both researchers independently assessed the content, categorizing information into relevant sections. Following this individual evaluation, a collaborative effort was made to consolidate findings and produce a summarized table, presenting a cohesive overview of the explored literature. The results were analyzed to identify research gaps and create frameworks or proposals for future research.

3 Results

WSUD is an important aspect of managing water in urban areas. It offers innovative solutions that combine environmental sustainability with urban development. This section will explain the crucial principles and factors to consider when implementing WSUD, focusing on rain gardens as a fundamental component of sustainable urban water management.

Each principle is tailored to address the specific needs of their respective areas, considering factors such as catchment characteristics, existing infrastructure, seasonal climate patterns, local water cycles, and societal expectations.

3.1 Principle of Water Sensitive Urban Design

The fundamental design principles of rain gardens are based on the Water Sensitive Urban Design (WSUD) principles. They comprise two main components that focus on enhancing water retention capabilities: rainfall collection structures and plant characteristics for rainwater absorption (Ristianti et al., 2022; Wanitchayapaisit et al., 2023).

3.1.1 Structure for Retaining Rainwater

The design criteria for the non-vegetative materials for rain gardens may include the following:

- **Ponding depth:** It is a key function in effectively collecting rainwater. Its primary function is to facilitate the accumulation of rainwater within a designated area(Dunnett & Clayden, 2007).
- **Soil mix/media:** Soil mix or media is a filtering medium for debris, preventing it from reaching the underlying layers. Typically composed of a combination of sandy loam, loamy sand, or loam, it acts as a filter and provides essential nutrients for the plants thriving above (Emanuel et al., 2010a).
- **The infiltration layer:** This layer is essential in enhancing the soil layer's permeability efficiency. It is specifically designed to promote better drainage and permeability. Typically composed of gravel or coarse sand, this layer not only facilitates effective water infiltration but also possesses the necessary strength to support the activities above it (Ristianti et al., 2022; Wahab et al., 2016).

3.1.2 Characteristics of Vegetation Used in Rain Gardens

The criteria for rain garden plants involve living or movable materials that change according to various environmental factors that consist of the following (Kasprzyk et al., 2022; Sharma & Malaviya, 2021; Wanitchayapaisit et al., 2023).

- **Groundcovers (mulch/pebbles/rocks)**: Groundcovers mitigate soil erosion and provide aesthetics for the garden.
- **Plant materials**: Planting is vital in enhancing the soil surface and contributing to the garden's aesthetics. The soil surface is covered by introducing various plants, creating a protective layer that helps prevent erosion and provides aesthetics for the garden.
- **Grass bumper**: It can supplement the design but is unnecessary for most rain gardens. It is an optional choice that depends on the designer's preferences.
- **Overflow system**: It is a supplemental or alternative system for draining overflow (Zhou, 2014), depending on the precipitation level in the area. The surface area of the rain garden is not dependent on the guide but rather the volume of stormwater it needs to slow and contain. This information needs to be calculated based on the design storm (the amount of storm it needs to contain) and the area of the sub-catchment (the surface of the runoff it is collecting) (Wanitchayapaisit et al., 2022).

The characteristics of vegetation used in rain gardens are shown in Fig. 2.

3.2 The Factors in Landscape Design that Impact Water Retention Ability

The literature review provides a summary of the factors that influence landscape design for rainwater management. Tables 1 and 2 present a synthesis of key principles (compare Sect. 4.1) extracted from various national guidelines of WSUD practices. The guidelines encompassed in this overview include WSUD, Low Impact Development (LID), Sustainable Urban Drainage Systems (SUDS), Low-Impact Urban Design and Development (LIUDD), Sponge Cities, and Active, Beautiful, Clean Waters (ABC Waters).

In landscape design, there are two primary elements: hardscapes and softscapes. Hardscape refers to inanimate, man-made structures that provide durability and define the spatial layout. In this research, the hardscape element is the structure used to retain rainwater. On the other hand, softscapes include living characteristics and horticultural plants such as trees and plants that provide beauty and benefits to the ecosystem (Pressman, 2007). In this research, the softscape element is the characteristics of vegetation. These elements work together to create a rain garden, each component having different water retention properties, which are explained below.

Under the category of "Structure to retain rainwater" or "Hardsacpe" (compare Sect. 4.1.1), the guidelines emphasize diverse approaches to achieving water retention objectives. WSUD guidelines advocate for soil porosity ranging from 20 to 40% and layer depths between 5 and 60 cm, while LID focuses on combining soil with local river sand

Table 1 Principle of each rain garden, guidance, and basic implementation: Hardscape

	Structure to retain rainwater		
	Ponding depth	Soil mixed/media	Infiltration layer
WSUD (Ahiablame et al., 2012; Rinchumphu et al., 2021) (Coombes et al., 2000; Zhang et al., 2019)	• Maximum depth of ponding in the detention basin of 260 mm with a residence time of 6 h • Ponding depth of 50 to 350 mm	• Planting soil layer depth of 10 to 20 cm	• Porous pavements have been shown to reduce runoff by an average of 50–93% • Soil porosity is 20–40%, with the layer depth ranging from 50 to 600 mm
LID (Jiang et al., 2022) (Baek et al., 2020) (Fassman-Beck et al., 2015; Gülbaz & Kazezyılmaz-Alhan, 2017)	• Important key design parameters: ponding depth and discharge ratio • Inflow rate at the entrance of bioretentions and ponding depth on bioretention columns are measured during selected artificial rainfall events and durations • Maximum depth 300–457 mm	• The soil is combined with local river sand and wood chips in a mass ratio 65:30:5 • Soil properties should be permeable and have a porosity similar to gravel	• Important key design parameters: infiltration
SUDS (Green, 2019) (Ali et al., 2021; Chen et al., 2021; Dell et al., 2021; Sañudo-Fontaneda et al., 2020)	• The design of bioretention cells aims to maintain a consistent presence of water, although the size of the ponding may fluctuate over time • Maximum depth 600 mm	• Media depth ranges from 400 to 1200 mm, and the depth also depends on the type of plant and pollutant concentration	• Reduced runoff through increased infiltration with limited effects • Gravel layer 400–750 mm
LIUDD (Ignatieva et al., 2008) (Akter, 2022; Wella-Hewage et al., 2016)	• Depth of surface ponding 350 mm	• The structure (planted bed) is a soil mix with deep, porous surfaces where water typically accumulates to a depth of 100 to 300 mm for several hours up to 36 h following rainfall • 600 mm deep amended soil layer	• 150 mm to 200 mm of drainage layer • 300 mm of storage layer
Sponge Cities (Wang et al., 2016). (Ali et al., 2021)	• The soil or media can be calculated from the depth of the ponding, the porosity, the rate of rainwater infiltration and the groundwater level • Maximum depth 110 mm	• The soil or media can be calculated from the depth of the ponding, the porosity, the rate of rainwater infiltration and the groundwater level	• The soil or media can be calculated from the depth of the ponding, the porosity, the rate of rainwater infiltration and the groundwater level
ABC Waters (H. S. Fowdar et al., 2022a, 2022b) (Heng, 2021)	• The influence of infiltration comes from the ponding depth, the hydraulic properties (porosity and hydraulic conductivity of the filtration medium), and its water content • There are a variety of designs for hardscapes. Prescribe layer specifications: a ponding depth layer (200–300 mm)	• Filter media layer 300 mm • Transition layer 100 mm • Materials include sand and sandy loam for the filter media, sand for transition	• Storage layer 300 to 400 mm • Mterial Large gravel for storage • Catchment characteristics produce different sizes and properties of runoff and quality for the design

Table 2 Principle of each rain garden, guidance and basic implementation: Softscape

	Characteristics of vegetation / Planting
WSUD (Wanitchayapaisit et al., 2022)	• Groundcover mitigates soil erosion whilst enhancing the visual appeal of the garden • To mitigate soil erosion, native plants are recommended
LID (Lee et al., 2022; Liu et al., 2021)	• Biodiverse plants are recommended • The evapotranspiration process carried reduced the outflow of stormwater • Suitable plant types include succulent plants and herbaceous grasses
SUDS (Green, 2019)	• Plants use transpiration to help move water from the soil to the air • Water can be retained on vegetative and soil surfaces, where it can be lost through evaporation or infiltration • Trees are highly recommended
LIUDD (Ignatieva et al., 2008)	• Medium to low-height vegetation, although the inclusion of trees may be considered based on the project's scale • Suggested diverse original native vegetation
Sponge Cities (Payne et al., 2018) (H. Fowdar et al., 2022a, 2022b) (Shao et al., 2023)	• The impact of plant dynamics on urban water storage remains unclear • Plant morphology and growth traits are crucial factors in determining the selection of biofilter plants • Alongside root features, the growth rate is also a significant consideration in plant design
ABC Waters (H. Fowdar et al., 2022a, 2022b)	• The flow rate is influenced by the density and height of the vegetation

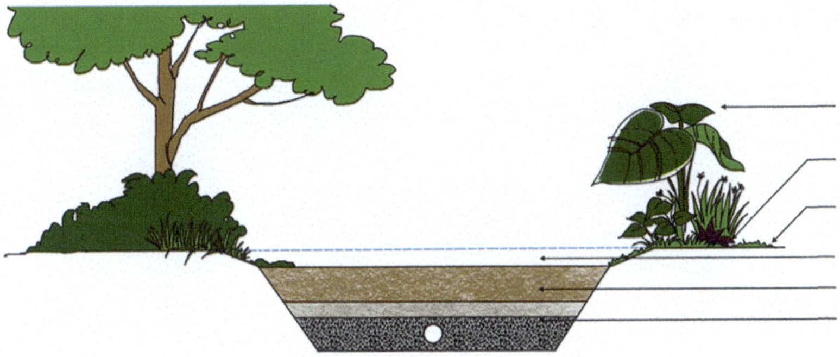

Fig. 2 Construction element of a rain garden via cross-section diagram. (diagram by author)

and wood chips in specific mass ratios. SUDS guidelines highlight the design of bioretention cells to maintain a consistent water presence, while LIUDD structures involve porous soil mixed with deep, porous surfaces for water accumulation. Sponge Cities emphasizes calculating soil or media properties based on various factors, and ABC Waters highlights the influence of catchment characteristics on runoff properties.

Regarding "Characteristics of vegetation" or "Softscape" (compare Sect. 4.1.2), the guidelines recommend different types of vegetation to enhance water retention and mitigate runoff. WSUD suggests using ground cover to prevent soil erosion and native plants to mitigate soil erosion. LID emphasizes the importance of biodiverse plants, succulent plants, and herbaceous grasses. SUDS guidelines recommend trees for water transpiration and loss reduction, while LIUDD suggests diverse native vegetation. Sponge Cities stresses the need for further research on plant dynamics and focuses on native Australian species, and ABC Waters highlights the influence of vegetation density and height on flow rate.

Overall, the summarized principles offer valuable insights into the varied strategies and considerations employed in national guidelines for adequate water-sensitive urban design implementations, as shown in Tables 1 and 2.

Table 1 provides comprehensive details regarding the hardscape structure, facilitating its practical application in design endeavours. This information serves as a constant parameter for computing water retention ability to suit the area. The calculation of water retention ability for rain gardens aligns with established conceptual frameworks that enable the quantification of runoff reception efficiency (Wanitchayapaisit et al., 2023). Table 2 offers an overview and a guide on selecting and implementing softscape elements, such as vegetation, specifically for the design of rain gardens. The guide focuses on characteristics of plants and planting strategies that can help manage water in urban environments. However, it is important to note that while Table 2 provides detailed guidance, there is still a need to establish principles for selecting plants in regions like Southeast Asia with unique climatic conditions.

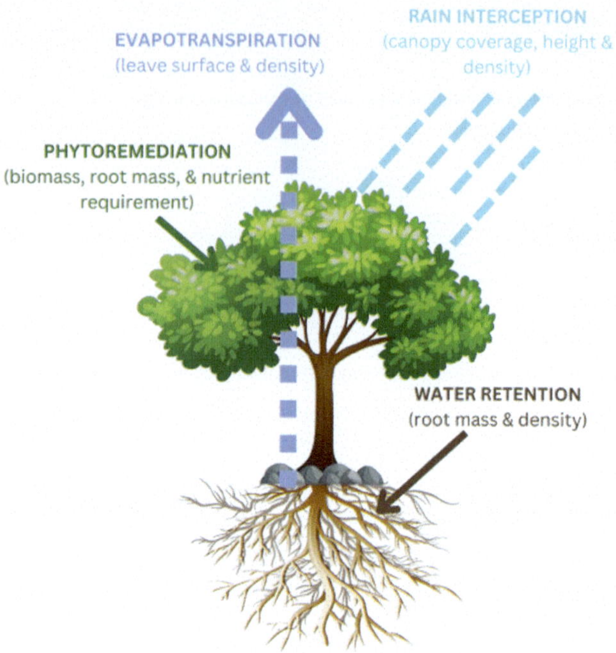

Fig. 3 Vegetation factors (image by author)

The next section of the study will investigate the factors that affect the performance of plants in GSI systems. The aim is to address existing gaps in knowledge and improve the effectiveness of rain garden designs in managing urban water challenges.

3.3 The Role of Plants and Design Factors

Plants alter the soil–water dynamics. For example, plants can enlarge and elongate soil pores following root turnover, reverse soil compaction, and create macropores (Yan et al., 2014; Yuan et al., 2017). Adopting vegetation enhances infiltration rates in rain garden media due to improved soil permeability and porosity. Despite the documented benefits, comparatively few studies dedicated their efforts to the role of vegetation in bioretention performance (Skorobogatov et al., 2020).

The results of the synthesis suggested that the factors considered in the initial assessment of a rain garden's water retention ability should include vegetation characteristics. These characteristics can be used to calculate various constants to determine the water retention ability of a rain garden (Dagenais et al., 2018; Fowdar et al., 2022a, 2022b; Hunt et al., 2015). The processes of evapotranspiration, phytoremediation, and water retention interact with non-vegetative factors such as soil porosity, groundwater level, wind speed, humidity, temperature, and solar radiation to influence the water retention ability of rain gardens (H. Fowdar et al., 2022a, 2022b; He et al., 2021; Körner, 2015), as shown in Figs. 3 and 4.

Furthermore, constant values obtained from plant data can be calculated, as demonstrated in Table 3. The table shows the factors and sources of different parameters

Fig. 4 Non-vegetative factors that affect plant characteristics (image by author)

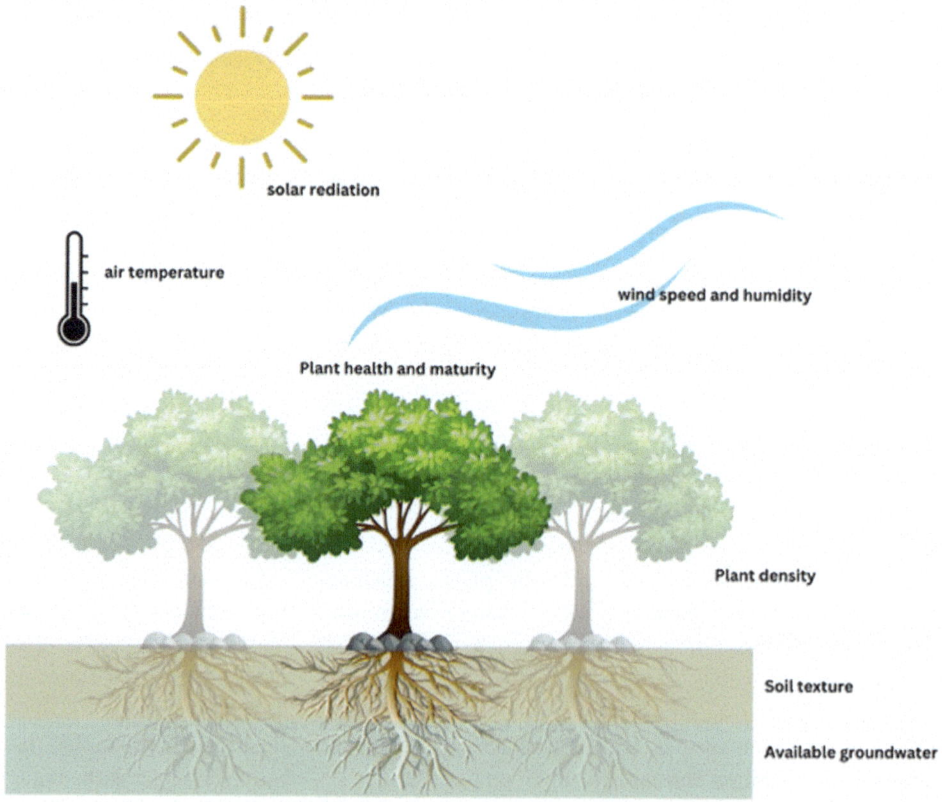

Table 3 Determination of vegetative factor constants

	Trunk/stem		Crown/leaf		Root	
	Girth	Total hight	Leave surface	Leave density	Root mass	Root density
Evapotranspiration	/	/			/	
Phytoremediation	/		/	/		
Water retention					/	/

Note Refers to factors that can be used to evaluate the potential of water absorption by plants

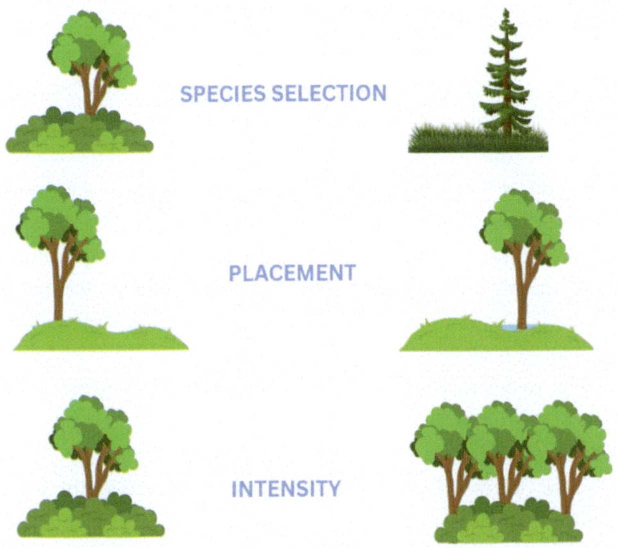

Fig. 5 Planting design strategies that might affect rain garden performance (image by author)

related to plant characteristics, which are likely to be used in the future to derive constants for calculating the water retention ability for WSUD.

These vegetation factors can be influenced by planting design factors (see Fig. 5), which determine how plants may interact with the ecosystem. Planting design factors can be simplified to include the following (Davar et al., 2021).

- **Plant species selection:** What species or communities of plants would be placed in the rain garden? Some plants have a more extensive leaf area index (LAF), contributing to a higher evapotranspiration rate (Fang et al., 2019). Some plants have larger biomass, meaning they might require and retain more water (Abino et al., 2014). Some species might survive better in rain garden conditions, providing continuous ecosystem services (Emanuel et al., 2010b). Special mechanisms in some plants bind certain minerals and nutrients, such as nitrogen fixation (Huang et al., 2000). Alternatively, some plants may require extra care, increasing the rain garden's costs in its life cycle (Jaber et al., 2012).
- **Placement:** Where would the plants be placed? Certain plants survive better in specific microclimates, and while they could be in the same garden, sun/shade and soil humidity, as well as underground water, can affect plant health and maturity (Emanuel et al., 2010b; Franti & Rodie, 2007; Yuan et al., 2017). Furthermore, the amount of solar radiation, relative humidity, and wind movement impacts plant growth and evapotranspiration rates, as previously discussed (He et al., 2021; Kim & Beard, 1988).
- **Intensity:** How dense should we place the plants? Some stormwater models, such as SWWM, only discuss intensity under the surface area coverage (Baek et al., 2020), which helps the designers know how much planting areas should be in the rain gardens. However, the biomass that comes with different intensities of plants would also affect the root mass, rainfall interruption, and evapotranspiration rate (Abino et al., 2014).

It should be noted that plants interact with other dynamic factors, such as survivability and maturity. Other complex environmental factors cannot yet be calculated in the current models (Dagenais et al., 2018; Fowdar et al., 2022a, 2022b; He et al., 2021).

The solution when considering the commonality between the three factors (plant species, placement, and intensity) is plant biomass. The more biomass, both trunk, leaf, and root density, can be used to improve the water retention ability of rain gardens. Thus, we propose that in the future, plant biomass should be included in the calculation aside from vegetation coverage, which looks only from the area perspective and not depth.

4 Discussion

The study analyzed the current literature and found that the factors used to measure the water retention ability of WSUD can be divided into two parts, including "hardscape" and "softscape".

Substantial research supports the hardscape component's use in calculating water retention ability, specifically regarding the width, depth, and porosity of the GSI structures. However, the softscape aspect needs more extensive data due to the involvement of variable plant factors (such as survivability and maturity) and insufficient information to calculate GSI effectiveness using plant-related factors. Key findings include:

(1) Research conducted on the same plant family in different climatic conditions may vary in results.
(2) Limited studies focus on Water Sensitive Urban Design (WSUD) in climates similar to Southeast Asia.
(3) Plant absorption efficiency varies in different environments.
(4) There is a need for more data on planting design details, such as species, positioning, and density while existing computational models mainly consider plant coverage. The research proposes that future models should calculate biomass for water retention calculations.

This research is consistent with existing publications calling for investigating the above-ground parts of rain gardens and how they might affect performance (Dagenais et al., 2018). The research also considers the physiology of the plants, which may be the part that people interact with the most in rain gardens (Suppakittpaisarn et al., 2020). The results add to previous studies focusing only on underground structural calculations (Baek et al., 2020), hopefully producing a more holistic picture and inviting more involvement in the design professions.

To enhance the accuracy of determining rain garden efficiency. Further research is needed to identify appropriate design factors specifically for rain gardens in tropical and subtropical climates. Additionally, establishing a standardized measurement unit to assess the performance of different plant species used in rain garden management in tropical and sub-tropical climates would enable immediate implementation by designers. These established constants can be utilized for future modelling calculations.

5 Conclusion

In conclusion, this exploratory literature review provides a comprehensive examination of the design factors that affect the water retention ability of water-sensitive urban design (WSUD) in the context of tropical and sub-tropical climates. The study highlights the limited focus on WSUD in climates similar to tropical and sub-tropical climates and a need for more information regarding the impact of planting design on the performance of rain gardens. The literature review identifies key principles of WSUD, such as Green Stormwater Infrastructure (GSI), implemented in various international contexts. Notably, the study underlines the need for examples and principles of GSI design in tropical and sub-tropical climates or Southeast Asia, specifically for rain gardens. The need for established principles for selecting plants and insufficient data on plant performance in GSI systems are also highlighted.

The reviewed literature emphasizes the crucial role of size and structure in implementing rain gardens, with specific parameters like the porosity of the infiltration layer and ponding depth being well-defined and quantifiable. However, some studies suggest that the potential of rainwater harvesting goes beyond these structural aspects. The importance of plants, especially large trees, in water absorption and evaporation is acknowledged as a significant factor. Despite these insights, the literature reveals a lack of universal or consistent measures to assess the efficiency of plants in rain gardens. Each plant group or species cannot be quantitatively evaluated in conjunction with the water retention structure. This research gap highlights the need for standardized measurement units and further exploration into the specific design factors influencing the performance of plants in the context of rain gardens in tropical and subtropical climates.

The research proposes the inclusion of plant biomass in future calculations for water retention ability, aiming for a more holistic perspective on rain gardens' efficiency. To advance knowledge in this field, the study recommends further investigations into appropriate design factors for WSUD in tropical and sub-tropical climates, establishing standardized measurement units, and developing constants for modelling calculations. This approach aims to enhance the accuracy of assessing different plant species' performance in rain garden management, facilitating immediate implementation by designers and contributing to the broader field of WSUD.

Moreover, future studies should explore additional aspects, such as investigating input values for modelling, to improve the capability of calculating water retention ability based on relevant plant factors. This could involve examining the water-absorbing performance of different tree species.

References

Abino, A. C., Castillo, J. A. A., & Lee, Y. J. (2014). Assessment of species diversity, biomass and carbon sequestration potential of a natural mangrove stand in Samar, the Philippines. *Forest Science and Technology, 10*(1), 2–8.

Ahiablame, L. M., Engel, B. A., & Chaubey, I. (2012). Effectiveness of low impact development practices: Literature review and suggestions for future research. *Water, Air, & Soil Pollution, 223*, 4253–4273.

Akter, A. (2022). Stormwater management. In *Rainwater Harvesting—Building a Water Smart City* (pp. 117–163). Springer.

Ali, W., Takaijudin, H., Yusof, K. W., Osman, M., & Abdurrasheed, A. S. I. (2021). The common approaches of nitrogen removal in bioretention system. *Sustainability, 13*(5), 2575.

Baek, S.-S., Ligaray, M., Pyo, J., Park, J.-P., Kang, J.-H., Pachepsky, Y., Chun, J. A., & Cho, K. H. (2020). A novel water quality module of the SWMM model for assessing low impact development (LID) in urban watersheds. *Journal of Hydrology, 586*, 124886.

Baird, J. B., Winston, R. J., & Hunt, W. F. (2020). Evaluating the hydrologic and water quality performance of novel infiltrating wet retention ponds. *Blue-Green Systems, 2*(1), 282–299.

Chen, S. S., Tsang, D. C., He, M., Sun, Y., Lau, L. S., Leung, R. W., Lau, E. S., Hou, D., Liu, A., & Mohanty, S. (2021). Designing sustainable drainage systems in subtropical cities: Challenges and opportunities. *Journal of Cleaner Production, 280*, 124418.

Coombes, P. J., Argue, J. R., & Kuczera, G. (2000). Figtree Place: A case study in water sensitive urban development (WSUD). *Urban Water, 1*(4), 335–343.

Coutts, A. M., Tapper, N. J., Beringer, J., Loughnan, M., & Demuzere, M. (2013). Watering our cities: The capacity for Water Sensitive Urban Design to support urban cooling and improve human thermal comfort in the Australian context. *Progress in Physical Geography, 37*(1), 2–28.

Dagenais, D., Brisson, J., & Fletcher, T. D. (2018). The role of plants in bioretention systems; does the science underpin current guidance? *Ecological Engineering, 120*, 532–545.

Davar, S. A., Rostami, P., & Moradi, M. (2021). Planting design utilizing phytoremediation of garden-rangeland ecotypes in urban green space (case study of the new city of Pardis). *PalArch's Journal of Archaeology of Egypt/egyptology, 18*(18), 547–558.

Dell, T., Razzaghmanesh, M., Sharvelle, S., & Arabi, M. (2021). Development and application of a SWMM-Based simulation model for municipal scale hydrologic assessments. *Water, 13*(12), 1644.

Dunnett, N., & Clayden, A. (2007). Rain gardens. *Managing water sustainably in the garden and designed landscape*.

Eckart, K., McPhee, Z., & Bolisetti, T. (2018). Multiobjective optimization of low impact development stormwater controls. *Journal of Hydrology, 562*, 564–576.

Emanuel, R., Godwin, D., & Stoughton, C. (2010a). The Oregon rain garden guide: a step-by-step guide to landscaping for clean water and healthy streams.

Emanuel, R., Godwin, D., & Stoughton, C. (2010b). *The Oregon Rain Garden Guide: Landscaping for Clean Water and Healthy Streams*. Oregon Sea Grant. https://www.portlandoregon.gov/bes/article/474026

Fang, H., Baret, F., Plummer, S., & Schaepman-Strub, G. (2019). An overview of global leaf area index (LAI): Methods, products, validation, and applications. *Reviews of Geophysics, 57*(3), 739–799.

Fassman-Beck, E., Wang, S., Simcock, R., & Liu, R. (2015). Assessing the effects of bioretention's engineered media composition and compaction on hydraulic conductivity and water holding capacity. *Journal of Sustainable Water in the Built Environment, 1*(4), 04015003.

Ferrans, P., Torres, M. N., Temprano, J., & Sánchez, J. P. R. (2022). Sustainable Urban Drainage System (SUDS) modeling supporting decision-making: A systematic quantitative review. *Science of the Total Environment, 806*, 150447.

Ferreira, J. C., Monteiro, R., & Silva, V. R. (2021). Planning a green infrastructure network from theory to practice: The case study of Setúbal, Portugal. *Sustainability, 13*(15), 8432.

Fowdar, H., Payne, E., Deletic, A., Zhang, K., & McCarthy, D. (2022a). Advancing the Sponge City Agenda: Evaluation of 22 plant species across a broad range of life forms for stormwater management. *Ecological Engineering, 175*, 106501.

Fowdar, H. S., Neo, T. H., Ong, S. L., Hu, J., & McCarthy, D. T. (2022b). Performance analysis of a stormwater green infrastructure model for flow and water quality predictions. *Journal of Environmental Management, 316*, 115259.

Franti, T. G., & Rodie, S. N. (2007). *Stormwater Management: Rain Garden Design for Homeowners*. BebGuide. http://extensionpublications.unl.edu/assets/pdf/g1758.pdf

Green, A. (2019). Sustainable drainage systems (SuDS) in the UK. *Urban Stormwater and flood management: Enhancing the Liveability of cities*, pp. 69–101.

Gülbaz, S., & Kazezyılmaz-Alhan, C. M. (2017). Experimental investigation on hydrologic performance of LID with rainfall-watershed-bioretention system. *Journal of Hydrologic Engineering, 22*(1), D4016003.

He, Y., Lin, E. S., Tan, C. L., Tan, P. Y., & Wong, N. H. (2021). Quantitative evaluation of plant evapotranspiration effect for green roof in tropical area: A case study in Singapore. *Energy and Buildings, 241*, 110973.

Heng, N. T. (2021). *Monitoring and Modelling of Active, Beautiful and Clean (ABC) Waters Design Features in Urban Singapore* National University of Singapore (Singapore)].

Huang, J., Reneau, R., & Hagedorn, C. (2000). Nitrogen removal in constructed wetlands employed to treat domestic wastewater. *Water Research, 34*(9), 2582–2588.

Hunt, W. F., Lord, B., Loh, B., & Sia, A. (2015). *Plant selection for bioretention systems and stormwater treatment practices*. Springer Nature.

Ignatieva, M., Meurk, C., Van Roon, M., Simcock, R., & Stewart, G. H. (2008). How to put nature into our neighbourhoods: Application of Low Impact Urban Design and Development (LIUDD) principles, with a biodiversity focus, for New Zealand developers and homeowners.

Jaber, F., Woodson, D., LaChance, C., & York, C. (2012). Stormwater management: Rain gardens. *The Department of Soil and Crop Sciences and Texas A&M AgriLife Communications, The Texas A&M System, USA*, 20.

Jiang, C., Lv, P., Li, J., & Gao, J. (2022). Rapid determination methods for extent parameters of typical LID facilities in Urban stormwater management. *Journal of Environmental Engineering, 148*(10), 04022064.

Kasprzyk, M., Szpakowski, W., Poznańska, E., Boogaard, F. C., Bobkowska, K., & Gajewska, M. (2022). Technical solutions and benefits of introducing rain gardens–Gdańsk case study. *Science of the Total Environment, 835*, 155487.

Kim, K., & Beard, J. (1988). Comparative turfgrass evapotranspiration rates and associated plant morphological characteristics. *Crop Science, 28*(2), 328–331.

Körner, C. (2015). Paradigm shift in plant growth control. *Current Opinion in Plant Biology, 25*, 107–114.

Lee, J., Bae, S., Lee, W. H., & Gil, K. (2022). Effect of surface area to catchment area ratio on pollutant removal efficiency in vegetation-type facilities. *Ecological Engineering, 179*, 106609.

Li, J., Sun, C., Chen, W., Zhang, Q., Zhou, S., Lin, R., & Wang, Y. (2022). Groundwater quality and associated human health risk in a typical basin of the Eastern Chinese Loess Plateau. *Water, 14*(9), 1371.

Liu, Q., Cui, W., Tian, Z., Tang, Y., Tillotson, M., & Liu, J. (2022). Stormwater management modeling in "Sponge City" construction: Current state and future directions. *Frontiers in Environmental Science, 9*, 721.

Liu, T., Lawluvy, Y., Shi, Y., & Yap, P.-S. (2021). Low impact development (LID) practices: A review on recent developments, challenges and prospects. *Water, Air, & Soil Pollution, 232*(9), 344.

Matthews, T., Lo, A. Y., & Byrne, J. A. (2015). Reconceptualizing green infrastructure for climate change adaptation: Barriers to adoption and drivers for uptake by spatial planners. *Landscape and Urban Planning, 138*, 155–163.

Neo, T. H., Xu, D., Fowdar, H., McCarthy, D. T., Chen, E. Y., Lee, T. M., Ong, G. S., Lim, F. Y., Ong, S. L., & Hu, J. (2022). Evaluation of active, beautiful, clean waters design features in tropical urban cities: A case study in Singapore. *Water, 14*(3), 468.

Norton, B. A., Coutts, A. M., Livesley, S. J., Harris, R. J., Hunter, A. M., & Williams, N. S. (2018). Planning for cooler cities: A framework to prioritise green infrastructure to mitigate high temperatures in urban landscapes. In *Planning for Climate Change* (pp. 233–249). Routledge.

Payne, E. G., Pham, T., Deletic, A., Hatt, B. E., Cook, P. L., & Fletcher, T. D. (2018). Which species? A decision-support tool to guide plant selection in stormwater biofilters. *Advances in Water Resources, 113*, 86–99.

Peel, M. C., Finlayson, B. L., & McMahon, T. A. (2007). Updated world map of the Köppen-Geiger climate classification. *Hydrology and Earth System Sciences, 11*(5), 1633–1644.

Pressman, A. (2007). *Architectural graphic standards*. John Wiley & Sons.

Radcliffe, J. C. (2019). History of water sensitive urban design/low impact development adoption in Australia and internationally. In *Approaches to water sensitive urban design* (pp. 1–24). Elsevier.

Reis, V. E. (2019). *Techniques for Investigating Causes of Green Stormwater Infrastructure Underperformance and Recommendations for Rehabilitation*. Villanova University.

Rinchumphu, D., Tangsongsuwan, R., Kridakorn Na Ayudhya, T., Irvine, K., Chua, L., Suppakittpaisarn, P., & Chakrawattana, N. (2021). *Potential Evaluation of Urban Bioretention Design* [Grant]. Chiang Mai, Thailand.

Ristianti, N. S., Bashit, N., Ulfiana, D., & Windarto, Y. E. (2022). Bioretention Basin, Rain Garden, and Swales Track Concepts through Vegetated-WSUD: Sustainable Rural Stormwater Management in Klaten Regency. IOP Conference Series: Earth and Environmental Science,

Sañudo-Fontaneda, L. A., Roces-García, J., Coupe, S. J., Barrios-Crespo, E., Rey-Mahía, C., Álvarez-Rabanal, F. P., & Lashford, C. (2020). Descriptive analysis of the performance of a vegetated swale through long-term hydrological monitoring: A case study from Coventry, UK. *Water, 12*(10), 2781.

Shao, R., Shao, W., & Wang, Y. (2023). Inferring the influence of urban vegetation on urban water storage capacity from evapotranspiration recession. *Journal of Hydrology, 620*, 129355.

Sharma, R., & Malaviya, P. (2021). Management of stormwater pollution using green infrastructure: The role of rain gardens. *Wiley Interdisciplinary Reviews: Water, 8*(2), e1507.

Skorobogatov, A., He, J., Chu, A., Valeo, C., & van Duin, B. (2020). The impact of media, plants and their interactions on bioretention performance: A review. *Science of the Total Environment, 715*, 136918.

Suppakittpaisarn, P., Chang, C.-Y., Deal, B., Larsen, L., & Sullivan, W. C. (2020). Does Vegetation Density and Perceptions Predict Green Stormwater Infrastructure Preference? *Urban Forestry & Urban Greening, 55*. https://doi.org/10.1016/j.ufug.2020.126842

Suppakittpaisarn, P., Jiang, X., & Sullivan, W. C. (2017). Green infrastructure, green stormwater infrastructure, and human health: A review. *Current Landscape Ecology Reports, 2*, 96–110.

Swedberg, R. (2020). Exploratory research. *The production of knowledge: Enhancing progress in social science*, pp. 17–41.

Tu, X., & Tian, T. (2015). Six questions towards a Sponge City–report on power of public policy: Sponge City and the trend of landscape architecture. *Landscape Architecture Frontiers, 3*(2), 22–32.

Valley, C. (2010). Low impact development stormwater management planning and design guide. In.

van Roon, M., Ignatieva, M., Meurk, C., Simcock, R., & Stewart, G. (2008). How to put nature into our neighbourhoods: Application of low impact urban design and development (LIUDD). principles, with a biodiversity focus, for New Zealand developers and homeowners. In: Manaaki Whenua Press.

Venvik, G., & C. Boogaard, F. (2020). Infiltration capacity of rain gardens using full-scale test method: effect of infiltration system on groundwater levels in Bergen, Norway. *Land, 9*(12), 520

Wahab, S., Mamtaz, R., & Islam, M. (2016). Applicability of water sensitive Urban Design (WSUD) in Dhaka City. In *Proceedings of the 3rd International Conference on Civil Engineering for Sustainable Development (ICCESD2016)*. Khulna, Bangladesh,

Wang, H., Cheng, X., Man, L., Li, N., Wang, J., & Yu, Q. (2016). *Challenges and future improvements to China's sponge city construction*. International Low Impact Development Conference China

Wanitchayapaisit, C., Charoenlertthanakit, N., Surinseng, V., Yaipimol, E., Rinchumphu, D., & Suppakittpaisarn, P. (2023). Enhancing water-sensitive urban design in Chiang Mai through a research-design collaboration. *Sustainability, 15*(22), 16127.

Wanitchayapaisit, C., Suppakittpaisarn, P., Charoenlertthanakit, N., Surinseng, V., Yaipimol, E., & Rinchumphu, D. (2022). Rain garden design for stormwater management in Chiang Mai, Thailand: A Research-through-Design Study. *Nakhara: Journal of Environmental Design and Planning, 21*(3), 222–222.

Wella-Hewage, C. S., Alankarage Hewa, G., & Pezzaniti, D. (2016). Can water sensitive urban design systems help to preserve natural channel-forming flow regimes in an urbanised catchment? *Water Science and Technology, 73*(1), 78–87.

Yan, H., Lipeme Kouyi, G., Gonzalez-Merchan, C., Becouze-Lareure, C., Sebastian, C., Barraud, S., & Bertrand-Krajewski, J.-L. (2014). Computational fluid dynamics modelling of flow and particulate contaminants sedimentation in an urban stormwater detention and settling basin. *Environmental Science and Pollution Research, 21*(8), 5347–5356.

Yuan, J., Dunnett, N., & Stovin, V. (2017). The influence of vegetation on rain garden hydrological performance. *Urban Water Journal, 14*(10), 1083–1089.

Zhang, K., Manuelpillai, D., Raut, B., Deletic, A., & Bach, P. M. (2019). Evaluating the reliability of stormwater treatment systems under various future climate conditions. *Journal of Hydrology, 568*, 57–66.

Zhou, Q. (2014). A review of sustainable urban drainage systems considering the climate change and urbanization impacts. *Water, 6*(4), 976-992.

Zuniga-Teran, A. A., Gerlak, A. K., Mayer, B., Evans, T. P., & Lansey, K. E. (2020). Urban resilience and green infrastructure systems: Towards a multidimensional evaluation. *Current Opinion in Environmental Sustainability, 44*, 42–47.

Adapting Modern Cities to the Evolved Needs of People

Amro Abdelalim

Abstract

People needs are considered to be a major indicator and director in their lives. Urban form of cities has been noticed to be affected and changed with the developed people needs. They are continuously changing all over hundreds of years according to the human development and scientific discoveries. Studies show that people tend to settle in cities because they know that urban settlements can offer better services and diversity in satisfying their diverse needs the best. That's why having a track over the arrangement and evolution of cities forms is an important domain and interesting subject to be studied. To have a livable city is to have its people needs studied very well and generally fulfilled. This paper focuses on finding a relation between the hierarchy of needs of people living in gathered settlements from ancient ages to present along with their cities forms. A city is considered to be successful and remarkable only when the services they present fits the most its diverse citizens needs. A city, in its spatial form, components and land uses, corresponds with citizens needs in some way. The way it's designed and planned gives a related overview to its people and how they live in the circumstances of space, resources and time. That's why tracking the needs of people and studying them recently is necessary to ensure that cities are suitable and proper for their citizens to live in them regarding the accelerated updated forms of cities and not contradicting with the digital age human history is entering.

Keywords

Hierarchy of needs · Suitability · Social needs

A. Abdelalim (✉)
Fayoum University, Fayoum, Egypt
e-mail: aka03@fayoum.edu.eg

1 Introduction

Having suitable cities is becoming more and more important to achieve sustainability. Living in cities requires to use resources in the best way due to scarcity of resources. This results in a situation where people interact with each other and get the full usage of the land they are living on without neglecting other generations rights. Services and work should be provided in a way that help them make their lives better. Having cities or living in urban settlements where people feel comfortable and relieved is the ultimate target of all urbanization process which has developed in different ways through ages to achieve goals of sustainability as in Fig. 1.

Cities are largely related and affected with the economic aspect where a successful economy results in a flourishing city and citizens welfare. A city, as an expression, has always been looked at as the dream place where there is everything in certain times of the last centuries. It is a place where one can find what he wants starting from daily and weekly groceries in supermarkets ending with already totally-made built-in kitchens in big stores. It is a place where one can't think of a product and doesn't find it there. "For centuries, the city has been the heart, the lifeblood, of various civilizations—the center of economic, political, and artistic events" (Macionis and Parrillo, 2017). A city is being imagined as the place of meeting and interaction where people gather and live life at its most. Imagined as the contemporary and modern expressionism of the society, people have always had that look to the city as the most newly developed and civilized world. People living in new cities are always tracked on how they eat, drink, walk, and even what they hold in their hands while walking, how they do the daily-done activities in their way.

Studying cities all over ages has been a point of interest and prerequisite for having better ways of living and fixing issues in developing cities. In old ages, needs were too simple to be studied and fulfilled where people only lived

B. He et al. (eds.), *Urban Climate Change Adaptation*, Advances in Science, Technology & Innovation,
https://doi.org/10.1007/978-3-031-65088-8_7

Fig. 1 The global goals for sustainable development *Source* United Nations

on basic and simple ones of just eating, drinking and finding some shelter to sleep. That's why most civilizations arouse around water features like the Nile in Egypt where it was the easiest and most applicable for housing units to be located beside the resource of water and start communities.

2 Maslow Pyramid of Needs

Human needs start from being basic to advanced and are listed in levels illustrated in Maslow pyramid of needs as in Fig. 2. Bottom of the pyramid includes basic psychological needs of breathing, food, water, shelter, clothing and sleep. These are considered to be the simple needs required for just survival without aspiring to achieve any advanced progress in life, where the person only fulfills them to pass the days without having to achieve pride or great achievements. This is obvious and clearly seen in the first settlements of people in the ancient ages where the human being used to live by the canals and water features in very simple huts as housing units for sleeping. The water feature was responsible for providing water and fish for eating where it was the simplest way of living to drink from the canal, eat some fish and sleep in the hut.

By time, it became obvious that sleeping in simple huts and relying on just fish as a source of food was not quite enough and it became possible to identify some opportunities of having diverse sources of food from trees and upgrading the housing unit to be more spacious, to include more living activities. This came after the invention of farming and knowing fire. This was considered to be a big leap in civilization where it became under control to live away from water features beside farming lands as farming became a golden key of providing permanent food resources all year. Using advanced materials helped in building more spacious housing units becoming wider and

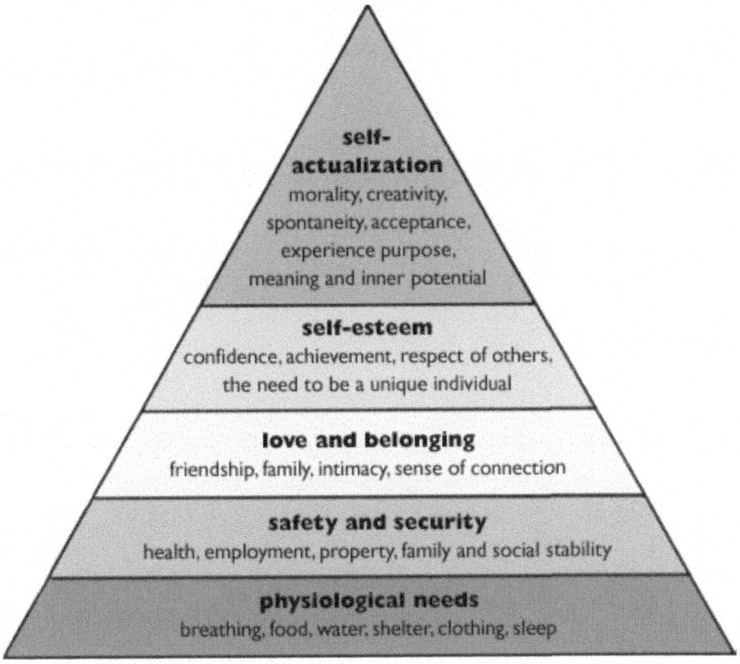

Fig. 2 Maslow Pyramid of needs
Source (Maslow, 1954)

higher to accommodate a larger number of people in the same place becoming called a tribe later. Living together on a land of some farms and housing units of huts represented the bottom level of needs of Maslow pyramid and can be identified to be the first settlement of civilizations which was able to fulfill needs of food, water and shelter.

Maslow's hierarchy offers five levels of needs for any human being. It is graduated from what is biological to the secondary ones. Needs grading starts from physiological needs passing through safety and security, love and belonging, self-esteem and ends with self-actualization. The first level of needs which is considered the level of biological life or death is clear. One can not live without oxygen or water or even without appropriate sleep. Having a shelter and clothes are considered necessary needs for life for achieving needs of sleeping well and being able to engage safely with others. Second level which comes directly after biological needs is safety and security. First level of needs defines if one can be alive or not, but when it comes to the second one, it defines where it is possible to keep living or not with the current living situations. A non-healthy man will not be alive except for a certain period of time, that's for his safety. Concerning security, employment, property, family and social stability and taking care of such needs make one secure and give some sense of guarantee for living. A level of social needs can be seen here. One lacks some sense of security without a family.

For the third level of needs, it is totally about social needs. It has a title of love and belonging, enlisted friendship, family, intimacy and sense of connection. In general, it describes the life cycle of any person. For example, a baby after getting born, cries for food and drink. After a while and after getting used to the milk of mother, he takes time to get used to the new home outside his mother's body. After taking some time to be a kid, he wants more than a home and a family and to discover more and more, as a need of exploring and aspiring what is next in life. Children search for other children to make friends after they had already found the safeness and security in their houses with their families. They search for the next level of needs as they aspire for belonging and intimacy, in a more advanced way. In this specific level, it is considerably right to say that a kid has a life. Children start to feel happy and joyful when they run after each other and when they do something together. They sense their presence and belonging together with each other.

It is not only about the children for the third level of needs. It can also be seen and observed clearly in the life of any teenager. When children get older and get ready for graduation, they feel a lack in security, a financial one. One aspires for a sense of financial security which translates in the form of searching for a job and having an income. Upon fulfilling this lack of security, one starts to look for the next, to get a higher level of love and belonging either coming as friendship or intimacy.

Fourth level of needs is said to be about achieving self-esteem. It is mainly concerned about the individual to achieve something, self-confidence, making something and to have the feeling of a successful and worthy life. Fifth level, self-actualization, is considered to be the highest level of needs. Some may not recognize it as a need but rather a level of welfare. So a graduation of five levels of needs, graduating from biological to social then self-satisfaction eventually explains how much social needs can be necessary for a person's life.

3 Origin of Cities and Their Development Across Ages

Cities emerge historically only where a food surplus can be extracted (Jacobs, 1969). They are said to have emerged in certain locations as in Fig. 3. The need of eating is the most primary factor of the first cities to be built where people could sleep knowing that there is a resource of food that they can rely on. The term of city has originated since ages from the beginning of history and agglomeration of some people together to form a gathered community, yet in small groups. It wasn't called cities by that time, but rather a town or a tribe. A gathered group of people living in the same place with shared culture and style of living was the very beginning of people settlement and first gathered communities urban form. Groups started to become larger in numbers by time and develop their abilities and living activities to form towns where people started having an administrative order and some heads to rule the settlement. Hereafter, progress of development of cities started to take place according to the upgrading of the needs for eating, shelter and protection. The term of city has taken ages along history so as to take the current form of cities in the twenty-first century. "It is documented that earliest urban centers appear to have emerged in the Middle East around 7500 BCE as modestly sized settlements such as Byblos (Scott & Storper, 2015)."

There was a common need for security in the time when there was the danger of being preyed on. Danger came in different levels along history. That's when people started to group together to form safer settlements when it was in the form of attacking wild animals. By time, it took other forms of invading enemies on settlements as in Fig. 4. However, another thing to catch sight of here is that people used to stay in cities aiming at a better life.

Cities started to upgrade their forms and take more developed ones after discovering of simple construction elements and grouping of people together in tribes. They could form territories by time and put some boundaries to provide more protection, making urban settlements bigger. Farms were included in the settlements along with simple walls and watching towers. This allowed for the settlements to be self-independent and be capable of generating their own food for their people without having to engage and interact with what is outside the settlement as seen in Fig. 5.

Defense was the first reason in building new towns of the medieval era, besides other reasons such as agriculture and trade (Kostof, 1991). The word bastion (known as city's fortification) itself came from the word bastide (new town in the Middle Ages). It was also evident that the defense

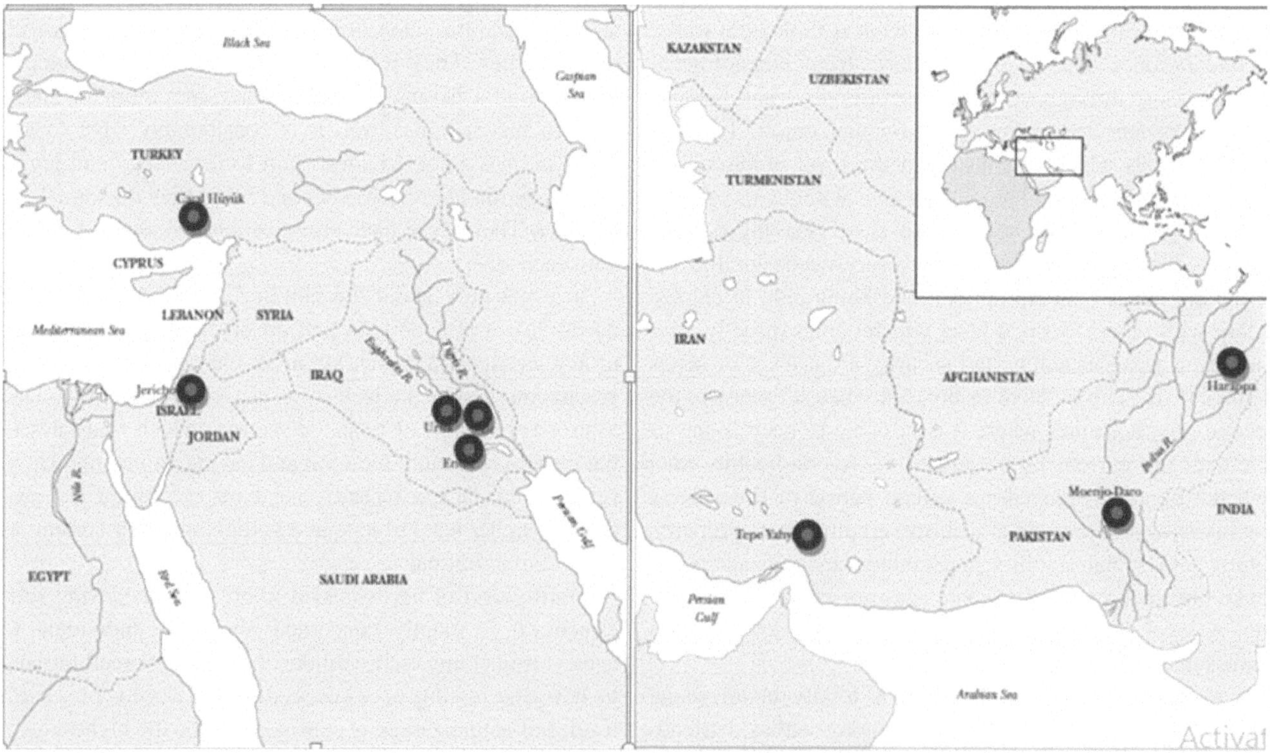

Fig. 3 Location of the earliest cities around the world *Source* (Macionis and Parrillo, 2017)

Fig. 4 View of Babylon, about 600 B *Source* (Frankfort, 1950)

system of cities grew more "sophisticated" in the medieval era. For example, in the fourteenth century, European cities reinforced their circuit wall by addition of second wall, so that "double curtain" (two layers) of city wall was created (Kostof, 1992). Moreover, the gateways in medieval era were flanked by strong towers and protected by mechanisms like the portcullis and the barbican making cities more developed in the mechanisms of protection and offering security.

Fig. 5 Aigues-Mortes (France), a Fortified Medieval city *Source* (Kostof, 1991)

Developed cities by time were improved in levels of protection through applying policy rules. Meanwhile, development in cities have turned to be internal in scopes of more varieties of food, diverse types of housing and better ways of transportation between different zones as in Figs. 6 and 7. For example, in the 1880s, the electric street trolley increased mobility both within the city and just beyond its boundaries, leading to the creation of "streetcar suburbs" and a new population of daily commuters to work in the city (Webb and Tee, 2011). Upon the invention of engines after the industrial age, railway was constructed for city users to transport longer distances in shorter times and people could be able to move between cities in several hours instead of days. Cities became consisting of double layers of networks for pedestrians and vehicles to transport people and its form became more complex.

Cities were taking the shape of towers distributed in many areas along with streets paved through them for transportation. Also, the layout of some cities started to take the shape of different zones distributed all over the city, with railway connecting them and linking them with other cities, in addition to other different modes of transportation as in Fig. 8. All of cities developments before the last century were mainly horizontal, in extending and increasing land uses and districts of more land. Upon entering the last century, it was able to have taller buildings. By the end of the nineteenth century, some buildings reached 30 stories (Still, 1999).

In the beginning of the twenty-first century, more towers were implemented in cities reaching larger heights providing residential, administrative and commercial uses as in Fig. 9. This resulted in the skyline of cities having different heights of buildings which accommodate diverse uses and cities taking more 3D versions. An elevated overview of capital and mega polis cities shows a dense number of buildings situated in different regions allowing for larger numbers of inhabitants on smaller areas as a reflection to the huge revolution in the technology of building materials and construction.

4 Shifting Towards Smart Cities

Smart cities are one of the newest trends in urbanism that started to be constructed along with the huge revolution in technology. Revolution in the field of technology has

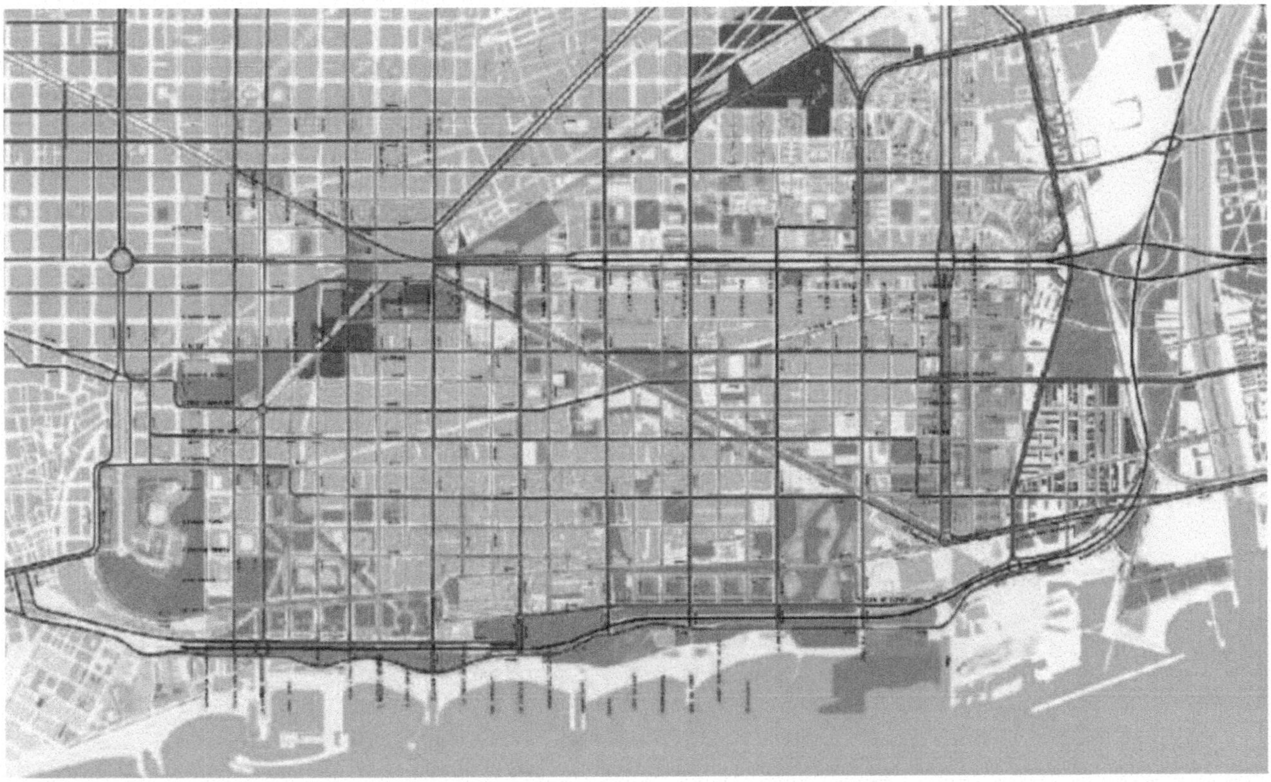

Fig. 6 Barcelona city *Source* (Busquets, 2006)

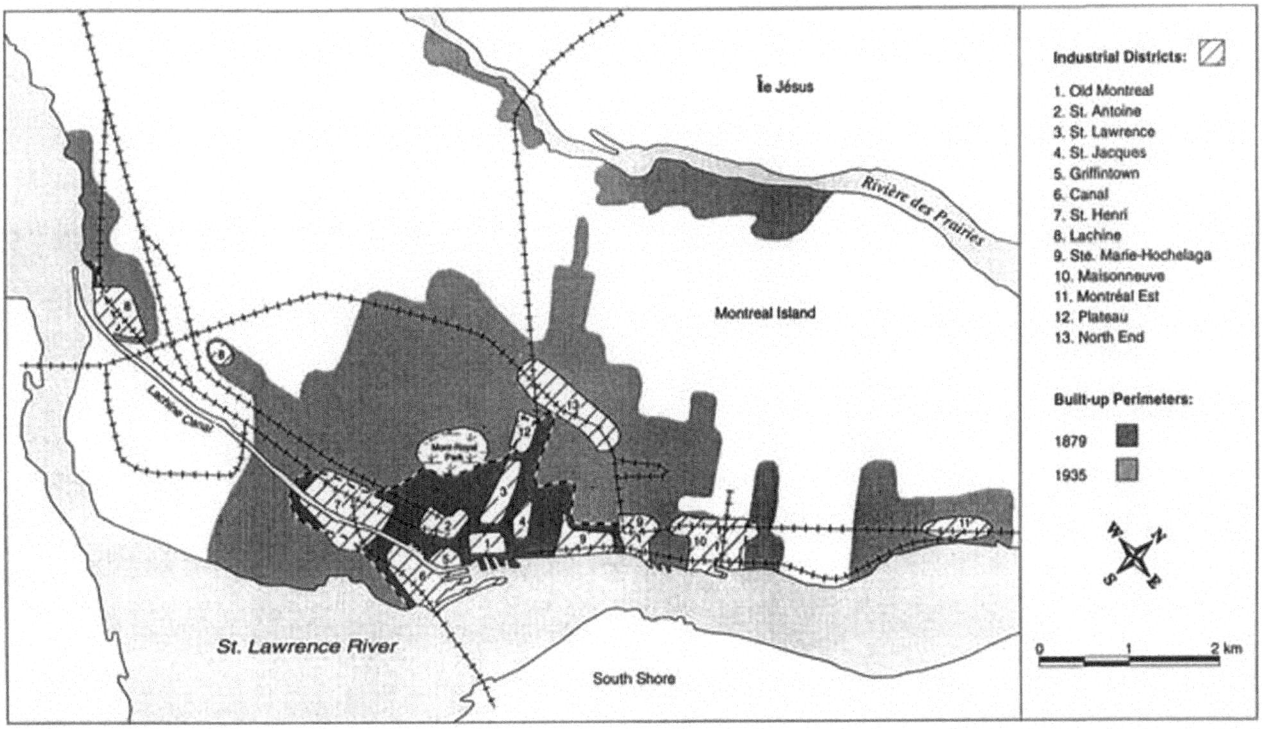

Fig. 7 Toronto in the 1920 *Source* (Harris and Lewis, 2001)

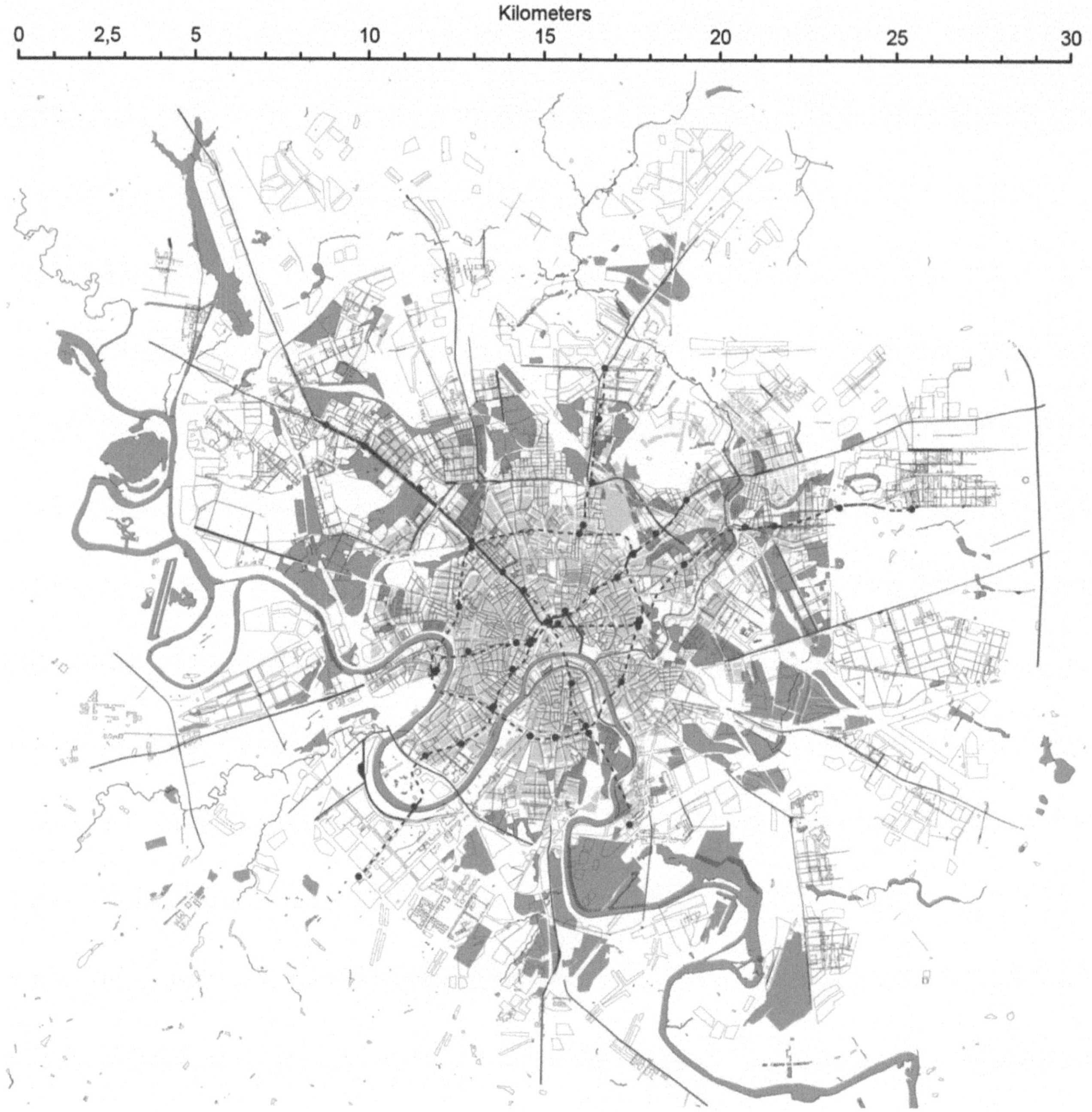

Fig. 8 Moscow city underground metro map (in red) *Source* (Bobkova, 2018)

obliged all people to use mobile phones and live among screens of laptops, computers, televisions and bundles of digital information transferring all the time as in Fig. 10. This resulted in types of cities that consist of lighted screens in the streets monitored with computers, all connected digitally and virtually.

Smart cities monitored by screens have been established through huge bundles of transferred data and abilities of computers. "The growth in complexity of both the hardware and software of information systems has generated communities of people, IT architects and software engineers, who are deeply familiar with methods to compose together vast networks of nodes, as well as tools for creating, managing, and navigating such networks" (Harrison and Donnelly, 2011). Cities, in the form of being smart have acquired the power of simulation, which has enabled city planners and architects to visualize, quantify and virtually see the massive flow of urban data happening daily in cities as in Fig. 11 of Venice carnival. This data is gathered and collected by softwares to be re-imagined and reconstructed virtually to evaluate the positive and negative impacts of different proposals.

Fig. 9 Petronas Towers *Source* (skyscrapercity.com)

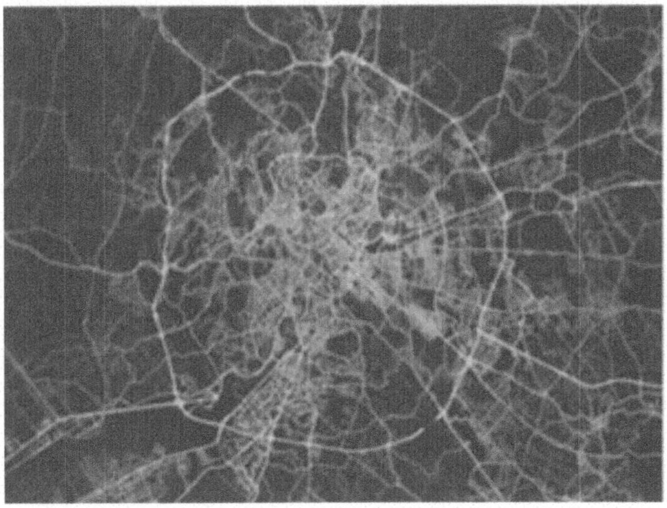

Fig. 10 GPS Volumes in Rome *Source* (Batty et al., 2012)

5 Urbanization Progress and Cities

Urbanization is considered to be one of the processes related with cities. Cities grow and become larger through the process of having its buildings constructed and streets built by time. This process of urbanization is the representation of citizens living in more housing units and making more transportation rates. One of the central features of urbanization has always been its efficiency-generating qualities via agglomeration (Scott and Storper, 2015). This is considered one of the big reasons why cities gain their power and brand from the people within them and how they act, co-live and respond to continuously happening changes.

In the current time, after inventing screens and being used in laptops, mobile phones and advertising tools in streets, they invaded the surrounding environment indoors and outdoors. Each apartment or administrative office in the twenty-first century contains at least 2 or 3 screens

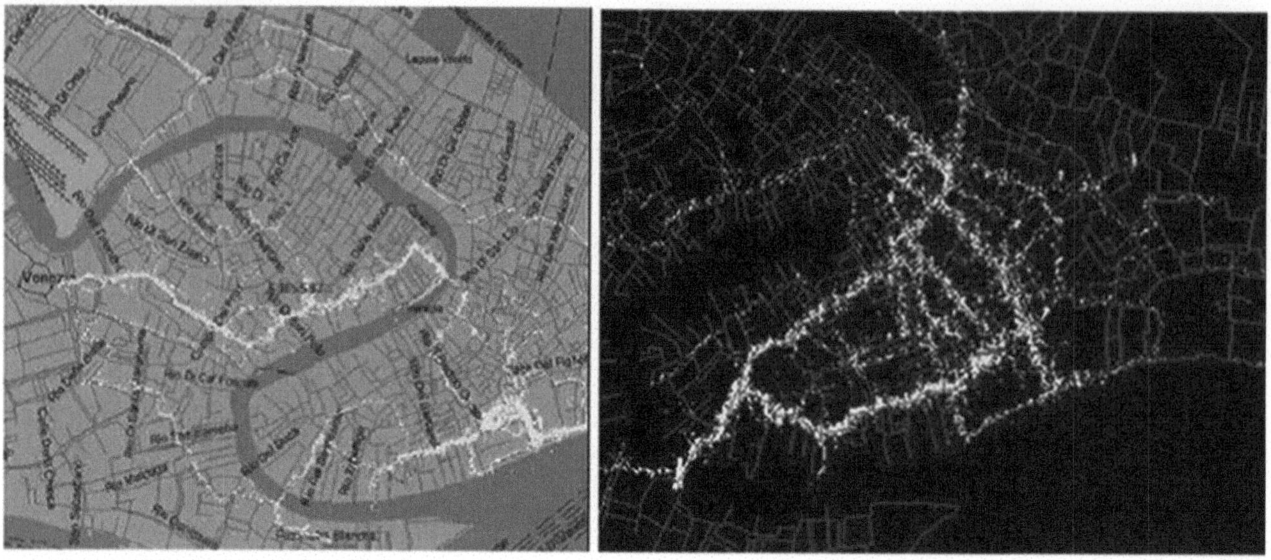

Fig. 11 Reconstruction of movement dynamics of pedestrian movements at different scales in the Venice carnival 2007 *Source* (Batty et al., 2012)

of televisions or computers in addition to mobile phones which transfer news and digital data continuously, making the human brain process most of the time.

In streets, screens of shops and advertising media work in the evening, making them full of light and data that work as a stimulating factor for brain processing as well. Cities developed in a way making them full of screens everywhere as a matter of evolution that makes them function to serve the needs of retails uses the most.

6 Understanding the Relation Between Cities and Needs of People

Cities, where people choose to live, are considered to have and provide what people need in their lives. One chooses the apartment that fits him the most, in terms of privacy, wideness, location and surrounding environment. Some want to in a live place where they pick that apartment in the city center. Others inspire for some calmness away from traffic noise preferring the far suburbs from the city, picking the aspects that fit them the most and provide what they need in their life.

A need can be explained hereby as the lack of something required to live well and more comfortably. People need to maintain their physical beings with material goods and services, but also have needs that are related to their status as members of a community. They need to work—not only to earn income but to participate in the community. They need to have a voice in the decisions that are made, and need a level of social well-being that comes from being cared for and caring for others in return. Generally speaking, living in a city or any settlement is meant to make human beings better.

A city says a lot about its people's needs as well as interests. Since cities are well-known for more diversity, good quality of life and wellness—besides maintaining the basic needs of people—they are expected to give and offer more services and opportunities and fulfill needs better. It is the very first reason cities were constructed for.

Needs are lacks and interests are different levels of needs. Nevertheless, human-beings are continuously looking for more and more interests. People always aspire for more and by time want more and more. Yet without the basic and primary needs, they will be seeking for interests. One will definitely want a fast car when he has a grand house, but if one doesn't have an apartment or a place to live, there will be no wanting for a car as much as for the house.

7 Results

Tracking the needs of people along time since ages beginning from the first settlements has shown different interesting results. People have invented cities as permanent settlements to fulfill their needs either becoming primary or secondary. The needs of people have shown to be in continuous change from the beginning of history till present. Designers and planners should be aware of that and track them repeatedly. In the end, cities are just another invention the same as cars, televisions and phones. They are liked as long as they are enjoyed and make peoples' lives easier and more comfortable and meet the needs of their people.

However, when they start being not comfortable, they shall become ghost cities and people head to more comfortable and suitable places.

The research shows that detailed description of humans needs can reflect very clearly on the cities form and human settlements type. Over the history and course of the time, needs have been developing and changing as humans kept civilizing. This can be observed clearly starting from the first settlements of people which were formed due to lack of security and ending with today's generations aspiring for welfare and interests. A photo of any ancient city will bigly reflect that era's lack of security and surrounded with walls and guarding towers to offer protection from invaders. On the contrary, any modern city will not reveal as much security as much as its modernization and the technology. Cities speak for the people's needs as they reflect for them and their way of living and life style.

References

Batty, M., Axhausen, K. W., Giannotti, F., Pozdnoukhov, A., Bazzani, A., Wachowisz, M., Ouzounis, G., & Portugali, Y. (2012). Smart cities of the future. *European Physical Journal Special Topics, 214*, 481–518. https://doi.org/10.1140/epjst/e2012-01703-3

Bobkova, E. (2018). Utopia and crisis of the post-socialist city: Historical transformations of the 19th and 20th century Moscow. In *Urban form and social context: From traditions to newest demands. Proceedings of the 25th international seminar on urban form* (pp. 221–229).

Busquets, J. (2006). *Barcelona: The urban evolution of a compact city*. Nicolodi & Harvard University Graduate School of Design.

Frankfort, H. (1950). Town planning in ancient Mesopotamia. *The Town Planning Review, 21*(2), 98–115. http://www.jstor.org/stable/40102125

Harris, R., & Lewis, R. (2001). The geography of North American cities and suburbs, 1900–1950: A new synthesis. *Journal of Urban History, 27*(3), 262–292. https://doi.org/10.1177/009614420102700302

Harrison, C., & Donnelly, I. A. (2011). A Theory of Smart Cities. In *Proceedings of the 55th annual meeting of the ISSS-2011, Hull, UK*, (Vol. 55, No. 1). Retrieved from https://journals.isss.org/index.php/proceedings55th/article/view/1703

Jacobs, J. (1969). *The economy of cities*. Vintage.

Kostof, S. (1991). *The city shaped* (pp. 108–111). Thames and Hudson.

Kostof, S. (1992). *The city assembled* (pp. 28–29). Thames and Hudson.

Macionis, J. J., & Parrillo, V. N. (2017). *Cities and urban life* (7th ed.). Pearson

Maslow, A. (1954). *Motivation and personality* (p. 236). Harper.

Scott, A. J., & Storper, M. (2015). The nature of cities: The scope and limits of urban theory. *International Journal of Urban Research., 39*, 1–15.

Still, B. (1999). *Mirror for Gotham: New York as Seen by Contemporaries from Dutch Days to the Present* (2nd ed.). Fordham University Press.

Webb, M., & Tee, J. (2011). *Jane's Urban Transport Systems 2011–2012*. Jane's Information Group.

Urban Planning and Climate Adaptation in Peru. Analysis of the Main Methods of Risk Prevention and Mitigation in Cities

Yeimis Milton Palomino Pichihua, Jimmy Delgado Roman, and Elizabeth Eliana Bautista Huaipar

Abstract

Peru is one of the South American countries that suffers from natural disasters as a result of the El Niño phenomenon, a climate cycle that alters the climate of the Pacific Ocean and is manifested with abnormal rainfall during the months of December and March. This phenomenon has a significant impact on coastal cities, causing a situation of urban fragility. In the face of this, the Peruvian government has implemented regulatory measures, which are the result of Law No. 29664 of the National Disaster Risk Management System (SINAGERD) of 2017. Its scope includes measures for risk prevention and assessment. As well as a series of technical manuals that allow a correct diagnosis of the territory. The present study delves into this methodological framework of risk management, analyzing the proposed schemes and their adaptation to the different urban realities of the country. In other words, it delves into the content of the different manuals with the intention of identifying both their positive and negative points. It also assesses the relevance of these processes as part of a preventive and not just reactive strategy in disaster situations. The results suggest that this methodological framework omits urban characteristics such as vulnerability factors such as proximity to services, access to housing, and transportation, among others. That is, it has a very open and generic approach, which allows discretion in the choice of risk indicators. That is why it produces diverse and even contradictory results. Also, these documents have to omit the geographical differences of Peru, which is a country with numerous ecological floors and problems arising from climate change. In conclusion, climate change adaptation policies omit the peculiarities of urban growth or the phenomenon of global urbanization.

Keywords

Climate adaptation · Urban planning · Latin America · Risk · Vulnerability

1 Introduction

In 1994, the first world conference on natural disaster reduction was held, the outcome of which was known as the Yokohama Strategy. Among the scopes of this document is the recognition of risk assessment as the first measure for disaster reduction. Per the most important measure, as it allows to identify the factors and characteristics to improve in human settlements exposed to natural hazards. Subsequently, the second world conference was held, the outcome of which was called the "Hyogo Framework of Action for 2005–2015". This new document reaffirms the mission of safeguarding people's lives. By increasing the resilience of nations to disasters, that is, it privileges the social component. Thus, the priorities for the decade were as follows:

- Ensure disaster risk reduction.
- Identify, evaluate, and monitor disaster risks.
- Use knowledge, innovation, and education to create a culture of security.
- Reduce underlying risk factors.
- Strengthening disaster preparedness.

It should be noted that strategies have improved over the years due to greater institutional commitment by nations

Y. M. P. Pichihua (✉) · J. D. Roman · E. E. B. Huaipar
Universidad Andina del Cusco, Cusco, Peru
e-mail: espacio.paralelo.es@gmail.com

J. D. Roman
e-mail: 015300118a@uandina.edu.pe

E. E. B. Huaipar
e-mail: 016200418b@uandina.edu.pe

B. He et al. (eds.), *Urban Climate Change Adaptation*, Advances in Science, Technology & Innovation,
https://doi.org/10.1007/978-3-031-65088-8_8

and states. This translates into the mobilization of financial resources for these purposes. Similarly, the issue of citizen participation was emerging as an axis of new policies on natural disasters.

There was a third conference that provided support for contemporary policies. This is the Sendai Framework for Disaster Risk Reduction 2015–2030. Peru is included in the list of 168 countries that have signed this document. The commitment of states is to reduce the risk of disasters, in parallel with increasing the resilience of communities. They also seek to integrate disaster risk reduction into policies, plans, programs, and budgets at all levels of government. For this purpose, four priorities of action are set: understanding disaster risk, strengthening disaster-risk governance, investing in risk reduction, and increasing disaster preparedness.

In 2010, Peru took an important step in the implementation of these strategies with the publication of the State Policies of the National Agreement, a document that, in Policy No. 32, indicates the need for a Disaster Risk Management Policy (GRD). Which text states:

> We are committed to promoting a policy of disaster risk management with the aim of protecting the life, health, and integrity of individuals, as well as public and private property, promoting and ensuring the location of the population and its equipment in areas of greater security, reducing vulnerabilities with equity and inclusion, under a process approach that includes risk assessment and reduction, emergency and disaster response, and reconstruction (CEPLAN, 2010).

This state policy seeks to strengthen the institutionality of disaster risk management at the three levels of government: national, regional, and local. Emphasizing the role of civil society and communities. It also identifies the need for the creation of appropriate regulations and technical guidelines for these purposes. This is the first document on the integration of risk into public policies.

It should be noted that urban growth is closely linked to natural disasters. As potential losses are high in urban areas compared to rural environments. That is, the process of urbanization increases the exposure and susceptibility of human groups. This is why urban factors should be incorporated into risk analysis and assessment. Because there are intrinsic conditions such as social inequality and environmental degradation that make cities less resilient to disasters (Brecht, 2012), In this context, proactive and preventive urban planning is needed, with the aim of reducing risk. Therefore, the task of the present study is to examine the standardized risk assessment methodology in Peru. From the hands of the official technical manuals, in order to evaluate their links with the urbanization phenomenon of the country and Latin America.

2 Background

2.1 Climate Adaptation in Peru

The impact of natural phenomena during the last two decades of the twentieth century has aroused international interest, especially in countries that presume a high degree of vulnerability. As is the case in developing countries, it is estimated that 80 percent of the lives lost in natural disasters belong to these territorial areas. (Brecht et al., 2013). As explained earlier, international strategies emphasize the increased resilience of nations and communities to disasters. In such a way that the losses caused by these phenomena have less consequences on the survival and livelihoods of human beings.

In this context, Peru has Law 29664 establishing the National Disaster Risk Management System (SINAGERD). Document materializing the objectives of international agendas. Emerging Risk Management (GRD) radically presents a vision that replaces loss management with risk management (PCM, 2016). That is, the likelihood of experiencing adverse effects resulting from changes in natural and/or climatic conditions. This conceptual advance allowed us to understand disasters as a transition. That is, not to understand it as "products" or accomplished events, but as a process, the end result of social processes that erode the capacity of society to adapt to its natural and built physical environment. (Lavell, 2005).

This new legal framework recognizes that disasters occur when there is social, institutional, or environmental vulnerability (Martínez, 2015). Thus, disaster is interpreted as a social event that unfolds as a result of the expression of a phenomenon of natural or anthropological origin in a society in conditions of vulnerability. In other words, it is a disruption to the "normal" circumstances of a society that paralyzes the performance of daily activities, causing material losses. The worst of the cases was the loss of human lives. This situation requires a necessary response of attention to the affected and the restoration of normal conditions in the community.

It should be noted that the catastrophe does not depend solely on the event that triggers it but also on the resilience (or absorption capacity) of the society that receives it. In general, a disaster is the result of a process of building risky situations in society. (De la Torre, 2011). Society is the main cause of disasters. In short, disasters are the product of processes of transformation and growth in society that do not guarantee an adequate relationship with the natural and built environment that supports them. In general, the majority of Peruvian institutions agree that disasters are defined by their effect on people; otherwise, they would simply be interesting geological or meteorological phenomena (Martínez, 2015).

2.2 Risk Assessment in Peru (GRD)

Disaster risk management is defined as a social process oriented towards disaster risk reduction and permanent control. To that end, one should become aware of the risk, analyze and understand it, and evaluate options to reduce it (Álvarez, 2016). In the Peruvian case, risk management has several dimensions of analysis. It is based on the analysis of hazards, vulnerabilities, and risks. Estimates are usually based on scientific research and reliable information recording. This enables action to be directed at the three levels of government: national, regional, and local.

By virtue of Law 29667, the National Disaster Risk Management System (Sinagerd) came into existence in 2011. This was the first legislative document to define disaster risk management as a complex social process. It consists of three (03) components of action: prospective, corrective, and reactive. It also considers four main processes, which are estimation, prevention and risk reduction, preparation, and reconstruction. Together, these approaches seek to be integrated into government actions to create the culture of prevention that is so needed in the country. The aforementioned law defines the roles and functions of the institutions linked to the GRD, such as the National Centre for Estimation, Prevention, and Reduction of Disaster Risk (CENEPRED) and the National Institute of Civil Defence (INDECI). They share the mission of developing technical documents (guides) that facilitate risk assessment and response processes.

Subsequently, Peru introduced the first urban policy derived from emergency situations. It is Reconstruction with Change (RRC). A set of measures was deployed in 2017, when the effects of the El Niño phenomenon produced multiple natural disasters. Many cities were affected by floods, slides, and collapses, among others. The social and economic impact was so strong that the government implemented actions from a ministerial-like institution called the RCC Authority. This initiative would end up changing the scope of Act 29664. Complementing it with the creation of services focused on adaptation to climate change and the development of technical manuals focusing on risk prevention, assessment, and mitigation. In other words, an exceptional situation led to a reformulation of the Peruvian legislative framework, perfecting the preceding documents (Ortiz Sánchez et al., 2022). In short, emergencies can be the trigger for structural change. The cyclicity of the event in the case of the El Nio phenomenon justifies this mobilization. I mean, in your likelihood of returning. Condition that does not occur in the pandemic, as it is an unusual event that is not threatening to return.

Finally, in 2021, the "National Risk Management Policy by 2050" will be published. Which is a more ambitious document since it is the result of a prospective process that begins with the identification of the public problem. Which is defined in the following terms:

> The geomorphology and location of the country, as well as its great climate diversity, make Peru and its population highly vulnerable, a country exposed to a wide variety of potential dangers that are throughout the territory and presented in an environment marked by a number of endogenous factors, peculiarities of the area, and the heterogeneity of the living conditions of the population (PCM, 2021).

The youth policy is structured around six (06) priority objectives, within which it emphasizes improving understanding of disaster risk for decision-making at the level of the population and state entities. In other words, the strengthening of the methodological framework of the RDG. It also proposes incorporating this into both public and private investment processes. In such a way that projects focused on natural disasters are accelerated. The policy structure is complemented by 17 guidelines of action and 27 specific services. Finally, the risk governance proposal is understood as a cross-cutting approach to territorial planning. It also incorporates dimensions of concerted planning, adaptation to climate change, water resource management, and other approaches that enrich risk management. This is in order to empower the communities or human groups involved. Especially in the phases of prevention or early warning in the event of disasters.

2.3 Risk Estimation Methods

Globally, there are numerous methodologies for estimating risks and losses from natural disasters. These differ in structure and scope; however, they share the approach of social resilience, or the strength of human groups to cope with abnormal natural events. (Sun et al., 2020). For this reason, most of them place special emphasis on the social dimension and the qualities of settlements. In other words, they deepen vulnerability as a counterpart to natural conditions. On the other hand, the strong mapping approach to risk assessment should be highlighted. That is, it is not an abstract calculation indifferent to physical space. Rather, it focuses on location and geographical conditions. In other words, it prioritizes the development of maps as a synthetic result of the prior analysis. Emerging methods view Geographic Information Systems (GIS) as indispensable tools (Brecht, 2012). It is possible to estimate the risk at the regional or subregional (continental) level, taking into account mortality rates as well as the economic risks arising from natural disasters. Contrasting these dimensions with natural conditions such as exposure to sea currents, geological faults, or flooded areas. That is, the risk is deduced from the place and the characteristics of the inhabitants who occupy it.

Peru is not alien to these trends, as it effectively incorporates this synthetic approach into its technical manuals. Therefore, in 2006, the "Basic Manual for Risk Assessment" developed by INDECI was published, a document that was characterized by its eminently qualitative approach based on the current conditions of the study site. It inaugurates the methodological scheme of hazards (P), vulnerability (V), and risk (R), understood as indispensable steps for estimation. As well as a stratification on four scales: low, medium, high, and very high. Thus, the level of risk is the result of the contrast between danger and vulnerability. In simple terms, Subsequently, in 2015, CENEPRED published "Manual for the Assessment of Risks Arising from Natural Phenomena". A second document perfects the previous methodology. Expanding the indicators needed for the calculation. It also proposes an extensive taxonomy of natural hazards, which allows for more specific analyses. This is a mixed calculation (quantitative and qualitative), which is used for the analysis of one hazard at a time. It preserves the schemes P, V, and R, enriching them with hierarchical calculations. It should be noted that this document is in force today and is the basis for the preparation of the so-called risk assessments (EVAR).

This methodology has been widely used in the country; it is a standard and a requirement of competent institutions. However, its implementation can be enriched by multiple elements, such as the extensive use of GIS. Also, participatory maps can be used for estimation, where the citizens involved can characterize the territory according to their perception of the place. Likewise, technological means such as applications and web forms, among others, can be used. Since the CENEPRED methodology is relatively flexible (Zeballos-Velarde, 2021).

However, the opening of this manual could make it a threat to citizens. Since it violates the privacy of families by exposing data that is relevant to the study but sensitive in the light of public information policies, This problem adds to the indifference of both the authorities and the citizens themselves before studies of this nature. A situation that becomes problematic because it prevents a real characterization of the place studied. Furthermore, the informal status of Peruvian urban settlements makes it difficult to process information, as these sectors are often not included in the cadastral records and are therefore inaccessible (Sánchez Tarrillo et al., 2020).

In summary, the methodology in Peru is valuable to the extent that the information is likely to be collected. It also prioritizes a social approach to calculations and estimates. At the country level, it shows the high vulnerability of urban settlements, resulting from the lack of social development and the absence of urban planning. Facts contrasting with Peru's macroeconomic development (Espinoza and Booker, 2023).

3 Research Design

This study is based on a documentary analysis aimed at reviewing the Peruvian methodology for estimating the risk of natural disasters. It has a qualitative approach and is focused on the aforementioned CENEPRED Manual, a mandatory implementation document in the country. It is developed in two stages: the first emphasizes the method itself (indicators and calculations), while the second deepens its connection with the urban legislation in force in the country.

4 Results

The CENEPRED Manual is the basis for estimating the risk of disasters, on the basis of which seven (07) manuals focused on specific natural hazards, such as earthquakes, slides, and floods, among others, were developed. However, the process and calculation are the same in these documents. Because these reproduce the same structure or sequence: risk analysis, vulnerability analysis, risk analysis, and risk communication and mitigation actions.

Since external factors are what create the danger, the first stage focuses on the circumstances. It is the probability that a natural event affects people's living conditions. It in turn consists of two calculations: susceptibility and parameters. In both cases, the characteristics of the phenomenon, such as soil type, slope, vegetation, and volume of precipitation, among others, are assessed. It is worth noting that the development analysis is done individually for each danger. Studies that combine several types of natural phenomena are not accepted, as their conditions of manifestation and impact are different.

The second stage is vulnerability. Where they are analyzed in terms of the conditions of the human groups that inhabit the area of study. It is in turn subdivided into social, economic, and environmental dimensions Fig. 1, and within each of them into variables of exposure, fragility, and resilience. Giving, as a result, an analysis matrix of nine (09) parts or sections. A hierarchical weighing process known as the SAATY method is used for quantification. In which the parameters and descriptors are chosen first, these are stratified into 5 categories, and after passing through a matrix, they acquire a specific value Table 1. The process continues while adding the variables to the dimensions and replicating this process for later stages. That is, it replicates the SAATY process for vulnerability and for danger.

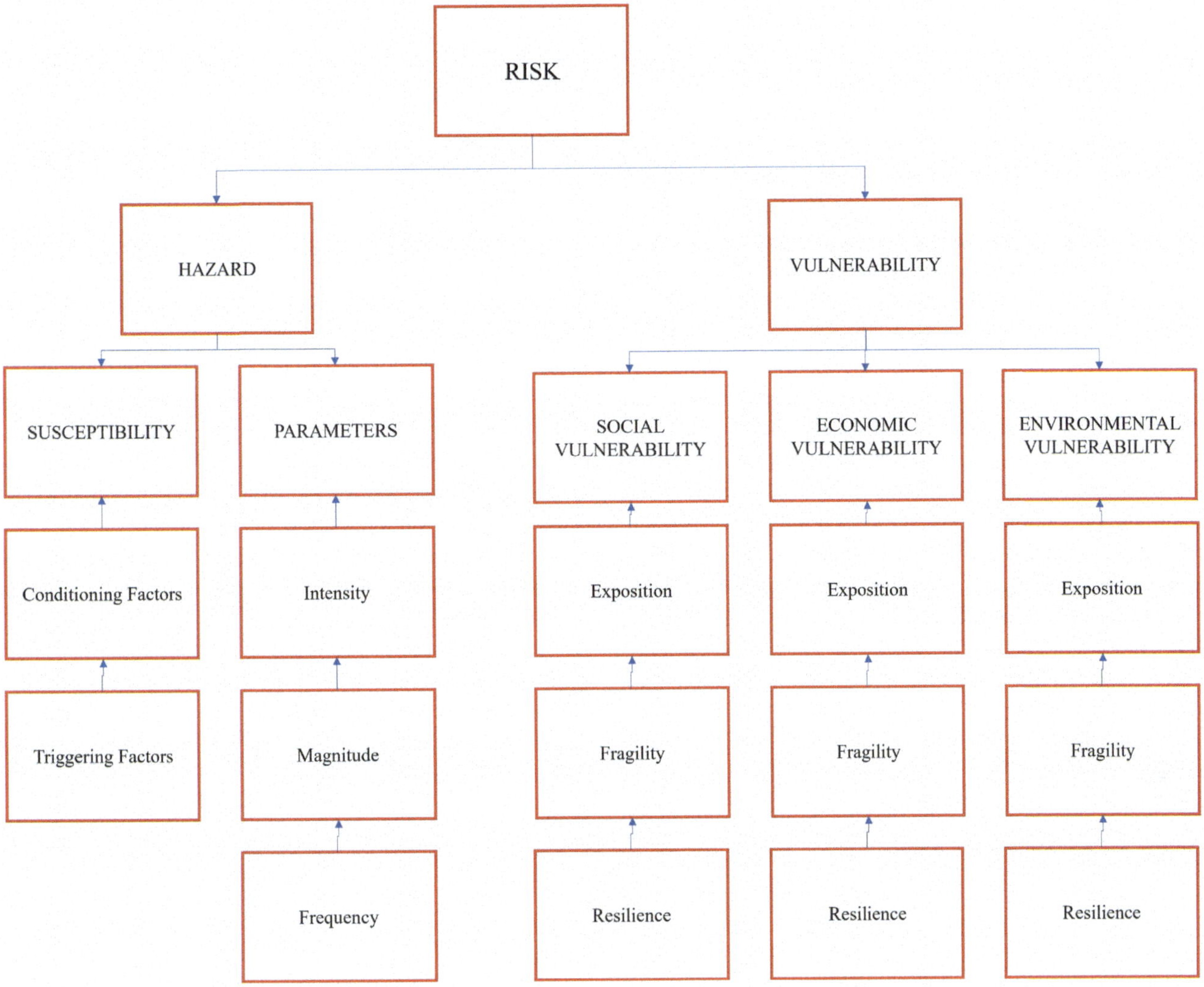

Fig. 1 Study dimensions and variables for risk assessment. *Source* Authors' elaboration based on Manual Cenepred

Table 1 Example of parameters and descriptors; weighting process. *Source* Authors' elaboration based on Manual Cenepred

Parameter		Flow rate	Weighting	
Describer	VD1	Extremely fast (v = 5 m/s)	PVD1	0.503
	VD2	Very fast (v = 0.05 m/s)	PVD2	0.26
	VD3	Fast (v = 0.0033 m/s)	PVD3	0.134
	VD4	Moderate ($v = 3.009 \times 10-^{4}$ m/s)	PVD4	0.068
	VD5	Slow to extremely slow ($v = 5.144 \times 10^{-8}$ m/year to $5.144 \times 10-^{10}$ m/year)	PVD5	0.035

It should be noted that the method of weighting is reasonable and logical, since it starts with knowledge of the conditions of the place and its inhabitants. Part of the observation, which allows to identify the minimum and maximum values; or, in their absence, the most frequent and less frequent values. However, it retains some discretion since the professionals developing the study have the power to modify this scheme voluntarily. That is, despite being a complex methodology, it possesses a certain degree of subjectivity. Where values are scientifically dependent on the perception of specialists.

As previously stated, the CENEPRED method gives the cartographic component top priority. In such a way that the main product of the estimate are the "Risk Maps," documents where the risk conditions are spatially shown. In this case, the exposed elements, which can be housing, estates, or urban blocks, are evaluated. And their risk level is a product of stratification into low, medium, high, and very high. The danger stage results in a map of its own; vulnerability also produces its own map. Accordingly, the risk map will be the result of the overlapping of both and the calculation or product of the values assigned to urban cells, sectors, and basins versus the values assigned to urban units. In such a way that the resulting mosaic shows different values Fig. 2.

Finally, it should be pointed out that, despite bankruptcies, this methodology is mandatory in the country. Its open structure allows it to adapt to various dangers as well as to various geographical realities. However, having no fixed parameters allows speculation or manipulation of trends in calculations. It also highlights the use of strictly urban qualities; the method is so generic in omitting the substantial difference between the countryside and the city. That, as mentioned earlier, is a condition of vulnerability in itself, since cities are not usually resilient spaces and are the ones that experience the most loss of both material and human lives. On the other hand, in 2021, Peruvian legislation managed to incorporate risk assessment into urban policies. In particular, the publication of the new Law on Urban Development (Law 31313) included an article requiring the preparation of these studies as part of the urban and territorial plans (Art. 92). That is, every plan must incorporate stratified risk maps as a mandatory input to continue the urban planning process. This achieved the articulation that both State Policy 32 and the GRD Policy raised at the time.

5 Discussion

In the Cenepred methodological framework, it should be noted that the threat is a set of external conditions, mostly natural. Thus, they remain stable unless structural interventions, such as infrastructure, reduce their impact. On the other hand, vulnerability is the synthesis of the conditions of the inhabitants, their housing, and their livelihoods. Thus, these elements are susceptible to change. Finally, risk is the product of both factors.

In this respect, it can be said that the calculation of risk depends to a large extent on the conditions of the population. Therefore, their characteristics, such as exposure, fragility, and resilience, are analyzed. However, this analysis is insufficient because it does not incorporate urban habitat conditions. It is limited to a description of the age distribution of the population, type of employment, disabilities, and land ownership, among others.

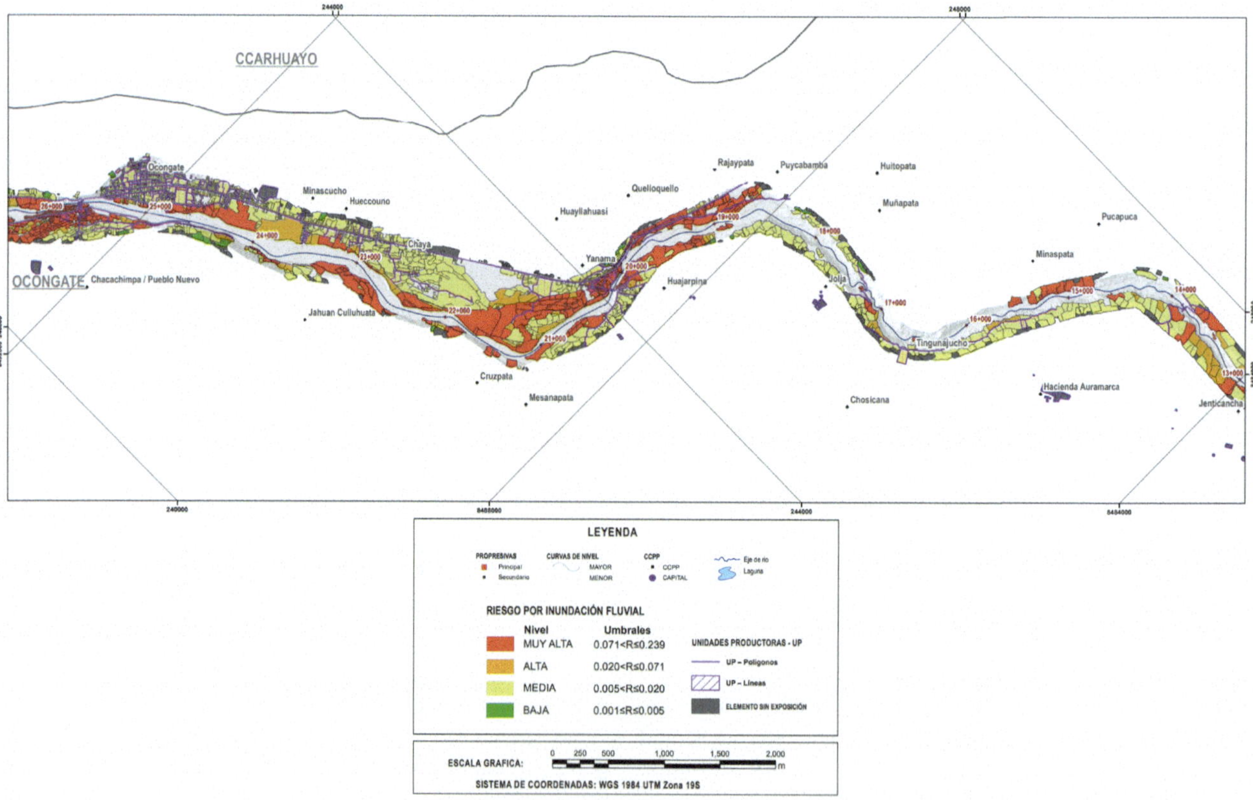

Fig. 2 Examples of risk maps and stratification. *Source* Authors

From a sociological point of view, there is an omission of the classic border between urban and rural. In other words, the method is applied, ignoring the occupation condition of the land. This is relevant in the Peruvian case, given that there is a large percentage of informal land as well as hybrid modalities such as communal land. This constitutes a methodological barrier because it automatically assigns greater vulnerability to non-urban populations and vice versa. In other words, it comes from the assumption that land ownership is a factor in vulnerability to natural disasters.

On the other hand, the family composition is analyzed at the age group level. It is implicitly assumed that a family with an aging population is a factor of high vulnerability. However, this situation varies depending on the area analyzed. In rural areas, aging is associated with poverty, as there is a population that does not receive pensions or livelihoods. In contrast, in cities, the population of the same age group can cover their expenses through a pension.

Finally, the risk analysis omits the conditions of proximity and the habits of residents' displacement. In other words, it is an analysis that abstracts the urban condition from a statistical analysis that ignores the urban space and the difficulties it imposes on its inhabitants. Elements of analysis such as proximity to food markets, schools, hospitals, parks, and gardens, among other urban services. As well as daily travel times or monthly transport costs, factors affecting the economy of families and indicating the degree of dependence of the population on urban centers.

6 Conclusion

Peru has incorporated natural disaster reduction guidelines from Yokohama, Hyoto, and the current Sendai framework. In other words, since the 1990s, it has made efforts to build a legislative and methodological framework in accordance with the most important international references.

Since 2006, the Peruvian government has been tasked with developing its own technical manuals for risk assessment. Highlighting the first manual of INDECI (qualitative) and the current manual of CENEPRED (mixed). In both cases, a similar methodology is used in other countries. Separating danger or threat from vulnerability. However, the calculations still show some inaccuracies inherent in a weighted method.

Strong geographical or cartographic bias should be highlighted in the risk assessment. Since the end product is the stratified risk map, Where spatial conditions and elements exposed to danger are assessed. The use of GIS media or platforms to systematize the calculations is also valued.

Over the past decade, Peru has strengthened its disaster risk management and climate change adaptation policies. Ministries and institutions have been established to oversee the entire implementation process through legislation and technical manuals, among which CENEPRED stands out. Today there are seven risk assessment manuals, very useful documents for characterizing the territory and identifying the risk areas that cannot be mitigated and therefore cannot be urbanized. However, the proposed methodology omits the distinction between rural and urban areas. It also ignores intrinsically urban factors such as land ownership and mobility. The analysis is therefore based on the assumption that there are disadvantaged populations, i.e., that they are inherently vulnerable to natural disasters.

References

Álvarez, S. (2016). Redes Sociales de Gestión del Riesgo de Desastres en el Perú. Pontificia Universidad Católica del Perú (PUCP)—Lima.

Brecht, H. (2012). *Natural Hazards, Risk Analysis and Emergency Preparedness: Applying Spatial Methods in Disaster Risk Management APPLYING SPATIAL METHODS IN DISASTER RISK MANAGEMENT* [Louisiana State University and Agricultural and Mechanical College]. https://doi.org/10.31390/gradschool_dissertations.2972

Brecht, H., Deichmann, U., Gun Wang, H. (2013). *A global urban risk index policy research working paper no. 6506 World Bank Washington DC*. License: CC BY 3.0 IGO. http://hdl.handle.net/10986/15865

CEPLAN, (2010). *Políticas de Estado del Acuerdo Nacional. Gobierno del Perú—Centro Nacional de Planeamiento Estratégico (CEPLAN)*. https://www.ceplan.gob.pe/wp-content/uploads/2019/02/35-POL%C3%8DTICAS-DE-ESTADO-actualizado-Feb.2019.pdf

De la Torre, A. (2011). *Rol del comité de Defensa Civil a nivel local, en la Gestión de las Políticas de prevención y atención de emergencias y desastres. El caso del distrito de la Molina—Lima*. Pontificia Universidad Católica del Perú (PUCP)—Lima.

Espinoza, A. J., & Booker, J. D. (2023). Building national disaster resilience: Assessment of ENSO-driven disasters in Peru. *International Journal of Disaster Resilience in the Built Environment, 14*(4), 423–433. https://doi.org/10.1108/IJDRBE-10-2022-0102

Lavell, A. (2005). Los conceptos, estudios y práctica en torno al tema de los riesgos y desastres en américa latina: evolución y cambio, 1980–2004: el rol de la red, sus miembros y sus instituciones de apoyo. In *La gobernabilidad en América Latina: evolución y cambio, 1980–2004. Balance reciente y tendencias a futuro. FLACSO, Facultad Latinoamericana de Ciencias Sociales*.

Martínez, M. (2015). *La construcción del conocimiento científico del riesgo de desastre: Epistemología, teorías y metodología de los estudios desde una perspectiva geográfica*. Universidad Pedagógica y Tecnológica de Colombia.

Ortiz Sánchez, I., Mayo D´Arrigo, M., Alvarez, J., Carlos, C., Bardales, E., Garaycochea, A., Christian, C., Calderón, I., Vivas, D., Pascuala Vásquez, H., Ramos Díaz, L. (2022) Los retos de las ciudades del Perú frente al cambio climático. Aproximaciones desde la Justicia Ambiental. In A. Castro, & M. Merino-Gómez (Eds.), *Desafíos y perspectivas de la situación ambiental en el Perú. En el marco de la conmemoración de los 200 años de vida republicana* (Vol. 1, pp. 414–440) INTE-PUCP. https://doi.org/10.18800/978-9972

PCM, (2021). *Política Nacional de Gestión del Riesgo de Desastres al 2050.*

PCM (2016) *Mecanismos de financiamiento para el Sistema Nacional de Gestión de Riesgo de Desastres (SINAGERD).* Lima—PCM—CARE. Septiembre.

Sánchez Tarrillo, S. J., Carrera Calderón, L. A., Aguinaga Vásquez, S. J. (2020). Vulneración de los derechos humanos ante los desastres naturales en el Perú. *EDUCARE ET COMUNICARE: Revista de Investigación de La Facultad de Humanidades, 8*(2): 88–95. https://doi.org/10.35383/educare.v8i2.474

Sun, R., Gao, G., Gong, Z., & Wu, J. (2020). A review of risk analysis methods for natural disasters. *Natural Hazards, 100*(2), 571–593. https://doi.org/10.1007/s11069-019-03826-7

Zeballos-Velarde, C. (2021). Participatory geographic information systems for integrated risk analysis: A case of Arequipa, Peru. In *Strengthening disaster risk governance to manage disaster risk* (pp. 99–106). Elsevier. https://doi.org/10.1016/B978-0-12-818750-0.00010-6

Understanding Farmers Utilization of Rice and Maize Residue in the Coastal Region of Togo: A Survey-Based Study

Mawunyo Simon Pierre Kitegi, Kperkouma Wala, Madjouma Kanda, and Folega Fousseni

Abstract

The coastal region of Togo underwent a comprehensive research study aimed at exploring the utilization of rice and corn residues by local farmers. To ensure accuracy and depth, the investigation involved direct collaboration with farmers' organisations operating in the area. The data pertinent to these organisations were diligently collected from the esteemed Ministry of Agriculture of Togo. For this study, surveys were conducted with direct interaction and engagement with farmers' organisations. The research specifically targeted maize growers in Ave, Bas Mono, Lacs, and Yoto, while focusing on rice producers in Zio, Yoto, and Bas-Mono. The compelling results derived from the study affirmed that rice paddy and maize are the most prominently cultivated cereal crops in the coastal region. However, it was found that both maize straw and stalks tend to be disposed of through burning. Similarly, rice paddy straw and husks are also subjected to burning, except for rice bran, which is utilized as animal feed due to its nutritional value. The research identified three primary reasons that compel farmers to resort to burning crop residues. Firstly, time constraints play a significant role, as farmers often face pressing schedules and labor demands, leaving them with limited options for residue management. Secondly, the lack of access to advanced machinery or appropriate equipment poses a notable hindrance in handling and processing crop residues efficiently. Lastly, economic constraints play a pivotal role, as farmers may perceive burning as a cost-effective and accessible means of disposing of crop residues. It is important to note that using a combine harvester for rice harvesting helps to increase efficiency while also preventing farmers from burning the residue on the farm. In conclusion, this comprehensive study sheds light on the prevailing practices and challenges related to crop residue management in the coastal region of Togo. Understanding the factors influencing farmers' decisions to burn residues provides valuable insights for formulating strategies to encourage more sustainable and environmentally friendly approaches in agricultural practices within the region.

Keywords

Agriculture residue · Maize residue · Rice residue · Farmers · Crop residue valorisation

1 Introduction

Togo's coastal region, known as the Maritime region, is primarily supported by agriculture and commerce. The agricultural system is the backbone of livelihoods in this region, with a heavy reliance on subsistence crops. These crops are grown using traditional production methods and basic tools, reflecting the local communities' cultural heritage. The production and trade of various food products such as maize, rice paddy, beans, sorghum, yams, and cassava account for a significant portion of economic activity. Furthermore, animal husbandry, particularly for chickens and small ruminants, is critical to many households' livelihoods (INSEED, 2023). However, effectively managing crop residues is one of the major challenges that farmers in this region face. Cereal crops, which are widely grown, produce a large amount of straw as a by product. Crop residue has traditionally been burned as a primary method of disposal by farmers. According to UNFCCC (2021), non-CO2 greenhouse gas emissions, particularly CH4 (methane) and N02

M. S. P. Kitegi (✉)
CERVIDA-DOUNEDON, University of Lomé, Lome, Togo
e-mail: simonpierrekitegi@gmail.com

K. Wala · M. Kanda · F. Fousseni
Department of Botanic, Faculty of Sciences, University of Lomé, Lome, Togo

B. He et al. (eds.), *Urban Climate Change Adaptation*, Advances in Science, Technology & Innovation,
https://doi.org/10.1007/978-3-031-65088-8_9

(nitrous oxide), increased by 25.89% and 25.86%, respectively, from 1995 to 2018. The UNFCCC study highlights that this increase in emissions may be traced mostly to the growth of cultivated fields and the widespread burning of agricultural leftovers. As the region develops and addresses the evolving challenges of sustaining its economy and livelihoods, adopting sustainable agricultural practices becomes increasingly important. To pave the way for a greener, more resilient future for Togo's coastal area, more efficient and eco-friendly crop residue management techniques must be implemented. To effectively address the issue of crop residue burning in Togo's coastal region, a thorough understanding of how farmers currently handle these residues and the reasons for their decision to burn them is required. Previous studies have found two primary factors that contribute to agricultural residue burning: time restrictions and financial limits (compare Ahmed & Ahmad, 2020; Ahmed et al., 2015; Choudhary et al., 2021; Grover et al., 2015; Thandra et al., 2021). Despite the region's substantial reliance on agriculture, there is an unusual dearth of current and thorough information on this practice. Recognizing this knowledge gap, this paper aims to fill it by providing valuable insights into the existing crop residue management practices in Togo's coastal area. We hope to unravel the complexities and motivations that underpin the widespread practice of burning crop residues through meticulous investigation and direct engagement with farmer organisations.

This paper aims to shed light on the multifaceted reasons behind farmers' choices by delving deeply into their perspectives and challenges. Time constraints, which are frequently caused by demanding schedules and labor-intensive farming practices, may drive farmers to the quick and seemingly convenient option of burning residues. Similarly, economic constraints may limit access to advanced machinery or efficient residue management alternatives. We intend to present a thorough picture of the situation on the ground and uncover the specific elements impacting crop residue management in this region through our research. Understanding these intricacies enables us to build localized strategies and treatments that address the key difficulties encountered by farmers. Furthermore, this research is compatible with wider worldwide initiatives to promote sustainable practices and lessen the environmental effects of agriculture. Alternatives to agricultural residue burning are becoming increasingly relevant as the world strives to prevent climate change and achieve sustainable development goals. Our study intends to contribute to this collaborative endeavor by offering evidence-based insights that pave the way for more ecologically friendly and efficient residue management procedures in Togo's coastal region.

1.1 Crop Residue Management Practices

Table 1 depicts some of the research findings on the management practices concerning crop residue. The table comprehensively presents various crop residue management practices, attributing each method to noteworthy authors who have significantly contributed to the discourse. These practices encompass a broad spectrum, including removing residues for livestock feed (Erenstein, 2002; Herrero et al., 2010; Valbuena et al., 2012), allowing livestock grazing (Erenstein, 2002), selling residues for animal fodder or biofuel (Erenstein, 2002), burning residues (Ahmed & Ahmad, 2020; Ahmed et al., 2015; Choudhary et al., 2021; Erenstein, 2002; Grover et al., 2015; Thandra et al., 2021), leaving residues on the soil surface (Wilhelm et al., 1986, 2007), incorporating residues into the soil (Rufino et al., 2011; Tisdale

Table 1 Review of crop residue management practices

Management practices	Authors
Removing residue for livestock	(Erenstein, 2002; Herrero et al., 2010; Valbuena et al., 2012)
Allowing livestock to graze on fields	(Erenstein, 2002)
Selling residue for animal fodder/biofuel	(Erenstein, 2002)
Burning residue	(Ahmed & Ahmad, 2020; Ahmed et al., 2015; Choudhary et al., 2021; Erenstein, 2002; Grover et al., 2015; Thandra et al., 2021)
Leaving residue on soil surface	(Wilhelm et al., 1986, 2007)
Incorporating residue into soil	(Rufino et al., 2011; Tisdale et al., 1985)
Conventional tillage (CT) practices	(Tisdale et al., 1985)
Winter ploughing	(Rufino et al., 2011)
Conservation tillage (chiseling, stubble mulching, no-till)	(Álvaro-Fuentes et al., 2009; Tisdale et al., 1985), (Baudron et al., 2012; Govaerts et al., 2006)
Composting	(Saharawat & Sidhu, 2014)
Biochar production	(Saharawat & Sidhu, 2014)
Mechanization	(Saharawat & Sidhu, 2014)

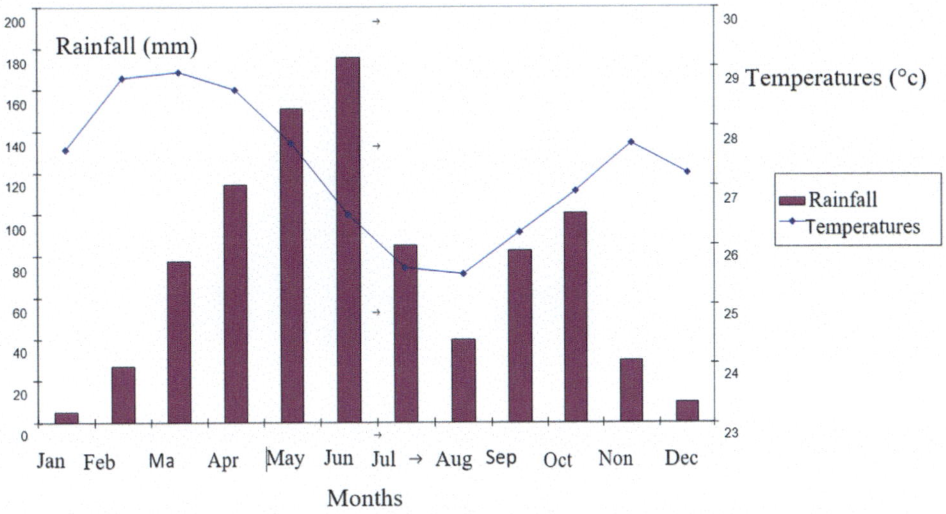

Fig. 1 Ombrothermic diagram for maritime region: monthly average rainfall and temperatures data from 1977 to 2017 (Agbewornu, 2018)

et al., 1985), conventional tillage practices (Tisdale et al., 1985), winter ploughing (Rufino et al., 2011), conservation tillage methods like chiseling, stubble mulching, and no-till (Álvaro-Fuentes et al., 2009; Baudron et al., 2012; Govaerts et al., 2006; Tisdale et al., 1985), and composting, biochar production and mechanization (Saharawat & Sidhu, 2014). This summay helps to understand different crop residue managements. It lacks the crop resiudes strategy in Togo's coastal region. To try to understand it, our study would like to contribute to know what techniques farmers use to manage crop residues and why they burn them.

2 Methodology

This section describes the research technique. It starts with agricultural output figures and finishes with the survey's methodology.

Study area

The research was conducted along the coastline of Togo(Fig. 2), a region situated in western Africa. The area by the coast is officially referred to as the Maritime Region within Togo's administrative division (Kodjovi Edjame, 2011). This area is located between latitudes 6°10 and 6°55 north, and longitudes 0°40 and 1°45 east. The Plateau Region to the north, the Atlantic Ocean to the south, the Republic of Benin to the east, characterized by the natural boundary of the Mono River, and Ghana to the west, define its geographical limits. The total area consists of around 6100 km^2. As projected by (INSEED, 2015), the region's population is anticipated to experience slight growth, with numbers rising from 2,199,396 in 2020 to 2,251,603 in 2021. According to the Togo population could reach 1.2 million people in 2028 (*Statista*, n.d.). The Maritime Region consists of eight distinct prefectures, each contributing its unique charm and diversity to the region's allure (Fig. 2). The climate in the region (Fig. 1.) is a tropical Guinean climate that boasts four distinct seasons. This climate is characterized by a climatic anomaly peculiar to southern Togo, offering two distinct rainy seasons: a long one stretching from March to July, and a shorter one occurring between September and December.

Agriculture and trade form the backbone of the Maritime Region's economy. Food crops take center stage in the agricultural sector, reflecting the region's preference for traditional industrial practices and the use of primitive tools. Maize, beans, sorghum, yam, cassava, and animal products like poultry and small ruminants dominate the economic activities here. Simultaneously, the manufacturing industry caters to the demand for various consumer goods. With its captivating coastal allure and dynamic economic activities, the Maritime Region remains a significant contributor to Togo's overall development and growth.

Crop production statistics

In this section, our objective is to emphasize the primary cereal crop cultivated in the area. To achieve this, we gathered data on agriculture production from the past 10 years, from 2013 to 2022, from the National Institute of Statistics and Economic Studies (INSEED). The collected data was then subjected to analysis by computing the average of all yield data. The purpose of this analysis was to gain a comprehensive understanding of the most prominently cultivated cereal crop in the region.

The visualizations was created using the R programming language with the aid of the packages "dplyr" and "tmap". By utilizing this comprehensive approach, we aim to offer a clearer picture of the agricultural landscape and the crop that holds the highest significance in the region's farming practices.

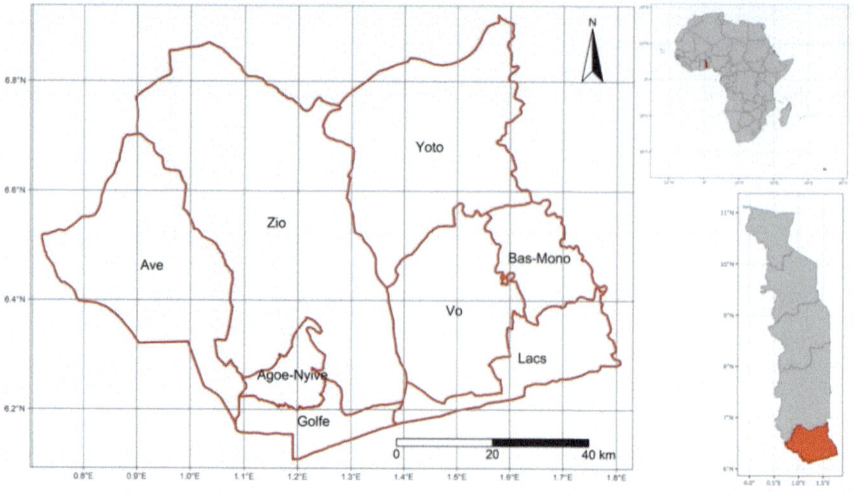

Fig. 2 Map of the Togo coastal area. *Source* Authors

The Survey

After conducting our initial investigation in the region, the results showed that rice paddy and maize stand out as the primary cereal crops cultivated along the Togolese coastline, significantly contributing to agricultural production. To delve deeper into the world of farmers' organisations involved in maize and rice paddy farming, we initiated a meticulous data collection process in the region from April to June 2022. We worked closely with the Ministry of Agriculture, Livestock, and Rural Development to guarantee the efficiency of data collecting. The majority of our data came directly from farmer groups actively involved in maize and rice farming. We gained vital information from these exchanges, including the names and locations of the organisations, the exact crops planted, and contact information for future communication. This extensive database provided us with useful insights into the regional distribution of maize and rice paddy cultivation. To ensure accuracy and comprehensiveness, we recruited data collectors who helped us to visit as many farmer groups as possible. To expedite the process and maintain consistency, we made use of digital technologies, specifically the Kobo tools and Kobo Collect software platforms. These tools proved to be highly effective for data entry, organization, and storage. To seamlessly integrate the survey questions into the Kobo tools platform, we sought the expertise of a digitizing software professional. This specialist efficiently converted the survey questions into a digital format, ensuring optimal performance within the Kobo tools platform. Additionally, the specialist conducted training sessions for the data collectors, familiarizing them with the digital questionnaire and empowering them to utilize the data gathering tools effectively. Given the linguistic diversity in the region, we took great care to provide the survey questions in both French and Ewe, the native language commonly spoken in the area. Our aim was to encourage open communication with farmer groups, ensuring their active participation and allowing them to express their perspectives and experiences in the language they felt most comfortable with. Our goal with this extensive data collection effort was to gather a wide range of reliable information on the farmer groups involved in maize and rice paddy production along the Togolese coastline. By analyzing this data, we aimed to gain a comprehensive understanding of the agricultural landscape in the region and develop effective strategies to promote sustainable farming practices. To achieve this, we employed various data analysis methods, including ratios to examine demographic characteristics and the range to assess farming experience and analyze land use patterns. Additionally, we used the dataset mode to investigate the factors that influence crop burning by farmers.

3 Results

3.1 Cereal Crop Production in Togo Coastal Area

The Togolese coastline is known for producing several important cereal crops that play a significant role in the region's agriculture. These crops include Maize, Millet, Sorghum, and Rice paddy. In the map (Fig. 3), we can observe the relative importance of each of these crops in the area. Among the eight prefectures that cover the Togolese coastline, six of them contribute significantly to cereal production. These prefectures are Zio, Ave, Bas-Mono, Lacs, Vo, and Yoto. The cultivation of Maize remains predominant in these six prefectures, demonstrating its significance in the local agricultural practices. Following Maize, Rice is the second most widely cultivated cereal crop, with its presence being more pronounced in four prefectures, namely Ave, Bas Mono, Zio, and Yoto. These areas seem to favor

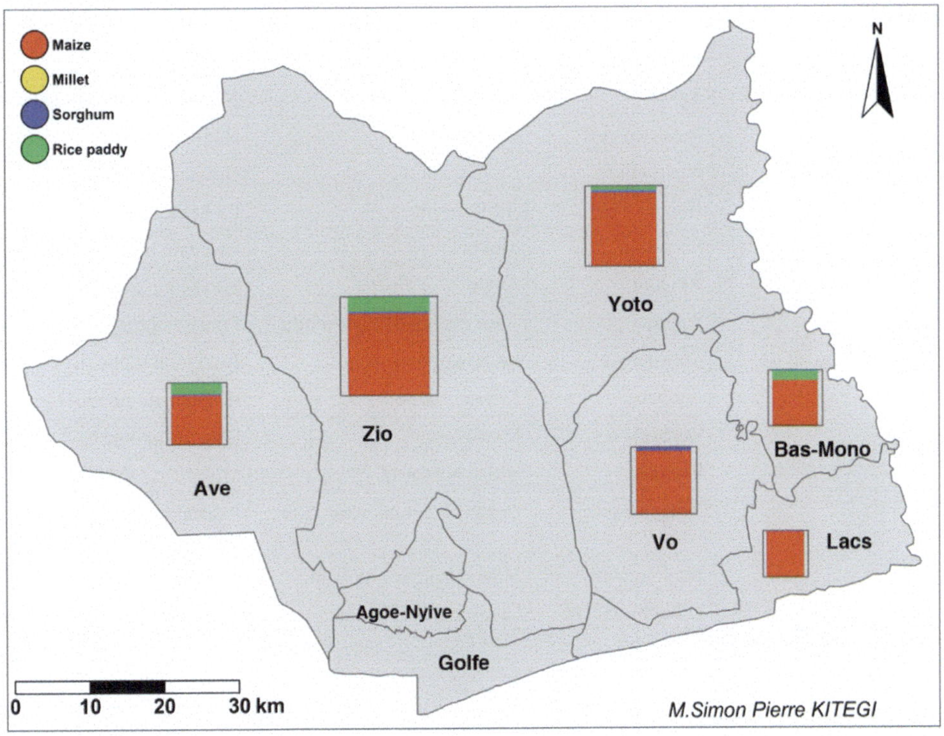

Fig. 3 Main cereal crop produced in the Togo coastline. *Source* Authors

Rice farming due to their conducive environmental conditions or other factors that support its growth. Millet, another essential cereal crop, is primarily cultivated in a single prefecture, which is Vo. This indicates that Millet has a localized importance in this specific area, possibly due to unique soil or climatic conditions that suit its growth.

3.2 *Famers' Demographics and Statistics*

Maize growers were polled in Ave, Bas Mono, Lacs, and Yoto, whereas rice producers were polled in Zio, Yoto, and Bas-Mono. A total of 19 maize farmers' organisations from various prefectures in the region and 56 rice farmers' organisations were included in the study. These were the only prefectures that were able to answer to the survey in the scope of this study. Due to lack of time and means we met one respondant per organization (Tables 2 and 3).

- **Maize producers**

Males make up 84% of those who respond to the poll, while females make up 16% of the sample. All of the participants have obtained land loans for farming purposes. The ages of the participants range from 35 to 75 years, indicating a diverse mix of age groups within the group. The participants had varying levels of maize growing expertise, ranging from 8 to 48 years. Each person cultivates an area ranging from 2 to 35 hectares, reflecting the different sizes of farming operations.

Table 2 Farmers organisations met for the survey on Maize production

Prefectures	Lacs	Ave	Yoto	Bas-mono
Farmers organisations	N'tsifafa	Mawuli	Dieu est grand	Alodo
	Douanefon	Avenir	Lonlongnon	
	Sainteté	Miwoenenyo	Bethel	
	Alafia	Tsivepe	Migabidjiwo	
	Du_courage	Kpedenou	Lebene	
	Djidjole		Novissi	
			Doenyeame	

Table 3 Farmers organisations met for the survey on Rice production

Prefectures	Bas-mono	Bas-Mono	Bas-Mono	Yoto	Zio
Farmers organisations	Kekeli	Vovossi	Mawougnigan gbandji	Motchouni	biova
	Assileassime	Dzidudu	N'tifafa gbandi	Chaine d'or	kekeli
	Novilonlon	Novissi agome	Novissi afomonou	Elayi	clémence
	Gajga	Assileassime	La lumiere agome		mokpokpo
	Dieu seul	Mokpokpo 2	Essenam midjresso kondji		nevame
	Apicum	Alodo	Elavagnon avegbo		persévérance
	Agepa	Assinesse midjresso kondji	Lom-nava midjresso kondji		volontos
	Mawuenam	Kpotowogbo zogbe	Novignon togbodji		agrivitale
	Baobab	Novissi zogbe	Didoula afomonou		force divine
	Degnigban	Mawussi	Elagnon agome		fraternité
	Eveame	Elatse hontowou	Milawé avegbo		la patience
	Nomahanou	Dekawowo hontowou	Alliance agome		main dans la main
	Lonlongnon	Mawukoenya hontowou	La sante avegbo		mawutodzi
	N'kogbe avegbo	Dieu est grand agome			

- **Rice paddy producers**

The poll was conducted in the cities of Zio, Yoto, and Bas-Mono. Males account for 82.93% of the Rice paddy group, while females account for 17.07%. Members in this group have all obtained land loans for agricultural purposes. Individuals range in age from 28 to 68 years old, indicating a large age range. The group's participants have agricultural experience spanning from 2 to 34 years and demonstrate a wide spectrum of expertise. Each cooperative cultivates an area ranging from 1 to 22 hectares, displaying a wide range of agricultural practices.

3.3 Different Use of Maize and Rice Residue

- **Rice residue**

In the three prefectures visited, Zio, Yoto, and Bas-Mono, rice straw is burned. The employment of rented combine harvesters discourages rice residue burning in the Bas-Mono region, notably in the canton of Agomeglouzou, as residues are scattered on the field (Fig. 4a and b). These combine harvesters can gather and cut rice straw while also dispersing it around the field. If this equipment did not exist, the rice straw would be burned. However, their availability is limited, and they are unable to serve the entire area. According to the study results, 56 respondents (100%) agree that rice husks are burned, resulting in their transformation into charcoal. Furthermore, the poll found that rice bran is marketed in sacks for animal nutrition, with 100% of respondents agreeing. All of the responders agree to sell all of the rice residue (straw and husk) if they find suitable market.

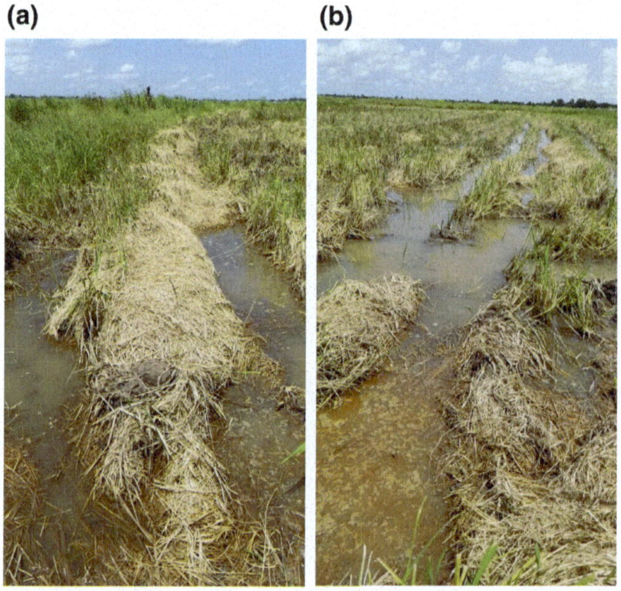

Fig. 4 Rice straw chop by the combine harvesters

- **Maize Residue**

The straw and stalks are entirely burned in 100% of the cases, and a%age of these leftovers are frequently left in the fields. These agricultural wastes are widely used as a useful source of energy for a variety of productive operations, such as the creation of earthenware pots, potash extraction, and fish smoking. Furthermore, the vast majority of responders (almost 99%) accept to commercialize all the residues created by their agricultural activities.

3.4 Drivers of Crop Burn the Residue According to Farmers

Short time

According to the farmers' unanimous statements, the timeframe between the harvest of crops and the subsequent planting season is perceived as insufficient for crop residues to adequately mineralize in the field. The low rate of residue mineralization presents challenges during soil preparation activities, whether performed manually or mechanically. The rice and maize cultivation campaigns occur twice a year, following the two rainy seasons that extend from June to August and from September to December. The limited time available to prepare the seedbed for the next season is a significant constraint faced by farmers. It is one of the primary reasons driving them to resort to the burning of agricultural residues. This practice allows them to quickly clear the field and prepare the soil for the next crop without having to wait for the natural decomposition process, which would require considerably more time. The urgency to prepare the land promptly for the next planting season, combined with the short period between harvest and planting, compels farmers to opt for residue burning as an expedient solution. By doing so, they can avoid delays in planting and ensure timely cultivation of the next crop. Waiting for crop residues to break down naturally would not align with their time-sensitive agricultural schedules.

Scarcity of agriculture machinery

The participants in the study also emphasized the significant absence of agricultural machinery for field operations. This lack of machinery leads to challenges in the mineralization of residues in the fields, making subsequent seedbed preparation difficult, irrespective of the availability of agricultural machinery. Interestingly, among all the surveyed individuals, only those from the prefecture of Bas Mono in Agomeglouzou benefitted from the activities of a private company that provides mechanical rice harvesting solutions while simultaneously crushing the straw residues. In contrast, other prefectures rely on manual methods, such as using the sickle for harvesting. The scarcity of agricultural equipment in the region poses a considerable challenge in addressing the issue of burning agricultural residues. Data obtained from INSEED reveals that the operational agricultural machinery in different prefectures along the coast is inadequate. The region only has access to 38 tractors, 21 tillers, 30 plows, 27 threshers, 4 combines, 6 winnowers, 5 seeders, 2 harvesters, 1 animal-drawn plow, 5 shellers, and 50 graters for a cultivable area of 360,503.7 hectares that were cultivated in 2022 for agricultural production.

This scarcity of equipment significantly impacts the agricultural productivity and sustainability of the region. Addressing this issue and enhancing access to modern agricultural machinery could prove to be instrumental in tackling the challenges related to residue management and promoting more efficient and sustainable agricultural practices in the area.

Economic constraints

Moreover, all the respondents also mentioned the lack of financial means as a significant factor that drives them to burn agricultural residues in the fields, particularly straw. Some participants argued that due to climate change and reduced yields, agricultural production is not profitable enough to afford agricultural machinery. In the survey, 99% of the participants expressed their willingness to sell all the residues generated from their production. This financial constraint adds to the challenges of adopting modern agricultural practices and investing in equipment that could help manage residues effectively. With the changing climate affecting crop yields, farmers find it difficult to justify the expenses associated with purchasing agricultural machinery. As a result, the practice of burning residues becomes a seemingly cheaper and more accessible option for many. However, the widespread willingness to sell agricultural residues indicates the potential for alternative solutions. Establishing systems for the collection and proper disposal of these residues could not only help mitigate the burning issue but also present an opportunity for generating income for farmers. Creating avenues for the commercial use of agricultural residues, such as for bioenergy production or other value-added products, could be explored to incentivize farmers to manage residues responsibly.

4 Discussion

This study identifies maize and rice paddy as the primary cereal crops in Togo's coastal area. Besides, it presents three main factors that lead farmers to opt for the practice of burning agricultural residues, a phenomenon that is detrimental in many ways. The first predominant factor relates to the limited time between the two cropping seasons. While farmers acknowledge the positive impact of agricultural residues on soil fertility, they find themselves with limited time to effectively and sustainably manage these residues. In addition to lack of time, another crucial factor is the lack of access to essential farm machinery for simultaneous harvesting and cutting of straw. One of the main reasons for burning residues is the lack of sufficient equipment. However, during the research conducted in Agomeglouzou Canton, we made an encouraging discovery: the availability of combine harvesters that can harvest rice and chop rice straw residues at the same time. In this respect, the results of this study are consistent with those of Bajracharya et al., 2021, who also highlighted the positive impact of the use of combine harvesters on reducing the burning of agricultural residues. The last major barrier identified in this study is economic. Many farmers indicated that they do not have sufficient financial resources to purchase the agricultural equipment needed to effectively manage agricultural residues. Some farmers in Agomeglouzou Canton also pointed out that it is often difficult to rent a combine harvester for rice harvesting, which adds to the economic constraint.

In addition, other potential reasons for burning agricultural residues that we were unable to include in our findings have been identified in subsequent studies. These include the scarcity of labour and the low acceptance of straw as a viable alternative (Dutta et al., 2022). It therefore seems essential to take concrete measures to address all these constraints related to the management of agricultural residues of rice and maize, the main cereals on the Togolese coast.

Positive examples from other countries can inspire us in this approach. In Nepal, for example, a study on the determinants of agricultural residue burning found that rice farmers who use combine harvesters are less likely to burn agricultural residues, as are those who combine livestock with production (Bajracharya et al., 2021). This demonstrates the potential effectiveness of integrated agricultural practices in reducing residue burning. As highlighted by Choudhary et al. (2020) the burning of agricultural leftovers emerges as a cost-effective and expeditious technique. Nevertheless, their findings underscore the adverse consequences it imposes on various aspects such as air quality, global warming, soil fertility, local economies, and human health. This highlights the urgent need for exploring more sustainable alternatives.

The financial limitations faced by farmers, scarcity of machinery and the time constraint combined with the effects of climate change on agricultural profitability, contribute to the persistence of residue burning. Nevertheless, the high level of openness to selling residues offers hope for implementing sustainable solutions that can benefit both farmers and the environment. By providing support, education, and financial incentives, policymakers and stakeholders can work together to promote more sustainable practices such as composting, or using residue for biofuel productions, and reduce the detrimental impacts of residue burning on agricultural lands and communities.

However, it is crucial to express the limitations of this study. For instance, data collection from rice producers' cooperatives in Avé Prefecture, another Togolese coast prefecture with abundant rice farming, was regrettably not feasible, as well as maize producers' cooperative in Vo and Zio. Additionally, due to financial and time constraints, we were unable to expand the number of cooperatives included in the study. Despite these challenges, the study presents notable strengths, particularly the comprehensive coverage of main drivers of maize and rice residue burning in prefectures along the coast. Through in-depth interviews with producers, we succeeded in gaining valuable insights into the underlying motivations behind the prevailing practice of burning agricultural leftovers.

Addressing these limitations in future research can further contribute to the development of sustainable alternatives and help mitigate the adverse effects on various ecological, economic, and health-related aspects to build sustainable cities. Finally, further research should also be conducted until such a strategy for managing residuals as economically and sustainably as possible can be identified in the coastal region.

5 Conclusion

This study represents a significant milestone in Africa as it explores the intricate factors that influence farmers' practices of burning crop residues. In the coastal region of Togo, farmers face various challenges that prompt them to resort to residue burning in order to maximize crop production throughout the seasons. These challenges include time constraints, limited access to machinery, and economic limitations. By focusing on rice paddy and maize residues, which are the primary cereal crops in the study area, this research sheds light on the specific agricultural context of the region. Understanding the reasons behind the widespread burning of the residues provides valuable insights into the larger agricultural landscape of Togo's coastal region. One potential solution to address this issue is to utilize the residues for the production of bioenergy or green hydrogen. This approach not only helps tackle energy challenges in the region but also offers an eco-friendly alternative to burning, thereby mitigating the environmental impact. Achieving sustainable agriculture and environmental preservation requires collaboration among various stakeholders, including government agencies, research institutions, and farmer communities. By working together and implementing supportive policies, viable solutions can be identified and implemented. Collaborative efforts to address the pressing issue of crop residue burning hold great promise for promoting responsible residue management and reducing the detrimental emissions associated with this practice. Furthermore, the significance of this research extends beyond the boundaries of Togo's coastal area. It serves as a compelling model for similar studies across Africa, encouraging a broader movement towards sustainable practices that prioritize the well-being of both farmers and the environment. As we work in tandem, embracing eco-friendly alternatives and fostering partnerships among stakeholders, we can collectively contribute to a greener and more prosperous future for farmers throughout the continent.

Acknowledgements The authors are deeply grateful to the World Bank for funding this research through CERViDA-DOUNEDON.

References

Agbewornu, K. (2018). Global warming and water resources variability in the maritime Region of Togo (West Africa). *Environment and Natural Resources Research, 8*, 49. https://doi.org/10.5539/enrr.v8n4p49

Ahmed, T., Ahmad, B., & Ahmad, W. (2015). Why do farmers burn rice residue? Examining farmers' choices in Punjab, Pakistan. *Land Use Policy, 47*. https://doi.org/10.1016/j.landusepol.2015.05.004

Ahmed, T., & Ahmad, W. (2020). Burning crop residue: Farmers' choice among various practices. *Pakistan Journal of Applied Economics, 29*, 265–290.

Álvaro-Fuentes, J., López, M. V., Arrúe, J. L., Moret, D., & Paustian, K. (2009). Tillage and cropping effects on soil organic carbon in Mediterranean semiarid agroecosystems: Testing the Century model. *Agriculture, Ecosystems and Environment, 134*(3), 211–217. https://doi.org/10.1016/j.agee.2009.07.001

Bajracharya, S., Mishra, A., & Maharjan, A. (2021). Determinants of crop residue burning practice in the Terai region of Nepal. *PLoS ONE, 16*. https://doi.org/10.1371/journal.pone.0253939

Baudron, F., Tittonell, P., Corbeels, M., Letourmy, P., & Giller, K. E. (2012). Comparative performance of conservation agriculture and current smallholder farming practices in semi-arid Zimbabwe. *Field Crops Research, 132*, 117–128. https://doi.org/10.1016/j.fcr.2011.09.008

Choudhary, A., Kadian, K., & Meena, M. S. (2021). *Reasons and awareness levels of farmers on residue burning in Indo-Gangetic plain of India: An exploratory research.*

Choudhary, A., Singh, K., Prashanth, C., & Bishnoi, R. (2020). *Crop residue burning and its harmful impacts a review.*

Dutta, A., Patra, A., Hazra, K., Nath, C., Kumar, N., & Rakshit, A. (2022). A state of the art review in crop residue burning in India: Previous knowledge, present circumstances and future strategies. *8*, 100581. https://doi.org/10.1016/j.envc.2022.100581

Erenstein, O. (2002). Crop residue mulching in tropical and semi-tropical countries: An evaluation of residue availability and other technological implications. *Soil and Tillage Research, 67*(2), 115–133. https://doi.org/10.1016/S0167-1987(02)00062-4

Govaerts, B., Sayre, K. D., & Deckers, J. (2006). A minimum data set for soil quality assessment of wheat and maize cropping in the highlands of Mexico. *Soil and Tillage Research, 87*(2), 163–174. https://doi.org/10.1016/j.still.2005.03.005

Grover, D., Kaur, P., & Sharma, H. (2015). Possible reasons and farmers awareness towards crop residue burning: An overview and a case study from Mirzapur village of Kurukshetra district, India. *Environment and We: An International Journal of Science and Technology, 10*, 75–85.

Herrero, M., Thornton, P. K., Notenbaert, A. M., Wood, S., Msangi, S., Freeman, H. A., Bossio, D., Dixon, J., Peters, M., van de Steeg, J., Lynam, J., Rao, P. P., Macmillan, S., Gerard, B., McDermott, J., Seré, C., & Rosegrant, M. (2010). Smart investments in sustainable food production: Revisiting mixed crop-livestock systems. *Science, 327*(5967), 822–825. https://doi.org/10.1126/science.1183725

INSEED. (2015). *Perspectives démographiques du Togo 2011–2031.*

INSEED. (2023). *Annuaire Statistique 2020–2021 De La Region Maritime.*

Kodjovi, E. (2011). Approche géophysique de définition du climat du littoral océanique et du bas-togo. *Revue de Géographie Tropicale et d'Environnement.* https://www.revues-ufhb-ci.org/fichiers/FICHIR_ARTICLE_1169.pdf

Rufino, R. D., Luna, J. M., Sarubbo, L. A., Rodrigues, L. R. M., Teixeira, J. A. C., & Campos-Takaki, G. M. (2011). Antimicrobial and anti-adhesive potential of a biosurfactant Rufisan produced by Candida lipolytica UCP 0988. *Colloids and Surfaces B: Biointerfaces, 84*(1), 1–5. https://doi.org/10.1016/j.colsurfb.2010.10.045

Saharawat, Y., & Sidhu, H. (2014). Management of cereal crop residues for sustainable rice-wheat production system in the indo-gangetic plains of India. *Proceedings of the Indian National Science Academy, 80*, 95. https://doi.org/10.16943/ptinsa/2014/v80i1/55089

Statista. (n.d.). Retrieved 10 February 2024, from https://www.statista.com/statistics/528816/total-population-of-togo/

Thandra, B., Sairam, M., Shankar, T., Maitra, S., & Praharaj, S. (2021). Crop residue burning in India: causes. *Impacts and Solutions, 12*, 37430–37437.

Tisdale, Nelson, W.L., & Beaton, J. D. (1985). *Soil and fertilizer nitrogen.*

UNFCCC. (2021). Rapport National D'inventaire Des Gaz a effet De Serre Togo.

Valbuena, D., Erenstein, O., Homann-Kee Tui, S., Abdoulaye, T., Claessens, L., Duncan, A. J., Gérard, B., Rufino, M. C., Teufel, N., van Rooyen, A., & van Wijk, M. T. (2012). Conservation agriculture in mixed crop–livestock systems: scoping crop residue trade-offs in Sub-Saharan Africa and South Asia. *Field Crops Research, 132*, 175–184. https://doi.org/10.1016/j.fcr.2012.02.022

Wilhelm, W. W., Doran, J. W., & Power, J. F. (1986). Corn and soybean yield response to crop residue management under no-tillage production systems[1]. *Agronomy Journal, 78*(1), 184–189. https://doi.org/10.2134/agronj1986.00021962007800010036x

Wilhelm, W. W., Johnson, J. M. F., Karlen, D. L., & Lightle, D. T. (2007). Corn stover to sustain soil organic carbon further constrains biomass supply. *Agronomy Journal, 99*(6), 1665–1667. https://doi.org/10.2134/agronj2007.0150

Climate Adaptation and Resilience Strategies

Exploring the Interplay Between Urban Landscape and Land Surface Temperature: Case Study of South Twenty Four Parganas

Sudarshana Sinha, Ankhi Banerjee, and Archana Patnaik

Abstract

The past decade's shifting urban landscapes and ecological dynamics have profoundly impacted land use, land cover (LULC), and land surface temperature (LST) in diverse ways across regions. However, in the context of areas undergoing urban transitions in India, a comprehensive study connecting LULC changes to LST remains limited. This study analyses the correlation between LULC types (specifically built-up areas, waterbodies, green spaces, and bare land) and LST. Utilising statistical methods like Correlation analysis, dimension reduction analysis like Principal Component Analysis, and one-way ANOVA, the research examines their interrelation, identifying the most influential factor. Data derived from Landsat 8 and MODIS datasets form the basis of this analysis. Findings confirm the impact of impervious surfaces on LST, aligning with existing research, while highlighting vegetation's stronger influence compared to water bodies. Elevated LST in bare land areas underscores the urgency of urban planning emphasizing green spacc cxpansion for sustainable thermal regulation. PCA was used to identify key sub-indicators across diverse domains and their influence on LST in different sections of the study region. Notable. Indicators such as UI (Urban Index), MNDWI (Modified Normalized Difference Water Index), NDVI (Normalised Vegetation Index), SAVI (Soil Adjusted Vegetation Index), and DBSI (Dry Bare Soil Index) emerge as dominant in their respective domains. While ANOVA analysis reveals variations among indicators, sectioning the region uncovers significant LST and LULC differences. This paper offers insights into nuanced LULC-LST relationships, emphasizing the need for tailored urban sustainability strategies within the region. Insights from this research could provide valuable guidance for policymakers and urban planners, facilitating the development of more sustainable urban strategies.

Keywords

LST · Built-up index · Vegetation index · UHI · Waterbody index · Bare land analysis · LULCC

1 Introduction

Land Surface Temperature (LST) results from the interplay of landscapes and urban interventions (Asgarian et al., 2015; Hamstead et al., 2016; Zhou et al., 2014a), notably influenced by urbanization's socioeconomic and ecological changes (Dutta et al., 2021; Guha et al., 2018; Naikoo et al., 2022). These alterations significantly impact LST, causing temperature variations and localized thermal anomalies (Kafy et al., 2021). India's urbanization surged from 27.8% in 2001 to 31.16% in 2011, mirroring West Bengal's rise from 28% in 2001 to 31.89% in 2011 (Census, 2001, 2011), albeit with diverse growth patterns across the state.

Urbanization disrupts both terrestrial and aquatic ecosystems, altering Land Use and Land Cover (LULC) domains, and profoundly impacting the remaining areas (Das et al., 2020; Rousta et al., 2018). This transformation involves converting barren, blue, and green spaces into

S. Sinha (✉) · A. Patnaik
Department of Humanities and Social Sciences, Indian Institute of Technology Kharagpur, Kharagpur, India
e-mail: sudarshanasinha1@gmail.com

A. Patnaik
e-mail: archana@hss.iitkgp.ac.in

A. Banerjee
Ranbir and Chitra Gupta School of Infrastructure Design and Management, Indian Institute of Technology Kharagpur, Kharagpur, India
e-mail: ankhi@infra.iitkgp.ac.in

B. He et al. (eds.), *Urban Climate Change Adaptation*, Advances in Science, Technology & Innovation,
https://doi.org/10.1007/978-3-031-65088-8_10

predominantly built-up areas, primarily driven by human influences (Boehme et al., 2015; Choudhury et al., 2019; Karakuş, 2019; Ogunjobi et al., 2018; Qu et al., 2020). Such changes strip natural land cover (LC) and modify indigenous land use (LU) into impermeable surfaces, mainly comprising grey infrastructure, thereby affecting regional environments and escalating Land Surface Temperature (LST) (Mathew et al., 2016; Pal & Ziaul, 2017). Urbanization disrupts the energy balance of a region, potentially increasing sensible heat flux while decreasing latent heat flux (Huang et al., 2022), leading to environmental degradation and loss of natural land cover in urbanized regions (Bharath et al., 2018; Daneshvar et al., 2019; Montazeri & Masoodian, 2020). Each facet of urban LULC distinctively influences LST (Dutta et al., 2019; Grover & Singh, 2015), shaping climatic patterns within built-up areas, water bodies, vegetation, and barren lands within LULC domains. Urban expansion plays a pivotal role in shaping Urban Heat Islands (UHI) and altering urban metabolism (Wang et al., 2021; Xiong & Zhang, 2021). Impermeable surfaces, driven by extensive grey infrastructure, significantly influence microclimate dynamics through soil sealing, enhancing the thermal retention capacity in built-up areas (Lee et al., 2021; Yang et al., 2021). This disruption in energy equilibrium is characterised by impervious surfaces storing thermal heat during the day and releasing it gradually at night (Nie et al., 2016; Ziaul & Pal, 2020). Variances in physical properties, radiation emissivity, absorption, and radiation capacity among urban built-up surfaces within grey infrastructure contribute to LST fluctuations (Imran & Mehmood, 2020; Imran et al., 2021; Kuang et al., 2015; Rehman et al., 2022; Ren et al., 2023). Albedo and heat retention capacity deviations further amplify LST variations in urban settings (Guha et al., 2018). The morphology, geometry, and spatial arrangement of built-up spaces wield significant influence on LST (Estoque et al., 2017; Ma et al., 2016). Disparities in construction material characteristics affect urban canopy and contribute to local LST elevation (Morais et al., 2019; Stache et al., 2022). Core urban areas consistently exhibit higher LST, even at night, contributing notably to Surface Urban Heat Island (SUHI) formation (Dissanayake et al., 2019; Mbuh et al., 2019; Tu et al., 2016). Intra-urban LST discrepancies significantly impact resident satisfaction and comfort levels (Kafy et al., 2021; Naim & Kafy, 2021; Sejati et al., 2019). Specific urban zones, like city cores and Central Business Districts, often experience elevated LST due to dense construction and increased concretization (Li et al., 2020). Notably, diurnal and nocturnal LST variations among built-up surfaces are closely associated with city size, densities, and urban morphological features (Ren et al., 2021).

Uncontrolled urban growth disrupts thermodynamics and exacerbates heat waves (Gazi et al., 2021; Kafy et al., 2021; Rahaman et al., 2021). Vegetation shows promise in mitigating this impact (Tepanosyan et al., 2021; Yan et al., 2020; Yang et al., 2017), with studies noting a negative correlation between LST and vegetation due to shading and evapotranspiration (Choudhury et al., 2019; Das & Angadi, 2020; John et al., 2020; Yan et al., 2018). Green space's capacity to moderate LST depends on its design and quality within urban areas (Guo et al., 2020; Yin et al., 2019). Conflicting perspectives exist: some advocate for strategic enhancement of green space coherence for better cooling, while others doubt the cooling impact of green space geometry alone (Guo et al., 2021; Shih, 2017). However, riparian forests often exhibit higher LST compared to surrounding areas (Tan et al., 2020), and the geographical variability of vegetation's impact on LST is evident (Wang et al., 2018). Tan et al. (2021) stated that fragmented blue and green spaces (BGs) are more effective in regulating LST than LCs. Nonetheless, BGs' cooling effects vary spatially, temporally, and seasonally (Amani-Beni et al., 2018; Du et al., 2016; Yu et al., 2017). Therefore, Rahman et al. (2021) advocate for further research using multispectral vegetation indices to comprehensively analyse urban environments, stressing the need to consider vegetation diversity and topographical variations in urban area studies.

The intricate interrelationship among LST, urban metabolism, and water bodies highlights a significant correlation. The proliferation of grey infrastructure has been associated with the depletion and degradation of water bodies (Nath et al., 2021; Patra et al., 2018). Water, while exerting a substantial influence on LST, demonstrates varied impacts contingent on its design and geometry (Manteghi et al., 2015; Syafii et al., 2017). Irregularly shaped water bodies, as underscored by Yun et al. (2020), tend to diminish their cooling effects. The most pronounced cooling effect of water bodies and vegetation is typically observed within a 500-m buffer zone (Chen et al., 2018; Jamei et al., 2019; Rahaman et al., 2021; Xie et al., 2022). This cooling magnitude is affected by factors such as waterbody depth, surface slope, and channel orientation (Lin et al., 2020). Additionally, the cooling effect varies based on regional geographical attributes and seasonal fluctuations (Solcerova et al., 2019). Temperature variations in neighboring regions' boundary layers also impact the effectiveness of blue spaces (Barlow, 2014; Gunawardena et al., 2017). Wang et al. (2018) posit a positive correlation between water bodies and mean night-time LST. Tan et al. (2020) observed that muddy beaches and marshy lands exhibit higher LST compared to the surrounding countryside, particularly during the summer months. They advocate for further research to comprehend the influence of diversity and geometry of urban blue spaces (Ampatzidis & Kershaw, 2020). Assumptions about their cooling potential may be oversimplified, necessitating a more nuanced investigation to avoid misleading conclusions.

Exposed bare surfaces, due to their increased solar radiation absorption, typically demonstrate higher LST in comparison to other land covers (Guha et al., 2018; Rehman et al., 2022; Song et al., 2014) highlights that micro-topography and soil composition significantly impact the LST of bare lands (Karakus, 2019; Swamy et al., 2017). Similarly, Madanian et al. (2018) note elevated LST in both bare soil and harvested agricultural areas, particularly during summer and winter seasons. Notably, LST tends to be higher over bare soil in tropical and subtropical areas compared to arid regions (Li et al., 2012; Zhao et al., 2016).

In the Indian context, as can be observed, the dynamics between LST and LULC have been investigated in the case of Kanyakumari district (Sam & Balasubramanian, 2022), Waynad district (John et al., 2020), Raipur City (Guha et al., 2020), Asansol sub-division (Das et al., 2020), Surat and Bharuch (Mukherjee & Singh, 2020), Asansol and Durgapur Municipal Corporation (Choudhury et al., 2019), Delhi (Dutta et al., 2019, 2021), English Bazar urban center (Pal & Ziaul, 2017), Ahmedabad city (Mathew et al., 2016). Since the correlation between LST and LULC is affected by several attributes, a generalization in this domain can underestimate niche subtleties and lead to misleading conclusions (Guha et al., 2018). This approach risks oversimplification, overlooking nuanced complexities in these associations. Previous studies often limited their scope to single LULC domains or specific urban contexts, hindering a comprehensive understanding of broader landscape dynamics. Particularly lacking in India is a thorough assessment of transitioning regions, including urban, peri-urban, and rural areas. To comprehend LST dynamics, an integrative approach considering multiple variables simultaneously is essential. Therefore, an inclusive multi-spectral analysis covering built-up spaces, greenspaces, water bodies, and bare land is crucial to unveil their intricate interplay. Tan et al. (2021) accentuate the necessity for further inquiry into differentiating green and blue spaces, underscoring the sway of external determinants like geographical locale, LC, and the extent of urbanization on LST dynamics. Siddique et al. (2020) underscored the criticality of exploring the interplay between LULC and LST for evaluating ecological health and adaptability, emphasizing the significance of updated data to monitor environmental dynamics.

Consequently, this study's primary aim is twofold: first, to analyze the LULC composition of the region, and second, to scrutinize the relationship between LST and built-up areas, water bodies, greenspaces, and bare land areas. This research diverges from prior research by encompassing diverse landscapes—urban, peri-urban, and rural areas—beyond the prevalent focus on the top-tier megalopolises. This approach enables a more nuanced understanding of the complex interplay between LST and LULC. This research employs a comprehensive multi-spectral analysis covering built-up areas, greenspaces, water bodies, and bare land. The methodology adopted ensures a comprehensive view of LST dynamics about various land cover types avoiding oversimplification and enhancing the accuracy of our conclusions. The findings contribute significantly to evaluating ecological well-being. Understanding the correlations between LST and diverse land cover types is crucial for assessing environmental health and adaptability, offering essential insights for sustainable urban development. The insights derived from our study hold practical significance for policymakers and urban planners. Identifying factors contributing to thermal stress and comprehending LST dynamics concerning different land cover types informs the development of effective and sustainable urban strategies. Through the diversity in the methodological framework and a broader focus on varied landscapes, our research contributes substantially to the field. The implications extend beyond academic realms, offering practical insights into urban planning and environmental management.

The paper consists of seven sections. The first section introduces the topic, while the second section provides a concise exploration of the research focus. The third section meticulously details methodologies, approaches, and presented data, including a subsection explaining the rationale for utilizing multiple sub-indicators. In the fourth section, various aspects, including LST variation, built-up area analysis, greenspace analysis, waterbody analysis, and bare land analysis, are comprehensively examined. The fifth section investigates the correlation between these categories and LST, employing PCA analysis to identify key contributing factors to variance. Additionally, one-way ANOVA analysis is conducted to highlight intra-region differences and inter- and intra-sub-indicator differentials. The paper provides a brief discussion in the sixth section and concludes succinctly in the seventh section.

2 Study Area

South Twenty-four Parganas spans latitudes 22°30′45″ N to 20°29′ N and longitudes 89°4′56″ E to 88°3′45″ E, bordered by Kolkata district, the Bay of Bengal, and rivers Bidya, Matla, and Hoogly (refer to Fig. 1). Its diverse land use can be classified into urban, peri-urban, rurban sectors, rural areas, and the Sunderbans marshlands. With 29 blocks, 5 sub-divisions, and 7 municipalities, the district's population surged by 18.17% (Census, 2011), surpassing West Bengal's 13.8% increase. From 2001 to 2011, the urban population in South Twenty-four Parganas increased by 9.85%, expanding urban regions by 241.96 km^2. Notably, Budge Budge-I and Thakurpukur Maheshtola recorded

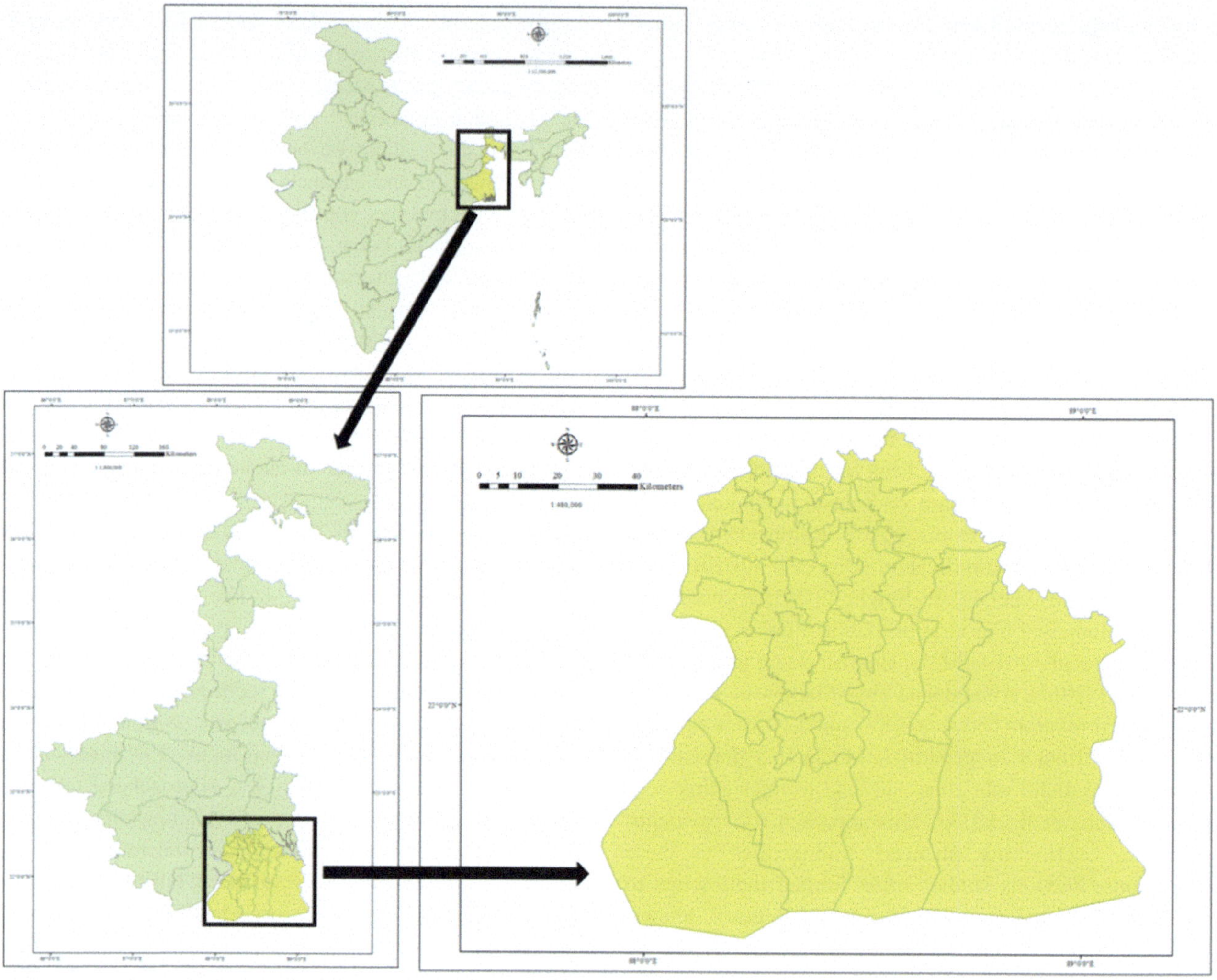

Fig. 1 Showing the location map of the study area. *Source* Computed by the authors

the highest urban population concentrations at 73.81 and 48.82%, respectively. Conversely, several areas reported no urban population. Diamond Harbour boasts the highest population density at 5673.86 persons/km^2.

3 Methods and Data Sources

Figure 2 elucidates the methodological framework employed in this investigation, entailing the collection of data from Landsat 8 and Terra MODIS satellite imagery. This involved a series of preprocessing stages, notably image cleansing and noise reduction, followed by the computation of diverse indices, (refer to Fig. 2). Land surface temperature retrieval commences with satellite sensors, such as thermal infrared sensors, essential for capturing Earth's surface radiation in the form of thermal infrared radiation. Commonly used satellite platforms include those from reputable. Agencies like the National Aeronautics and Space Administration (NASA) or the European Space Agency (ESA), such as the Moderate Resolution Imaging Spectroradiometer (MODIS) dataset. The data curation process involves several key steps. Firstly, radiative transfer models, mathematical simulations of electromagnetic radiation interactions with the atmosphere and Earth's surface, are employed to correct for atmospheric effects on observed thermal infrared radiation. Secondly, atmospheric correction is conducted to address the absorption and scattering of thermal infrared radiation by the atmosphere, ensuring accurate land surface temperature values. Radiative transfer models assist in mitigating atmospheric interference. Thirdly, emissivity correction is initiated, accounting for variations in surface emissivity—the measure of a surface's efficiency in emitting thermal radiation. Different surfaces exhibit distinct emissivity values, requiring corrections for accurate land surface temperature retrieval. Subsequently, brightness temperature calculation is performed, defining the temperature of a black body emitting the same total

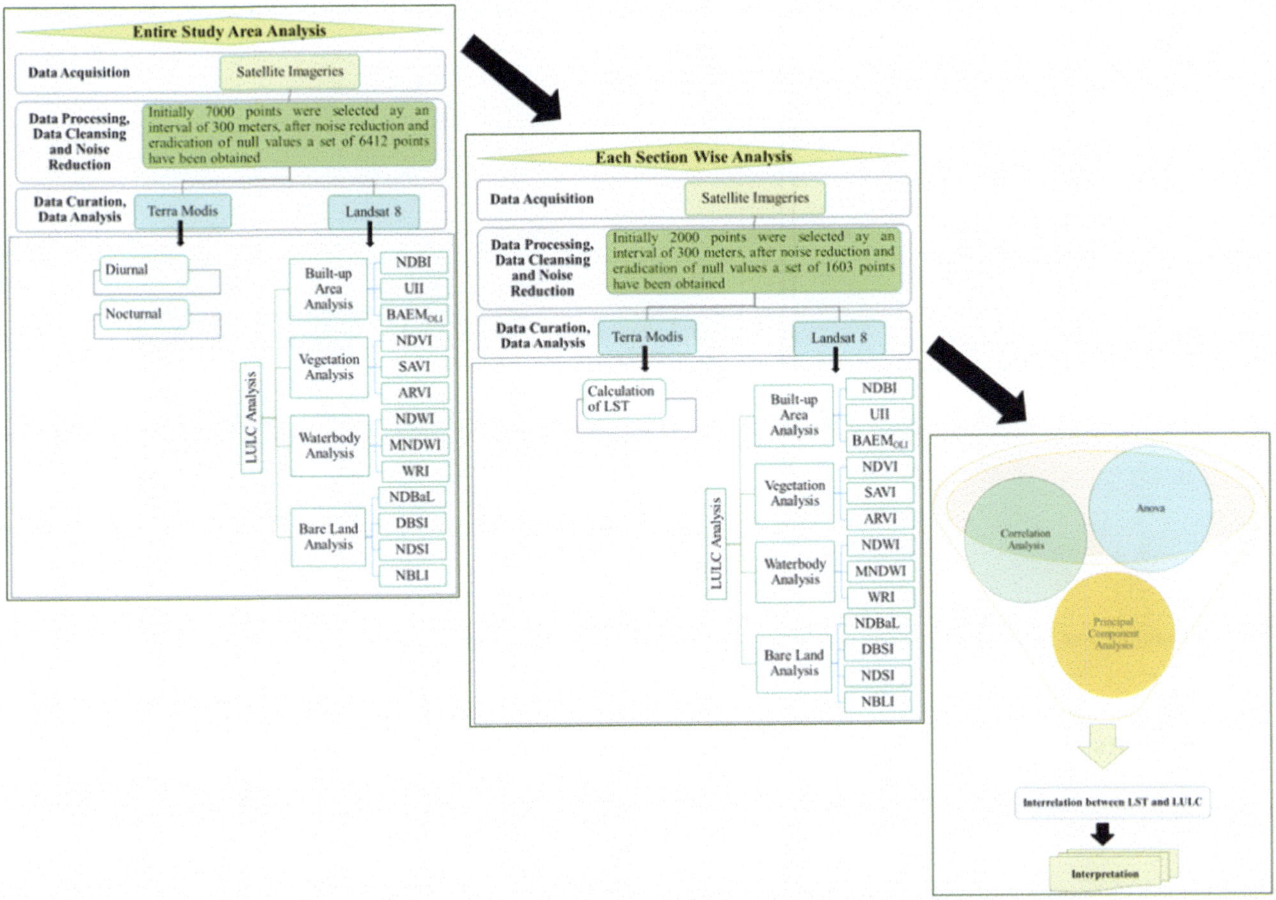

Fig. 2 Showing the methodological framework of the paper. *Source* Computed by the authors

radiant energy as the object in a specific spectral band. It serves as a crucial intermediate step in LST retrieval. The fourth step involves the initiation of the Temperature Retrieval Algorithm, specialized algorithms converting brightness temperature into land surface temperature. These algorithms consider factors such as atmospheric conditions, emissivity, and sensor characteristics. Subsequently, validation and calibration ensure consistency and accuracy over time. Temperature values are converted from Kelvin to degrees Celsius, followed by data analysis and visualization through the construction of interpretation and land surface temperature maps.

The study concentrated on four LULC domains: Built-Up Area Analysis, Vegetation Analysis, Waterbody Analysis, and Bare Land Analysis. Sub-indices were meticulously computed within these domains to ensure a comprehensive assessment. To discern the intricate association between these domains and Land Surface Temperature (LST), a fishnet technique was applied, selectively gathering 7000 points at 300-m intervals. Post-refinement procedures, eliminating noise and null values, resulted in a refined dataset of 6412 points, instrumental in conducting correlation analyses. Identification of pivotal sub-indicators necessitated the utilization of Principal Component Analysis (PCA). The study area underwent segmentation into four distinct sections, each contributing 2000 points at 300-m intervals through fishnet analysis. Subsequent data refinement encompassed noise reduction and null value elimination, culminating in each section yielding a dataset of 1603 points. These datasets were subjected to One-way ANOVA analysis (refer to Fig. 3) to reveal nuanced inter and intra-group variations within the study area.

3.1 Rationale Behind the Selection of the Indices

The rationale behind selecting the various indices used to gauge the LULCC has been stated below (refer to Table 1).

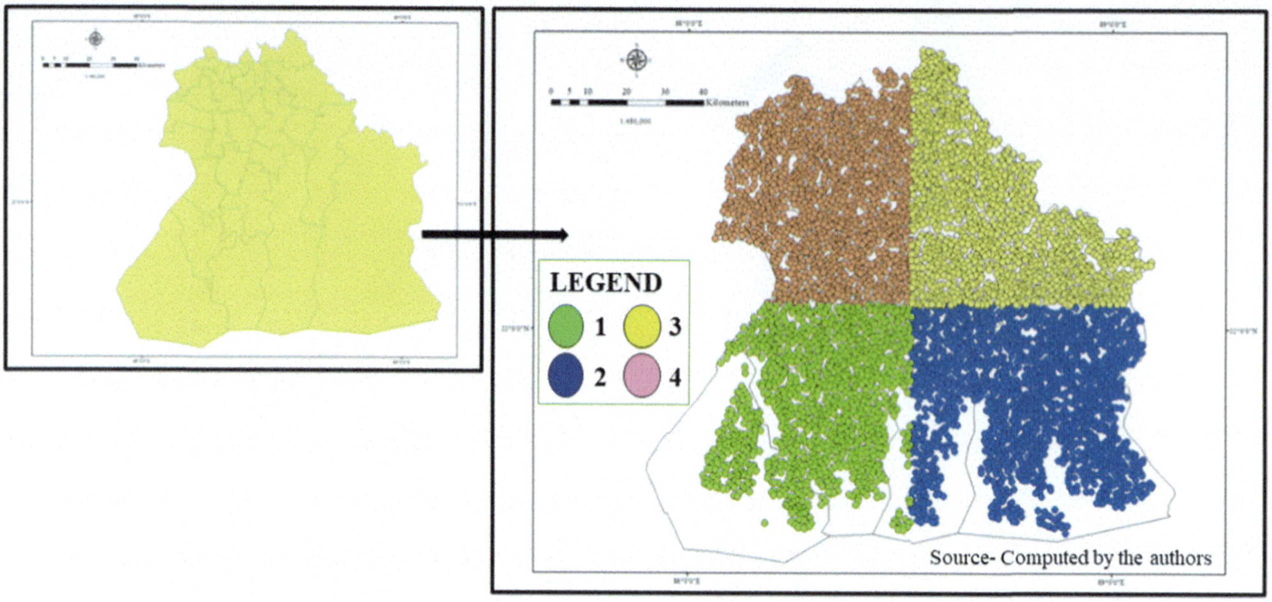

Fig. 3 Showing the four parts and the points as per the fishnet technique. *Source* Computed by the authors

4 Results

4.1 LST Analysis

The computed LST within the study area displays marked variations both diurnally and nocturnally. Daytime temperatures peaked at 40.81 °C, notably dropping to 20.61 °C during the nocturnal period, (refer to Table 2, Fig. 4). These variations reveal a significant temperature differential of 14.2 °C during the day and 7 °C at night, consistently observed diurnally and nocturnally, spanning between maximum and minimum temperatures ranging from 13.2 to 6 °C. Elevated daytime temperatures were notably concentrated near the river, extending through the southern and central areas, while the northernmost region registered notably higher temperatures. Conversely, despite high daytime temperatures, the central and southeastern regions experienced unusually low nighttime temperatures. Notably, the northern part of the study area maintained substantially higher temperatures even during the night.

This region was selected for the study to scrutinize an urban transition area reflecting analogous dynamics. Results revealed that regions characterized by dense urban structures exhibited notably higher LST compared to areas near BGs. These urbanized forms are predominantly clustered in the northern, northwestern, and central regions, whereas the southern area displayed dense mangrove vegetation. These findings resonate with prior research, indicating that coastal regions often exhibit lower LST owing to the cooling potential of these regions. Significantly, a large water body delineates the southern boundary of the study area and contributes to significantly lower temperatures in those regions. Consistently with past studies, our findings reaffirm that bare spaces tend to register higher LST. Table 2 provides a descriptive analysis of LST characteristics, illustrating distinct temperature variations between day and night across different areas within the study region.

4.2 LULC Analysis

A comprehensive multi-spectral analysis was conducted across four major domains, namely built-up areas, vegetation, waterbodies, and bare land (refer to Table 3 and 7). The examination aimed to understand the intricate nuances within the research area. Notably, among the built-up indices, $BAEM_{OLI}$ exhibited the widest range between maximum and minimum values (1.393), contrasting with NDBI having the narrowest range (0.895). In the realm of waterbody indices, WRI displayed the highest value (1.203), while MNDWI displayed the lowest (0.882). Within vegetation indices, NDVI stood out with the highest value (0.734), in contrast to SAVI with the lowest (1.101). Finally, among bare land indices, NDBaL recorded the highest value (1.108), while NBLI displayed the lowest (0.766). Notably, $BAEM_{OLI}$ exhibited the most significant range between maximum and minimum values among the sub-indicators, emphasizing the substantial intra-region differences across all domains.

Table 1 Showing the rationale behind the selection of the various indices

Abbreviations	Name	Details	Sources
Built-up indices			
NDBI	Normalised built-up index	This algorithm leverages medium spatial and spectral resolution satellite images to delineate built-up areas $\mathrm{NDBI} = \frac{(\mathrm{SWIR1-NIR})}{(\mathrm{SWIR1+NIR})}$	Das et al. (2020), Guha et al. (2018), Kafy et al. (2021), Zha et al. (2003)
$\mathrm{BAEM_{OLI}}$		It diverges from binary image methods by utilizing a continuous NDBI algorithm, enhancing accuracy by 20% compared to prior approaches $\mathrm{BAEMOLI} = \left\{\left[\frac{(\mathrm{SWIR1-NIR})}{(\mathrm{SWIR1+NIR})} - \frac{(\mathrm{NIR-RED})}{(\mathrm{NIR+RED})}\right]\right\}$	He et al. (2010), Zhou et al. (2014b)
UI	Urban index	This algorithm helps in identifying built-up areas $\mathrm{UI} = \frac{(\mathrm{SWIR2-NIR})}{(\mathrm{SWIR2+NIR})}$	Kaur and Pandey (2022)
Vegetation indices			
NDVI	Normalised vegetation index	It gauges vegetation density $\mathrm{NDVI} = \frac{(\mathrm{NIR-RED})}{(\mathrm{NIR+RED})}$	Gascon et al. (2016), Grover and Singh (2015), Hussain and Karuppannan (2021), Kant et al. (2009), Mukherjee and Singh (2020)
SAVI	Soil adjusted vegetation index	It incorporates a soil brightness correction factor (L) to reduce soil luminance influence. It aims to minimize soil effects in vegetation analysis $\mathrm{SAVI} = \left[\frac{(\mathrm{NIR-RED})}{(\mathrm{NIR+RED+L})}\right] * (1 + \mathrm{L})$	Candiago et al. (2015)
ARVI	Atmospherically resistant vegetation index	It counters atmospheric influences, especially in areas with high aerosol content $\mathrm{ARVI} = \frac{(\mathrm{NIR-RB})}{(\mathrm{NIR+RB})}$ Where $\mathrm{RB} = \mathrm{RED} - \gamma(\mathrm{BLUE} - \mathrm{RED})$	Xue and Su (2017)
Waterbody indices			
NDWI	Normalized difference water index	It is used for the analysis of water bodies $\mathrm{NDWI} = \frac{(\mathrm{NIR-SWIR})}{(\mathrm{NIR+SWIR})}$	Abutaleb et al. (2015), Kafy et al. (2021), McFeeters (1996)
MNDWI	Modified normalized difference water index	It enhances open-water detection while minimizing influences from built-up areas, vegetation, and soil. It is widely used in areas with water bodies surrounded by developed land, and it achieves an accuracy of 96.9% $\mathrm{MNDWI} = \frac{(\mathrm{GREEN-SWIR})}{(\mathrm{GREEN+SWIR})}$	Singh et al. (2014), Xu (2006)
WRI	Water ratio index	It utilizes an Adaboost algorithm with four spectral reflectance bands to identify waterbodies, achieving an impressive overall accuracy of 97.30% $\mathrm{WRI} = \frac{(\mathrm{GREEN+RED})}{(\mathrm{NIR+SWIR1})}$	Acharya et al. (2018), Rokni et al. (2014), Shen and Li (2010)

(continued)

Table 1 (continued)

Abbreviations	Name	Details	Sources
Bare land indices			
NDBaL	Normalized difference bareness index	It specializes in accurately mapping barren land within humid regions $NDBaL = \frac{(SWIR1-TIR)}{(SWIR1+TIR)}$	Rehman et al. (2022)
DBSI	Dry bare soil index	It is tailored for discriminating between urban areas and bare soil using Landsat 8 data and achieves an accuracy of 92% $DBSI = \frac{(SWIR1-GREEN)}{(SWIR1+GREEN)} - \frac{(NIR-RED)}{(NIR+RED)}$	Rasul et al. (2018)
NDSI	Normalized difference soil index	It is used to extract impervious surfaces from satellite imagery $NDSI = \frac{(SWIR1-NIR)}{(SWIR1+NIR)}$	Xu (2010)
NBLI	Normalized difference bare land index	It is primarily employed for distinguishing bare land surfaces in satellite imagery and achieves an overall accuracy rate of 92% $NBLI = \frac{(RED-TIR)}{(RED+TIR)}$	Li et al. (2017)

Table 2 Descriptive analysis of LST

Category	Maximum	Minimum	Mean	S.D.	C.V.
Day	40.810	26.610	33.310	3.064	0.092
Night	27.610	20.610	23.449	1.118	0.048
Temperature difference between day and night	13.200	6.000	9.861	1.946	0.044

Source Computed by the authors

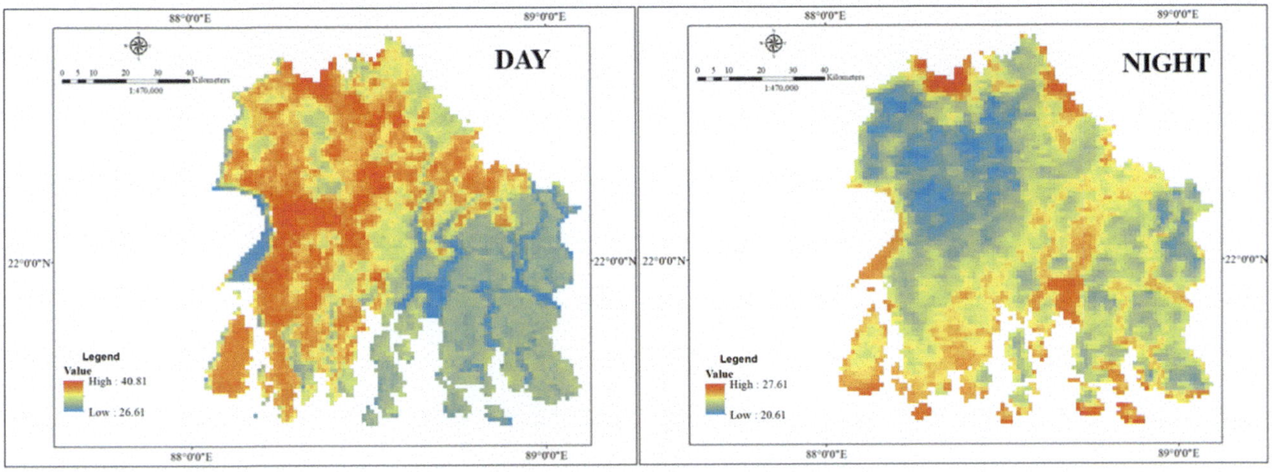

Fig. 4 Showing the LST of the study region during the day and night. *Source* Computed by the authors

Table 3 Showing the descriptive analysis of the various indicators

Name	Maximum	Minimum	S.D.	C.V.	Kurtosis	Skewness
Built-up indices						
NDBI	0.477	− 0.418	0.075	− 0.443	− 0.284	− 0.097
UI	0.756	− 0.555	0.102	− 0.397	− 0.504	0.219
$BAEM_{OLI}$	0.449	− 0.944	0.241	− 0.764	− 0.426	0.54
Waterbody indices						
NDWI	0.418	− 0.477	0.075	0.443	− 0.284	0.097
MNDWI	0.347	− 0.535	0.164	2.584	− 0.115	0.962
WRI	1.632	0.429	0.377	0.389	1.448	1.716
Vegetation indices						
NDVI	0.557	− 0.177	0.193	1.331	0.609	− 1.327
SAVI	0.835	− 0.266	0.289	1.335	0.609	− 1.327
ARVI	0.529	− 0.37	0.218	2.804	0.747	− 1.391
Bare area indices						
NDBaL	0.399	− 0.709	0.095	− 0.173	− 0.772	− 0.315
NBLI	0.131	− 0.635	0.036	− 0.067	− 0.304	0.749
DBSI	0.448	− 0.478	0.088	− 0.421	− 0.339	− 0.125
NDSI	0.477	− 0.418	0.075	− 0.443	− 0.284	− 0.097

Source Computed by the authors

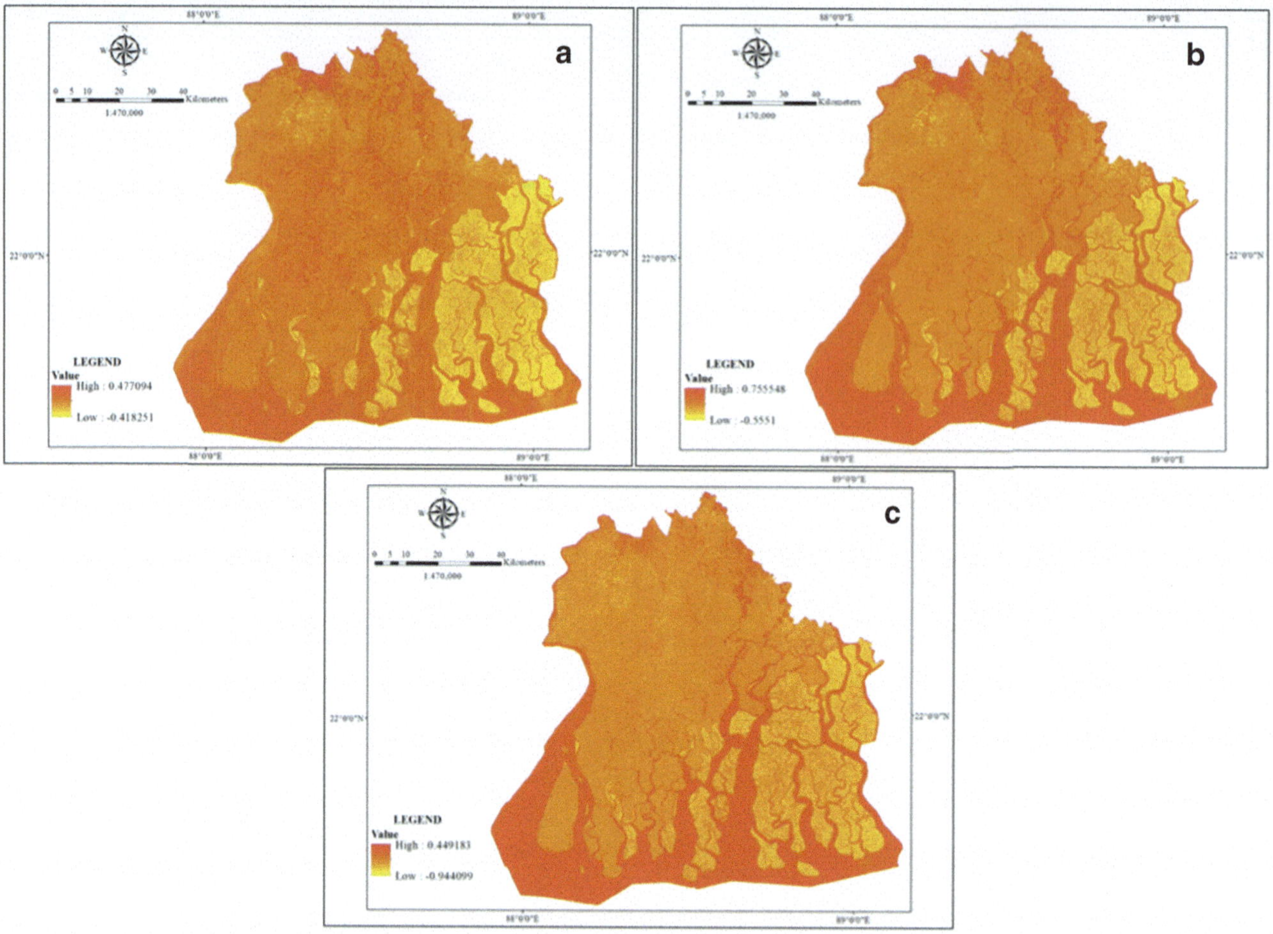

Fig. 5 Showing the built-up area analysis (**a** NDBI, **b** UI, **c** $BAEM_{OLI}$). *Source* Computed by the authors

4.2.1 Built-Up Area Analysis

Three analyses, namely NDBI, UI, and BAEMOLI, were conducted to assess urban growth morphology and configuration (refer to Fig. 5). The NDBI, UI, and $BAEM_{OLI}$ recorded a mean value of 0.895, 1.311, and 1.393 respectively. While values varied across these indices, the study area's northern tip consistently exhibited the highest density of built-up areas across all analyses. In the NDBI analysis, the northern section displayed the highest values, followed by the middle and western sectors, with the eastern and south-eastern edges reflecting the lowest values. Diverging slightly, the UI analysis exhibited a similar trend in high and low-density urban areas, while the $BAEM_{OLI}$ configuration displayed a range from moderate to low concentrations of built-up spaces. The southeastern region displayed the lowest concentration, contrasting with the highest in the northern area, while the rest depicted a moderate concentration. These disparities stem from distinct topographical features limiting intensive urbanization and influencing infrastructure complexities within the study region.

4.2.2 Vegetation Analysis

Three analyses namely NDVI, ARVI, and SAVI were conducted to assess vegetation status (refer to Fig. 6), with mean values recorded for NDVI, ARVI, and SAVI analysis being 0.734, 0.899, and 1.101 respectively. The northern tip of the study region displayed sparse vegetation, while the southeastern part exhibited significantly denser vegetation due to variations in built-up and urban areas. A considerable portion of the region comprises water bodies and marshy lands of differing densities. Overall, vegetation density ranged from moderate to low across most of the study area. Despite the lowest value at the northernmost point, the remainder of the area depicted moderate vegetation. The region's elevated aerosol content prompted an ARVI analysis to evaluate vegetation health. This demonstrates that

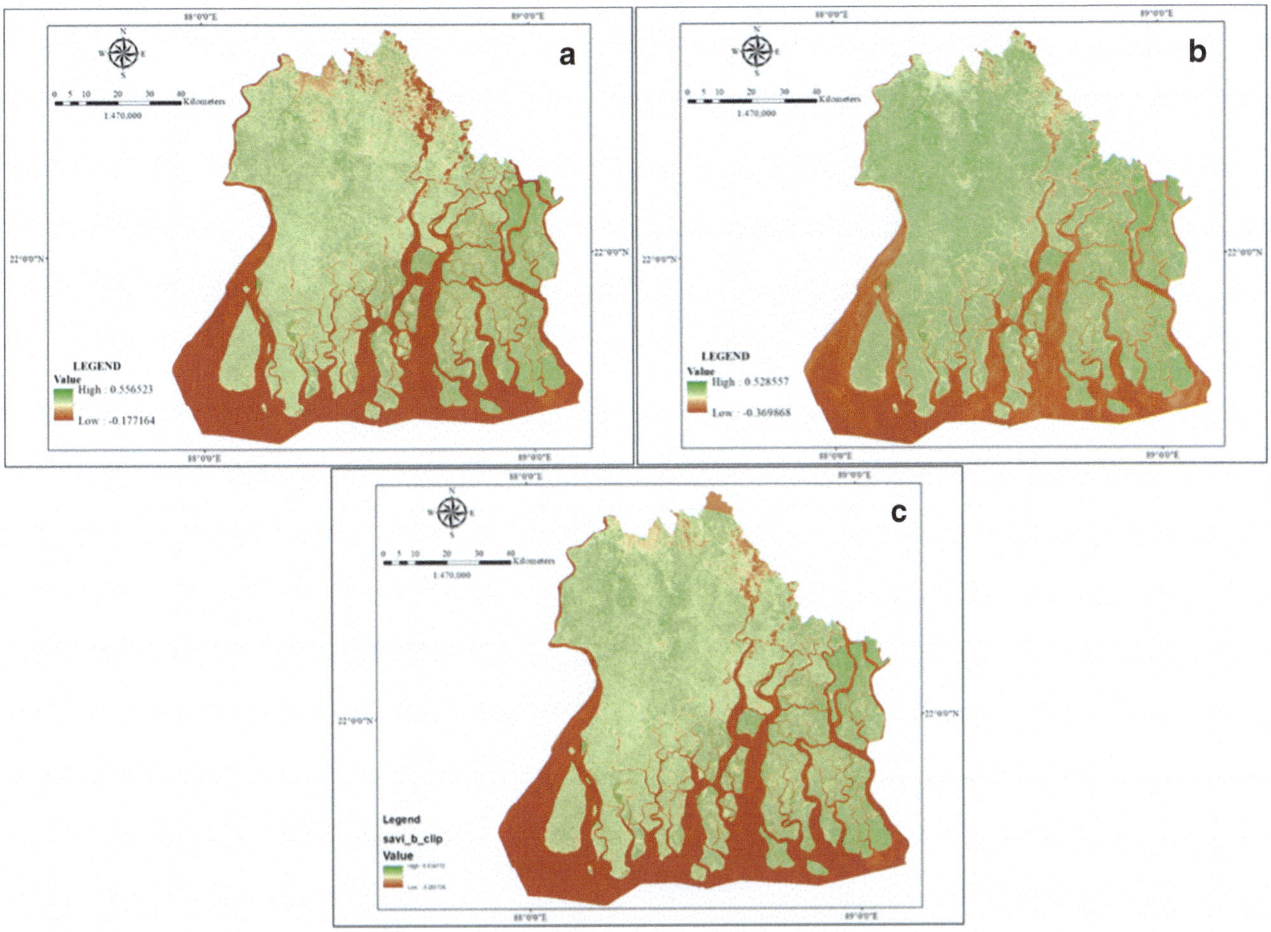

Fig. 6 Showing the vegetation analysis (**a** NDVI, **b** ARVI, **c** SAVI). *Source* Computed by the authors

vegetative health could be classified between the poor and to average category.

4.2.3 Water Body Analysis

Three analyses, NDWI, MNDWI, and WRI, were conducted to assess the status of water bodies (refer to Fig. 7). The NDWI, MNDWI, and WRI recorded the mean values of 0.895, 0.882, and 1.203 respectively. While NDWI and MNDWI exhibit some similarities, WRI displays notably higher values. However, all assessments consistently highlight abundant water bodies in the southern region of the study area, attributable. To the presence of the Bay of Bengal. Furthermore, the unique topography of the southern expanse, characterized by deltaic areas, alongside distributaries traversing the remainder of the study territory, contributes significantly to the prevalence of high values across a majority of locations.

4.2.4 Bare Land Analysis

Four analyses namely NDBaL, NBLI, DBSI, and NDSI were conducted to assess the area covered by bare lands (refer to Fig. 8). These indices recorded mean values of 1.108, 0.766, 0.926, and 0.895 respectively. Notably, the availability of bare land is not uniformly distributed across the entire study region and most indices depict a moderate to very low range of values. The southeastern zone registers comparatively lower bare land due to prevalent mangrove forests and water bodies. At the northern tip, analyses by NDBaL, NBSI, and DBSI indicate the presence of bare land, yet only the NBLI analysis contradicts this by showing no discernible evidence of bare ground in that area.

5 Interplay Between LST and LULC

To unveil the inherent relationships, a correlation analysis between LST and various sub-indicators was conducted (refer to Table 4), with correlation coefficients assessed at a significance level of 0.01. Positive correlations were identified between LST and built-up indices, such as NDBI (0.465), UI (0.135), and BAEMOLI (0.70), possibly attributed to the higher radiation absorption capacity of

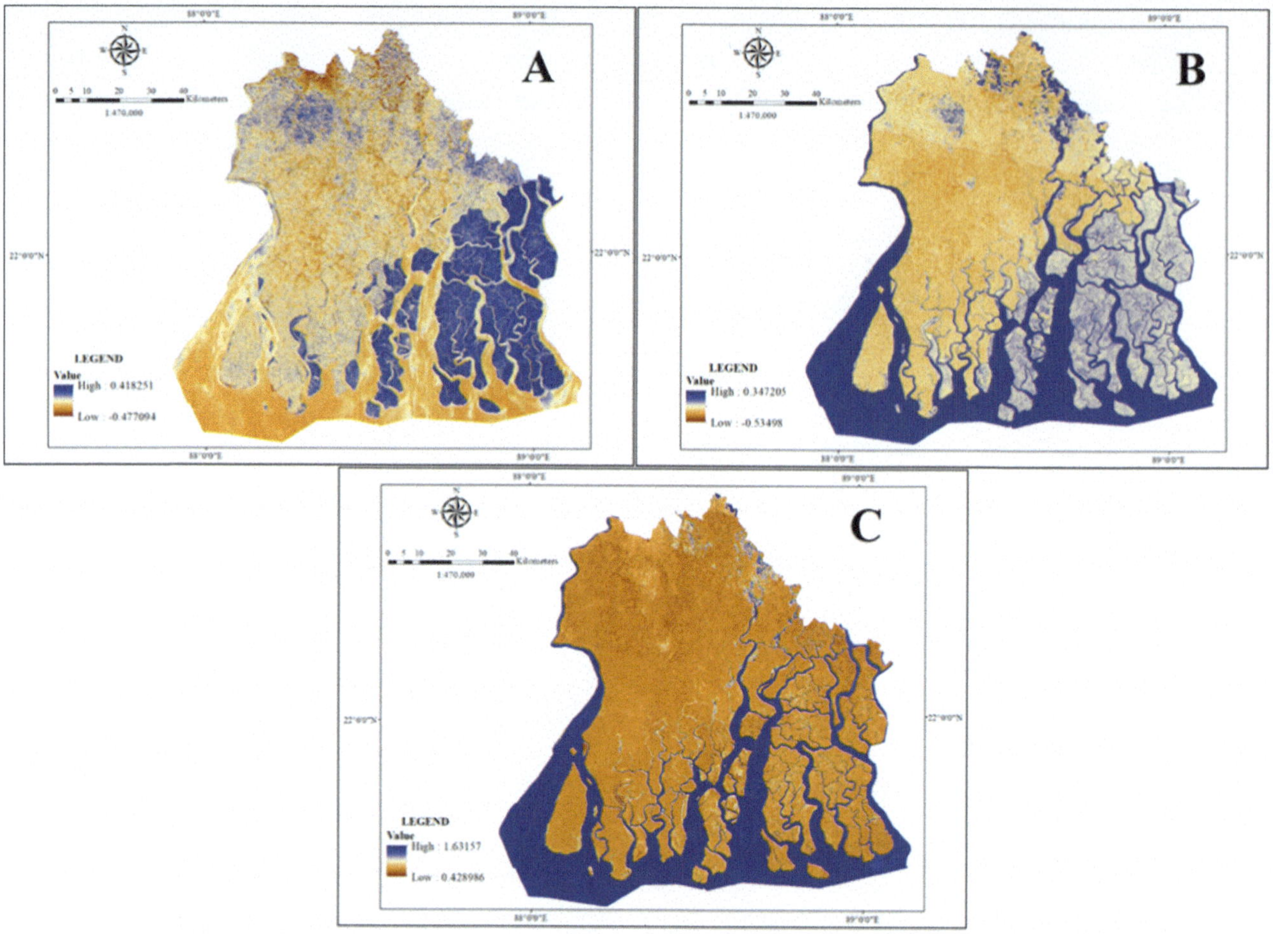

Fig. 7 Showing the waterbody analysis (**a** NDWI, **b** MNDWI, **c** WRI). *Source* Computed by the authors

impermeable surfaces. Conversely, built-up indices exhibited an inverse correlation with vegetation and waterbody indices, indicating a negative relationship between LST and these domains. All waterbody and vegetation sub-indicators displayed varying degrees of negative correlation with LST, reflecting differences in their cooling effects. Even within the broader category of vegetation, NDVI (− 0.342), ARVI (− 0.382), SAVI (− 0.342), and LST exhibited significant correlations at the 0.01 level. Similarly, within the waterbodies category, various sub-indicators such as WRI (− 0.494), NDWI (− 0.465), and MNDWI (− 0.682) demonstrated distinct coefficients. Notably, a positive correlation emerged between bare land indices and LST. Sub-indicators, including DBSI (0.437), NBLI (0.064), NDBaL (0.712), and NDSI (0.465), revealed significant positive correlations with LST at the 0.01 level, attributed to the geographical location, micro-topography, and increased solar radiation exposure of such surfaces.

PCA was computed to identify the most significant sub-indicator capable of illustrating the highest proportion of variance across all broad domains (refer to Table 5). Regarding the built-up index, the first component can elucidate 90.16% of the variance. Among the sub-indicators within the built-up domain, UI (0.995) was identified as the first component in the rotated component matrix. Concerning the waterbody index, the first component can clarify 63.69% of the variance among the various sub-indicators. MNDWI (0.985) is listed as the first component in the rotated component matrix. Among the various sub-indicators used to analyse vegetation, the first component can elucidate 99.7% of the variation, and the rotated component matrix jointly designates NDVI (0.999) and SAVI (0.999) as the first component. In the case of the bare land indices, the first component can clarify 64.03% of the variance in the sub-indicators used to analyse the influence of bare land on LST, with DBSI (0.988) being identified as the first component in the rotated component matrix.

A one-way ANOVA analysis was conducted among various indicators to assess inter and intra-indicator differences (refer to Table 6). However, the overall analysis results were

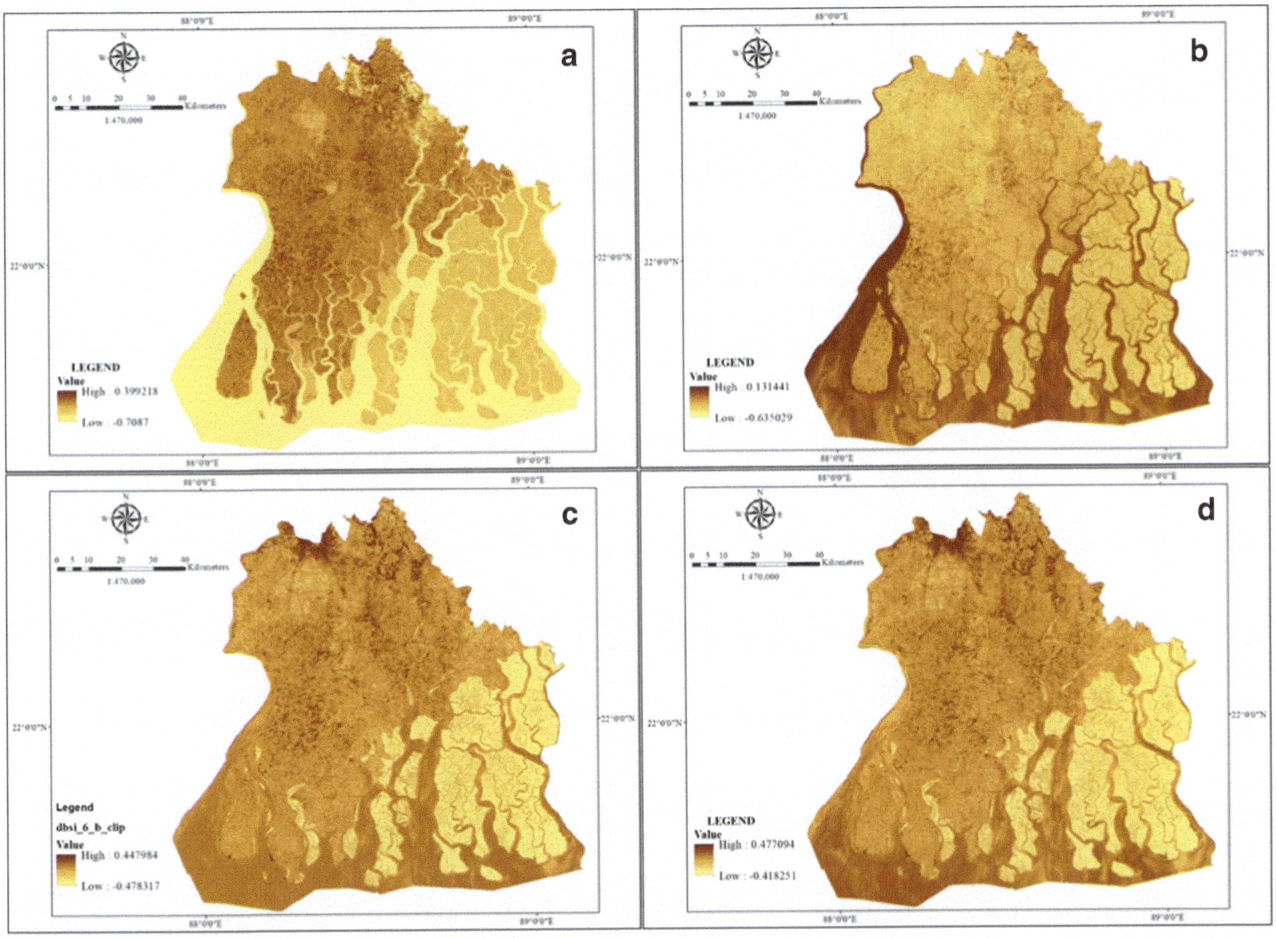

Fig. 8 Showing the bare land analysis (**a** NDBaL, **b** NBLI, **c** DBSI, **d** NDSI). *Source* Computed by the authors

insufficient to account for niche differences between the various areas within the study region. To address this limitation, the study area was divided into four sections for a comprehensive analysis (refer to Table 7). Notably, the fourth section exhibited the highest differential in the LST range. Within the built index, the third section recorded the highest LST of 40.81 °C, closely followed by the fourth section at 40.75 °C. The third section demonstrated the most significant differences in NDBI (0.444) and UI (0.494) values, potentially attributed to concentrated urban areas in specific locales. Conversely, the second part displayed less pronounced variation in the built-up space pattern due to the fragmented nature of its topography. The most considerable differences in water indices, WRI (1.088), MNDWI (0.533), and NDWI (0.444), were observed in Sects. 4 and 3, respectively. This diverse pattern of built-up space concentration also impacted vegetation, with Sect. 4 exhibiting the most significant disparities in NDVI (0.624), SAVI (0.937), and ARVI (0.662) values. Notably, the range between the maximum and minimum values over the presence of open spaces peaked in the third section, likely due to a similar concentration reason. Given the region's distinct micro-topography, generalizations become impractical. Therefore, a single-attribute examination or employing a sole technique for determining its impact on LST may lead to erroneous conclusions. Hence, our approach of utilizing multi-spectral indices to explore the interaction between LULC and LST is justified.

6 Discussion

After a comprehensive assessment, a close relationship between LST and the identified broad domains has been established. Our findings align with Xu's (2010) assertion that increasing impervious surfaces contribute to the acceleration of LST and the emergence of UHI. Interestingly, our results differ from Tan et al.'s (2021) findings, which suggested that water bodies exhibit slightly higher cooling efficacy and significantly broader cooling ranges than

Table 4 Showing the correlation between LST and various indicators

Category		LST	NDBI	UI	$BAEM_{OLI}$	WRI	NDWI	MNDWI	ARVI	NDVI	SAVI	DBSI	NBLI	NDBaL	NDSI
LST	Pearson's correlation	1.000													
NDBI		0.465**	1.000												
UI		0.135**	0.884**	1.000											
$BAEM_{OLI}$		0.070**	0.722**	0.946**	1.000										
WRI		− 0.494	0.170**	− 0.583	− 0.801	1.000									
NDWI		− 0.465	− 1.000**	− 0.884	− 0.722	0.170**	1.000								
MNDWI		− 0.682	− 0.242	− 0.221	− 0.495	0.908**	0.242**	1.000							
ARVI		− 0.382	− 0.357	− 0.733	− 0.9	− 0.973	− 0.357	− 0.819	1.000						
NDVI		− 0.342	− 0.412	− 0.77	− 0.928	− 0.963	− 0.412	− 0.781	0.993**	1.000					
SAVI		− 0.342	− 0.412	− 0.77	− 0.928	− 0.963	− 0.412	− 0.781	0.993**	1.000**	1.000				
DBSI		0.437	0.991**	− 0.893	− 0.75	− 0.216	− 0.991**	− 0.202	− 0.389	− 0.454	− 0.454**	1.000			
NBLI		0.064**	0.537**	− 0.696	− 0.793	− 0.658	− 0.537**	− 0.394	− 0.684	− 0.755	− 0.755**	0.622**	1.000		
NDBaL		0.712**	0.393**	− 0.063**	− 0.326	− 0.797	− 0.393**	− 0.965	− 0.7	− 0.64	0.640**	0.373**	0.147**	1.000	
NDSI		0.465**	1.000**	− 0.884	− 0.722	− 0.17	− 1.000**	− 0.242	− 0.357	− 0.412	− 0.412**	0.991**	0.537**	0.393**	1.000

Source Computed by the authors

** Correlation is significant at the 0.01 level (2-tailed)

Table 5 Showing the PCA among the various indices clubbed under the different domains

Total variance explained							
Component	Initial eigenvalues			Extraction sums of squared loadings			Loadings
	Total	% of variance	Cumulative %	Total	% of variance	Cumulative %	Total
Built-up indices							
1	2.705	90.159	90.159	2.705	90.159	90.159	
2	0.283	9.427	99.586				
3	0.012	0.414	100.000				
Waterbody indices							
1	1.911	63.693	63.693	1.911	63.693	63.693	1.909
2	1.082	36.080	99.773	1.082	36.080	99.773	1.088
3	0.007	0.227	100.000				
Vegetation indices							
1	2.991	99.695	99.695	2.991	99.695	99.695	
2	0.009	0.205	100.000				
3	0.008	0.100	100.000				
Bare land indices							
1	2.561	64.028	64.028	2.561	64.028	64.028	2.495
2	1.140	28.491	92.519	1.140	28.491	92.519	1.282
3	0.297	7.413	99.932				
4	0.003	0.068	100.000				

Source Computed by the authors

tree-based green spaces. Cai et al. (2018) attribute this variance to water's superior cooling capacity compared to air and other ground-surface elements. Our study indicates that vegetation demonstrates a stronger negative correlation with LST compared to water bodies, a trend attributed to the unique micro-topography and geographical location of the study area. Nevertheless, parallels could be drawn between our findings and those of Estoque et al. (2017) as both studies identify a significant positive correlation between LST and impervious surfaces, coupled with a negative correlation between green and blue surfaces and LST. Our results are also consistent with Rehman et al. (2022), who observed higher LST in built-up and bare land areas. Madanian et al. (2018) similarly reported elevated LST in barren terrain across all seasons. In tropical areas, as observed in our investigation, higher LST is prevalent (Li et al., 2012). The greater LST in bare land areas in our study suggests that the conversion of native LULC to impermeable surfaces could lead to a continuous rise in LST, emphasizing the need for more effective planning focused on expanding BGs to modify the region's thermodynamics.

Investigating the interplay between LULC and LST holds critical significance in evaluating a location's ecological health and overall viability. Staying up-to-date with the latest data is equally imperative to monitor and understand new environmental changes occurring within the area. Moreover, the outcomes derived from this analysis have the potential to aid policymakers and urban planners in formulating robust strategies to foster a more sustainable urban environment.

7 Conclusion

In summary, the intricate relationship between LULC and LST warrants a nuanced, multi-dimensional approach. Understanding specific land cover types, spatial distribution, and regional variations is crucial for effective policymaking and urban planning. This comprehensive understanding enables tailored strategies for sustainable urban development, emphasizing the pivotal role of green spaces not only in mitigating temperature rise but also in fostering ecological resilience, improving air quality, promoting biodiversity, and enhancing overall liveability in urban areas. Prioritizing the preservation and expansion of green spaces stands as a cornerstone for sustainable urban growth and the well-being of communities. In summary, the intricate relationship between LULC and LST warrants a nuanced, multi-dimensional approach. Understanding

Table 6 Showing the inter and intra-indicator ANOVA analysis

		Sum of squares	df	Mean square	F	Sig.
LST	Between Groups	27,993.865	3	9331.288	1872.927	≤0.05
	Within Groups	31,920.920	6407	4.982		
	Total	59,914.785	6410			
NDBI	Between Groups	10.020	3	3.340	746.911	≤0.05
	Within Groups	28.650	6407	0.004		
	Total	38.670	6410			
UI	Between Groups	3.516	3	1.172	155.774	≤0.05
	Within Groups	48.204	6407	0.008		
	Total	51.720	6410			
$BAEM_{OLI}$	Between Groups	3.259	3	1.086	30.431	≤0.05
	Within Groups	228.739	6407	0.036		
	Total	231.998	6410			
WRI	Between Groups	42.059	3	14.020	213.301	≤0.05
	Within Groups	421.118	6407	0.066		
	Total	463.178	6410			
NDWI	Between Groups	10.020	3	3.340	746.911	≤0.05
	Within Groups	28.650	6407	0.004		
	Total	38.670	6410			
MNDWI	Between Groups	29.537	3	9.846	775.677	≤0.05
	Within Groups	81.324	6407	0.013		
	Total	110.861	6410			
ARVI	Between Groups	6.660	3	2.220	96.478	≤0.05
	Within Groups	147.426	6407	0.023		
	Total	154.086	6410			
NDVI	Between Groups	4.166	3	1.389	68.528	≤0.05
	Within Groups	129.820	6407	0.020		
	Total	133.986	6410			
SAVI	Between Groups	9.372	3	3.124	68.527	≤0.05
	Within Groups	292.081	6407	0.046		
	Total	301.453	6410			
DBSI	Between Groups	13.126	3	4.375	677.502	≤0.05
	Within Groups	41.378	6407	0.006		
	Total	54.504	6410			
NBLI	Between Groups	0.618	3	0.206	222.114	≤0.05
	Within Groups	5.939	6407	0.001		
	Total	6.557	6410			
NDBaL	Between Groups	13.884	3	4.628	942.346	≤0.05
	Within Groups	31.465	6407	0.005		
	Total	45.349	6410			
NDSI	Between Groups	10.020	3	3.340	746.911	≤0.05
	Within Groups	28.650	6407	0.004		
	Total	38.670	6410			

Source Computed by the authors

Table 7 Showing the descriptive analysis of the various parts

Part	LST	NDBI	UI	$BAEM_{OLI}$	WRI	NDWI	MNDWI	ARVI	NDVI	SAVI	DBSI	NBLI	NDBaL	NDSI
1	11.760	0.355	0.394	0.795	0.995	0.355	0.511	0.637	0.554	0.831	0.425	0.141	0.371	0.355
2	11.780	0.309	0.391	0.864	0.986	0.309	0.431	0.659	0.571	0.857	0.352	0.114	0.281	0.309
3	13.420	0.444	0.494	0.839	1.021	0.444	0.525	0.657	0.587	0.880	0.524	0.179	0.366	0.444
4	13.700	0.368	0.457	0.791	1.088	0.368	0.533	0.662	0.624	0.937	0.420	0.148	0.351	0.368

Source Computed by the authors

specific land cover types, spatial distribution, and regional variations is crucial for effective policymaking and urban planning. This comprehensive understanding enables tailored strategies for sustainable urban development, emphasizing the pivotal role of green spaces not only in mitigating temperature rise but also in fostering ecological resilience, improving air quality, promoting biodiversity, and enhancing overall liveability in urban areas. Prioritizing the preservation and expansion of green spaces stands as a cornerstone for sustainable urban growth and the well-being of communities.

This research stands out for its inclusive landscape assessment, spanning urban, peri-urban, and rural areas, providing a more comprehensive understanding than studies focusing solely on major cities. The uniqueness lies in considering diverse landscapes and their LULC dynamics, avoiding the limitations of single LULC domains or specific urban contexts. The study employs a multi-spectral analysis covering built-up areas, greenspaces, water bodies, and bare land, ensuring a nuanced understanding of LST dynamics. This integrative approach incorporates external determinants like geographical locale, land cover, and urbanization extent, offering a sophisticated perspective on the complex interplay between LST and LULC. Addressing a literature gap, the research focuses on transitioning regions, contributing valuable insights often overlooked in studies on highly urbanized areas. Additionally, the study critically analyzes the interplay between LST and differentiating green and blue spaces, aligning with contemporary trends and emphasising the importance of understanding environmental elements' cooling effects on thermal dynamics.

This research employs a holistic approach to investigate urban, peri-urban, and rural areas, analyzing the interplay between Land Use and Land Cover (LULC) and Land Surface Temperature (LST). The methodology involves multi-spectral analysis and indices for a robust assessment of LST dynamics. Addressing literature gaps, the findings inform urban planning and policymaking, offering insights into sustainable strategies amid urbanization. Focusing on South Twenty Four Parganas enhances regional applicability. The study illuminates environmental implications, aiding ecological evaluation and informing sustainable urban development. In broader terms, understanding the correlation between land cover types and LST is crucial for effective urban planning. Strategies like increasing green spaces mitigate heat island effects, contributing to climate resilience and public health. This assessment informs decisions on building design, energy use, and water management, fostering sustainable, resilient, and liveable urban environments, benefiting both the environment and the community.

References

Abutaleb, K., Ngie, A., Darwish, A., Ahmed, M., Arafat, S., & Ahmed, F. (2015). Assessment of urban heat island using remotely sensed imagery over Greater Cairo, Egypt. *Advances in Remote Sensing, 4*, 35. https://doi.org/10.4236/ars.2015.41004

Acharya, T. D., Subedi, A., Yang, I. T., & Lee, D. H. (2018). Combining water indices for water and background threshold in Landsat image. *Proceedings, 2*(3), 143, 1–6. https://doi.org/10.3390/ecsa-4-04902

Amani-Beni, M., Zhang, B., Xie, G.-d., & Xu, J. (2018). Impact of Urban Park's tree, grass and waterbody on microclimate in hot summer days: A case study of Olympic park in Beijing, China. *Urban Forestry & Urban Greening, 32*, 1–6.https://doi.org/10.1016/j.ufug.2018.03.016

Ampatzidis, P., & Kershaw, T. (2020). A review of the impact of blue space on the urban microclimate. *Science of the Total Environment, 730*, 139068. https://doi.org/10.1016/j.scitotenv.2020.139068

Asgarian, A., Amiri, B. J., & Sakieh, Y. (2015). Assessing the effect of green cover spatial patterns on urban land surface temperature using landscape metrics approach. *Urban Ecosystem, 18*(1), 209–222. https://doi.org/10.1007/s11252-014-0387-7

Barlow, J. F. (2014). Progress In Observing And Modelling The Urban Boundary Layer. *Urban Climate, 10*, 216–240. https://doi.org/10.1016/j.uclim.2014.03.011

Bharath, H. A., Chandan, M. C., Vinay, S., & Ramchandra, T. V. (2018). Modelling urban dynamics in rapidly urbanising Indian cities. *The Egyptian Journal of Remote Sensing and Space Science, 21*(3), 201–210. https://doi.org/10.1016/j.ejrs.2017.08.002

Boehme, P., Berger, M., & Massier, T. (2015). Estimating the building based energy consumption as an anthropogenic contribution to urban heat islands. *Sustainable Cities and Society, 19*, 373–384. https://doi.org/10.1016/j.scs.2015.05.006

Cai, Z., Han, G., & Chen, M. (2018). Do water bodies play an important role in the relationship between urban form and land surface temperature? *Sustainable Cities and Society, 39*, 487–498. https://doi.org/10.1016/j.scs.2018.02.033

Candiago, S., Remondino, F., De Giglio, M., Dubbini, M., & Gattelli, M. (2015). Evaluating multispectral images and vegetation indices

for precision farming applications from UAV images. *Remote Sensing, 7*(4), 4026–4047. https://doi.org/10.3390/rs70404026
Census. (2001). Census of India. Census Digital Library. Office of the Register General & Census Commissioner, India, Ministry of Home Affairs, Government of India.
Census. (2011). Census of India. Census Digital Library. Office Of The Register General & Census Commissioner, India, Ministry Of Home Affairs, Government Of India.
Chen, L., Zhang, M., Zhu, J., Wang, Y., & Skorokhod, A. (2018). Modeling impacts of urbanization and urban heat island mitigation on boundary layer meteorology and air quality in Beijing under different weather conditions. *Journal of Geophysical Research: Atmospheres, 123*(8), 4323–4344. https://doi.org/10.1002/2017JD027501
Choudhury, D., Das, K., & Das, A. (2019). Assessment of land use land cover changes and its impact on variations of land surface temperature in Asansol-Durgapur Development Region. *The Egyptian Journal of Remote Sensing and Space Sciences, 22*(2), 203–218. https://doi.org/10.1016/j.ejrs.2018.05.004
Daneshvar, M. R. M., Rabbani, G., & Shirvani, S. (2019). Assessment of Urban Sprawl effects on regional climate change using a hybrid model of factor analysis and analytical network process in the Mashhad city. *Environmental Systems Research, 8*(23), 1–12. https://doi.org/10.1186/s40068-019-0152-2
Das, S., & Angadi, D. P. (2020). Land Use-Land Cover (LULC) transformation and its relation with land surface temperature changes: A case study of Barrackpore Subdivision, West Bengal, India. *Remote Sensing Applications: Society and Environment, 19*, 100322. https://doi.org/10.1016/j.rsase.2020.100322
Das, N., Mondal, P., Sutradhar, S., Ghosh, R. (2020). Assessment of variation of land use/land cover and its impact on land surface temperature of Asansol Subdivision. *Egyptian Journal of Remote Sensing and Space Science*, 1–12. https://doi.org/10.1016/j.ejrs.2020.05.001
Dissanayake, D. M. S. L. B., Morimoto, T., Ranagalage, M., & Murayama, Y. (2019). Land-use/land-cover changes and their impact on surface urban heat islands: Case study of Kandy City, Sri Lanka. *Climate, 7*, 1–20. https://doi.org/10.3390/cli7080099
Du, H., Song, X., Jiang, H., Kan, Z., Wang, Z., & Cai, Y. (2016). Research on the cooling island effects of water body: A case study of Shanghai China. *Ecological Indicators, 67*, 31–38. https://doi.org/10.1016/j.ecolind.2016.02.040
Dutta, D., Rahman, A., Paul, S. K., & Kundu, A. (2019). Changing Pattern of urban landscape and its effect on land surface temperature in and around Delhi. *Environmental Monitoring and Assessment, 191*, 551. https://doi.org/10.1007/s10661-019-7645-3
Dutta, D., Rahman, A., Paul, S. K., & Kundu, A. (2021). Impervious surface growth and its inter-relationship with vegetation cover and land surface temperature in Peri-Urban Areas of Delhi. *Urban Climate, 37*, 100799. https://doi.org/10.1016/j.uclim.2021.100799
Estoque, R. C., Murayama, Y., & Myint, S. W. (2017). Effects of landscape composition and pattern on land surface temperature: an urban heat island study in the megacities of Southeast Asia. *Science of the Total Environment, 577*, 349–359. https://doi.org/10.1016/j.scitotenv.2016.10.195
Gascon, M., Cirach, M., Martínez, D., Dadvand, P., Valentín, A., Plasència, A., & Nieuwenhuijsen, M. (2016). Normalized Difference Vegetation Index (NDVI) as a marker of surrounding greenness in epidemiological studies: The case of Barcelona City. *Urban Forestry & Urban Greening, 19*, 88–94. https://doi.org/10.1016/j.ufug.2016.07.001
Gazi, M. Y., Rahman, M. Z., Uddin, M. M., & Rahman, F. M. A. (2021). Spatio-temporal dynamic land cover changes and their impacts on the urban thermal environment in the Chittagong Metropolitan Area, Bangladesh. *GeoJournal, 86*, 2119–2134. https://doi.org/10.1007/s10708-020-10178-4
Grover, A., & Singh, R. B. (2015). Analysis of Urban Heat Island (UHI) in relation to Normalized Difference Vegetation Index (NDVI): A comparative study of Delhi and Mumbai. *Environments, 2*, 125–138. https://doi.org/10.3390/environments2020125
Guha, S., Govila, H., Dey, A., & Gill, N. (2018). Analytical study of land surface temperature with NDVI and NDBI using Landsat 8 OLI and TIRS data in Florence and Naples City, Italy. *European Journal of Remote Sensing, 51*(1), 667–678. https://doi.org/10.1080/22797254.2018.1474494
Guha, S., Govil, H., Dey, A., & Gill, N. (2020). A case study on the relationship between land surface temperature and land surface indices in Raipur City, India. *Geografisk Tidsskrift-Danish Journal of Geography, 120*(1), 35–50. https://doi.org/10.1080/00167223.2020.1752272
Gunawardena, K. R., Wells, M. J., & Kershaw, T. (2017). Utilising green and bluespace to mitigate urban heat island intensity. *Science of the Total Environment*, 584–585. https://doi.org/10.1016/j.scitotenv.2017.01.158
Guo, G., Wu, Z., Cao, Z., Chen, Y., & Yang, Z. (2020). A multilevel statistical technique to identify the dominant landscape metrics of greenspace for determining land surface temperature. *Sustainable Cities and Society, 61*, 102263. https://doi.org/10.1016/j.scs.2020.102263
Guo, G., Wu, Z., Cao, Z., Chen, Y., & Zheng, Z. (2021). Location of greenspace matters: A new approach to investigating the effect of the greenspace spatial pattern on urban heat environment. *Landscape Ecology, 36*, 1533–1548. https://doi.org/10.1007/s10980-021-01230-w
Hamstead, Z. A., Kremer, P., Larondelle, N., McPhearson, T., & Haase, D. (2016). Classification of the heterogeneous structure of urban landscapes (STURLA) as an indicator of landscape function applied to surface temperature in New York City. *Ecological Indicators, 70*, 574–585. https://doi.org/10.1016/j.ecolind.2015.10.014
He, C., Shi, P., Xie, D., & Zhao, Y. (2010). Improving the normalized difference built-up index to map urban built-up areas using a semi-automatic segmentation approach. *Remote Sensing Letters, 1*, 213–221. https://doi.org/10.1080/01431161.2010.481681
Huang, H., Li, Y., Zhao, Y., & Zhai, W. (2022). Analysis of the impact of urban summer high temperatures and outdoor activity duration on residents' emotional health: Taking hostility as an example. *Frontiers in Public Health, 10*, 955077. https://doi.org/10.3389/fpubh.2022.955077
Hussain, S., & Karuppannan, S. (2021). Land use/land cover changes and their impact on land surface temperature using remote sensing technique in District Khanewal, Punjab Pakistan. *Geology, Ecology, and Landscapes*, 1–14 https://doi.org/10.1080/24749508.2021.1923272
Imran, H. M., Hossain, A., Islam, A. K. M. S., Rahman, A., Bhuiyan, M. A. E., Paul, S., & Alam, A. (2021). Impact of land cover changes on land surface temperature and human thermal comfort in Dhaka city of Bangladesh. *Earth Systems and Environment, 5*, 667–693. https://doi.org/10.1007/s41748-021-00243-4
Imran, M., & Mehmood, A. (2020). Analysis and mapping of present and future drivers of local urban climate using remote sensing: A case of Lahore, Pakistan. *Arabian Journal of Geosciences, 13*, 1–10. https://doi.org/10.1007/s12517-020-5214-2
Jamei, Y., Rajagopalan, C., & Sun, Q. C. (2019). Spatial structure of surface urban heat island and its relationship with vegetation and built-up areas in Melbourne, Australia. *Science of the Total Environment, 659*, 1335–1351. https://doi.org/10.1016/j.scitotenv.2018.12.308
John, J., Bindu, G., Srimuruganandam, B., Wadhwa, A., & Rajan, P. (2020). Land use/land cover and land surface temperature analysis

in Wayanad District, India, using satellite imagery. *Annals of GIS, 26*(4), 343–360. https://doi.org/10.1080/19475683.2020.1733662

Karakus, C. (2019). The impact of Land Use/Land Cover (LULC) changes on land surface temperature in Sivas City Center and its surroundings and assessment of Urban Heat Island. *Asia-Pacific Journal of Atmospheric Sciences, 55*, 1–16. https://doi.org/10.1007/s13143-019-00109-w

Kafy, A.-A., Dey, N. N., Rakib, A. A., Rahaman, Z. A., Nasher, N. M. R., & Bhatt, A. (2021). Modeling the relationship between land use/land cover and land surface temperature in Dhaka, Bangladesh, using CA-ANN Algorithm. *Environmental Challenges, 4*, 100190. https://doi.org/10.1016/j.envc.2021.100190

Kant, Y., Bharath, B. D., Mallick, J., Atzberger, C., & Kerle, N. (2009). Satellite-based analysis of the role of land use/land cover and vegetation density on surface temperature regime of Delhi, India. *Journal of the Indian Society of Remote Sensing, 37*, 201–214. https://doi.org/10.1007/s12524-009-0030-x

Karakuş, C. B. (2019). The impact of land use/land cover (LULC) changes on land surface temperature in Sivas City Center and its surroundings and assessment of urban heat Island. *Asia-Pacific Journal of Atmospheric Sciences, 55*, 1–16. https://doi.org/10.1007/s13143-019-00109-w

Kaur, R., & Pandey, P. (2022). A review on spectral indices for built up area extraction using remote sensing technology. *Arabian Journal of Geosciences, 15*(391), 1–22. https://doi.org/10.1007/s12517-022-09688-x

Kuang, W., Liu, Y., Dou, Y., Chi, W., Chen, G., Gao, C., Yang, T., Liu, J., & Zhang, R. (2015). What are hot and what are not in an urban landscape: quantifying and explaining the land surface temperature pattern in Beijing, China. *Landscape Ecology, 30*, 357–373. https://doi.org/10.1007/s10980-014-0128-6

Lee, Y. Y., Md Din, M. F., Iwao, K., Lee, Y. H., & Anting, N. (2021). Impact of thermal behaviour of different environmental conditions on ambient environment and thermal discomfort in Malaysia. *Indoor and Built Environment, 30*, 520–534. https://doi.org/10.1177/1420326X19897956

Li, H., Wang, C., Zhong, C., Su, A., Xiong, C., Wang, J., & Liu, J. (2017). Mapping urban bare land automatically from landsat imagery with a simple index. *Remote Sensing, 9*(249), 1–10. https://doi.org/10.3390/rs9030249

Li, X., Zhou, W., Ouyang, Z., Xu, W., & Zheng, H. (2012). Spatial pattern of greenspace affects land surface temperature: evidence from the heavily urbanized Beijing Metropolitan Area, China. *Landscape Ecology, 27*, 887–898. https://doi.org/10.1007/s10980-012-9731-6

Li, Y., Schubert, S., Kropp, J. P., & Rybski, D. (2020). On the influence of density and morphology on the Urban Heat Island intensity. *Nature Communications, 11*(1), 1–9. https://doi.org/10.1038/s41467-020-16461-9

Lin, Y., Wang, Z., Jim, C. Y., Li, J., Deng, J., & Liu, J. (2020). Water as an urban heat sink: Blue infrastructure alleviates urban heat island effect in Mega-City Agglomeration. *Journal of Cleaner Production, 262*, 121411. https://doi.org/10.1016/j.jclepro.2020.121411

Ma, Q., Wu, J., & He, C. (2016). A hierarchical analysis of the relationship between urban impervious surfaces and land surface temperatures: Spatial scale dependence, temporal variations, and bioclimatic modulation. *Landscape Ecology, 31*, 1139–1153. https://doi.org/10.1007/s10980-016-0356-z

Madanian, M., Soffianian, A. R., Koupai, S. S., Pourmanafi, S., & Momeni, M. (2018). The study of thermal pattern changes using landsat-derived land surface temperature in the central part of Isfahan Province. *Sustainable Cities and Society, 39*, 650–661. https://doi.org/10.1016/j.scs.2018.03.018

Manteghi, G., Limit, H. B., & Remaz, D. (2015). Water bodies an urban microclimate: A review. *Modern Applied Science, 9*, 1–10. https://doi.org/10.5539/mas.v9n6p1

Mathew, A., Sreekumar, S., Khandelwal, S., Kaul, N., & Kumar, R. (2016). Prediction of surface temperatures for the assessment of urban heat island effect over Ahmedabad city using linear time series model. *Energy and Buildings, 128*, 605–616. https://doi.org/10.1016/J.ENBUILD.2016.07.004

Mbuh, M. J., Wheeler, R., & Cook, A. (2019). Spatiotemporal analysis of urban heat island intensification in the city of Minneapolis-St. Paul and Chicago Metropolitan Areas using Landsat data from 1984 to 2016. *Geocarto International*, 1–40. https://doi.org/10.1080/10106049.2019.1655802.

McFeeters, S. K. (1996). The use of the normalized difference water index (NDWI) in the delineation of open water features. *International Journal of Remote Sensing, 17*(7), 1425–1432. https://doi.org/10.1080/01431169608948714

Montazeri, M., & Masoodian, S. A. (2020). Tempo-spatial behavior of surface urban heat island of Isfahan Metropolitan Area. *Journal of the Indian Society of Remote Sensing, 48*, 263–270. https://doi.org/10.1007/s12524-019-01059-6

Morais, M. V. B. d., Urbina Guerrero, V. V., Freitas, E. D. d., Marciotto, E. R., Valdés, H., Correa, C., Agredano, R., & Vera-Puerto, I. (2019). Sensitivity of radiative and thermal properties of building material in the urban atmosphere. *Sustainability, 11*, 6865. https://doi.org/10.3390/su11236865

Mukherjee, F., & Singh, D. (2020). Assessing land use-land cover change and its impact on land surface temperature using LANDSAT data: A comparison of two urban areas in India. *Earth Systems and Environment, 4*, 385–407. https://doi.org/10.1007/s41748-020-00155-9

Naikoo, M. W., Rihan, M., Ishitaque, M., & Shahfahad. (2022). Analyses of Land Use Land Cover (LULC) change and built-up expansion in the suburb of a Metropolitan City: Spatio-temporal analysis of Delhi NCR using landsat datasets. *Journal of Urban Management, 9*(3), 347–359. https://doi.org/10.1016/j.jum.2020.05.004

Naim, M. N. H., & Kafy, A. Al. (2021). Assessment of urban thermal field variance index and defining the relationship between land cover and surface temperature in Chattogram City: A remote sensing and statistical approach. *Environmental Challenges, 100107*.https://doi.org/10.1016/j.envc.2021.100107

Nath, B., Ni-Meister, W., & Choudhury, R. (2021). Impact of urbanization on land use and land cover change in Guwahati City, India and its implication on declining groundwater level. *Groundwater for Sustainable Development, 12*, 100500. https://doi.org/10.1016/j.gsd.2020.100500

Nie, Q., Man, W., Li, Z., & Huang, Y. (2016). Spatiotemporal impact of urban impervious surface on land surface temperature in Shanghai, China. *Canadian Journal of Remote Sensing, 42*(6), 680–689. https://doi.org/10.1080/07038992.2016.1217484

Ogunjobi, K. O., Adamu, Y., Akinsanola, A. A., & Orimoloye, I. R. (2018). Spatiotemporal analysis of land use dynamics and its potential indications on land surface temperature in Sokoto Metropolis, Nigeria. *Royal Society Open Science, 5*, 180661. https://doi.org/10.1098/rsos.180661

Pal, S., & Ziaul, S. K. (2017). Detection of land use and land cover change and land surface temperature in English bazar urban centre. *The Egyptian Journal of Remote Sensing and Space Science, 20*, 125–145. https://doi.org/10.1016/j.ejrs.2016.11.003

Patra, S., Sahoo, S., Mishra, P., & Mahapatra, S. C. (2018). Impacts of urbanization on land use/cover changes and its probable implications on local climate and groundwater level. *Journal of Urban Management, 7*(2), 70–84. https://doi.org/10.1016/j.jum.2018.04.006

Qu, S., Wang, L., Lin, A., Yu, D., & Yuan, M. (2020). Distinguishing the impacts of climate change and anthropogenic factors on vegetation dynamics in the Yangtze River Basin, China. *Ecological Indicators, 108*, 105724. https://doi.org/10.1016/j.ecolind.2019.105724

Rasul, A., Balzter, H., Ibrahim, G. R. F., Hameed, H. M., Wheeler, J., Adamu, B., Ibrahim, S., & Najmaddin, P. M. (2018). Applying built-up and bare-soil indices from landsat 8 to cities in dry climates. *Land, 7*, 81. https://doi.org/10.3390/land7030081

Rahman, M. N., Rony, M. R., Jannat, F. A., Chandra Pal, S., Islam, M. S., Alam, E., & Islam, A. R. (2021). Impact of urbanization on urban heat island intensity in major districts of Bangladesh using remote sensing and geo-spatial tools. *Climate, 10*(1), 3. https://doi.org/10.3390/cli10010003

Rahaman, S., Jahangir, S., Haque, S. Md., Chen, R., & Kumar, P. (2021). Spatio temporal changes of green spaces and their impact on urban environment of Mumbai, India. *Environment, Development and Sustainability, 23*, 6481–6501.https://doi.org/10.1007/s10668-020-00882-z

Rehman, A., Qin, J., Shafi, S., Khan, M. S., Ullah, S., Ahmad, K., Rehman, N. U., & Faheem, M. (2022). Modelling of land use/cover and LST variations by using GIS and remote sensing: A case study of the Northern Pakhtunkhwa Mountainous Region, Pakistan. *Sensors, 22*, 4965. https://doi.org/10.3390/s22134965

Ren, J., Yang, J., Wu, F., Sun, W., Xiao, X., & Xia, J. C. (2023). Regional thermal environment changes: Integration of satellite data and land use/land cover. *iScience, 26*(2), 105820. https://doi.org/10.1016/j.isci.2022.105820.

Ren, T., Zhou, W., & Wang, J. (2021). Beyond intensity of urban heat island effect: A continental scale analysis on land surface temperature in Major Chinese Cities. *Science of the Total Environment, 791*, 148334. https://doi.org/10.1016/j.scitotenv.2021.148334

Rokni, K., Ahmad, A., Selamat, A., & Hazini, S. (2014). Water feature extraction and change detection using multitemporal landsat imagery. *Remote Sensing, 6*, 4173–4189. https://doi.org/10.3390/rs6054173

Rousta, I., Sarif, M. O., Gupta, R. D., Olafsson, H., Ranagalage, M., Murayama, Y., Zhang, H., & Mushore, T. D. (2018). Spatiotemporal analysis of land use/land cover and its effects on surface urban heat island using Landsat data: A case study of metropolitan city Tehran (1988–2018). *Sustainability, 10*, 4433. https://doi.org/10.3390/su10124433

Sam, S. C., Balasubramanian, G. (2022). Spatiotemporal detection of land use/land cover changes and land surface temperature using Landsat and MODIS data across the coastal Kanyakumari district, India. Geodesy and Geodynamics, 1–10.https://doi.org/10.1016/j.geog.2022.09.002

Sejati, A. W., Buchori, I., & Rudiarto, I. (2019). The spatio-temporal trends of urban growth and surface urban heat islands over two decades in the Semarang Metropolitan Region. *Sustainable Cities and Society, 46*, 101432. https://doi.org/10.1016/j.scs.2019.101432

Shen, L., Li, C. (2010). Water body extraction from Landsat ETM+ imagery using Adaboost Algorithm. In *2010 18th International Conference on Geoinformatics* (pp. 1–4). https://doi.org/10.1109/GEOINFORMATICS.2010.5567762

Shih, W. (2017). Greenspace patterns and the mitigation of land surface temperature in Taipei Metropolis. *Habitat International, 60*, 69–80. https://doi.org/10.1016/j.habitatint.2016.12.006

Siddique, M. A., Dongyun, L., Li, P., Rasool, U., Khan, T. U., Farooqi, T. J. A., Wang, L., Fan, B., & Rasool, M. A. (2020). Assessment and simulation of land use and land cover change impacts on the land surface temperature of Chaoyang District in Beijing, China. *PeerJ, 8*, e9115. https://doi.org/10.7717/peerj.9115

Singh, K. V., Setia, R., Sahoo, S., Prasad, A., & Pateriya, B. (2014). Evaluation of NDWI and MNDWI for assessment of waterlogging by integrating digital elevation model and groundwater level. Geocarto International, 1–14.https://doi.org/10.1080/10106049.2014.965757

Solcerova, A., van de Ven, F., & van de Giesen, N. (2019). Nighttime cooling of an urban pond. *Frontiers in Earth Science, 7*, 156. https://doi.org/10.3389/feart.2019.00156

Song, J., Du, S., Feng, X., & Guo, L. (2014). The relationships between landscape compositions and land surface temperature: Quantifying their resolution sensitivity with spatial regression models. *Landscape and Urban Planning, 123*, 145–157. https://doi.org/10.1016/j.landurbplan.2013.11.014

Stache, E. E., Schilperoort, B. B., Ottelé, M. M., & Jonkers, H. H. (2022). Comparative analysis in thermal behaviour of common urban building materials and vegetation and consequences for urban heat island effect. *Building and Environment, 213*, 108489. https://doi.org/10.1016/j.buildenv.2021.108489

Swamy, G., Nagendra, S. M. S., & Schlink, U. (2017). Urban Heat Island (UHI) influence on secondary pollutant formation in a tropical humid environment. *Journal of the Air & Waste Management Association, 67*(10), 1080–1091. https://doi.org/10.1080/10962247.2017.1325417

Syafii, N. I., Ichinose, M., Kumakura, E., Jusuf, S. K., Chigusa, K., & Wong, N. H. (2017). Thermal environment assessment around bodies of water in urban canyons: A scale model study. *Sustainable Cities and Society, 34*, 79–89. https://doi.org/10.1016/j.scs.2017.06.012

Tan, J., Yu, D., Li, Q., Tan, X., & Zhou, W. (2020). Spatial relationship between land-use/land-cover change and land surface temperature in the Dongting Lake Area, China. *Scientific Reports, 10*, 9245. https://doi.org/10.1038/s41598-020-66168-6

Tan, X., Sun, X., Huang, C., Yuan, Y., & Hou, D. (2021). Comparison of cooling effect between green space and water body. *Sustainable Cities and Society, 67*, 102711. https://doi.org/10.1016/j.scs.2021.102711

Tepanosyan, G., Muradyan, V., Hovsepyan, A., Pinigin, G., Medvedev, A., & Asmaryan, S. (2021). Studying spatial-temporal changes and relationship of land cover and surface urban heat island derived through remote sensing in Yerevan, Armenia. *Building and Environment, 187*, 107390. https://doi.org/10.1016/j.buildenv.2020.107390

Tu, L., Qin, Z., Li, W., Geng, J., Yang, L., Zhao, S., Zhan, W., & Wang, F. (2016). Surface urban heat island effect and its relationship with urban expansion in Nanjing, China. *Journal of Applied Remote Sensing, 10*, 026037. https://doi.org/10.1117/1.JRS.10.026037

Wang, C., Wang, Z., Kaloush, K. E., & Shacat, J. (2021). Perceptions of urban heat Island Mitigation and implementation strategies: Survey and gap analysis. *Sustainable Cities and Society, 66*, 102687. https://doi.org/10.1016/j.scs.2020.102687

Wang, Y.-C., Hu, B. K. H., Myint, S. W., Feng, C.-C., Chow, W. T. L., & Passy, P. F. (2018). Patterns of land change and their potential impacts on land surface temperature change in Yangon, Myanmar. *Science of the Total Environment, 643*, 738–750. https://doi.org/10.1016/j.scitotenv.2018.06.209

Xu, H. (2006). Modification of normalised difference water index (NDWI) to enhance open water features in remotely sensed imagery. *International Journal of Remote Sensing, 27*(14), 3025–3033. https://doi.org/10.1080/01431160600589179

Xu, H. (2010). Analysis of impervious surface and its impact on urban heat environment using the normalized difference impervious surface index (NDISI). *Photogrammetric Engineering & Remote Sensing, 76*, 557–565. https://doi.org/10.14358/PERS.76.5.557

Xie, C., Zhang, X., Zhuang, L., Zhu, R., & Guo, J. (2022). Analysis of surface temperature variation of Lakes in China using MODIS land surface temperature data. *Scientific Reports, 12*, 2415. https://doi.org/10.1038/s41598-022-06363-9

Xiong, Y., & Zhang, F. (2021). Effect of human settlements on urban thermal environment and factor analysis based on multi-source data: A case study of Changsha City. *Journal of Geographical Sciences, 31*, 819–838. https://doi.org/10.1007/s11442-021-1873-5

Xue, J., & Su, B. (2017). Significant remote sensing vegetation indices: A review of developments and applications. *Journal of Sensors*, 1–17. https://doi.org/10.1155/2017/1353691

Yan, C., Guo, Q., Li, H., Li, L., & Qiu, G. Y. (2020). Quantifying the cooling effect of urban vegetation by mobile traverse method: A local-scale urban heat island study in a subtropical megacity. *Building and Environment, 169*, 106541. https://doi.org/10.1016/j.buildenv.2019.106541

Yan, H., Wu, F., & Dong, L. (2018). Influence of a large urban park on the local urban thermal environment. *The Science of the Total Environment, 622*, 882–891. https://doi.org/10.1016/j.scitotenv.2017.11.327

Yang, J., Menenti, M., Wu, Z., Wong, M. S., Abbas, S., Xu, Y., & Shi, Q. (2021). Assessing the impact of urban geometry on surface urban heat island using complete and Nadir temperatures. *International Journal of Climatology, 41*, E3219–E3238. https://doi.org/10.1002/joc.6919

Yang, J., Sun, J., Ge, Q., & Li, X. (2017). Assessing the impacts of urbanization-associated green space on urban land surface temperature: A case study of Dalian, China. *Urban Forestry & Urban Greening, 22*, 1–10. https://doi.org/10.1016/j.ufug.2017.01.002

Yin, J., Wu, X., Shen, M., Zhang, X., Zhu, C., Xiang, H., Shi, C., Guo, Z., & Li, C. (2019). Impact of urban greenspace spatial pattern on land surface temperature: A case study in Beijing Metropolitan Area, China. *Landscape Ecology, 34*, 2949–2961. https://doi.org/10.1007/s10980-019-00932-6

Yu, Z., Guo, X., Jørgensen, G., & Vejre, H. (2017). How can urban green spaces be planned for climate adaptation in subtropical cities? *Ecological Indicators, 82*, 152–162. https://doi.org/10.1016/j.ecolind.2017.07.002

Yun, G. Y., Ngarambe, J., Duhirwe, P. N., Ulpiani, G., Paolini, R., Haddad, S., Vasilakopoulou, K., & Santamouris, M. (2020). Predicting the magnitude and the characteristics of the urban heat island in coastal cities in the proximity of desert landforms. The case of Sydney. *Science of the Total Environment, 709*, 136068. https://doi.org/10.1016/j.scitotenv.2019.136068

Zha, Y., Gao, J., & Ni, S. (2003). Use of normalized difference built-up in automatically mapping urban areas from TM imagery. *International Journal of Remote Sensing, 24*, 583–594. https://doi.org/10.1080/01431160304987

Zhao, H., & Chen, X. (2005). Use of normalized difference bareness index in quickly mapping bare areas from TM/ETM+. In *Proceedings of 2005 IEEE international geoscience and remote sensing symposium, IGARSS'05* (Vol. 3, pp. 1666–1668). https://doi.org/10.1109/IGARSS.2005.1526319

Zhao, W., Li, A., & Zheng, J. (2016). A study on land surface temperature terrain effect over mountainous area based on Landsat 8 thermal infrared data. *Remote Sensing. Technol. Appl., 31*, 63–73.

Zhou, W., Qian, Y., Li, X., Li, W., & Han, L. (2014a). Relationships between land cover and the surface urban heat island: Seasonal variability and effects of spatial and thematic resolution of land cover data on predicting land surface temperatures. *Landscape Ecology, 29*, 153–167. https://doi.org/10.1007/s10980-013-9950-5

Zhou, Y., Yang, G., Wang, S., Wang, L., Wang, F., & Liu, X. (2014b). A new index for mapping built-up and bare land areas from landsat-8 OLI data. *Remote Sensing Letters, 5*, 862–871. https://doi.org/10.1080/2150704X.2014.973996

Ziaul, S., & Pal, S. (2020). Modeling the effects of green alternative on heat island mitigation of a Meso Level Town, West Bengal, India. *Advances in Space Research, 65*(7), 1789–1802. https://doi.org/10.1016/j.asr.2019.12.031

Six Hidden Changes in Mobility During COVID-19 and Their Impact on Future Transport Technologies and Research Direction

Pooja Rao and Washington Y. Ochieng

Abstract

Transportation is integral to our daily activities, not only for social well-being but also economic success. Transformative technologies such as automation and ride pooling services due to comfort and convenience should enhance our use of mobility and complement public transit. However, restrictions on movement due to the COVID-19 pandemic have raised questions related to transport resilience and its long-term societal impact. Alternative mode choice for many individuals in the United States, for example, may not be an option given the low transit and land-use density distribution outside the urban core. This combined with inadequate funding, limits job growth spearing inequality and prejudice. Spatio-temporal distribution of cities that shape activity and transport accessibility have not fully benefited some population segments socioeconomically. Considerable research has been dedicated to understanding the impacts on public transit ridership, general trip frequency, distance and mode purpose, along with active health and the environment. Many existing studies around the pandemic's global effect on mobility have largely investigated salient impacts on people and sectors loosely and in isolation, rather than collectively. It is therefore imperative to fully establish from evidence, key dimensions and changing patterns of transport and mobility since the start of the pandemic in March 2020, their implications, and lessons learned. This paper explores a thought leadership through guidance and insight, the hidden salient impacts of the pandemic in six key areas: transport sector employment, essential workers, personal vehicle ownership, telecommuting, ridesharing, and gender equality including ethnic minorities. Further research recommendations given the evolution of transport technologies in the context of future pandemics are also provided.

Keywords

COVID-19 · Travel behavior · Emerging transport technologies · Shared mobility · Gender inequality · Racial inequality · Policy · Future research

1 Introduction

The impact of COVID-19 on all aspects of passenger transportation, including active mobility has been profound. The detailed report from IEA (2020) noted big differences across all transport systems with aviation taking the biggest hit with passenger reductions by almost 75% in April 2020. The competing challenge to decarbonize had net positive effects as a consequence of stay-at-home orders imposed by several governments. On the other hand, these mandates including social distancing measures required to mitigate transmission (Thakkar et al., 2020) led to an overall reduction in global transport activity. Noticeable declines in personal vehicle travel by as much as 50% compared to the previous year (IEA, 2020) meant that trip frequency and travel distance were severely depressed.

Variations across Europe were comparable, Hungary reported a drop in daily trip frequency by half to approximately 4 million, with larger impacts faced by public transit reductions (Bucsky, 2020). Elsewhere, as seen in Fig. 1, the mobility trends for the U.K. showed steep declines from March 2020 in driving and public transit, and for the 12-month period, walking was up 24%, and driving and public transit only 4 and 6% respectively. U.K. Public transit including buses and the underground rail network

P. Rao (✉) · W. Y. Ochieng
Imperial College London, London, England
e-mail: pooja.rao16@imperial.ac.uk

W. Y. Ochieng
e-mail: w.ochieng@imperial.ac.uk

B. He et al. (eds.), *Urban Climate Change Adaptation*, Advances in Science, Technology & Innovation,
https://doi.org/10.1007/978-3-031-65088-8_11

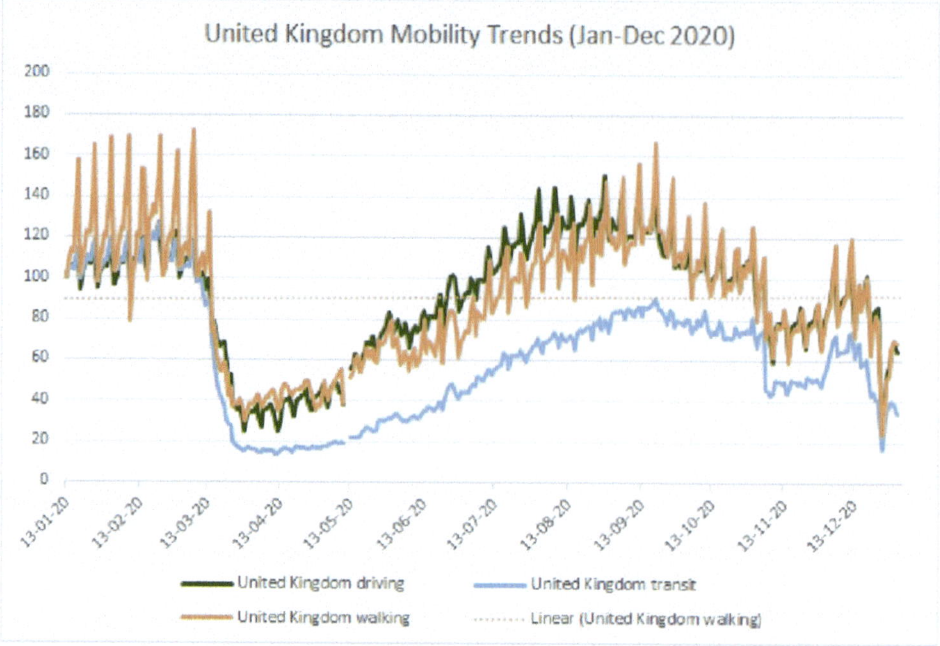

Fig. 1 U.K. Mobility Trends (author's own analysis)—*Data source* Apple COVID-19 Mobility Trends Reports (2020)

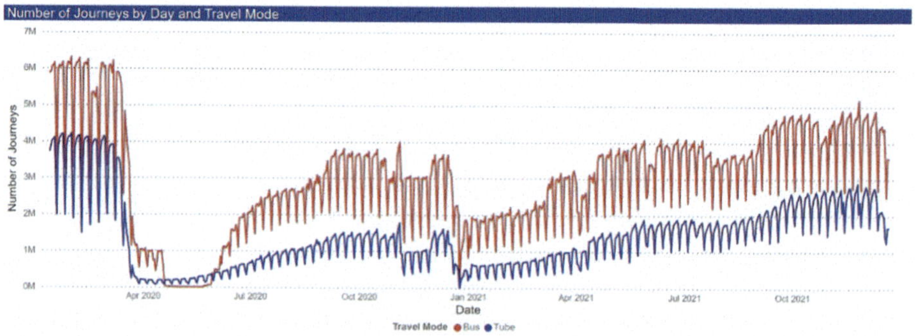

Fig. 2 London, U.K. Tube and Bus Journeys—*Source* Transport for London (2021)

remained at low levels throughout 2021, not regaining its pre-pandemic volumes as depicted in Fig. 2.

Private vehicle use in the U.S. decreased by 60% with comparable trip distances (Klein et al., 2020), and by 37% among very young socially geared adults and ethnic minorities (Stavrinos et al., 2020). But generally elsewhere, people appeared to be driving more often and recklessly with reported frequent collisions at top speeds (Katrakazas et al., 2020; Ledsom, 2020; NHTSA, 2021). This change, albeit not long-lasting has returned to pre-pandemic levels as of September 2021 with an 80% increase in personal vehicle use (Harrington & Hadjiconstantinou, 2022) with sustained transit ridership declines, as seen in London, U.K. (UK Parliament, 2021).

This is contrary to some poorer nations. For example, in India, most people rely heavily on public transit as witnessed during the pandemic, and have very few options for cheaper alternatives (Pawar et al., 2020).

The general decline in trip frequency and distance traveled meant people with better socio-economic characteristics were telecommuting by 40% (de Haas et al., 2020) and unaffected by the pandemic. In turn, these low ridership volumes directly impacted transit operators who experienced large fare-based revenue losses. They were, therefore, unable to provide reliable services during the pandemic, with below-optimum post-pandemic ridership numbers still persistent (UK Parliament, 2021).

Embodied in much of the literature covered thus far on the pandemic's impact on mobility systems, existing studies have primarily focused on public transit ridership distribution with limited details on key user groups. This is principally important for essential workers who rely on transit, whose services were largely reduced. There are limited details on the pandemic's unemployment impact levels perceived in various sectors of transport, and the types of workers that were specifically affected.

Studies have revealed insights into the reduction of vehicle use during the pandemic and fewer trips made. However, few have examined new changing travel patterns associated with booming new vehicle ownership from rising safety concerns. Increased new vehicle sales may be good for consumer spending growth, and is a predictor of greater car use, but raises questions of sustainability given the need to reduce transport emissions. It is believed that widespread electrification and future automation of vehicles will remedy this problem, but on its own, cannot be the oracle in mitigating other competing transport problems. Increased congestion may result from increased vehicle ownership and naturally into single occupancy use, creating a vicious cycle.

A key to solving these intertwined problems which is seen as another gap in the literature and not fully explored is ridesharing. Being a significant yet seldom used component of mobility, it has huge potential in reducing the number of vehicles and trips. Generally, in previous years adoption rates for ridesharing have been very low. Transportation Network Companies (TNC) garnered worldwide popularity through its convenient app-based on-demand ridehailing services, which have in turn displaced public transit services. More recently, TNCs have reported a significant drop in their ride pooling service due to passenger proximity fears from COVID-19 (Dalia Leven, 2020), which has not been fully addressed in previous pandemic-related transport papers.

TNCs such as Uber and Lyft were reported to shuttle essential workers, i.e. health workers to their workplaces as public transit services experienced disruptions. This is understood to be particularly important, and therefore covered in more detail in the relevant section.

One notable outcome of the pandemic is the shift to remote working, and much of the UK and parts of the developed world, for instance, have adopted hybrid work styles for better wellbeing. How this has impacted travel and transit operators has been only moderately reviewed. This paper aims to fulfill this gap by considering a thought leadership on the pandemic's impact on mobility, and to understand the role of gender equality and ethnic minorities in transportation during the pandemic, especially focusing on women and the obstacles they faced during this period. This paper will also consider what this means for emerging transport technologies and future research direction as a cohesive pandemic-related transport study.

This study provides a thought leadership into six hidden effects of the pandemic, specifically on transportation employment and sector-specific groups, essential workers, vehicle ownership, telecommuting, ridesharing, gender equality, and ethnic minorities, that form the basis of related work in Sect. 2. This is followed by future considerations in the context of emerging transport technologies in Sect. 3, future research direction including policy initiatives in a post COVID-19 era in Sect. 4, followed by the conclusion in Sect. 5.

2 Related Work and Thought Leadership

Several studies have disentangled the effects of the COVID-19 pandemic on the transportation sector. This paper encompasses a more focused and tightly-knit thought-leadership review of the six hidden impacts of the pandemic raised in the previous section and post-pandemic travel trends from a survey we carried out across London, UK.

2.1 Overview of Post-pandemic Travel Trends Across London

We conducted a recent motivation study using a stated preference survey across a sample population in London to examine travel behavior attributes and mode choice preferences as part of ongoing research into the likely future adoption of shared automated electric vehicles (SAEVs).

The topography of London is used in the appraisal of transport-related decisions as individual trip characteristics are dependent on the choice of mode and perceptions of alternatives given in Table 1.

Results indicate that private vehicle use is still widely dominant in the capital despite multimodal travel options available, approximately 36% of respondents use their cars daily. Walking as the only active mobility form since the pandemic remains positive at 35% contrary to cycling. This is expected since the sociodemographic characteristics of the sample base are over 55 years old with aged mobility impairments.

2.2 Transportation Employment and Specific Worker Groups

Transportation employment was one of the hardest hit sectors during the pandemic, yet little is known about it and its impacts across sub-sectors. The wave of job losses across the U.S from mid-March to mid-April 2020 was the largest in the country's history, surpassing the great recession (Şahin et al., 2020) with unemployment benefits exceeding 40 million by mid-June 2020 (Lambert, 2020). These effects were pronounced by heavy job losses across industries within the or, with many pivotal to a globally functioning market.

The divergence between essential key workers and non-essential workers was impactful. Leisure and hospitality employees, namely minorities, were more likely to

Table 1 Mode choice travel characteristics of the sample population in London

Mode used	Never	Less than once a year	Once or twice a year	Less than monthly, but more than twice a year	Once or twice a month	Less than weekly, but more than twice a month	Once or twice a week	3 or more days per week	Everyday
Private car	8.07	4.48	2.91	3.14	2.02	3.36	11.21	28.92	35.87
Bus/coach	20.18	10.99	14.35	16.14	11.21	7.17	9.19	8.97	1.79
Train	18.39	13.00	20.18	19.51	11.21	6.73	6.95	2.91	1.12
Tube	19.06	13.68	18.39	19.73	10.76	8.07	6.28	3.36	0.67
Taxi/Uber	28.92	17.26	23.32	17.49	7.17	2.24	2.02	0.67	0.45
Walking	10.54	2.24	2.02	3.81	5.16	5.83	14.57	20.85	34.98
Cycling	74.22	3.81	4.04	5.16	2.91	2.24	3.14	2.91	1.57
Motobike	87.44	0.67	0.90	0.90	2.91	1.12	0.90	2.02	1.12
E-scooter	97.53	0.00	0.45	0.00	0.67	0.00	0.22	0.45	0.67

be unemployed (Fairlie et al., 2020) than their essential worker counterparts who cannot work from home. The latter makes up around 53% of workers in the U.S. (Kearney & Pardue, 2020). Therefore, the importance of providing a robust and reliable transport network cannot be more emphasized. Even now, the recovery is not near pre-pandemic levels, being 2% higher in 2020 than the previous year (Bredemeier, 2021). This has been exacerbated by worldwide restrictions which impacted consumer spending resulting from closures of many businesses and job losses. Figure 3 provides a snapshot of impacted retail and recreational businesses (Ritchie et al., 2020) for a handful of randomly selected countries.

Lower economic activity in leisure, hospitality, retail, and other non-essential sectors, meant higher costs to businesses as a consequence of various lockdown measures. This in turn severely impacted the transportation sector globally.

In summary, choice modeling studies have shown the heterogeneous unemployment effects of the pandemic on various transport sectors, leading to global supply chain disruptions, and an e-commerce boom. The worst affected sector exposed fragilities of unemployment for essential and non-essential workers, particularly those faced by women and ethnic minorities. The disproportionate imbalance between workers from higher socio-economic groups, and those less educated lower income workers was stark. These imbalances are further widened given this sector's inability to telecommute (Dingel & Neiman, 2020), resulting in long-term unemployment when combined with a slow post-pandemic recovery.

2.3 Essential Workers

During the pandemic, when governments imposed stay-at-home orders, essential workers who were not able to work remotely have remained critical to the functioning of daily business operations. These essential workers according to Kearney and Pardue (2020) are predominantly low-income non-Caucasian males with poor educational qualifications. Through a validation study using multiple data sources,

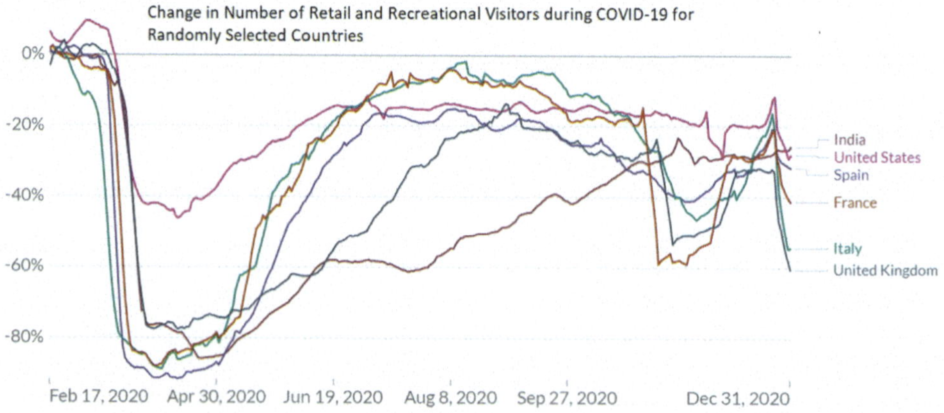

Fig. 3 Retail and recreational change since baseline Jan 3–Feb 6 (2020). *Note* some seasonal variation may exist. *Source* Google COVID-19 Community Mobility Trends/ourworldindata.org/coronavirus (2020)

Mongey et al., (2021) revealed that undocumented workers were unable to work remotely, and as their jobs required physical presence, which social distancing measures could not be satisfied, i.e. hospitality and retail (Gabe & Florida, 2021) were disproportionately affected by public transit disruptions.

Transport inaccessibility given the heterogeneous distribution of transit, has exposed system-wide inequalities and left many essential workers who rely on bus and train services unable to fulfil their roles and made redundant as a result. The lack of investment in urban transit systems in the United States, a lifeline for many Americans, only highlighted racial inequalities that led more to perish during 2020 (Planetizen, 2021). Somehow this is not new. The United States has a history of struggling with equitable access to reliable transportation, particularly in urban areas. Factors contributing to this include declining tax revenues, the flight of high-wage earners, and underinvestment in urban centers. Additionally, policymakers often prioritize highway and roadway improvements over public transit infrastructure, which disproportionately harms marginalized communities who cannot afford a vehicle and denies them access to good jobs (Urban Institute, 2020). Possible remediation would be to consider building community relationships early to help identify at-risk groups. However, it's important to consider what impact community engagement will have on those without access to transport if benefits are to be widely received.

In summary, lessons learned from transit failure support the need to deliver equitable and reliable access to mobility. Doing so would help steer policies toward better employment prospects and improve the well-being of key workers and other low-income marginalized groups alike. The combined force of strong leadership and infrastructure funding is desired when helping vulnerable communities transition to new automated transport technologies.

2.4 Vehicle Ownership

A global surge in vehicle ownership was witnessed during the pandemic and looks to continue as a trend, Fig. 4.

In the U.S. about 78% of transport is through personal vehicle use, versus approximately 58% in Europe, used primarily for work and leisure purposes (Dingil & Esztergar-Kiss, 2021). But as the pandemic unfolded, negative shifts in public transit and its causality on essential workers covered in detail previously, have led to a growing demand for personal vehicles given the anxiety of train and bus travel. A survey carried out by E&Y (2020), shows 50% of all respondents would purchase a vehicle in the first half of 2020, up 17% from the previous year.

The fear of contamination has seen increased car sales amid declining public transport amongst Millennials, who according to E&Y (2020) are expected to represent 45% of first-time owners globally within half a year. Their analysis was conducted by a global mobility survey of 3300 individuals in the latter half of 2020. To expand further, the report's survey results approximate 33% of 24–39 year old's (Millennials) will purchase their first vehicle within 6-months from the survey date, as a consequence of the pandemic. 20% of existing car owners may opt for an additional purchase. Of first-time buyers, only a meager 6% would consider an Electric Vehicle (EV) purchase. This low percentage of EV uptake is disappointing. Furthermore, younger consumers are driving new sales of ICE vehicles, and as such the value and cost benefits of owning an EV are either poorly misguided or misunderstood. The likely consequence of the upsurge in new vehicle purchases may negate the environmental benefits of increased congestion. This will no doubt depress transit use and delay urban capital investment, contributing further to equitable inaccessibility.

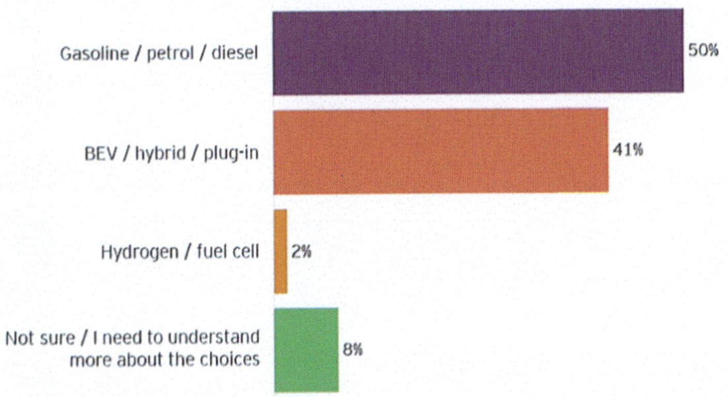

Fig. 4 New global vehicle purchases as of July 2021. *Source* E&Y (2021)

Similarly, the pandemic has shifted people's behavioral travel preferences out of fear and stigma of transit to private transport. Reversing a previously declining vehicle ownership trend, by fueling a more car-dependent society, from COVID-19 fears to oil-dollar price elasticities, would negate any positive net effects in decarbonizing our transport system by 2050.

2.5 Telecommuting

The effect of the pandemic on employment and how people work, hybrid, or fully remote has been significant, and the body of literature on telecommuting in specific sectors of the population provides interesting findings.

Before 2020, only 5% of the U.K. workforce worked remotely (Watson, 2020). By early Q2 2020 when the country was in lockdown, working patterns largely shifted to telecommuting with 47% of certain occupation groups working from home (Cameron, 2020). A classification breakdown of these groups would be later helpful in informing social equity and long-term mobility issues. Furthermore, the results from Fig. 5 highlight a large percentage of the skilled labor market able to telecommute versus unskilled groups. This has significant meaning not only for the heterogeneous effects remote working has on declining transit travel and loss of fare-based revenue but also for employee wellbeing and job satisfaction. The first half of this group equates telecommuting with flexible working practices allowing more freedom leading to better work productivity, job satisfaction, and wellbeing (Wheatley, 2016, 2020). Contrastingly, this can also lead to work-base exclusion, work-life imbalance, and poor future employment (Blundell et al., 2020; Kelliher & Anderson, 2009).

An Australian study (Hensher et al., 2021) examined the effects of remote working on travel activity patterns and found support for hybrid work styles when commuting is essential. However, a lack of homogeneity in telecommuting can lead to increased inequality and social mobility, particularly for those in precarious circumstances. The variability in equitable distribution between high-income educated remote workers and low-income often female key workers is significant, which influences trip density and travel demand as mobility patterns evolve over the next few years (Gezici & Ozay, 2020; Hensher et al., 2021).

2.6 Ridesharing

Ridesharing that facilitates travel between users of similar origin–destination pairs was negatively impacted, a further blow to preexisting declines in developed countries. The emergence of Transportation Network Companies (TNCs) such as Lyft and Uber has been rapid and of magnitude given its global popularity as an app-based Mobility-on-Demand (MoD) service. The pandemic's impact on

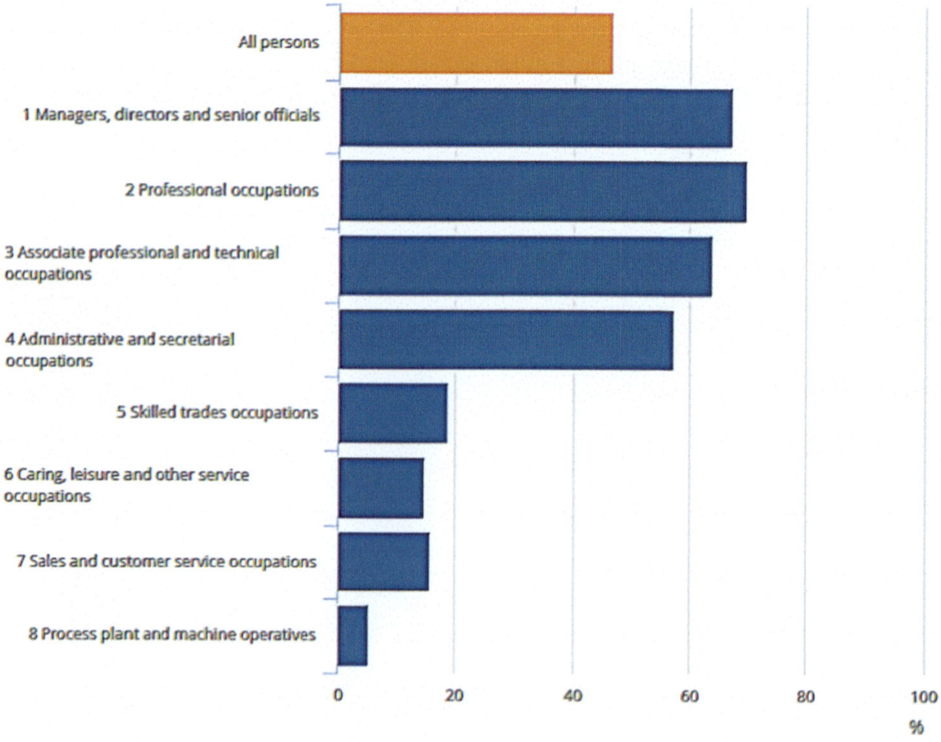

Fig. 5 Occupations of UK labor force remote working. *Source* Office for National Statistics (ONS) (2020)

Table 2 Frequency distribution of Londoners who previously experienced ridesharing e.g. Uberpool

	Frequency	Percent
Ridesharing		
Never	423	98.44
Less than once a year	7	1.57
Once or twice a year	4	0.90
Less than month, but more than twice a year	5	1.12
Once or twice a month	3	0.67
Once or twice a week	1	0.22
3 or days per week	2	0.45
Every day	1	0.22

ridesharing is therefore significant. Declining transit ridership as people opt for personal vehicles, as covered in Sect. 2.3, eschew ridesharing with strangers as a major contributing factor.

We carried out a recent survey across London as part of a larger assessment of the psychological attributes of ridesharing, and unsurprisingly from Table 2 only a small fraction have previously experienced a ridesharing service at the most a little more than twice annually, indicating more needs to be done to bring public awareness of the social and environmental benefits of shared mobility when automated vehicles are introduced.

Our survey results also include prejudice and bias, along with female safety as elements that hinder ridesharing over cost and convenience. However, unlike other studies (Sarriera et al., 2017), our results reveal cost and service time are more important deciding factors than negative social synergism depending on the time of day, despite social and behavioral judgments are still problematic generally.

During the pandemic, travel patterns from transportation network companies (TNCs) have helped people in need and non-profit agencies reach essential services and supplies. However, in a post-pandemic world with heterogeneous travel distributions, TNCs must reconsider their ridesharing business models to address inclusivity for vulnerable members of the population, including those with disabilities who cannot use public transport.

2.7 Gender Travel and Ethnic Minorities

The pandemic exacerbated gender inequality, particularly in developing countries where disparities among women were larger than in developed nations. In Sierra Leone, African women living in perilous conditions were particularly affected by the lack of proper information and misguidance on the state of the pandemic, which led to a greater risk of infection and transmission (Levine et al., 2021).

Mobility restrictions increased gender poverty as women took on primary caregiving activities, including homeschooling and other non-routine tasks which hampered their labor force participation (Madgavkar et al., 2020). This resulted in reduced work hours and potential job loss, diminishing spatial equality in paid and unpaid work, including job recovery in the UK, Spain, and India (Chung et al., 2022; Deshpande, 2021; Farre et al., 2020). The effects of the U.S. workforce within transportation and warehousing is one example of gender disparity. There has been evidence of strong job loss variation amongst worker groups within this sector. Figure 6 shows women experienced a 15% disproportionate rate of unemployment (Bluedorn et al., 2021) in July-Aug 2020 to men in the transport and warehouse sector, which peaked again higher in March 2021 at 10%.

The pandemic had regressive outcomes in Southeast Asian countries, with heterogeneous effects of mobility during lockdowns that were stark amongst women in informal

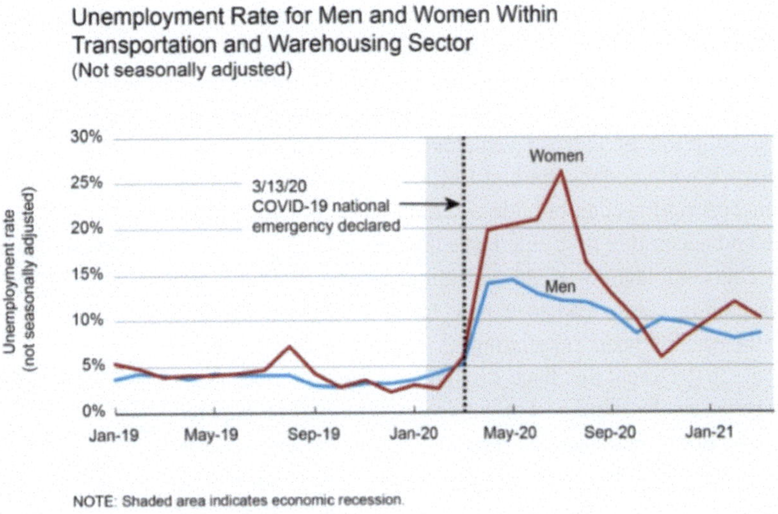

Fig. 6 Transport and warehouse sector job losses by gender (US Labor Market). *Source* Bureau of Transportation Statistics (BTS) (2021)

employment, which makes up over 85% of women in this bracket. Service disruptions and suspensions left many impoverished women out of work, but the recovery of services in India resulted in reduced passenger capacity, fare rises, and the digitization of public transit, exacerbating technology-led gender imbalances (Shah et al., 2023).

Evidence of similar variation is also revealed using mobile phone data in parts of Europe, including the U.K., where gaps in inequality amongst women and particularly younger people were affected. Hospitality venues and other non-essential retail closures imposed travel restrictions meant younger people who serve these businesses were made redundant (Blundell et al., 2020; Caselli et al., 2021). Generational divide shows younger people with unstable job contracts as opposed to senior full-time employees have insufficient savings, and thus suffer multiple financial and health-related problems.

3 Consideration of Future Transport Technologies

Attention should be drawn to the long-term effects of these six hidden impacts in the context of future transport technologies, and the recommendations considered to enable a more resilient and equitable transport framework. The pandemic has brought changes to travel and work patterns, which have benefited higher-income groups. This raises questions about the future of passenger transport and how people will adapt to changes in transport modes and decarbonization efforts. Effective policy intervention is needed to drive change in the post-recovery era. The shift to remote work has positive and negative impacts on mobility and accessibility, as well as environmental sustainability.

Better policy initiatives and training can help reduce income inequality and promote hybrid working patterns. To achieve gender parity, primary caregivers should be allowed to telecommute. Travel behavior changes resulting from the pandemic have implications for the future of mobility and transport, and dynamic ridesharing and public–private partnerships can be used to reduce congestion and emissions. The adoption of electric vehicles must be balanced with network-wide congestion and traffic flow, and multi-modal transport options can provide essential workers with equitable and accessible first-mile/last-mile access.

The combination of electrification, shared mobility, and automation could lead to a new form of microtransit that could revolutionize ridesharing. A Singapore study (Ongel et al., 2019) found that fleets of automated electric microtransit buses can be more cost-effective in the long run compared to traditional bus services. Private companies and startups are already funding such services in the automated vehicle space. The public sector needs to catch up and compete by introducing similar shared automated vehicles (SAVs) as a dynamic on-demand public transport service. Cost reduction is a key consideration in bringing new technologies to market, which typically requires competitive tension.

4 Future Research Direction

Current challenges facing mobility are wide and complex. The gradual transition to automated transport combined with traditional vehicles could have beneficial environmental impacts, but further research is needed on how new technologies blend into the existing urban ecosystem. Fleet deployment of shared automated vehicles may increase congestion, reducing the consumer experience, but dynamic ridesharing offered through TNC or public transit agencies could help relieve road congestion and increase average vehicle occupancy. The pandemic has highlighted the need for equitable transport access, especially for low-income and underrepresented groups, and public–private partnerships could help deliver better access to areas with restrictive coverage. Further research is needed to understand travelers' willingness to ride with strangers and to improve survey design. The effects of remote working on transit revenues negatively impact current and future transit services, highlighting the need for a rethink of on-demand automated mobility in an evolving urban ecosystem.

5 Conclusion

This paper discusses the impact of the COVID-19 pandemic on the transportation system, particularly on essential workers and low-income groups, and the need for effective and reliable transport. The paper suggests that emerging electric automated vehicles can either benefit or hinder equitable accessibility and hinder climate objectives. The paper also emphasizes the need for public–private partnerships and realistic policy measures to steer the three transport revolutions toward a more fair and sustainable future. Lastly, the paper discusses the inevitable consequences of automation across every sector, including the widening of societal shifts, job access, and equity.

Acknowledgements This research was supported by the Imperial College London scholarship.

References

Bluedorn, J., Hansen, N.-J., Shibata, I., Caselli, F., & Mendes Tavares, M. (2021). Gender and employment in the COVID-19 recession: Evidence on "she-cessions." *IMF Working Papers*, *2021*(095), https://doi.org/10.5089/9781513575926.001

Blundell, R., Costa Dias, M., Joyce, R., & Xu, X. (2020). Covid-19 and inequalities. *Fiscal Studies, 41*(2), 291–319. https://doi.org/10.1111/1475-5890.12232

Bredemeier, K. (2021, September 16). *US jobless benefit claims increase, but still near pandemic low*. Voice of America. https://www.voanews.com/a/us-jobless-benefit-claims-increase-but-still-near-pandemic-low/6231112.html

Bucsky, P. (2020). Modal share changes due to COVID-19: The case of budapest. *Transportation Research Interdisciplinary Perspectives, 8*, 100141. https://doi.org/10.1016/j.trip.2020.100141

Cameron, A. (2020, July 7). *Coronavirus and homeworking in the UK: April 2020*. Office for National Statistics. https://www.ons.gov.uk/employmentandlabourmarket/peopleinwork/employmentandemployeetypes/bulletins/coronavirusandhomeworkingintheuk/april2020

Caselli, F., Grigoli, F., Sandri, D., & Spilimbergo, A. (2021). Mobility under the COVID-19 pandemic: Asymmetric effects across gender and age. *IMF Economic Review, 70*(1), 105–138. https://doi.org/10.1057/s41308-021-00149-1

Chung, H., Seo, H., Birkett, H., & Forbes, S. (2022). Working from home and the Division of Childcare and housework among dual-earner parents during the pandemic in the UK. *Merits, 2*(4), 270–292. https://doi.org/10.3390/merits2040019

de Haas, M., Faber, R., & Hamersma, M. (2020). How covid-19 and the Dutch 'Intelligent Lockdown' change activities, work and travel behaviour: Evidence from longitudinal data in the Netherlands. *Transportation Research Interdisciplinary Perspectives, 6*, 100150. https://doi.org/10.1016/j.trip.2020.100150

Deshpande, A. (2021). The COVID-19 pandemic and Gendered Division of paid work, domestic chores and leisure: Evidence from India's first wave. *Economia Politica, 39*(1), 75–100. https://doi.org/10.1007/s40888-021-00235-7

Dingel, J. I., & Neiman, B. (2020). How many jobs can be done at home? *Journal of Public Economics, 189*, 104235. https://doi.org/10.1016/j.jpubeco.2020.104235

Dingil, A. E., & Esztergár-Kiss, D. (2021). The influence of the COVID-19 pandemic on mobility patterns: The first wave's results. *Transportation Letters, 13*(5–6), 434–446. https://doi.org/10.1080/19427867.2021.1901011

E&Y. (2020, November 12). *Millennials to lead covid-induced car ownership boom–EY survey*. https://www.ey.com/en_gl/news/2020/11/millennials-to-lead-covid-induced-car-ownership-boom-ey-surve

E&Y. (2021, July 20). *How did a global crisis pave the way for EV sales?* https://www.ey.com/en_tr/automotive-transportation/how-did-a-global-crisis-pave-the-way-for-ev-sales

Fairlie, R., Couch, K., & Xu, H. (2020). *The impacts of Covid-19 on minority unemployment: First evidence from April 2020 CPS microdata*. https://doi.org/10.3386/w27246

Farre, L., Fawaz, Y., Gonzalez, L., & Graves, J. (2020). How the COVID-19 lockdown affected gender inequality in paid and unpaid work in Spain. *SSRN Electronic Journal*. https://doi.org/10.2139/ssrn.3643198

Gabe, T., & Florida, R. (2021). Impacts of jobs requiring close physical proximity and high interaction with the public on U.S. industry employment change during the early stages of the COVID-19 pandemic. *The B.E. Journal of Economic Analysis & Policy, 21*(3), 1163–1172. https://doi.org/10.1515/bejeap-2021-0052

Gezici, A., & Ozay, O. (2020). How race and gender shape covid-19 unemployment probability. *SSRN Electronic Journal*. https://doi.org/10.2139/ssrn.3675022

Harrington, D. M., & Hadjiconstantinou, M. (2022). Changes in commuting behaviours in response to the COVID-19 pandemic in the UK. *Journal of Transport & Health, 24*, 101313. https://doi.org/10.1016/j.jth.2021.101313

Hensher, D. A., Beck, M. J., & Wei, E. (2021). Working from home and its implications for strategic transport modelling based on the early days of the COVID-19 pandemic. *Transportation Research Part a: Policy and Practice, 148*, 64–78. https://doi.org/10.1016/j.tra.2021.03.027

IEA. (2020). *Changes in transport behaviour during the COVID-19 crisis*. IEA. https://www.iea.org/articles/changes-in-transport-behaviour-during-the-covid-19-crisis

Katrakazas, C., Michelaraki, E., Sekadakis, M., & Yannis, G. (2020). A descriptive analysis of the effect of the COVID-19 pandemic on driving behavior and Road Safety. *Transportation Research Interdisciplinary Perspectives, 7*, 100186. https://doi.org/10.1016/j.trip.2020.100186

Kearney, M. S., & Pardue, L. (2020). *Exposure on the job: Who are the essential workers who likely cannot work from home?*. Pesquisa. https://pesquisa.bvsalud.org/global-literature-on-novel-coronavirus-2019-ncov/resource/pt/grc-740397

Kelliher, C., & Anderson, D. (2009). Doing more with less? flexible working practices and the intensification of work. *Human Relations, 63*(1), 83–106. https://doi.org/10.1177/0018726709349199

Klein, B., LaRock, T., McCabe, S., Torres, L., Friedland, L., Privitera, F. (2020). Reshaping a nation: Mobility, commuting, and contact patterns during the covid-19 outbreak. *Northeastern University-Network Science Institute Report*.

Lambert, L. (2020, June 11). *Over 44.2 million Americans have filed for unemployment during the coronavirus pandemic*. Fortune. https://fortune.com/2020/06/11/us-unemployment-rate-numbers-claims-this-week-total-job-losses-june-11-2020-benefits-claims/

Ledsom, A. (2020, October 14). *Riskier behaviour during pandemic is causing more accidents*. Forbes. https://www.forbes.com/sites/alexledsom/2020/10/14/emptier-roads-but-traffic-fatalities-soaring-why/

Leven, D. (2020). *Disrupting the disruptors: Part 4: Covid-19 impacts on ride-sharing/cambridge systematics*. Disrupting the Disruptors: Part 4 | COVID-19 Impacts on Ride-sharing/Cambridge Systematics. https://camsys.com/blog/covid-19-impacts-on-ride-sharing

Levine, M., Meriggi, N., Mushfiq Mobarak, A., Ramakrishna, V., & Voors, M. (2021, March 8). *How covid-19 is making gender inequality worse in low-income countries-and what to do about it*. https://insights.som.yale.edu/insights/how-covid-19-is-making-gender-inequality-worse-in-low-income-countries-and-what-to-do

Madgavkar, A., White, O., Krishnan, M., Mahajan, D., & Azcue, X. (2020). *Covid-19 and gender equality: Countering the regressive effects*. McKinsey Global Institute.

Mongey, S., Pilossoph, L., & Weinberg, A. (2021). Which workers bear the burden of social distancing? *The Journal of Economic Inequality, 19*(3), 509–526. https://doi.org/10.1007/s10888-021-09487-6

NHTSA.(2021, June 3) *2020 fatality data show increased traffic fatalities during pandemic*. https://www.nhtsa.gov/press-releases/2020-fatality-data-show-increased-traffic-fatalities-during-pandemic

Ongel, A., Loewer, E., Roemer, F., Sethuraman, G., Chang, F., & Lienkamp, M. (2019). Economic assessment of autonomous electric microtransit vehicles. *Sustainability, 11*(3), 648. https://doi.org/10.3390/su11030648

Pawar, D. S., Yadav, A. K., Akolekar, N., & Velaga, N. R. (2020). Impact of physical distancing due to novel coronavirus (SARS-COV-2) on daily travel for work during transition to lockdown. *Transportation Research Interdisciplinary Perspectives, 7*, 100203. https://doi.org/10.1016/j.trip.2020.100203

Planetizen. (2021). *What Covid taught about mobility justice*. https://www.planetizen.com/news/2021/10/115093-

Ritchie, H., Mathieu, E., Rod´es-Guirao, L., Appel, C., Giattino, C., Ortiz-Ospina, E., & Roser, M. (2020, March 5). *Coronavirus pandemic (COVID-19)*. Our World in Data. https://ourworldindata.org/coronavirus

Şahin, A., Tasci, M., & Yan, J. (2020). The unemployment cost of COVID-19: How high and how long? *Economic Commentary*

(Federal Reserve Bank of Cleveland), 1–7. https://doi.org/10.26509/frbc-ec-202009

Sarriera, J. M., Álvarez, G. E., Blynn, K., Alesbury, A., Scully, T., & Zhao, J. (2017). To share or not to share. *Transportation Research Record: Journal of the Transportation Research Board, 2605*(1), 109–117. https://doi.org/10.3141/2605-11

Shah, S., Rajiv, R. M., & Lokre, A. (2023, April). *Moving toward gender-equitable. Transportation in post-covid-19 urban South Asia*. Transportation research record. https://www.ncbi.nlm.nih.gov/pmc/articles/PMC10152222/

Stavrinos, D., McManus, B., Mrug, S., He, H., Gresham, B., Albright, M. G., Svancara, A. M., Whittington, C., Underhill, A., & White, D. M. (2020). Adolescent driving behavior before and during restrictions related to covid-19. *Accident Analysis & Prevention, 144*, 105686. https://doi.org/10.1016/j.aap.2020.105686

Thakkar, N., Burstein, R., Hu, H., Selvaraj, P., & Klein, D. (2020). Social distancing and mobility reductions have reduced covid-19 transmission in king county, wa. *Institute for Disease Modeling*.

Transport for London. (2021). *Every journey matters*. Coronavirus Publications. https://tfl.gov.uk/corporate/publications-and-reports/coronavirus-publications

UK Parliament. (2021). *Decarbonisation: The impact of coronavirus on land transportation*. https://commonslibrary.parliament.uk/research-briefings/cbp-9266/

Urban Institute. (2020, October 6). *The unequal commute*. https://www.urban.org/features/unequal-commute

Watson, B. (2020, March 24). *Coronavirus and homeworking in the uk labour market: 2019. UK Office of National Statistics*. https://www.ons.gov.uk/employmentandlabourmarket/peopleinwork/employmentandemployeetypes/articles/coronavirusandhomeworkingintheuklabourmarket/2019

Wheatley, D. (2016). Employee satisfaction and use of flexible working arrangements. *Work, Employment and Society, 31*(4), 567–585. https://doi.org/10.1177/0950017016631447

Wheatley, D. (2020). Workplace location and the quality of work: The case of urban-based workers in the UK. *Urban Studies, 58*(11), 2233–2257. https://doi.org/10.1177/0042098020911887

An Evaluation of Sustainable Development Goal Seven (Affordable and Clean Energy) in Ghana

Simon Ofori Ametepey, Clinton Aigbavboa, Wellington Didibhuku Thwala, and Hutton Addy

Abstract

The Sustainable Development Goals (SDGs), which were unanimously approved by all member nations of the United Nations, are expected to be implemented by all. One of the critical goals developed was SDG 7, the all-inclusive access to affordable and clean energy for all by 2030. This study critically evaluates the progress of SDG 7 implementation in Ghana. This research paper heavily relies on document analysis and secondary data from sources such as gathering information from credible and up-to-date reports such as the 2021 Voluntary National Review (VNR) and the Medium-Term Expenditure Framework (MTEF) reports etc. Through the combination of policies, projects, and interventions, Ghana has made commendable progress toward realizing SDG 7. However, it is essential to highlight that only one of the five targets associated with SDG 7 has made significant strides. Distinctly, Target 7.1, which seeks to ensure all-inclusive access to affordable, reliable, and modern energy services by 2030, has achieved an 86.3% attainment nationwide. This rcscarch discovered that though substantial advancements have been achieved in the power generation and distribution sector, the same cannot be said for establishing and integrating Renewable Energy sources (REs). The country continues to face significant challenges in adopting Renewable Energy Technologies (RETs) and systems, usually stemming from high purchase and installation costs. To attain all the targets outlined in SDG 7 by 2030, stakeholders in the energy sector, particularly the government, must double up efforts in soliciting investment and allocating adequate resources towards the sector.

Keywords

Affordable energy · Clean energy · Electricity · Ghana · Renewable energy · SDG 7

1 Introduction

Universal accessibility to cost-effective, dependable, and sustainable energy is the mandate of Sustainable Goal 7 (Kuamoah, 2020). Energy has been a fundamental element supporting life since ancient times, with the sun serving as the earliest source of heat and light (National Grid Group, 2023). Sustainable energy is pivotal in improving the quality, accessibility, and reliability of essential services required for survival, development, and well-being in households, health centers, and schools (UNICEF, 2022.). Energy sources are generated from the environment for renewables and non-renewables, such as hydropower and uranium. Over the past two decades, a remarkable increase of 1.7 billion people worldwide have gained access to electricity (The World Bank, 2017). However, as the global population grows, the demand for affordable energy escalates, necessitating a shift towards green energy sources such as wind and solar (UNEP, n.d.). The reliance on fossil fuels and the consequent rise in greenhouse gas emissions

Simon Ofori Ametepey, Clinton Aigbavboa, Wellington Didibhuku Thwala, Hutton Addy are contributed equally to this study

S. O. Ametepey (✉) · H. Addy
Centre for Sustainable Development (CenSuD), Koforidua Technical University, Koforidua, Ghana
e-mail: simon.ametepey@ktu.edu.gh

H. Addy
e-mail: hutad@outlook.com

S. O. Ametepey · C. Aigbavboa
Department of Construction Management and Quantity Surveying, Faculty of Engineering and Built Environment, University of Johannesburg, Johannesburg, South Africa

S. O. Ametepey · W. D. Thwala
Department of Civil Engineering, College of Science Engineering and Technology, University of South Africa, Pretoria, South Africa

B. He et al. (eds.), *Urban Climate Change Adaptation*, Advances in Science, Technology & Innovation,
https://doi.org/10.1007/978-3-031-65088-8_12

have triggered substantial climate changes, requiring investments in renewable energy technologies (RET) and adopting cost-effective standards (UNEP, n.d.). Such measures could potentially reduce worldwide electricity consumption by 14% for buildings and industries. Modernizing technology and extending infrastructure to supply green energy sources in developing countries is a vital objective that promotes economic growth and environmental sustainability. Unfortunately, recent data from the International Energy Agency (IEA) suggests an alarming trend, projecting a rise of nearly 20 million people without electricity in 2022, particularly in sub-Saharan Africa (IEA, 2022). According to IEA (2022), access to electricity stands at 86.3% as of 2021, with variations between urban and rural areas in Ghana.

Nonetheless, the government is working to boost Ghana's economic growth and is giving priority to increasing energy output, facilitating access to conventional power sources, and introducing sustainable energy sources. With renewable energy projected to produce 152.34 GWh of energy in Ghana in 2021, the nation's dependency on imported methane for electricity generation could be reduced (IEA, 2022). The absence of transition mechanisms and energy poverty continue to impede socioeconomic development. The capacity to recruit and retain instructors, as well as academic achievement and attendance rates, are all correlated with access to power. Furthermore, using renewable energy sources greatly improves children's health by lowering indoor air pollution (UNEP, n.d.). Ghana's quest to meet its energy demands cannot be well achieved without considering the affordable and clean energy goal of the sustainable development goals (SDG 7) with its associated targets. The government of Ghana has consistently reiterated its commitment to achieving this goal. As a result, this paper sought to assess Ghana's progress in achieving SDG 7 and government, non-governmental organizations, and other stakeholder contributions towards the goal. By conducting a comprehensive analysis of Ghana's progress and ongoing efforts, the study made valuable discussions, provided recommendations, and drew meaningful conclusions regarding the improvements of contemporary standards and commitments by stakeholders to achieve the objectives outlined in SDG 7.

2 SDG 7-Affordable and Clean Energy for All

Over the past ten years, there has been a notable increase in the percentage of people globally who have access to electricity and clean cooking fuels (Whitby et al., 2020). According to Whitby et al. (2020), the percentage of people using electricity worldwide increased from 82 percent in 2008 to almost 90% in 2018. Additionally, 6.3% of people used contemporary cooking appliances and clean fuels, up from 55% in the preceding ten years. But it's important to recognize that a large number of people still do not have access to power, with Sub-Saharan Africa suffering disproportionately from this issue. In the West African region, the access rate increased from 34% in 2010 to 47% in 2018 (Whitby et al., 2020). The countries with the most prominent electricity access deficits, 14 out of 20 in Sub-Saharan Africa, constituted 78% of individuals worldwide without electricity availability in 2018 (Whitby et al., 2020). Innovative strategies are needed to optimize limited public resources and mobilize investments from private businesses to address the challenge of electricity access in Africa. The Economic Commission for Africa (ECA) introduced the "SDG 7 Initiative for Africa" as a response to the inadequacy of electricity access (UNECA, 2020). This initiative centers around three key pillars: finance, governance, and sustainability. Its goal is to bring together governments, donors, and developers of clean energy projects to expedite private sector funding for the implementation of clean energy solutions in Africa. The program is essential in helping nations close the gaps in their energy access, satisfy the rising energy demand, and take action against climate change in line with the updated Nationally Determined Contributions (NDCs) of the Paris Agreement (UNECA, 2020). The United Nations General Assembly requested high-level talks in 2021 in recognition of the significance of energy-focused goals in the 2030 Agenda for Sustainable Development and the United Nations Decade of Sustainable Energy for All (2014–2024) (UN, 2021). This dialogue marked a significant milestone as the inaugural international assembly on energy under the General Assembly since the event presented a unique opportunity to advance the implementation of the Paris Agreement and drive transformative action in the initial years of the SDG Decade of Action. As a result of the High-level Dialogue on Energy, a worldwide action plan was established to attain universal energy accessibility and facilitate the energy transition (UNDESA & DSDG, 2021). The dialogue also generated over $400 billion in new commitments to support these goals (United Nations, 2021).

Ghana has achieved significant progress in its electricity sector, positioning itself as a leader in Africa with a reliable power supply and many grid-connected households (Kukah et al., 2022; Murray, 2015). The country's commitment to national electrification has increased its power transmission and distribution system over the years (Kukah et al., 2022; Singh et al., 2015; Tengey et al., 2022). The evolution of Ghana's electricity generation can be traced through various stages, starting with the use of diesel generators by industries, mines, and factories as initial power sources. The establishment of the Aksombo Dam marked the beginning of the hydroelectric phase (Kumi, 2017). The generation mix was later diversified to include thermal power plants utilizing crude oil or gas and renewable energy sources such as solar and mini-hydro (Afful-Dadzie et al., 2022; Energy Commission, 2020). Ghana also had plans to incorporate coal and nuclear power into its energy mix (Nyasapoh et al., 2022). According to Power Africa, Ghana has achieved an 84% nationwide access rate to power connections, with 93% in urban areas and 73% in rural areas (2022). The country has also made a slight effort in renewable energy generation, with an anticipated output of 152.34 GWh from renewable sources in 2021 (Tulashie et al., 2022). However, thermal and hydroelectric generation remain the primary energy sources for Ghana, with the country also exporting electricity to neighboring countries such as Burkina Faso, Benin, and Togo. Ghana is making efforts to develop its grid line further, including installing transmission lines and Bulk Supply Points (BSPs), to expand its power exports in the sub-region. The hydroelectric portion of Ghana's generation phase is managed by the Volta River Authority (VRA) and Bui Power Authority (BPA), while the transmission system is overseen by the state-owned enterprise, Ghana Grid Company (GRIDCO) (IEA, 2022; Kumi, 2017). The Electricity Company of Ghana (ECG) and the Northern Electricity Distribution Company (NEDCO) are responsible for the final segment of electricity distribution (IEA, 2022). Unfortunately, Ghana still faces challenges in achieving its industrialization and agricultural modernization goals due to financial deficits and an unreliable and costly power supply (Power Africa, 2022). Despite an installed generation capacity of approximately 5300 MW, the access rarely exceeds 2400 MW due to fluctuating hydrological conditions, insufficient fuel sources, and ageing infrastructure. Nevertheless, the country is well-positioned to overcome these limitations with its abundant natural gas reserves and renewable energy resources (Power Africa, 2022).

Ghana has encountered many obstacles in achieving the aim of affordable and clean energy. Ghana has long struggled with problems related to the electricity supply (Kukah et al., 2022). These problems include low voltage, blackouts, and instability, and they provide real problems for customers who are industrial and residential (Ackah, 2021). The nation has encountered irregular power supply cycles as a result of industrialization, urbanization, and population expansion. Ghana is highly dependent on fossil fuels for electricity generation, even with its significant potential for renewable energy. As a result, it is vulnerable to changes in global fuel costs as well as economic issues such as exchange rates (Agyekum, 2020). The industry suffers from several critical flaws, such as the laborious and governmental permit application procedure and the high upfront expenses of renewable energy projects. Inadequate studies and innovation pose the most significant peril to the sector's sustainability, subsequently an imbalanced competitive landscape and the impact of shifting climatic conditions (Agyekum, 2020). Ghana encounters challenges in generating sufficient electricity resulting from diminished water reserves in hydropower plants like Akosombo and issues with gas delivery for gas-fired power plants (Nyasapoh et al., 2022). Despite multiple attempts to establish sustainable energy generation, various obstacles have impeded progress (Cobbold et al., 2019; Kumi, 2017; Nyasapoh, 2018). Ghana's slow pace of renewable energy penetration can be attributed to unfavorable market conditions and insufficient dedication to legislative and regulatory frameworks (Afful-Dadzie et al., 2022). Inadequate infrastructural investment, the absence of energy security diversification, high distribution losses, and consumer revenue non-payments further compound the challenges the electricity sector faces (Debrah et al., 2020; Cobbold et al., 2019). These difficulties have led to recurring electricity generation and supply obstacles, negatively impacting Ghana's economic growth and quality of life (Ackah, 2015; Clerici et al., 2016; Merem et al., 2017; Murray, 2015). Addressing these issues is essential for Ghana's achievement of SDG 7. Improving the regulatory environment, investing in research and development, and diversifying the energy mix are crucial steps toward achieving a more reliable and resilient power supply system. Additionally, enhancing infrastructure, reducing distribution losses, and addressing consumer revenue non-payments are vital to improving Ghana's energy sector's efficiency and financial stability.

3 Research Methodology

The study aimed to comprehensively evaluate the progress of the Sustainable Development Goal (SDG) seven implementation in Ghana. A research design based on document analysis was employed to achieve this objective, incorporating data from various secondary sources. This approach ensured the cohesiveness, reliability, and depth of the research. Document analysis is a qualitative research method to interpret and extract meaningful insights from electronic and physical documents. Through a meticulous examination of published international and national documents, the study sought to gain a comprehensive understanding of the advancements made towards SDG 7. In evaluating the progress of Sustainable Development Goal (SDG) seven in Ghana, secondary data played a pivotal role. The research design hinged on sourcing and analyzing preexisting data from various authoritative entities. The targets set by the United Nations for SDG 7 served as the benchmark for measuring progress, thus enabling an evaluation of Ghana's performance in implementing this goal. To ensure a robust analysis, data collected over 15 years is utilized.

However, particular emphasis is placed on data collected within the past five years as it provides crucial insights into the recent progress of SDG 7 implementation in Ghana. To evaluate the effectiveness of each target in achieving SDG 7 by 2030, the research gathered secondary data from reputable sources across multiple domains. These sources include national and international reports from entities like United Nations agencies, the World Bank, the Energy Commission, the World Bank, the Ghana Statistical Service, private sector sources, and other credible platforms. The meticulously collected data was evaluated for relevance and reliability, and then systematically analyzed using descriptive statistics to assess Ghana's strides toward SDG 7. Secondary data was also instrumental in examining the Ghanaian government's financial and strategic contributions to the goal, drawing from national documents such as the 2021–2022 medium-term expenditure framework (MTEF), the 2022 Voluntary National Report (VNR), the 2021 Population and housing census report, Energy Commission annual report, and others. By featuring these sources, the research established a comprehensive overview of the government's commitment to fulfilling the targets of Goal 7. However, Ghana has challenges in measuring the progress of the SDGs due to data limitations in specific sectors. Data scarcity can be attributed to factors such as limited facilities, inadequate funding, and the lack of expertise and resources within the relevant government institutions responsible for periodic data collection. To guide the analysis, the study adopted the targets and metrics of SDG 7, discussing each target's progress made in Ghana as well as the government and relevant stakeholder contributions to Goal 7. This comprehensive framework serves as the basis for evaluating Ghana's progress, facilitating a systematic assessment of the level of achievement for each target as outlined in Table 1.

4 Results

Some credible international and national reports provided valuable insights into Ghana's recent advancements toward achieving Sustainable Development Goal 7 (SDG 7). While the government's national report did not report the progress of some indicators due to the non-availability of data, this study relied on other credible sources to estimate Ghana's progress on all the metrics of SDG 7. The ensuing sections present a thorough assessment of the indicators to provide a holistic understanding of Ghana's progress in implementing SDG 7.

4.1 Indicator 7.1.1: Proportion of Population with Access to Electricity

In 2021, Ghana made considerable progress in expanding access to electricity, with 86.3% of the population expected to be connected to the national grid (National Development Planning Commission, 2022a, 2022b). The urban population witnessed a notable increase in electricity usage, with approximately 95.2% relying on electricity, compared to 84.6% in 2010 and 74.66% in 2000. In rural settlements, the percentage of people using electricity has more than quadrupled, rising from 16.1% in 2000 to 72.6% in 2021 (Fig. 1). The Greater Accra region has the lowest coverage at 57% (Fig. 2). Generally, the country's northern regions face more significant challenges in electricity access due to low population densities and long distances from major transmission lines. To address this, the Rural Electrification Project (REP) and the Self-Help Electrification Program (SHEP) have played a crucial role in connecting new communities to the national grid, with 1111 communities connected between 2018 and 2020 and an additional 843 communities connected between 2020 and 2021.

4.2 Indicator 7.1.2: Proportion of Population with Primary Reliance on Clean Fuels and Technology

According to the National Development Planning Commission (2022a, 2022b), there has been a notable increase in the percentage of households using gas (LPG) as their primary cooking fuel. From 2010 to 2021, this percentage rose from 18.2 to 36.9%. However, there is a significant disparity between rural and urban areas, with rural settlements having a much lower percentage of 14.8% compared to urban areas with 51.3%. The Greater Accra region has the highest percentage of LPG users at 68.2%, while the Northeast Region has the lowest at 4.2%. On the other hand, the proportion of households using electricity as their primary cooking fuel experienced a slight decline from 0.5% in 2010 to 0.4% in 2021 (Fig. 3). Despite this decline, wood and charcoal remain the primary cooking fuel for over half of households accounting for 54.3% of usage. The government has taken steps such as establishing the Ghana Cylinder Manufacturing Company, the Rural LPG Promotion (RLP) program, and others to promote the use of LPG and reduce reliance on fuelwood.

Table 1 SDG 7 and its targets and indicators identified in Ghana

Targets	Metrics	Data collection status
7.1 Universal access to modern energy: ensuring universal access to affordable, reliable, and modern energy services by 2030	Proportion of the population with access to electricity	Data was collected from UN Agencies such as IRENA VNR report, MFET energy, population and housing census, SDG tracker, energy outlook, and the energy commission annual reports
	Proportion of population with primary reliance on clean fuels and technology	Data was collected from the VNR report, MFET energy, population and housing census
7.2 Increase global percentage of renewable energy: increasing substantially the share of renewable energy in the global energy mix by 2030	Renewable energy shares in the total final energy consumption	Data was collected from UN Agencies such as IRENA, VNR report, energy outlook, and the energy commission annual reports
7.3 Double the improvement in energy efficiency: double the global rate of improvement in energy efficiency by 2030	Energy intensity measured in terms of primary energy and Gross Domestic Product (GDP)	Data was collected from the United Nation's sustainable energy for all initiative and the Ghana energy database system
7.a Promote access to research technology and investments in clean energy: enhance international cooperation to facilitate access to clean energy research and technology, including renewable energy, energy efficiency and advanced and cleaner fossil-fuel technology, and promote investment in energy infrastructure and clean energy technology by 2030	International financial flows to developing countries in support of clean energy research and development and renewable energy production, including in hybrid systems	Data was collected from UN Agencies such as IRENA, VNR report, MFET energy and the energy commission annual reports
7.b Expand and upgrade energy services for developing countries: expanding infrastructure and upgrade technology for supplying modern and sustainable energy services for all in developing countries, least developed countries, small island developing states and landlocked developing countries, in accordance with their respective programs of support by 2030	Installed renewable energy-generating capacity in developing countries (in watts per capita)	Data was collected from UN Agencies such as IRENA, VNR report, energy outlook, and the energy commission annual reports

Source Author, 2023

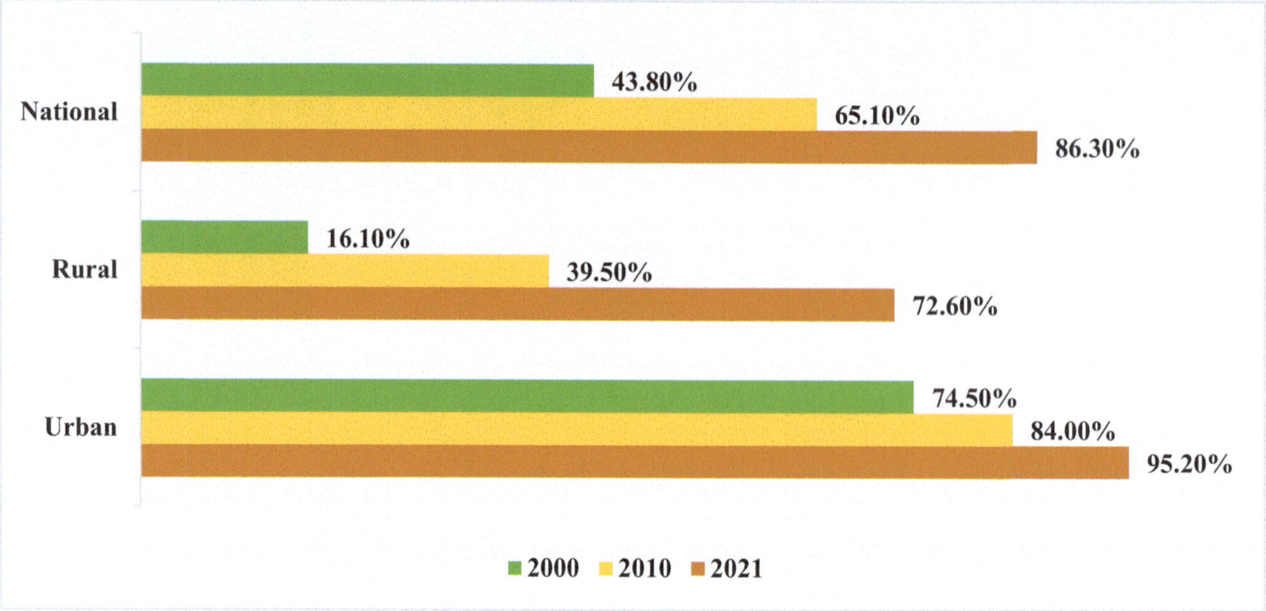

Fig. 1 Proportion of population with access to electricity, 2000, 2010, and 2021 (%). *Source* Author, 2023

Fig. 2 Population with access to electricity by region, 2021 (%). *Source* 2021 Population and Housing Census, 2022

4.3 Indicator 7.2.1: Renewable Energy Share in the Total Final Energy Consumption

Despite the various initiatives aimed at increasing the use of renewable energy in electricity generation, there has been a concerning trend. The share of renewable energy in Ghana's total final energy consumption has decreased over the years. In 2017, the percentage stood at 46.9%, but it declined to 42.7% in 2019 and further dropped to 39.6% in 2021 (National Development Planning Commission, 2022a, 2022b). This decline is depicted in Fig. 4, highlighting the need for further attention and efforts to reverse this trend and promote the use of renewable energy sources in Ghana's energy consumption. The outcome revealed by this indicator points to suboptimal performance, underscoring the need for heightened focus and action. This intensified effort is crucial to achieve all objectives of Sustainable Development Goal 7 (SDG 7).

4.4 Indicator 7.3.1: Energy Intensity Measured in Terms of Primary Energy and GDP

According to Sustainable Energy for All organization, Ghana's 5-year Energy Intensity rate improved between 2013 to 2018 by 2% and had a 5-year average GDP growth from 2013 to 2018 of 5% (SEforALL, 2022). This result is showcased in Fig. 5. The total energy supply in 2018 was 9872 kilo Tonne of oil Equivalent (ktoe). According to the

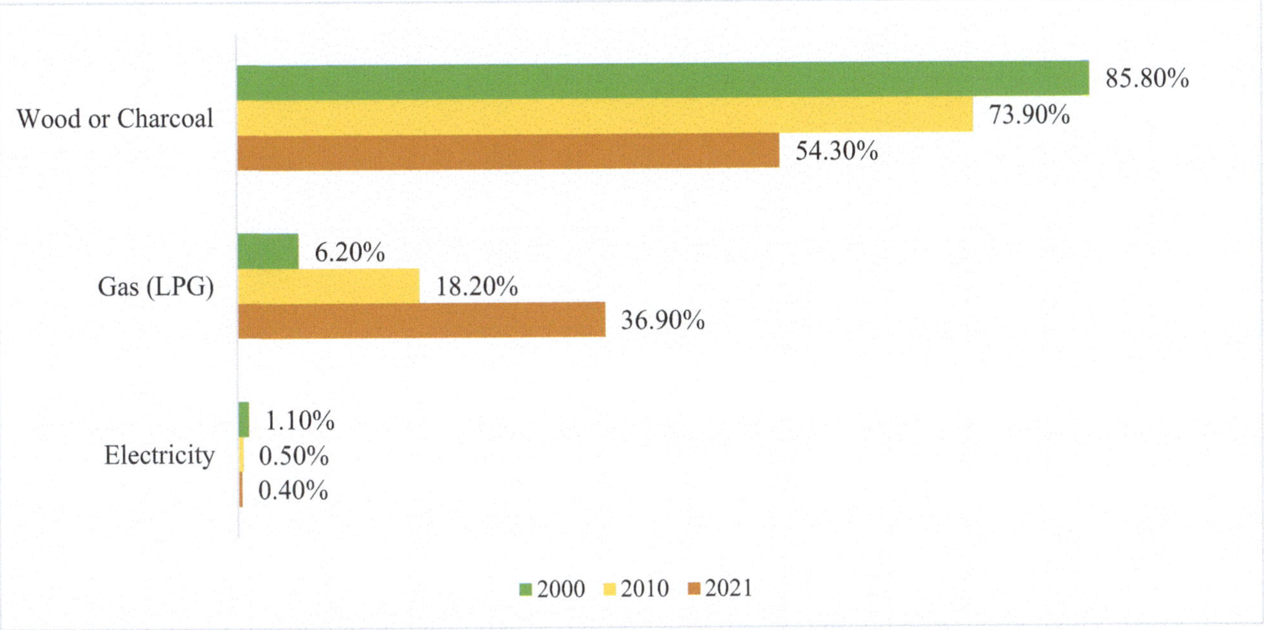

Fig. 3 Percentage of population with primary reliance on clean fuels and technology, 2000, 2010, and 2021 (%). *Source* National Development Planning Commission (2022a, 2022b)

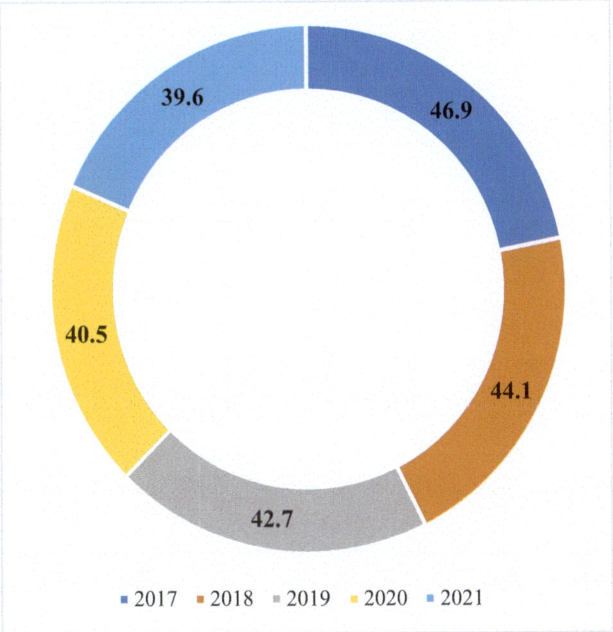

Fig. 4 Renewable energy share in the total final energy consumption, 2017–2021 (%). *Source* National Development Planning Commission (2022a, 2022b)

Energy Commission (2020), since 2020, there has been a gradual fall in energy intensity in terms of primary energy and GDP. In 2020 had about 71.22 tonnes of oil equivalent per million United States dollars (TOE/million USD), with 2021 and 2022 falling to 67.23 and 66.43 TOE/million USD (Energy Commission, 2022a). This visualized result is shown in Fig. 6, taken from the Ghana Energy Database System.

4.5 Indicator 7.a.1: International Financial Flows to Developing Countries in Support of Clean Energy Research and Development and Renewable Energy Production, Including Hybrid Systems

According to the IRENA (2020), the public flows to renewables, in conjunction with international aid from other countries, have contributed to the integration and development of RE production and hybrid systems. However, as shown in Fig. 7 below, about five hundred million dollars of funding was given to the country in 2016 as starting capital to implement RE development and integrate RE systems into the grid.

4.6 Indicator 7.b.1: Installed Renewable Energy-Generating Capacity in Developing Countries (in Watts Per Capita)

Continual cost declines and concerted efforts to diversify the energy mix contribute to a gradual growth in the proportion of renewable energy-generating capacity. Approximately 34.1%

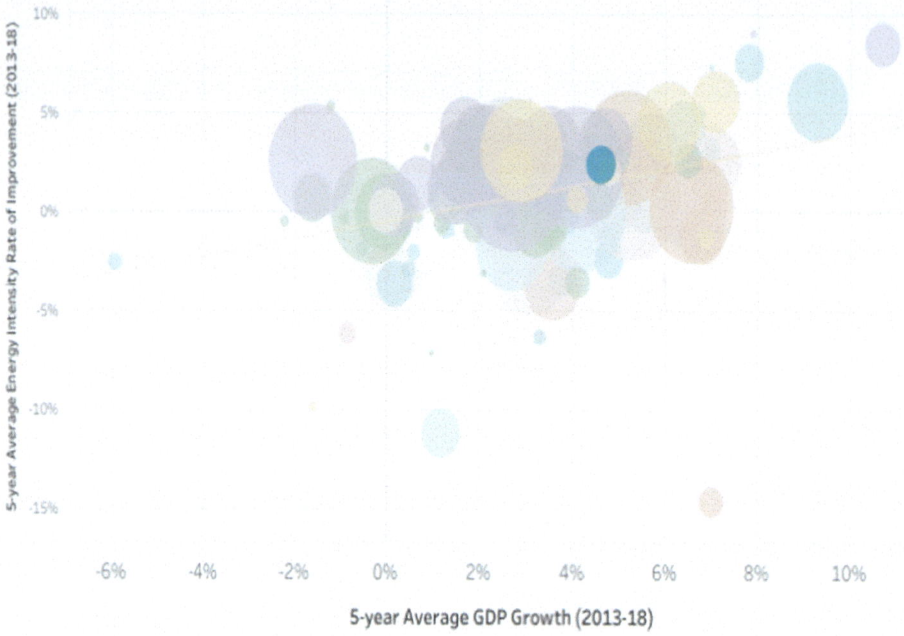

Fig. 5 Relationship between economic growth (GDP) and the rate of improvement in energy efficiency (using a 5-year average from 2013 to 2018). *Source* Sustainable energy for all

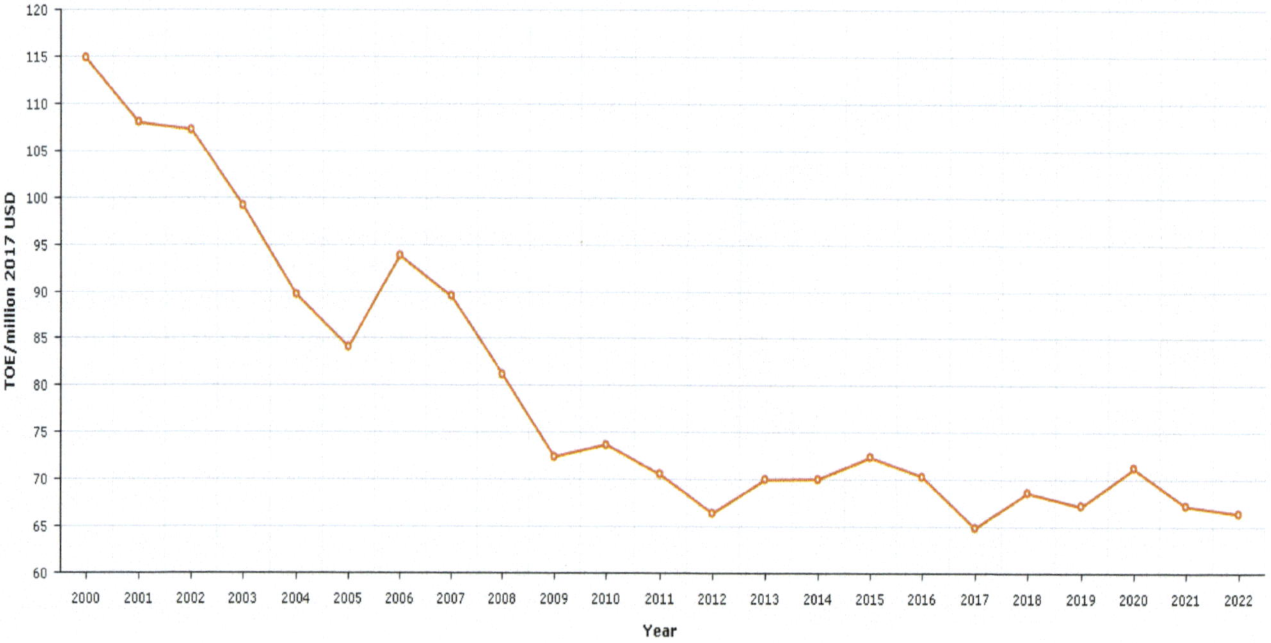

Fig. 6 Energy Intensity measured in terms of primary energy supply. *Source* Ghana energy database system

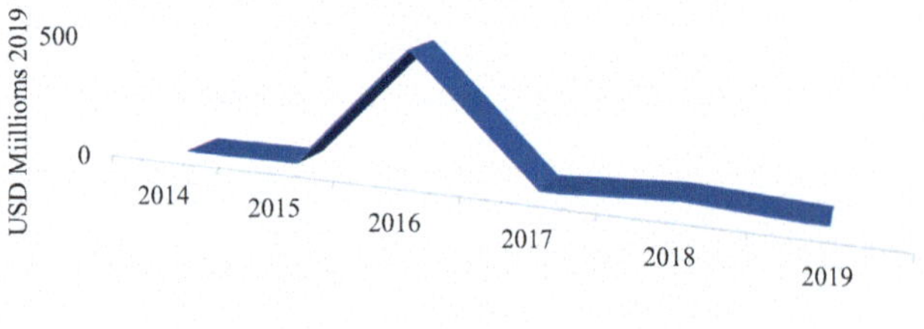

Fig. 7 Public and international flows to renewables. *Source* IRENA (2020)

of Ghana's electricity generation is hydro, compared to 65.3% thermal and 0.6% renewables, as shown in Fig. 8 (National Development Planning Commission, 2022a, 2022b). The renewable energy-generating capacity in Ghana has substantial growth from 0.79 watts per head in 2017 to 1.91 watts per person in 2020 and to 3.64 watts per capita in 2021, as shown in Fig. 9 (National Development Planning Commission, 2022a, 2022b). For further insights, refer to Figs. 8, 9, and 10.

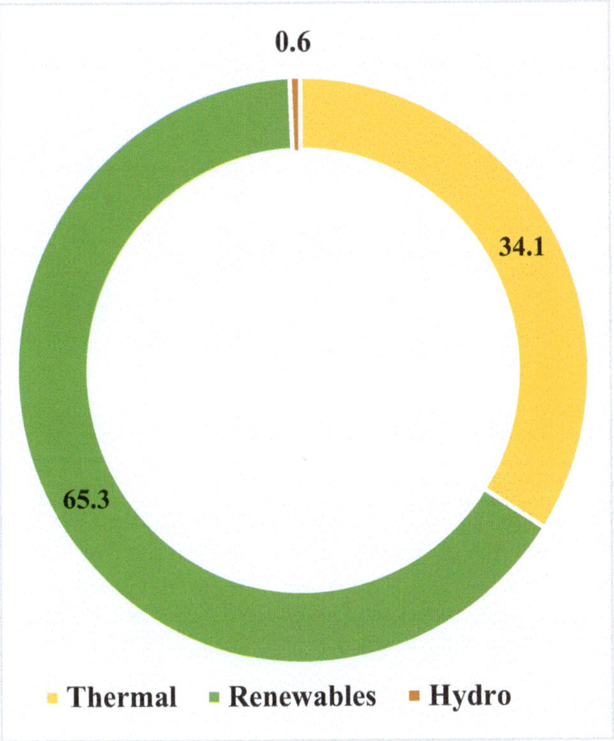

Fig. 8 Grid electricity supply by generation type, 2021 (%). *Source* National Development Planning Commission (2022a, 2022b)

4.7 Results of Indicators from Other Sources (Government, SOEs, NGOs, and Others)

The MFET report provides valuable insights into these outcomes, focusing on recent years and projecting trends for 2020 (Table 2). To gather relevant data for the 2020 outcomes, sources such as the Energy Commission Annual Reports were considered, ensuring a comprehensive analysis. In addition to reviewing past achievements, the report outlines planned targets for 2021 and 2025, indicating the government's commitment to advancing sustainable energy practices. However, the report acknowledges certain limitations in estimating actual results for 2021. The 2022 voluntary national report presented higher outcome values than the estimates in the MFET report (Ministry of Energy, 2022). This disparity stems from the non-inclusion of private and self-supply power or energy generation initiatives for households. To ensure an accurate assessment of Ghana's progress in SDG 7, it is essential to consider the insights provided by the MFET and Voluntary National Report (VNR). Both reports give a more vivid picture of Ghana's advancements in affordable and clean energy can be obtained.

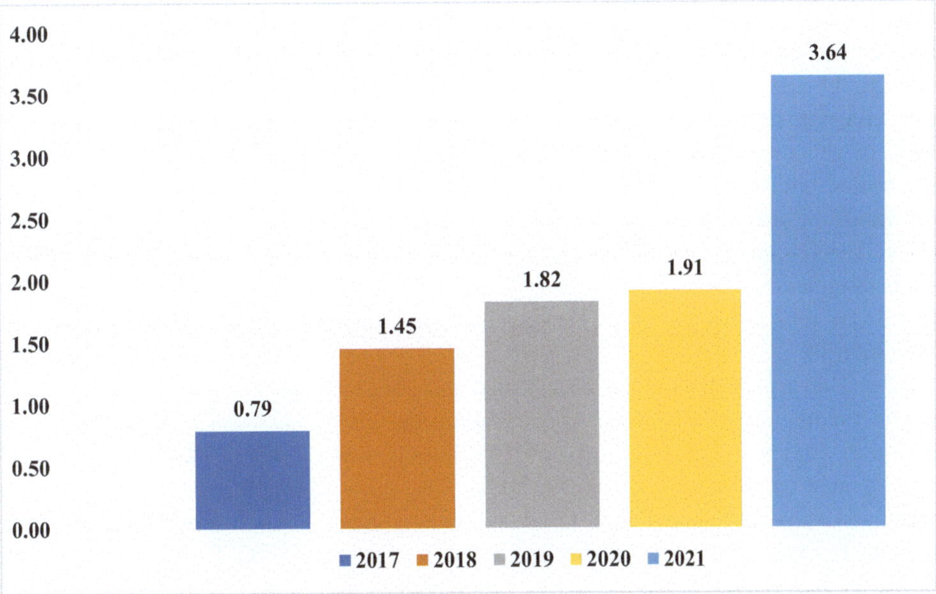

Fig. 9 Installed renewable energy-generating capacity, 2017–2021 (watts per capita). *Source* National Development Planning Commission (2022a, 2022b)

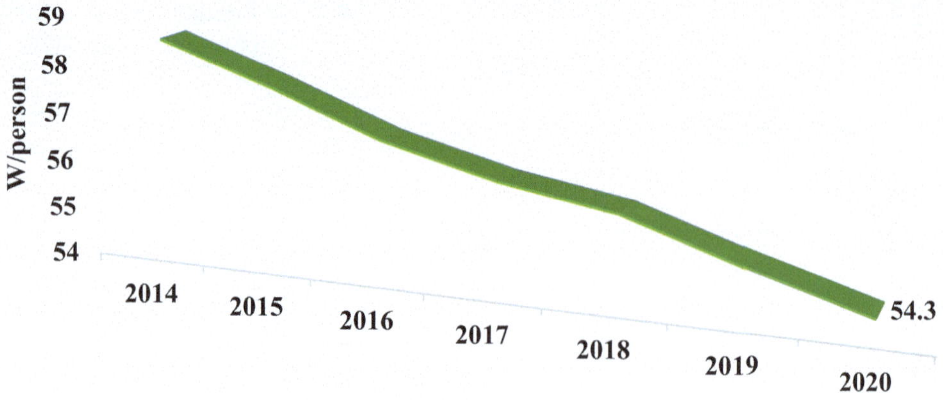

Fig. 10 Per capita renewable capacity from 2014 to 2020. *Source* National Development Planning Commission (2022a, 2022b)

4.7.1 Contributions of Government, SOEs, NGOs, and Private Sector Towards Goal 7

The Government of Ghana (GoG) is deeply committed to improving the energy and electricity access in the country. Despite financial challenges in funding and formidable obstacles to expanding access to affordable and clean energy sources due to rapid urbanization, the Government and other crucial stakeholders have paved the way for development within the energy sector. In 2021, hydropower and thermal plants were projected to generate 32.9% and 66.4% of the total electricity supply. Renewables were to constitute 0.7% of the total generation mix. Due to the progress in achieving these targets, the projections for 2022 increased to hydropower, thermal plants, and renewables, having total energy mix contributions of 31.4, 67.9, and 0.8% (Energy Commission, 2022b). These are shown in Figs. 11 and 12, respectively. These projections ensure that access to electricity is increased nationwide through thermal generation, the highest contributor to electricity generation in Ghana, by 1.5% and expanding clean energy utilization by increasing the renewable generation mix by 0.1%. The Government of Ghana has set further ambitious goals to increase its generation capacity from 2.6W in 2012 to 5.5 GW by 2026 as part of the Strategic National Energy Plan (SNEP).

The Ministry of Energy in 2022 stated that the total public funding allocated for Power generation and distribution in 2021 was estimated to be 551 million Ghanaian cedis (Table 3) (Ministry of Energy, 2022). The government of Ghana's commitment to renewable energy development is exemplified through the launch of the Scaling up Renewable Energy Program (SREP) with a budget allocation of USD 230 million. However, the oil and gas sector experienced a decline of 20.9% in its contribution to Ghana's GDP during Q3 2020, impacting overall economic growth. Despite the challenges posed by the COVID-19 pandemic, the energy sector in Ghana showed modest progress in implementing measures to mitigate its impact. Moreover, electricity played a vital role in Ghana's GDP, contributing close to 2.2 billion Ghanaian cedis (GHS) or 360 million U.S. dollars in 2020 as indicated in Fig. 13 (Sasu, 2022). This reflects an increase compared to the previous year, indicating the sector's growing importance. Additionally, Ghana's electricity generation exceeds domestic demand, exporting 167 kilowatt hours of electricity to neighboring countries.

Stakeholders in the energy sector, including the Electricity Company of Ghana (ECG) and Meinergy, have significantly contributed to Sustainable Development Goal 7 (SDG 7) in Ghana. ECG has worked diligently to provide reliable electricity distribution services, implementing infrastructure enhancements and capital investment projects to improve operational performance. Meanwhile, Meinergy has been actively expanding its renewable energy business, collaborating with Huawei to promote green development (Huawei, 2022). Their efforts align with Ghana's strategic goal of increasing the proportion of renewable energy in the country's energy mix to 10% by 2030. They have achieved remarkable results through utility-scale PV plants, PV-hydropower integration, and other initiatives. Their partnership aims to develop further PV & ESS plants, data centers, eLTE, and public cloud services, driving the nation's transition to a greener energy future.

5 Discussion

The study sought to evaluate the progress of implementing SDG 7 (Affordable and clean energy). The results showed that SDG 7 achievement in Ghana has a million miles of steps to take to achieve the goal. Ghana has not made any significant headway in achieving the targets. Target 7.1.1, the connection of the population to the national grid, received an 86.3% succession score through the number of households connected to the grid. This goal has seen that urban and rural vicinities have increased access to

Table 2 Policy outcome indicators and targets for the ministry of energy

Outcome indicator description	Unit of measurement	Baseline		Latest status		Target	
		Year	Value	Year	Value	Year	Value
Access to electricity increased (SDG Target 7.b)	Percentage of connection	2020	85.17	2021	86.63	2025	99
Electricity extended to deprived communities (SDG Target 7.b)	Number of communities connected to the national grid		11,287		162		559
Expansion of generation capacity (SDG Targets 7.1, 7.a, 7.b)	MW		5212.45		50 (Bui Solar)		360 (Bui, Early Power, Pwalugu)
Increased penetration of renewable energy in the national energy supply mix (SDG Targets 7.2, 7.3, 7.a)	RE capacity addition to the grid		1.50%		2%		26.40%
Mini-grid electrification using renewable energy technologies in island and lakeside communities installed (SDG Targets 7.2, 7.3, 7.a)	Number of mini grids installed	2020	5	2021	0 (Construction of three mini grids is 40% complete)	2025	3
Low reliance on wood fuels (SDG Targets 7.)	Number of improved charcoal cook stoves distributed		54,000		2,30,000		2,95,000
National LPG promotion policy implemented	Current LPG marketing model replaced with the cylinder recirculation model		Pilot implementation was launched in the Kwaebibiri, Denkyembuor and Adansi North Districts and the Obuasi Municipal		All LPG intervention activities have been reviewed and embedded in the LPG for Development (LPG4D) Program		203,160 cook stoves to be distributed under the LPG for development program (LPG4H & LPG4C)
Increased production from Jubilee field (SDG Target 9.b)	Volume of oil produced	2020	312.46 mmbbls	2021	19.94 mmbbls	2025	27.68 mmbbls
	Gas export		144.71 Bcf		26.55 Bcf		37.51 Bcf
Increased production from TEN field (SDG Target 9.b)	Volume of oil produced		89.45 mmbbls		9.48 mmbbls		13.09 mmbbls
	Gas export		14.93 Bcf		2.5 Bcf		16.47 Bcf
Increased production from SGN field (SDG Target 9.b)	Volume of oil produced		46.54 mmbbls		12.12 mmbbls		8.89 mmbbls
	Gas export		95.40 Bcf		48.14 Bcf		62.42 Bcf

Source Ministry of Finance and Economic Planning (2022)

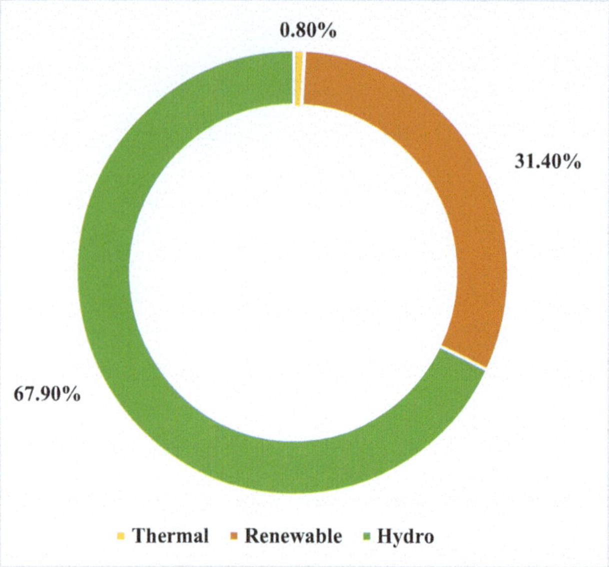

Fig. 11 Share of electricity supply by generation type for 2021. *Source* Ministry of Energy (2022)

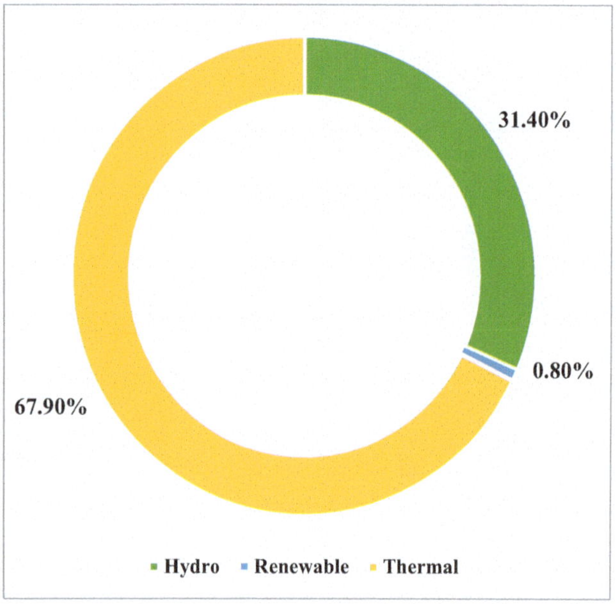

Fig. 12 Share of electricity supply by generation type for 2022. *Source* Ministry of Energy (2022)

electricity through the works of ECG, the Government of Ghana, the Ministry of Energy, and other stakeholders collectively working towards the goal. With the assistance of renewable energy, Ghana aims to achieve total household connection to the grid by 2030. Existing power generation centers are expected to expand their capacities to accommodate the rise in population per the projections of the MTEF. With continuous and consistent investments in the power generation and distribution sector, all relevant energy space stakeholders can contribute to attaining target 7.1.1.

Target 7.1.2 has also seen a significant rise, with about 18.7% of households using LPG gas and other clean fuels. Many urban and city centers adopt clean fuels; however, many people residing in rural areas still depend on wood and wood fuels such as charcoal which, when not appropriately managed, can damage nature by encouraging deforestation and enhancing climate change. In conjunction with the Liquefied Petroleum Gas (LPG) Marketers Association of Ghana, the government has agreed to achieve a 50% penetration of LPG in households by 2030. However, reaching this target could be challenging due to the constraints of limited distribution outlets across the country and the persistently soaring prices of LPG (Asante et al., 2018). Considering this perspective, the government must contemplate offering investment incentives to motivate Oil Marketing Companies (OMCs) and other potential investors to establish additional LPG storage and distribution facilities throughout the country. This would help enhance accessibility and affordability. Furthermore, the government must prioritize the speedy implementation of measures to promote the availability of LPG cylinders at subsidized prices through their Cylinder Recirculation Model (CRM) Policy to enable households to access LPG fuels, especially within rural areas.

Target 7.2 has faced regression since 2017 by reducing the percentage of renewable energy in the overall final energy consumption. In 2021, renewable energy share in total consumption dropped by 7.3% in 2021, which is quite detrimental to achieving the goal. The GoG, the Ministry of Energy, and other stakeholders have undergone myriads of projects such as the Solar Irrigation System at Tamalugu and mini grids at Pediatorkope to improve Ghana's reliance on renewable sources as well as providing energy for socio-economic growth for the people (Takouleu, 2019). According to the MTEF report for 2022–2025, renewable energy has experienced higher adoption and integration into the country's energy mix, from 1.5% in 2020 to 2% in 2021. Five mini-grids were installed in 2020 for Mini-grid Electrification using Renewable Energy technologies in the island and lakeside communities. In 2021, no additional mini grids were installed. Despite all these developments in the RE sector, there is still a decline in the adoption of RE sources. From pricing to reliance, citizens still have questions regarding RE sources and their usage, with only a few fully understanding their benefits. Other plausible external factors, such as COVID-19 and economic strife between 2019 and 2023, may have reduced prior investments pledged to the green energy cause (IRENA, 2020). According to the IRENA (2020), the projections for RE to the total energy supply by generation for 2022 was 0.8%, a

Table 3 MFET expenditure estimates for 2022–2025

Sector for funding under energy	Approximate expenditure estimates per year in a million cedis			
	2022	2023	2024	2025
Management and administration	300	300	300	300
Power generation, transmission, and distribution (power management)	140	140	140	140
Petroleum development	6	6	6	6
Renewable energy development	20	20	20	20
Energy sector registration	85	85	85	85
Grand total	551	551	551	551

Source Ministry of Energy (2022)

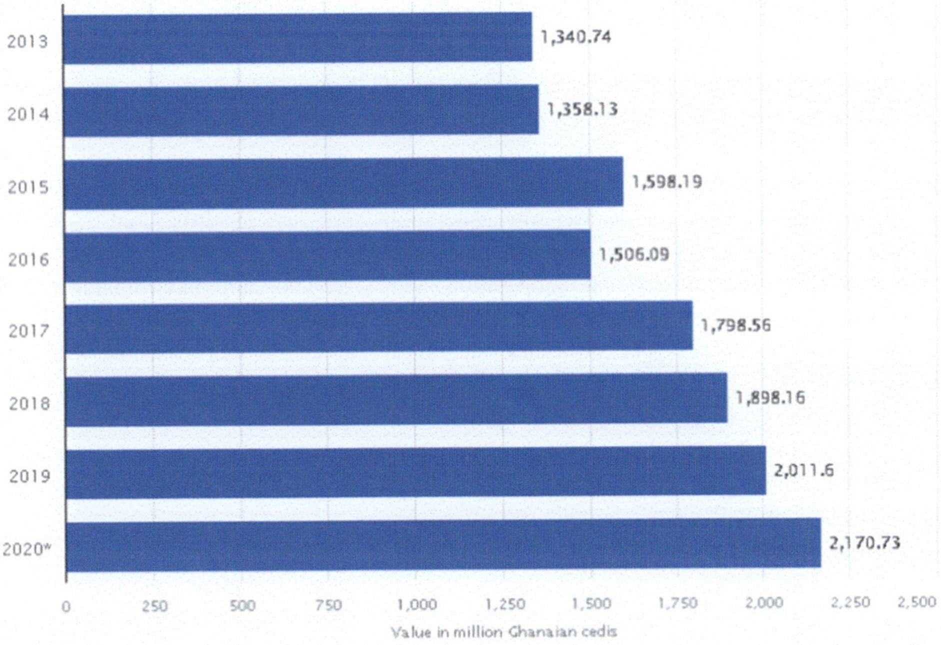

Fig. 13 Ghana's Contribution Electricity to GDP 2013–2020. *Source* Sasu (2022)

step up from 2017, where RE's contribution to power consumption was 0.5%. In 2020, the projected energy supply by generation by Renewables for 2021 was 0.7% (Energy Commission, 2021). Unfortunately, the government and other energy players achieved only 0.6% supply by generation (National Development Planning Commission, 2022a, 2022b).

Target 7.3 has shown satisfactory progress, although much improvement is still required. Doubling the country's energy efficiency has received some progress but still needs more investment by energy sector players. The energy intensity rate between 2013 to 2018 increased by 2% with a GDP growth rate of 5%. This result shows Ghana's readiness and progress towards achieving goal seven; however, data in recent times (2019–2023) has not been taken for the progress on the goal, especially showing the country's progress on this target during and after the pandemic. In 2020, the primary energy intensity per GDP was 72.62 TOE/million USD, an increase of 9.94 TOE/million USD. However, recent data (2019–2023) on the progress of indicator 7.3.1 has not been gathered. Hence it is challenging to understand the performance of energy efficiency, especially after the COVID-19 pandemic and the global economic crisis.

Target 7.a, which is the funding for Renewable Energy development and implementation, needs more attention from the international community and energy sector players to achieve the goal. Aside from 500 million USD invested in the country in 2016 for the critical startup of the green energy and renewables sectors, less than one hundred million dollars have been invested into developing RE systems. The inadequate funding is not expediting actions to increase the installation and utilization of Renewables in the country. RE technologies (RETs) and systems are costly and are often capital intensive such as wind turbine purchasing and installation. Hence massive investments to either purchase or subsidize prices of RETs for the nation and individuals is still quite challenging. Suppose investments and funding are not given to target 7.a. In that case, the country may still rely on thermal generation, which gives off massive carbon emissions, which is a leading factor towards global warming. With only twenty million Ghana cedis dedicated to RE by the Ministry of Energy (Ministry of Energy, 2022), international bodies and other private and business players must step in to improve access to REs across all the regions of Ghana.

Target 7.b is influenced by target 7.a since the installed capacity depends on acquiring RETs and RE systems, which are expensive and require a lot of funding. Ghana has less than 1% supply generation from RE sources. Although projections were made for the increase in installations per year, they have not been fulfilled. The Government of Ghana and other relevant stakeholders must continuously address all the challenges facing the respective targets and Ghana's energy sector holistically. Since funding seems to be a significant predicament in clean energy generation and grid integration, the government must utilize private businesses and allow for international cooperation with organizations and agencies like Power Africa to continue the country's renewable energy generation and integration growth.

The progress made by Ghana in implementing SDG 7 is not encouraging besides Target 7.1.1 and 7.1.2. If a deliberate and concerted effort is not made towards the improvement and achievement of the other targets, the affordable and clean energy goal of the sustainable development goals cannot be achieved by 2030.

5.1 The Way Forward

To move towards achieving Goal 7 of making affordable and easily accessible energy, with a focus on environmental sustainability, several vital actions must be taken based on the contributions of stakeholders in Ghana's energy sector. Firstly, it is crucial to strengthen the implementation of existing policies and programs, such as the Renewable Energy Act and the National Medium-Term Development Policy Framework. In close collaboration with the Energy Commission, the government should monitor the progress of these initiatives, address challenges that arise, and make necessary adjustments to ensure their effective implementation. The Energy Commission should focus on refining and enforcing regulations that promote sustainable energy practices. This includes creating a favorable environment for private sector investment, ensuring compliance with quality and safety standards, and facilitating the incorporation of sustainable energy sources into the existing energy grid. Regular reviews and updates of regulatory frameworks will enable the sector to adapt to changing market dynamics and emerging technologies. There is a need to boost the capacity of renewable energy sources. Meinergy and other private sector players have demonstrated their commitment to expanding renewable energy in Ghana. To build on this momentum, it is essential to encourage further investments in renewable energy projects, such as utility-scale PV plants and energy storage systems. The government can provide incentives and favorable policies to attract more private sector participation in the renewable energy sector, thereby diversifying the energy mix and reducing reliance on traditional fossil fuels. Collaboration between stakeholders is also crucial for success. The government, private sector entities, and regulatory bodies should work together to share knowledge, expertise, and resources. This collaboration can facilitate the development of innovative solutions, such as hybrid systems combining PV and hydropower, and accelerate the adoption of sustainable energy practices in Ghana. By taking these steps, Ghana can make noteworthy progress towards Goal 7, ensuring affordable, accessible, and sustainable energy for all while promoting economic growth and environmental stewardship. The involvement of key stakeholders, combined with a comprehensive approach, will contribute to the country's sustainable energy future.

6 Conclusions

The study sought to evaluate the implementation of SDG 7, which is focused on affordable and clean energy in Ghana. Ghana has made significant strides in achieving universal access to electricity and providing clean cooking fuel, in alignment with Target 7.1 of SDG 7. The objective of supplying affordable energy for all Ghanaian citizens is attainable, as evidenced by most national reports and data. However, there are challenges in achieving all the targets of Goal 7, particularly in terms of the renewable energy (RE) mix in the nation's energy consumption, which necessitates huge investment and stronger dedication from all stakeholders, particularly the government. Ghana's dependence on RE as a stable energy source has regressed, indicating a decline in its contribution to the total energy mix compared to thermal systems. The investment and supply generation of renewable energy technologies (RETs) and RE systems have not shown positive progress. Insufficient funding from both local and international stakeholders has hindered the effective integration of RE into Ghana's energy system, limiting the reduction of carbon emissions associated with thermal systems, which are significant contributors. Increased investments are necessary to improve the supply generation capacity to utilize renewable energy sources achieving at least a 1% share in the total energy supply generation mix by 2030. With the current interventions and approach toward SDG 7 implementation, achieving this goal by 2030 will be difficult. To drive progress and contribute to the achievement of SDG 7, the Government of Ghana (GoG) and all relevant stakeholders must prioritize investments and attention toward the advancements of RE sources, RETs, and systems. With adequate support and resources, Ghana can progress towards cost-effective and clean energy for all.

References

Ackah, C. (2015). Electricity insecurity and its impact on micro and small businesses in Ghana. In *A study presented at the Institute of Statistical, Social and Economic Research (ISSER) workshop, "Electricity Insecurity and Its Impact on the Economy of Ghana"* (Vol. 14). University of Ghana.

Ackah, I. (2021). *The latest power supply challenges in Ghana*. Energy for Growth Hub. Retrieved January 13, 2023, from https://www.energyforgrowth.org/memo/the-latest-power-supplychallenges-in-ghana/

Afful-Dadzie, A., Mensah, S. K., & Afful-Dadzie, E. (2022). Ghana renewable energy master plan: The benefits of private sector participation. *Scientific African, 17*, e01353.

Agyekum, E. B. (2020). Energy poverty in energy-rich Ghana: A SWOT analytical approach for the development of Ghana's renewable energy. *Sustainable Energy Technologies and Assessments, 40*, 100760. https://doi.org/10.1016/j.seta.2020.100760

Asante, K. P., Afari-Asiedu, S., Abdulai, M. A., Dalaba, M. A., Carrión, D., Dickinson, K. L., Abeka, A. N., Sarpong, K., & Jack, D. W. (2018). Ghana's rural liquefied petroleum gas program scale-up: A case study. *Energy for Sustainable Development, 46*, 94–102. https://doi.org/10.1016/j.esd.2018.06.010

Clerici, C., Taylor, M. S., & Taylor, K. (2016). *Dumsor: The electricity outages leaving Ghana in the dark*. Al Jazeera.

Cobbold, E. Y., Owusu, D., & Bathuure, I. (2019). An assessment of Ghana's electricity sector, challenges, and remedies. *International Journal of Multidisciplinary Research and Development, 6*(11), 242–247.

Debrah, S. K., Nyasapoh, M. A., Ameyaw, F., Yamoah, S., Allotey, N. K., & Agyeman, F. (2020). Drivers for nuclear energy inclusion in Ghana's energy mix. *Journal of Energy* (2020).

Energy Commission. (2020). *National Energy Statistics 2000–2019*. Energy Commission.

Energy Commission. (2021). (rep.). *2021 Energy Outlook for Ghana* (Ser. 12, pp. 33–34). Energy.

Energy Commission. (2022a). *SDG|Ghana Energy Database*. Ghana Energy Database System. Accra, Ghana. http://gheatoolkit.energycom.gov.gh/Analysis/SDG.

Energy Commission. (2022b). (rep.). *2022 energy outlook for Ghana*. Energy.

Huawei. (2022). *Meinergy signs agreement with Huawei on a 1 GW and 500 mwh project to facilitate green development of Ghana*. Retrieved April 18, 2023, from https://solar.huawei.com/eu/news/eu/2022/3/Meinergy-Signs-Agreement-with-Huawei-on-a1GW-and-500MWh-Project-of-Ghana.

IEA. (2022). *For the first time in decades, the number of people without access to electricity is set to increase in 2022*. IEA. https://www.iea.org/commentaries/for-the-first-time-in-decadesthe-number-of-people-without-access-to-electricity-is-set-to-increase-in-2022

IRENA. (2020). *Global renewables outlook: Energy transformation 2050*. International Renewable Energy Agency.

Kuamoah, C. (2020). Renewable energy deployment in Ghana: The hype, hope and reality. *Insight on Africa, 12*(1), 45–64. https://doi.org/10.1177/0975087819898581

Kukah, A. S., Anafo, A., Kukah, R. M. K., Blay Jnr, A. V. K., Sinsa, D. B., Asamoah, E., & Korda, D. N. (2022). Exploring innovative energy infrastructure financing in Ghana: Benefits, challenges and strategies. *International Journal of Energy Sector Management, 16*(2), 248–264. https://doi.org/10.1108/IJESM-12-2020-0010

Kumi, E. N. (2017). *The electricity situation in Ghana: Challenges and opportunities* (p. 30). Center for Global Development.

Merem, E. C., Twumasi, Y., Wesley, J., Isokpehi, P., Shenge, M., Fageir, S., & Nwagboso, E. (2017). Regional assessment of energy trends in West Africa using GIS. *International Journal of Energy Engineering, 7*(1), 1–27.

Ministry of Energy. (2022). *Medium Term Expenditure Framework (MTEF) for 2022–2025*. Retrieved December 17, 2022.

Murray, C. (2015). *IMF and USA Set to Ruin Ghana*. Uncategorized by Craig.

National Development Planning Commission. (2022). (rep.). *Ghana Voluntary National Report* (pp. 61–65).

National Grid Group. (2023). *Energy explained*. Retrieved January 11, 2023, from https://www.nationalgrid.com/stories/energy-explained

Nyasapoh, M. A., Debrah, S. K., Twerefou, D. K., Gyamfi, S., & Kholi, F. K. (2022). An overview of energy resource and future concerns for Ghana's electricity generation mix. *Journal of Energy, 2022*, 1–16. https://doi.org/10.1155/2022/1031044

Nyasapoh, M. A. (2018). *Modelling energy supply options for long-term electricity generation—A case study of Ghana power system* (Doctoral dissertation, University of Ghana).

Power Africa. (2022, December 9). *Power Africa in Ghana: Power Africa. U.S. Agency for International Development*. Retrieved January 13, 2023, from https://www.usaid.gov/powerafrica/ghana.

Sasu, D. D. (2022). *Ghana: Contribution of electricity to GDP in Ghana from 2013 to 2020*. https://www.statista.com/statistics/1272261/annual-contributions-of-electricity-to-gdp-in-ghana/#statisticContainer

SEforALL. (2022). *SEFORALL analysis of SDG7 progress*. https://www.seforall.org/data-stories/seforall-analysis-of-sdg7-progress.

Singh, G., Nouhou, S. A., & Sokona, M. Y. (2015). *Ghana renewables readiness assessment*. IRENA.

Takouleu, J. M. (2019). *Ghana: BPA Commissions a mini hydroelectric power plant in Hohoe*. Afrik 21. Retrieved January 12, 2023, from https://www.afrik21.africa/en/ghana-bpacommissions-a-mini-hydroelectric-power-plant-in-hohoe/

Tengey, C., Nwulu, N. I., Adepoju, O., & Longe, O. M. (2022). Analysis of the productivity dynamics of electricity distribution regions in Ghana. *Energies, 15*(24), 9414. https://doi.org/10.3390/en15249414

The World Bank. (2017). (rep.). *The state of electricity access report*. https://documents.worldbank.org/curated/en/285651494340762694/pdf/114841-ESM-PUBLIC-P148200-32pFINALSEAR-OverviewWEB.pdf

Tulashie, S. K., Odai, R., Dahunsi, A. M., Atisey, S., & Amenakpor, J. (2022). Feasibility study of wave power in Ghana. *International Journal of Sustainable Engineering, 15*(1), 299–311.

UN. (2021). *High-level dialogue on energy*. SDG Knowledge Hub. Retrieved January 12, 2023, from https://un-energy.org/high-level-dialogue-energy/.

UNDESA & DSDG. (2021). *Stakeholders thematic e-consultation for the United Nations high-level dialogue on Energy Summary Report|Department of Economic and Social Affairs*. United Nations. Retrieved January 12, 2023, from https://sdgs.un.org/publications/stakeholders-thematic-e-consultation-HLDE-summary-report

United Nations. (2021). *New commitments at UN energy summit*. United Nations. Retrieved January 12, 2023, from https://www.un.org/en/hlde-2021/page/new-commitments-un-energy-summit

UNECA. (2020). *On sustainable development goal 7: The "SDG7 initiative for Africa": Accelerating clean energy investments for access and climate ambition in Africa*. Addis Ababa. © UN. ECA. https://hdl.handle.net/10855/43684

UNEP. (n.d.). *Goal 7: Affordable and clean energy*. https://www.unep.org/explore-topics/sustainable-development-goals/why-do-sustainable-development-goals-matter/goal-7

UNICEF. (2022). *A brighter life for every child with sustainable energy*. UNICEF. https://www.unicef.org/media/127626/file/A%20brighter%20life%20for%20every%20child%20with%20sustainable%20energy.pdf

Whitby, A., Besnard, J., Mathew, P. P., & Portale, E. (2020). *The dawning promise of energy for all. SDG Atlas 2020*. Retrieved January 11, 2023, from https://datatopics.worldbank.org/sdgatlas/goal-7-affordable-and-clean-energy/

Assessing the Progress of Clean Water and Sanitation (SDG 6) Implementation in Ghana

Simon Ofori Ametepey, Clinton Aigbavboa, Wellington Didibhuku Thwala, and Hutton Addy

Abstract

In 2015, the United Nations set forth 17 Sustainable Development Goals (SDGs) to be realized by 2030, one of which is universal access to water and sanitation (SDG 6). This study critically evaluates the progress of Sustainable Development Goal 6 implementation in Ghana. The assessment relied on document analysis and secondary data from various sources, including the population and housing census report (2021), the 2021 Voluntary National Review (VNR), and the Medium-Term Expenditure Framework (MTEF) report, among others. Through a combination of policies, projects, and interventions, Ghana has made commendable progress toward realizing SDG 6. However, it is essential to highlight that only one of the eight targets associated with SDG 6 has made significant strides. Distinctly, Target 6.1, which seeks to ensure universal and equitable access to affordable and safe drinking water, was achieved by approximately 92.2% of all households. This research revealed that although substantial advancements have been achieved in the water industry, the same cannot be said for sanitation. The country continues to face significant challenges in delivering adequate and high-quality water and sanitation services to urban and rural communities. To meet all the targets outlined in SDG 6 by 2030, stakeholders in the water and sanitation sector, particularly the government, must intensify their efforts and allocate resources towards the sector. By undertaking comprehensive evaluations and addressing the existing constraints, Ghana can navigate the path toward the successful execution of Goal 6 targets.

Keywords

SDG6 · Water and sanitation · UNICEF · Open defecation · Sustainable development goals · Ghana

1 Introduction

Water is a life-preserving and life-sustaining resource, essential for the survival of humans, animals, and the environment. It plays a vital role in human existence, from household activities such as cooking and cleaning to industrial undertakings like farming and food and medicine production. Despite water covering about 70% of the planet's surface, many countries face challenges accessing safe and potable water sources. Over the past decade, significant steps have been taken globally to ameliorate access to clean drinking water sources (Mosello, 2017). However, discrepancies in the access to and usage of refined water sources persist, particularly in developing cities (Alba et al., 2020; Cha et al., 2021; Deshpande et al., 2020; Mutono et al., 2022). The availability of clean and safe drinking water and sufficient sanitation resources is crucial to the

Simon Ofori Ametepey, Clinton Aigbavboa, Wellington Didibhuku Thwala, Hutton Addy are contributed equally to this study.

S. O. Ametepey (✉) · H. Addy
Centre for Sustainable Development (CenSuD), Koforidua Technical University, Koforidua, Ghana
e-mail: simon.ametepey@ktu.edu.gh

H. Addy
e-mail: hutad@outlook.com

S. O. Ametepey · C. Aigbavboa
Department of Construction Management and Quantity Surveying, Faculty of Engineering and Built Environment, University of Johannesburg, Johannesburg, South Africa

S. O. Ametepey
Department of Building Technology, Faculty of Built Environment, Koforidua Technical University, Koforidua, Ghana

W. D. Thwala
Department of Civil Engineering, College of Science Engineering and Technology, University of South Africa, Johannesburg, South Africa
e-mail: wdthwala@wsu.ac.za

B. He et al. (eds.), *Urban Climate Change Adaptation*, Advances in Science, Technology & Innovation,
https://doi.org/10.1007/978-3-031-65088-8_13

population's health, safety, and well-being. Both the United Nations (UN) General Assembly and the Human Rights Council have declared that universal access to clean water and sanitation is a fundamental human right for all member countries of the UN (Roaf et al., 2018). Leading international organizations such as the UN and the World Health Organization (WHO) set targets and help ensure that everyone has access to affordable, good-quality, and safe drinking water. The equitable distribution of basic water and sanitation services is a crucial aspect of sustainable development and indirectly stated by SDG 10 (Targets 10.2, 10.3, 10. b), which focuses on improving equality within and between nations (Bayu et al., 2020; Duran et al., 2015). There is a close correlation between poverty and access to clean water (UNICEF, 2022). Individuals living in poverty are over twenty times more likely to spend more than 30 min fetching water than affluent individuals. The availability of clean water is paramount for human existence and overall quality of life, regardless of whether one resides in urban or rural settlements. Regrettably, many developing nations suffer from inadequate access to clean water, leading to poor sanitation in commercial and residential areas. The UNICEF Joint Monitoring Program and WHO reported that 844 million people worldwide still have inadequate access to basic water services, and 2.3 billion lack access to basic sanitation (Gelave, 2018). Moreover, over 673 million individuals excrete outdoors, predominantly in developing countries. Access inequality remains pervasive, particularly regarding basic sanitation services, where impoverished populations are often excluded from accessing these essential facilities (Bayu et al., 2020). In Ghana, numerous regions face challenges accessing clean water and adequate sanitation facilities. However, insubstantial research has been conducted to examine the contributions and performance of governmental and private organizations in addressing this issue. Therefore, this study assesses Ghana's progress in achieving SDG 6 over recent years. Additionally, the study intends to determine the commitment of the government, state-owned enterprises, non-governmental organizations, and companies operating in the water and sanitation industry on a private basis toward accomplishing this goal. Through a comprehensive analysis of the sector's performance, this research sought to provide discussions, recommendations, and conclusions regarding the advancement made in attaining SDG 6 and the roles these entities can play in enhancing current standards and commitment towards accomplishing the targets of SDG 6.

2 SDG 6—Clean Water and Sanitation for All

SDG 6 is part of the UN's 17 SDGs, which evolved from the Millennium Development Goals in 2015 and target achievement by 2030 (Chopra et al., 2022). This goal aims to ensure universal access to water and sanitation as a fundamental human right, contributing to Goal 10's focus on reducing inequality (United Nations, 2022). SDG 6 and SDG 10 experienced significant transformation from their origins in the Millennium Development Goals. SDG 6 adopted an all-inclusive and holistic approach to water and sanitation access, while SDG 10 directly tackled disparities within and among countries, reflecting their respective goals' expanded and nuanced focus (Rambla & Langthaler, 2016). SDG 6 was developed by the work of UN-Water (2015a) and further elaborated by Fukuda et al. (2019). The principle of "access fairness" is emphasized in SDG 6 and 10, as they are closely interconnected with various other SDGs such as SDG 1, 2, 3, 4, 5, 11, 13, 15 and 17 (Bayu et al., 2020; UN-Water, 2015b). The Global South, however, faces challenges in water governance due to weak and underfunded water institutions and managers (Dinar & Saleth, 2005). Water governance is a system that regulates and controls the availability of water and sanitation sources (OECD, 2015a; OECD, 2015b; Neef et al., 2009). Water governance refers to the interconnection of political, social, economic, and administrative systems to develop and regulate water resources and distribute water services among various segments of society (Bayu et al., 2020). Past studies have indicated that inequality stems from systemic factors rather than development practices (Bartels, 2016; Bayu et al., 2020; Ezcurra & Rodrguez-Pose, 2014). This system often leads to an inadequate understanding of water and sanitation as fundamental human rights. The quantification of inequality could direct development objectives to consider the most underprivileged individuals and groups more accurately (Bayu et al., 2020; Cowell, 2015). To monitor the progress of SDG 6, the WHO and UNICEF Joint Monitoring Program for Water Supply, Sanitation, and Hygiene launched a new status report and SDG baselines in 2017. This monitoring system utilizes service ladders and wealth quintile estimates to assess the growth of national, urban, and rural populations (Bayu et al., 2020). SDG 6 has 11 indicators or metrics that are universally accepted as a system of measurement across its eight targets. Each of these targets is set to be achieved by 2030 through the rigorous determination, tracking, and monitoring of their corresponding indicators using up-to-date data. This process is essential for making informed amendments and ensuring continuous development and improvements towards these goals.

2.1 Progress and Challenges of SDG 6 in Ghana

Ghana has made significant economic advancements in recent decades, improving household conditions, such as enhanced accessibility to refined water and sanitation facilities and adopting cleaner cooking fuels (Moulds et al., 2022). Sanitation issues in Ghana have received attention, and efforts have been made to address them structurally. The Government of Ghana has reaffirmed its commitment to the United Nations 2010 declaration that recognizes the accessibility of water and sanitation as fundamental rights. Ghana is committed to gradually ensuring universal basic sanitation access, as outlined in the directive principles of state policy in the 1992 Constitution (Sallah-Phillips, 2006). Development can be enabled if current national policies recognize sanitation services as social goods with reasonable economic value (Ametepey, 2023). Efforts have been made to develop sanitation policies and plans for Ghana's rural and urban areas. The 2010 Environmental Sanitation Policy (ESP) is founded on seven broad principles, such as capacity building and education, and incorporates associated principles that affect the implementation of improved environmental sanitation (Boamah, 2011). In addition, the policy provides a clear framework for achieving the sector's goals through training and workshops, education and communication, sustainable financing, cost recovery, monitoring, and evaluation, among others (Akoto-Bamfo, 2020). The 2010 ESP specifies how it should be implemented and which agencies are responsible for monitoring projects established at the household and community level. Ghana has also adopted Community-Led Total Sanitation (CLTS) as its basic sanitation promotion strategy for rural areas. At the same time, the Local Government Act assigns responsibility for sanitation services to metropolitan, municipal and district assemblies (Appiah-Effah et al., 2019). However, coordination and monitoring of urban sanitation approaches and activities remain ineffective due to the complexities of urban communities (Osumanu, 2010). Challenges such as inadequate water and sanitation databases, financial constraints, lack of private sector interest in rural sanitation, and limited innovative financing options for disadvantaged households pose obstacles to achieving SDG 6 in Ghana. Additionally, the construction of shared toilets discourages households from installing their own, particularly in wet regions prone to natural disasters (UNICEF Ghana, 2022). Ghana faces substantial challenges in providing adequate water and sanitation to rural and urban residents (Nyika & Dinka, 2023). Ghana Water Company Limited's performance in the urban water supply is poor, with a contested coverage of approximately 60% and unaccounted-for water continuing to account for approximately 50% of the company's water production (Awuah et al., 2009). Open defecation remains a challenge, with merely 23% of the population having access to enhanced sanitation premises. According to Akoto-Bamfo (2020), financial support, human indiscipline, carelessness, infrastructure and funding issues, climate change, water resource availability, socioeconomic inequalities, inefficient water use, and management, rapid urbanization, and political commitment all impede Ghana's progress in implementing SDG 6. Urbanization, with approximately 58% of the population residing in urban areas, has placed significant pressure on the sanitation and water management services of municipalities in Ghana (Ametepey, 2023). Discrepancies in basic sanitation services persist, necessitating effective targeting of underprivileged communities, schools, and health centers. Inadequate technical capacity, high capital expenditures, and a lack of priority given to sanitation services by households, public sectors, and commercial sectors contribute to Ghana's prevailing poor sanitation services (Akoto-Bamfo, 2020). Despite advancements in SDG 6, accessibility to areas of convenience remains a predicament in rural and metropolitan areas, with a low percentage of people having access to safely managed sanitation facilities in metropolitan areas. Low allotment, lack of planning and coordination, and inadequate resources from capacity building and program implementation are some of the credible barriers to supplying necessary sanitation services, particularly in rural areas (UNICEF Ghana, 2022). The heavy reliance on donor funding, low public investment, and limited private sector involvement create a significant bottleneck in addressing WASH (Water, Sanitation, and Hygiene) inequalities and achieving universal access. Present-day sector financing of 140 million US dollars yearly generates a significant financial divide in providing all-inclusive access. Forecasts regarding available financial resources indicate larger allotments and more dependable funding streams derived from taxes. Nevertheless, relying solely on taxes will not be enough to ensure the ambitious progress needed to meet the Sustainable Development Goals. Also, there is room for improving the contemporary revenue contribution towards operation and maintenance by providing exceptional service quality. The integration of returns, more effective budgeting, and the application of accessible funding are required. Furthermore, drawbacks to funding the SDGs must be eradicated, especially those generated by weak functioning, with operations and high expenses involved, to incentivize the expansion of sanitation services and ensure their long-term viability. Ghana's progress in implementing SDG 6 has various challenges, including inadequate infrastructure, funding, coordination, and limited capacity. The government's commitment to addressing these challenges still and beyond 2030, along with enhanced financial support, improved governance, and effective targeting, is crucial for gaining all-inclusive delivery of clean water and sanitation sources and services for Ghanaians (Ametepey, 2023).

3 Research Methodology

This investigation endeavored to conduct an exhaustive assessment of Ghana's implementation of SDG 6. The study employed a meticulous document analysis research design, integrating a diverse array of secondary data sources. This approach significantly bolstered the research's coherence, reliability, and depth. The methodological process for conducting the research is shown in Fig. 1.

Conducting this study on the contemporary development of Goal 6 in Ghana required outlining accurately all targets and metrics for assessing Goal 6 by each member state of the UN (as shown in Table 1). An extensive search for each of the indicator was then performed online to search for reliable data and information on Ghana's trending progress. In meticulously assessing each target's realization within the 2030 timeframe, the study harnessed secondary data from a spectrum of reputable sources. These included authoritative entities like UN agencies, the World Bank, the Ghana Statistical Service Population and Housing Census Report (2021), alongside contributions from the private sector. The research delved into an extensive array of secondary data, including notable government reports from the Ministry of Finance and Economic Planning (MFET) and the Voluntary National Review (VNR). A detailed document analysis, recognized as a pivotal qualitative research technique, was adeptly employed to distill, and interpret meaningful insights from an array of electronic documents, facilitating a comprehensive understanding of the strides made toward SDG 6. These critical resources were systematically analyzed, augmenting the document analysis approach, and providing a multidimensional view of Ghana's efforts. The benchmarks established by the United Nations for SDG 6 were instrumental as the definitive standard for appraising progress, thereby enabling a nuanced evaluation of Ghana's execution of these essential goals.

The 2022 VNR by the National Development Planning Commission serves as the foundational document for assessing Ghana's progress towards Goal 6 of the 2030 Agenda. This report provides a crucial benchmark in evaluating the nation's advancements. However, it is important to note that the VNR does not encompass all indicators for the targets, and there are notable discrepancies when compared with other reports, such as the Medium-Term Expenditure Framework (MTEF). Indicators such as 6.3.2, 6.4.1, 6.5.2, 6.6.1, and 6.a.1 were not considered by the VNR report nor any other National report in Ghana. Additionally, the UN bodies have updated data for all indicators relevant to Ghana on the official UN-Water platform in 2022, further contributing to the comprehensive assessment of progress by supplying information for missing data.

The analysis of the progress of achieving the targets of SDG 6 utilized the most recent available data between 2021 and 2020 as the baseline. Emphasis was placed on data and publications over the past five years to highlight current

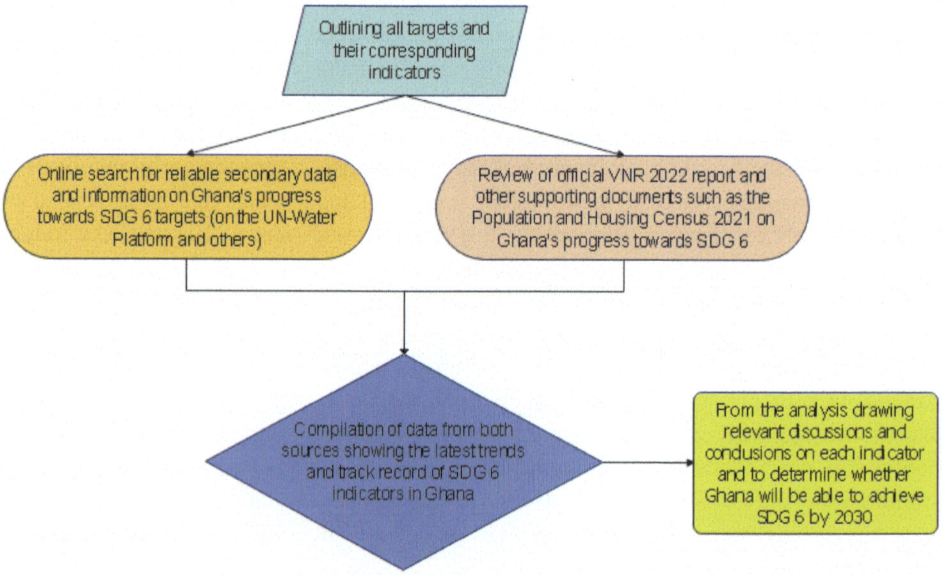

Fig. 1 Methodological procedure for the study, (illustration by authors)

Table 1 SDG 6, its targets and indicators

Targets	Indicators	Data collection status
6.1 By 2030, achieve universal and equitable access to safe and affordable drinking water for all	6.1.1 The proportion of the population using safely managed drinking water services	Data was collected from the VNR report and Population and Housing Census. Updated by WHO/UNICEF on UN-Water Platform
6.2 By 2030, achieve access to adequate and equitable sanitation and hygiene for all and end open defecation, paying particular attention to the needs of women and girls and those in vulnerable situations	6.2.1 The proportion of the population using (a) safely managed sanitation services and (b) a hand-washing facility with soap and water	Data was collected from the VNR report and Population and Housing Census. Updated by WHO/UNICEF on UN-Water Platform
6.3 By 2030, improve water quality by reducing pollution, eliminating dumping and minimizing release of hazardous chemicals and materials, halving the proportion of untreated wastewater and substantially increasing recycling and safe reuse globally	6.3.1 A proportion of domestic and industrial wastewater flows safely treated 6.3.2 The proportion of bodies of water with good ambient water quality	Data on 6.3.1 was collected from the VNR report and Population and Housing Census. Data was updated by WHO/UNEP on UN-Water Platform for indicator 6.3.1. For 6.3.2, data has not been assessed on this indicator in Ghana from documentation, UN- Water, and other assessment bodies. However, the water quality index could be determined from the VNR Report through the Ministry of Sanitation and Water
6.4 By 2030, substantially increase water-use efficiency across all sectors and ensure sustainable withdrawals and supply of freshwater to address water scarcity and substantially reduce the number of people suffering from water scarcity	6.4.1 Change in water-use efficiency over time 6.4.2 Level of water stress: freshwater withdrawal as a proportion of available freshwater resources	No data for 6.4.1 within any of the official documents. However, data was available on UN-Water platform. Data for 6.4.2 was collected from the VNR report and available on the UN-Water platform by FAO
6.5 By 2030, implement integrated water resources management at all levels, including through transboundary cooperation as appropriate	6.5.1 Degree in integrated water resources management 6.5.2 Proportion of transboundary basin area with an operational arrangement for water cooperation	Data for 6.5.1 was collected from the VNR report and Population and Housing Census and updated by UNEP on UN-Water Platform. No data for 6.5.2. within any of the official documents. However, data was available on UN-Water platform provided by UNESCO/UNECE

(continued)

Table 1 (continued)

Targets	Indicators	Data collection status
6.6 By 2030, protect and restore water-related ecosystems, including mountains, forests, wetlands, rivers, aquifers, and lakes	6.6.1 Change in the extent of water-related ecosystems over time	No data for 6.6.1 within any of the official documents. However, data was available on UN-Water platform by UNEP
6.a. By 2030, expand international cooperation and capacity-building support to developing countries in water- and sanitation-related activities and programs, including water harvesting, desalination, water efficiency etc	6.a.1 Amount of water- and sanitation-related official development assistance that is part of a government-coordinated spending plan	No data for indicator 6.a.1 within any of the official documents. However, data was available on UN-Water platform provided by the WHO/ OECD
6.b. Support and strengthen the participation of local communities in improving water and sanitation management	6.b.1 The proportion of local administrative units with established and operational policies and procedures for the participation of local communities in water and sanitation management	No data for indicator 6.b.1 within any of the official documents. The WHO have data on only the level of participation by communities in Ghana. Hence, this indictor has not been assessed in Ghana

trends and advancements in achieving the goals. This comprehensive approach was designed to garner a holistic perspective on Ghana's advancement toward each SDG 6 target. This research employed meta-analysis, a technique combining data from multiple studies and documents, to draw holistic conclusions. Commonly used in systematic reviews, this approach allowed for a thorough examination of each target, offering an in-depth understanding of Ghana's progress and emerging trends in its journey towards achieving SDG 6. Furthermore, the research extended its scope to include other salient government documents, such as the 2021–2022 MTEF. This was instrumental in ascertaining the financial commitments and strategic undertakings of the Ghanaian government in its quest to fulfill SDG 6. Such an analysis rendered a detailed overview of the government's dedication to achieving the targets of Goal 6. The study underscored the importance of employing universally accepted metrics or indicators as critical tools for monitoring the progression of the SDGs. Nonetheless, it acknowledged the prevalent challenges in tracking and quantifying SDG progress in numerous developing countries, including Ghana. To organize the analysis effectively, the study constructed a table that clearly outlined the metrics for SDG 6, along with their associated targets. This comprehensive framework played a crucial role in supporting the assessment of Ghana's progress, enabling a systematic and thorough evaluation of the achievement levels for each specified target.

4 Results

The results section presents the analysis carried out on data and information gathered from secondary sources and is presented according to the indicators of SDG 6. The progress of achievement and implementation of each indicator of SDG 6 is presented under this section.

4.1 Indicator 6.1.1: Proportion of Population Using Safely Managed Drinking Water Services

The results for Ghana's proportion of the population using safely managed drinking water services show a positive trend, according to the data sourced from WHO/UNICEF in 2022 as depicted in Fig. 2. In 2010, only 25% of Ghana's population had access to safely managed drinking water. This number saw a significant increase over the next decade, reaching 42% by 2020. By 2022, the proportion further improved to 44%. This steady growth indicates Ghana's ongoing efforts and progress in enhancing access to safe drinking water, a vital component of public health and well-being.

In 2021, approximately 92.2% of homes in Ghana had access to improved drinking water sources (Ghana Statistical Service, 2022; WHO/UNICEF, 2022). This number represents a seven percent increase from 2010, when 85.3% of Ghanaians had basic access to drinking water as presented in Fig. 3. Ghana is progressing towards its national objective of providing fair and equal availability of essential drinking water services by 2025. Compared to rural households, a significant percentage of urban homes have refined drinking water sources and basic drinking water services. According to WHO/UNICEF (2022), Ghana has seen a further increase of 2% in access to improved drinking sources, accumulating to 94% of the population with access to improved water services. The primary water sources in urban areas are sachet water (51.5%) and pipe-borne water (33.6%), while in rural areas, it is borehole/tube well water (33.6%) and pipe-borne water (28.8%) (Ghana Statistical Service, 2022). Six regions in Ghana have achieved nearly universal access to improved water sources, while in four regions (Northern, Northeast, Oti, and Savannah) more than one-fifth of the population still lacks access to such sources as depicted in Fig. 4.

4.2 Indicator 6.2.1: Percentage of Ghanaians Using (a) Safely Managed Sanitation Services and (b) A Hand-Washing Facility with Soap and Water

The results for Ghana's proportion of the population using safely managed sanitation services indicate a positive trend, according to data sourced from WHO/UNICEF. In 2010, the percentage of the population with access to safely managed sanitation services was 9%. Over the next decade, this Figure increased, reaching 15% by 2020. By 2022, the proportion had slightly improved to 16% as indicated in Fig. 5. Although the progress is modest, it signifies Ghana's

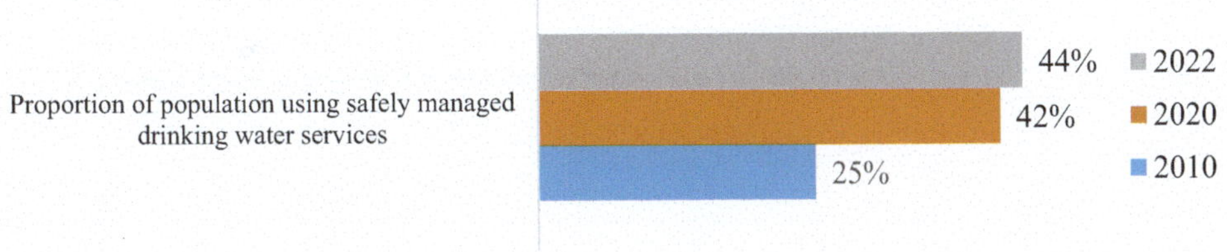

Fig. 2 Accessibility to safely managed water services 2010, 2020 and 2022, (%). *Source* National Development Planning Commission (2022) and WHO/UNICEF

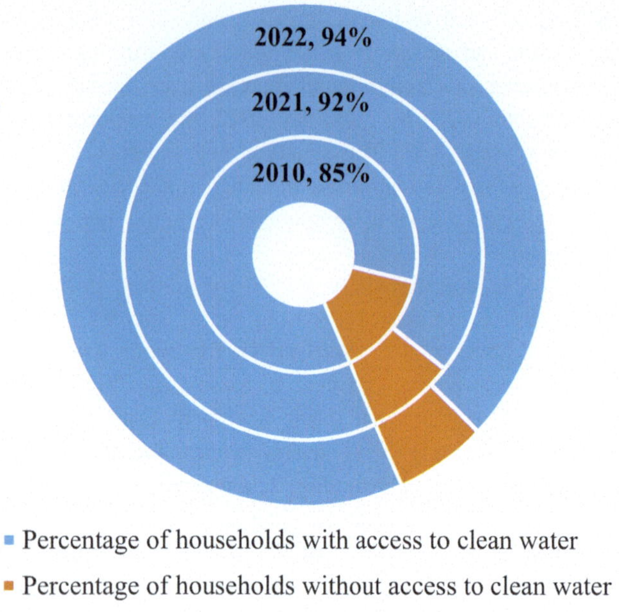

Fig. 3 Accessibility to enhanced water services 2010, 2021 and 2022, (%). *Source* National Development Planning Commission (2022) and WHO/UNICEF

ongoing efforts to enhance sanitation facilities. Regarding improved sanitation services, the data reflects a more substantial level of access among the population. In 2010, 65% of Ghanaians were using improved sanitation services. This number grew to 72% by 2020 and saw a slight increase to 73% by 2022 as indicated in Fig. 6. The consistent improvement over the years shows that a larger proportion of the population has access to better sanitation services, which are crucial for public health and well-being.

In 2021, 59.3% of households in Ghana had access to a toilet, an increase from 46% in 2010 (Ghana Statistical Service, 2022). Urban households have higher toilet access rates compared to rural households. The percentage of households using public restrooms decreased by twelve percentage points to 23% as of 2021 (see Fig. 7). Moreover, open defecation decreased by 2.3% in 2021, reducing to 17.7%. To eradicate open defecation by 2030, Ghana ought to minimize the population devoid of toilets by an average of 2% per annum (National Development Planning Commission, 2022). Data from the 2021 JMP report indicated the percentage of people using hand-washing facilities with soap and water showed a marginal increase from 41.3% to 41.53% between 2017 and 2020 (see Fig. 8). In 2020, approximately 47% of households in urban areas had available hand-washing facilities with soap and water, compared to 35% of the population in rural settings (Ghana Statistical Service, 2022; National Development Planning Commission, 2022).

4.3 Indicator 6.3.1: A Proportion of Domestic and Industrial Wastewater Flows Safely Treated

According to WHO (2022), the proportion of wastewater flow safely treated was 12% domestically in Ghana. A substantial percentage of homes (about 70.6%) in Ghana dispose of their wastewater on the ground, street, or outside, as stated by the 2021 PHC report. This practice is more prevalent in rural regions (88.9%) than in urban regions (58.7%) (Ghana Statistical Service, 2022). The percentage

Fig. 4 The Percentage of homes with accessibility to drinking water sources by region, 2021, (%). *Source* National Development Planning Commission (2022)

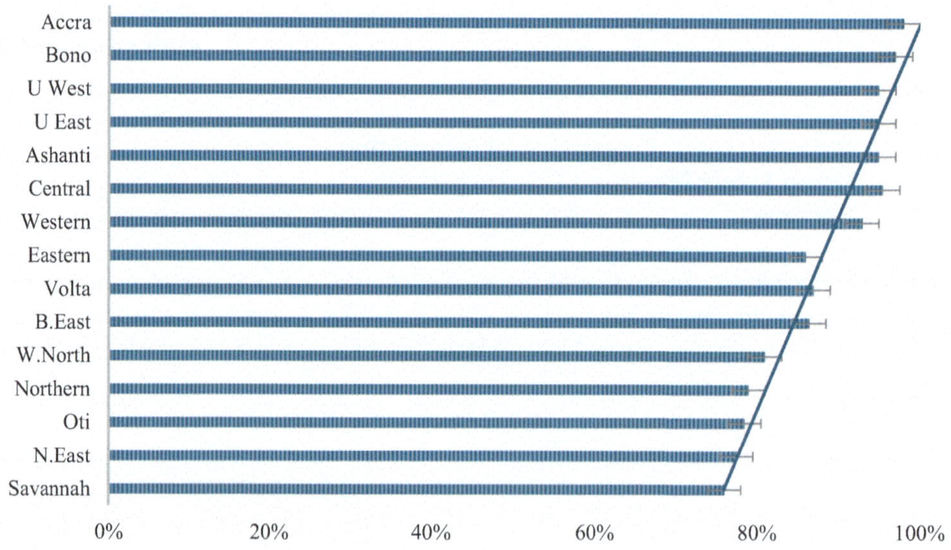

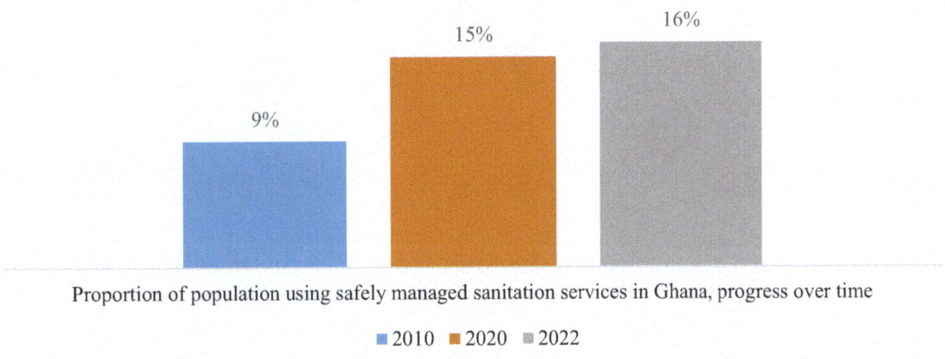

Fig. 5 Proportion of population using safely managed sanitation services 2010, 2020, 2022, (%). *Source* WHO/UNICEF

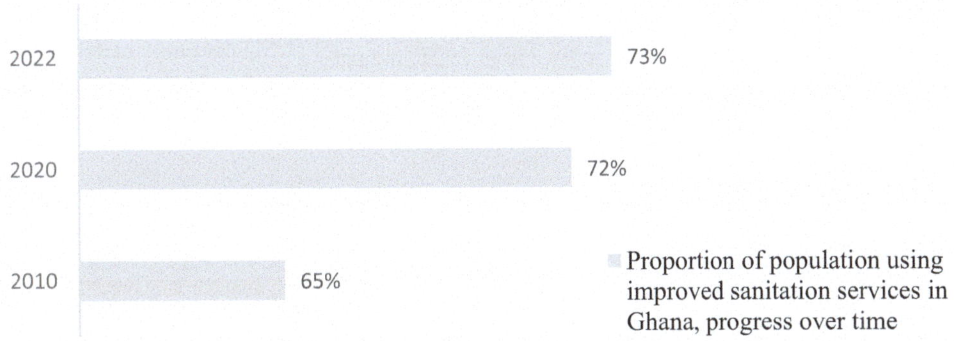

Fig. 6 Proportion of population using improved sanitation services in Ghana, 2010, 2020, 2022. *Source* WHO/UNICEF

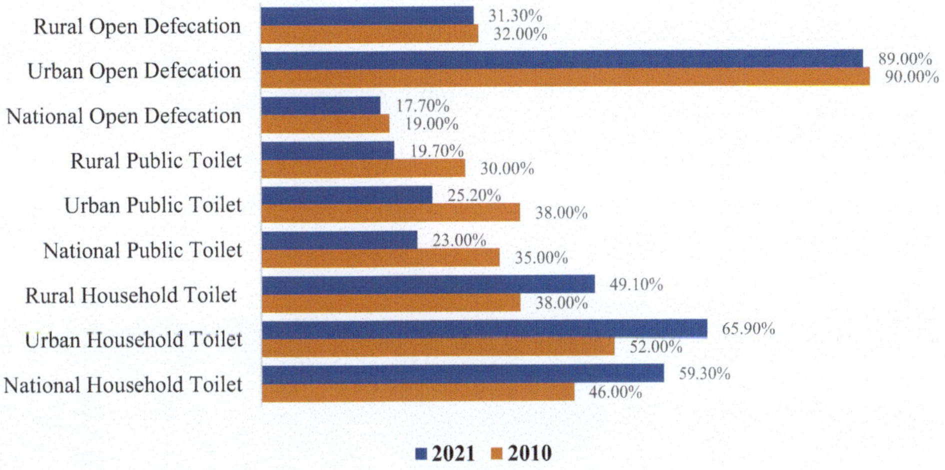

Fig. 7 Toilet facilities utilized by homes by type of vicinity 2010 and 2021, (%). *Source* 2010 and 2021 Population and Housing Census

of the population using sewage systems is more prevalent in metropolitan, town, and city areas (3.2%) than in rural settings (0.9%), representing a difference of 2.3% (Refer to Fig. 9). Additionally, there are regional variations in wastewater disposal methods. The Greater Accra Region shows a significant prevalence of soak-away wastewater disposal (17.8%). At the same time, the Bono East Region primarily disposes of wastewater on the ground, street, or outside (89.3%).

4.4 Indicator 6.3.2: Proportion of Bodies of Water with Good Ambient Water Quality

The proportion of bodies of water with good ambient quality in terms of the water quality index of Ghana has been assessed by the Ministry of Sanitation and Water. In 2021, Ghana implemented the National Water Quality Monitoring Program to improve the effectiveness of water pollution

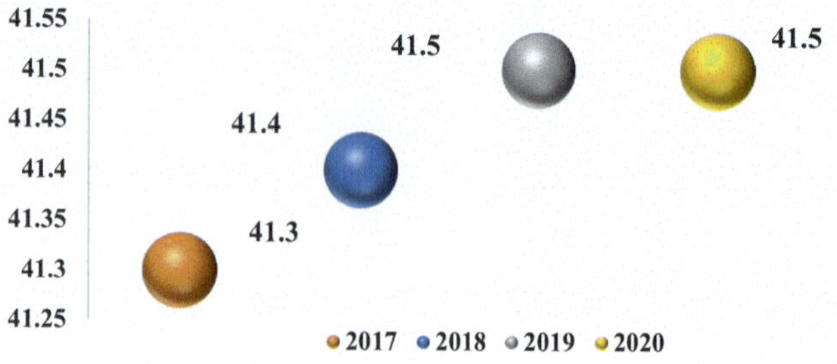

Fig. 8 The percentage of homes using hand-washing spaces with soap and water, 2017–2021, (%). *Source* JMP Report 2021

prevention (National Development Planning Commission, 2022). The water quality in major river basins improved from an index of 57.8% in 2019 to 58.2% in 2021, indicating good water quality (see Fig. 10). However, despite these slight improvements, the basins continue to face challenges from uncontrolled pollution, illegal mining, and waste disposal.

4.5 Indicator 6.4.1: Change in Water-Use Efficiency Over Time

In Ghana, the change in water-use efficiency from 2010 to 2020 has been marked by an increase in the cost of water per cubic meter. In 2010, the price stood at 18 USD per cubic meter. By 2015, there was a noticeable rise to 24 USD per cubic meter, reflecting a change in the economic or environmental factors affecting water resources. By the end of the decade, in 2020, the cost had risen further to 33 USD per cubic meter as indicated in Fig. 11. This ascending trend indicates a significant change in water-use efficiency in the country over a period of ten years.

Fig. 9 Household wastewater disposal by method and locality, 2021, (%). *Source* 2021 Population and Housing Census, 2022

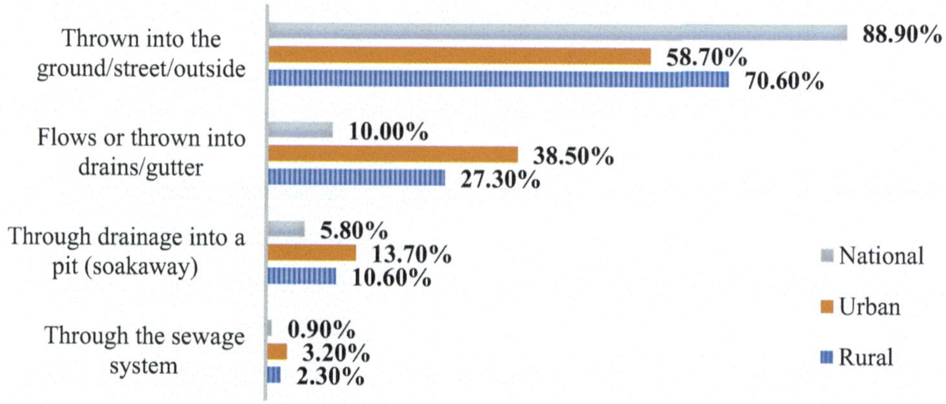

Fig. 10 Water Quality Index, 2017–2021. *Source* Ministry of Sanitation and Water Sector Medium-Term Development Plan, 2022–2025

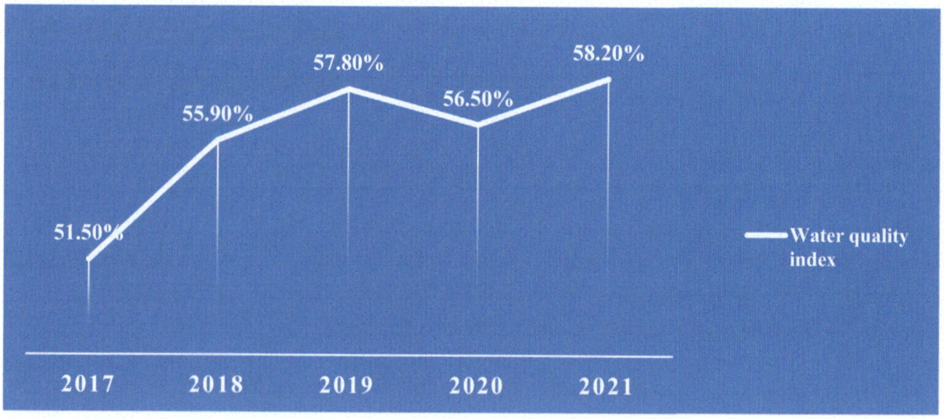

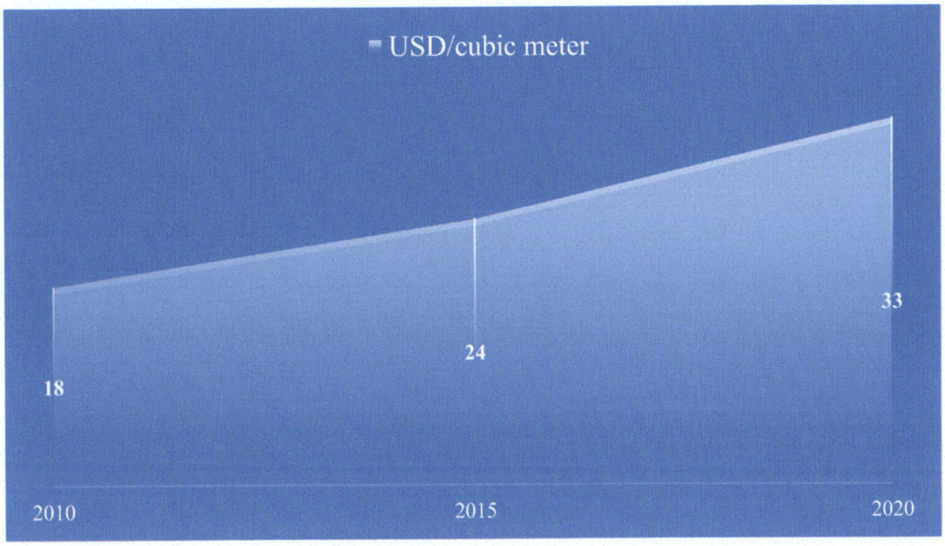

Fig. 11 Change in water-use efficiency over time, 2010, 2015 and 2020. *Source* FAO

4.6 Indicator 6.4.2: Level of Water Stress—Freshwater Withdrawal as a Proportion of Available Freshwater

According to the Water Resources Profile Series (USAID, 2021), Ghana is endowed with ample water resources and does not face significant water stress. The country's freshwater withdrawal as a proportion of available freshwater resources hit approximately 6% in 2020 and 6.3% in 2021 (FAO, 2022; USAID, 2021). This percentage is below the threshold of 25%, showing Ghana has a safe level of water stress. Additionally, its per capita renewable water resource of 1949 m^3 exceeds the Falkenmark Index for water stress. However, the accessibility of water per person in Ghana has decreased. It decreased from 1926 m^3/person from 2017 to 1916 m^3/person in 2020 and farther to 1725 m^3/person in 2021 (see Fig. 12) (National Development Planning Commission, 2022). The water stress levels are in proximity to the suggested lowest threshold of 1700 m^3 per head based on the 2021 population (National Development Planning Commission, 2022). To mitigate the adverse effects of human activities on river bodies, Ghana planted 16,937 tree seedlings through buffer zone restoration programs in specific communities located along the Black and White Volta, Tano, Offin, and Densu river basins (Water Resource Commission, 2013). This initiative aimed to restore degraded buffers along river bodies.

4.7 Indicator 6.5.1: Degree of Integrated Water Resources Management

The extent of management of water resources increased from 49% in 2017 to 51% in 2019 and 57% in 2021 (National Development Planning Commission, 2022; Water Resources Commission, 2021). The position in Ghana is near the classification of moderately high (60%); however, it remains significantly distant from being classified as very high (100%). As part of the Volta Basin project, Integrating Flood and Drought Management and Early Warning for Climate Change Adaption (VFDM), an evaluation of existing capacities and requirements for creating risk maps were conducted (Water Resources Commission, 2021). This initiative aimed to assist the six countries of the Volta Basin (Benin, Burkina Faso, Côte d'Ivoire, Ghana, Mali, and Togo) in implementing collaborative and unified efforts to enhance the existing Early Warning Systems and management strategies for floods and droughts while also bolstering regional, national, and local capacities resilience (Water Resources Commission, 2021). Furthermore, in partnership with essential stakeholders such as the Ministry of Water and Sanitation, the Water Resources Commission created an online platform to distribute early warning information to communities.

4.8 Indicator 6.5.2: Proportion of Transboundary Basin Area with an Operational Arrangement for Water Cooperation

In 2020, Ghana demonstrated commendable performance on the proportion of transboundary river and lake basins and aquifers with operational arrangements for water cooperation (UNESCO/UNECE, 2022). Specifically, Ghana achieved a high level of 91%, indicating the establishment of effective mechanisms for water cooperation within its transboundary river and lake basins at 88% and aquifers at 96% (UNESCO/UNECE, 2022).

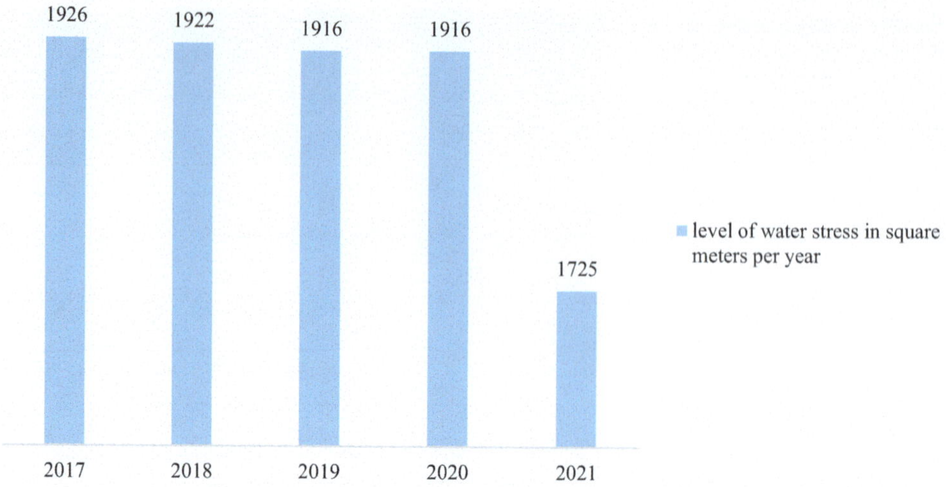

Fig. 12 Level of water stress, 2017–2021 (m^3 per capita). *Source* Water Resources Commission (2021)

4.9 Indicator 6.6.1: Change in the Extent of Water-Related Ecosystems Over Time

Ghana has observed significant changes in the spatial extent of surface water bodies, encompassing both expansions and contractions (UNEP, 2022). These changes are often associated with climate-related phenomena, including flooding and droughts. During the baseline period of 2001–2005, the total area covered by surface water in Ghana amounted to 5564 km^2 (UNEP, 2022). However, during the latest five-year period from 2011 to 2015, this area expanded to 7233 km^2, representing a notable gain of 30% compared to the baseline (UNEP, 2022). These findings underscore the dynamic nature of surface water resources in Ghana and their susceptibility to environmental fluctuations.

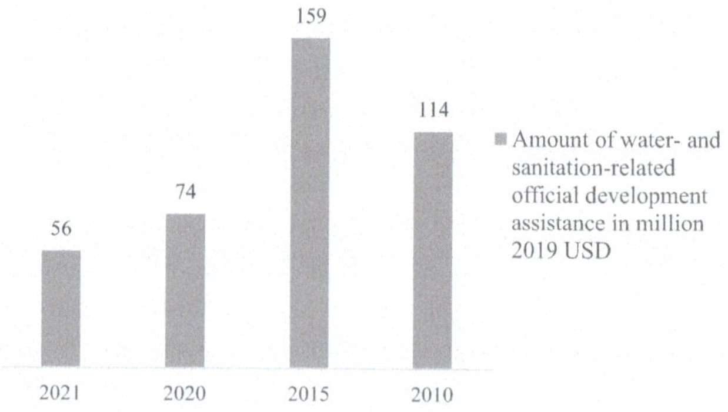

Fig. 13 Amount of water- and sanitation-related official development assistance that is part of a government-coordinated spending plan. *Source* WHO/OECD

4.10 Indicator 6.a.1: Amount of Water- and Sanitation-Related Official Development Assistance that is Part of a Government-Coordinated Spending Plan

The data on water- and sanitation-related Official Development Assistance (ODA) for Ghana reveals variations in funding over the last two decades, as reported by WHO/OECD. In 2010, the ODA commitment stood at 114 million USD, which indicates a substantial investment in the sector. By 2015, the assistance had increased significantly to 159 million USD, reflecting the highest point of aid during the reported period. However, post-2015, there was a downward trend. In 2020, the amount of ODA decreased to 74 million USD, almost half of what it was in 2015. This decline continued into 2021, with the amount further decreasing to 56 million USD (Refer to Fig. 13).

4.11 Indicator 6.b.1: The Proportion of Local Administrative Units with Established and Operational Policies and Procedures for the Participation of Local Communities in Water and Sanitation Management

The proportion of local administrative units with established and operational policies and procedures for the participation of local communities in water and sanitation management in Ghana has not been assessed within official documents and on UN-Water platform and other assessment bodies. However, the WHO determines the level of participation by communities in Ghana in water and sanitation management in three ranks. Participation levels at 1 are very low, at 2 are moderate, and at 3, they are high. In Ghana, procedures in law or policy for participation by communities are clearly defined. In 2017, the level of participation in water resources planning and management was moderate (level 2). However, in 2019 and 2021, there was a high level of community participation in water resources planning and management nationwide (level 3).

4.12 Results of Ministerial Sector Indicators from the Ministry of Water Resources and Sanitation

The Ministry of Water Resources and Sanitation plays a pivotal role in managing Ghana's water resources and ensuring the provision of clean water and sanitation services to the population. The following overview presents key findings from the latest set of Ministerial Sector indicators as displayed in the Medium-Term Expenditure Framework for 2022–2025. These indicators are instrumental in gauging the progress, challenges, and impact of policies and initiatives undertaken by the Ministry. This is indicated in Table 2. The analysis of these indicators is critical for informing strategic decisions and aligning efforts with the national development agenda to achieve sustainable management of water and sanitation resources.

The targets from the Ministry of Finance and Economic Planning for 2021 aimed for improvements in water and sanitation, with goals set for safely managed drinking water, water-use efficiency, and integrated water management. By end of year 2021, actual results from the VNR, WHO and other sources, reported a slight shortfall in drinking water access at 44% against 44.26%, and a level of water stress at 1725 compared to 1910 from the MFET. However, the goal for integrated water resource management was surpassed, achieving 57% against 55%. These outcomes highlight progress but also reflect the challenges in data scope and reporting timelines, emphasizing the need for continual strategy refinement to meet Ghana's water and sanitation ambitions.

4.13 Contributions of Government, SOEs, NGOs, and Private Sector Towards Goal 6

The Government of Ghana (GoG) is deeply committed to improving the availability of water and sanitation services and sources through various sectoral laws, strategies, and schemes (UNICEF, 2022). However, financing these services remains a significant challenge. The expansion of access to adequate and safe water and sanitation in Ghana is hindered by formidable obstacles, primarily due to rapid urbanization, which strains urban infrastructure, including the delivery of water and sanitation services (UNICEF, 2022). Meeting Ghana's SDG 6 water and sanitation targets would require an estimated annual investment of over six billion Ghanaian cedis, which is 0.82% of the nation's GDP (WASH budget brief, 2021). According to the Ghana Clean Water, Access 2000–2022 report, the proportion of the population with safe water availability in Ghana has steadily increased by approximately 1.6–1.75% per year. This is further backed by the VNR report and the WHO/UNICEF. Table 3 illustrates this trend from 2017 to 2022.

The Ministry of Sanitation and Water Resources (2022) stated that the total public funding allocated for water and sanitation in 2021 was estimated to be 560 million Ghanaian cedis, less than 10% of the total funding required. The WASH budget summary reveals that Ghana's domestic public expenditure on sanitation is estimated to be less than 0.04% of GDP in 2021. As per the eThekwini Declaration of 2008, the government has committed to allocating at least 0.5% of GDP to WASH. The Ministry has prepared a report on the appropriations bill based on the Medium-Term Expenditure Framework (MTEF) for 2022–2025. The summary of the bill's outcomes is presented in Table 4.

According to Statista, Ghana has observed a constant price Quarterly GDP contribution from water and sewage. The water supply and sewage industry, in the second quarter of 2021, contributed approximately 211 million Ghanaian cedis (GHS), equivalent to around 30.7 million USD, to the country's gross domestic product (GDP) (Sasu, 2022). The first quarter of 2021 recorded the highest value at current prices since 2019, increasing from the previous quarter's value of approximately 598 million GHS. However, the second and third quarters of 2021 witnessed the lowest contribution of water and sewage to the nation's GDP. Figure 14 shows the economic contribution made by the water and sewage sector to Ghana's GDP, as measured in current prices, from Q1 2019 to Q2 2021.

The UNICEF-Ghana WASH (Water, Sanitation, and Hygiene) program provides funding for water, sanitation, and hygiene initiatives in developing nations. The total WASH budget for Ghana in 2021 is GHS 558 million, marking a 74% increase compared to 2020. However, this amount falls significantly short of the estimated GHS 1750 million needed to ensure essential services in the sector (WASH budget brief, 2021). The potential future reduction (shown by the forecasted estimation of 496.4 million for 2022–2023) in the budget from this level raises concerns as it could undermine the progress made and impede effective planning and implementation within the sector. The projected WASH expenditures for 2021 are expected to amount to 0.13% of GDP (WASH budget brief, 2021), as depicted in Fig. 15.

All stakeholders in the water and sanitation field have contributed to enhancing accessibility to pure and hygienic water and sanitation sources. The government's "One Village, One Dam" policy and the efforts of organizations such as Water Aid and the Community Water and Sanitation

Table 2 Ministerial policy outcome indicators and targets

Outcome indicator description	Unit of measurement for outcome indicator	Baseline		Latest status		Target	
		Year	Value	Year	Value	Year	Value
Improve access to safe and reliable water supply services for all	Percentage of the population with basic access to drinking water sources	2020	71.5%	2021	75%	2024	85.51%
	Percentage of distribution losses	2020	49.98%	2021	49%	2024	45%
	Percentage of the population with access to safely managed drinking water sources (compare Indicator 6.1.1)	2020	41.4%	2021	44.26%	2024	55.7%
	The proportion of metered customers	2020	86.2%	2021	90%	2024	95.0%
	Billing & Collection ratio	2020	65.83%	2021	75%	2024	85%
Enhance access to improved and reliable environmental sanitation services	Percentage of the population with access to improved liquid waste management	2020	23.7%	2021	31.33%	2024	61.85%
	Number of communities achieving open defecation-free (ODF) status	2020	2148	2021	5498	2024	13,500
	The proportion of solid waste properly disposed of (major towns/cities)	2020	85%	2021	87%	2024	95%
	The proportion of liquid waste (faecal matter) safely disposed of on-site or properly collected, transported, and treated off site	2020	13.3%	2021	19.64%	2024	45%
Promote sustainable water resources development and management	The proportion of bodies of water with good ambient water quality (Water quality index) (compare Indicator 6.3.2)	2020	56.6%	2021	57.85%	2024	75%
	Level of Water Stress (compare Indicator 6.4.2)	2020	1916	2021	1910	2024	1900
	The extent to which integrated water resources management (IWRM) is implemented. It considers the various users and uses of water to promote social equity, economic efficiency, and environmental sustainability at all levels (compare Indicator 6.5.1)	2020	51%	2021	55%	2024	60%

Source Ministry of Finance and Economic Planning, 2022

Table 3 Water access in Ghana from 2017 to 2020

Year	Population percentage access to safe drinking water (%)	Increase/decrease from the previous year (%)
2020	41.47	1.71
2019	39.7	1.69
2018	38.01	1.67
2017	36.34	1.64

Source Ghana Clean Water Access 2000–2022

Table 4 Funding under water and sanitation

Sector for funding under water and sanitation	Approximate expenditure estimates per year in a million cedis			
	2022	2023	2024	2025
Management and administration	24	24	24	24
Water management	676	835	835	835
Sanitation management	102	102	102	102
Grand Total	802	960	960	960

Source Ministry of Sanitation and Water Resources (2022)

Agency that install boreholes within remote areas have performed a vital part in improving population accessibility to water Zoomlion Ghana Limited, a Ghanaian-owned company specializing in comprehensive waste management services, has also made noteworthy contributions. The company offers solutions for waste collection, transportation, transfer, sorting, recycling, and disposal. Zoomlion has collaborated with the government, WASH programs, and other enterprises and non-governmental organizations in the SDG 6 sector to establish projects that align with Goal 6.

5 Discussion

Ghana's journey towards ensuring sustainable water and sanitation for all, as outlined in the Sustainable Development Goals (SDGs), particularly SDG 6, reveals a landscape of significant challenges intertwined with notable progress. Starting with the indicator on safely managed drinking water services, we observe a commendable increase from 25% in 2010 to 44% in 2022 (Indicator 6.1). This nearly 20% rise over twelve years underscores the Ghanaian authorities' commitment to improving water access, although it also highlights the persisting challenges. The distinction between the broad availability of improved drinking water sources, which reached 94% in 2022 up from 85.3% in 2010, and safely managed services is pivotal. While the former suggests widespread access to water sources that are protected from contamination, the latter emphasizes the safety, reliability, and direct accessibility of water services in homes, which remain free from contamination. The existing gap between access to improved sources and safely managed services points to issues such as water quality, reliability, and accessibility that need comprehensive addressing to fully meet the SDG target 6.1. In the case of sanitation (Indicator 6.2), the proportion of the population using safely managed wastewater disposal services rose from 9% in 2010 to 16% in 2022.Although modest, the increase signals a positive shift towards improved public health conditions. This improvement indicates significant potential for enhancing access to sanitation facilities that effectively manage waste. The period also saw growth in access to improved sanitation services from 65 to 73%, indicating a broader trend of improving sanitation infrastructure. Yet, the transition from improved to safely

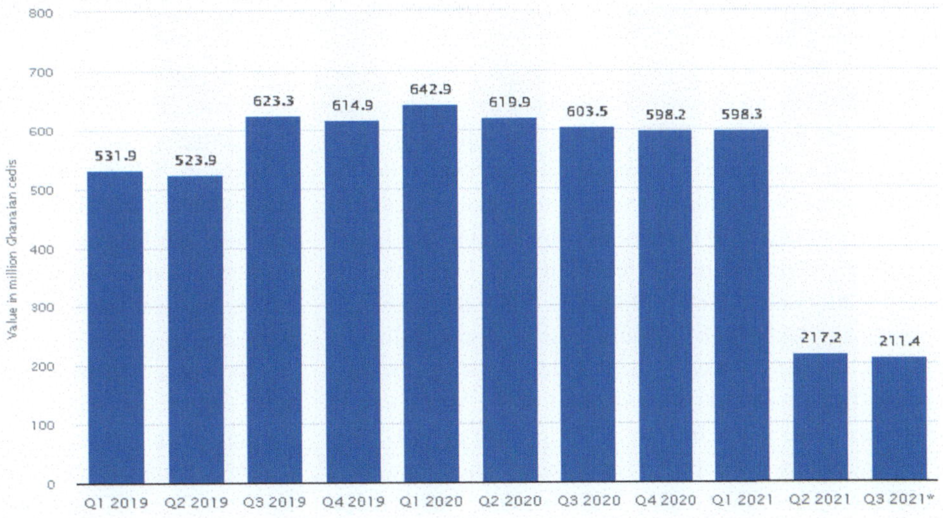

Fig. 14 Value added of water and sewerage to GDP at current prices in Ghana from Q1 2019 to Q3 2021. *Source* Sasu (2022)

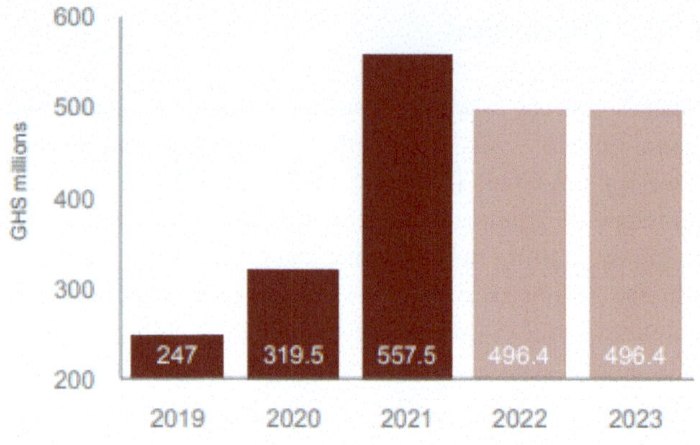

Fig. 15 WASH sector historical and forecasted budgets, 2019–2023. *Source* WASH Budget Brief (2021)

managed services, which involves the complete sanitation chain, including the treatment, disposal, or reuse of human waste, remains a lingering challenge. In the treatment of domestic and industrial wastewater (Indicator 6.3.1), Ghana faces a significant hurdle with only 12% of its wastewater being safely treated. This stark Figure indicates a major gap in wastewater management, with over 70% of homes resorting to inappropriate disposal methods such as dumping on the ground, street, or outside—a practice more prevalent in rural areas. The low utilization of sewage systems, particularly in rural locales, underscores an urgent need for enhanced wastewater management systems. These systems are crucial for mitigating environmental pollution and safeguarding public health. Moreover, the regional disparities in sanitation infrastructure demand targeted investments to bolster sanitation services, aiming for a comprehensive improvement in wastewater management essential for achieving SDG 6.3.1. Furthermore, the proportion of bodies of water with good ambient water quality, as per Indicator 6.3.2, showcases modest progress with an increase in the water quality index from 57.8% in 2019 to 58.2% in 2021. This slight improvement, achieved under the National Water Quality Monitoring Program, is tempered by ongoing challenges from pollution, illegal mining, and improper waste disposal that continue to menace Ghana's water bodies. These issues necessitate stronger regulatory measures, deeper community engagement in conservation efforts, and increased public awareness regarding the importance of water conservation. Such efforts are vital for protecting community health and ensuring the sustainability of Ghana's water resources. The indicator for water-use efficiency (Indicator 6.4.1), as evidenced by the rising cost of water per cubic meter from 18 USD in 2010 to 33 USD in 2020, marks a significant shift in the economic valuation of water resources in Ghana. This increase may reflect improved efficiency or elevated costs stemming from scarcity, treatment, and distribution challenges. It potentially signals environmental pressures and the urgent need for substantial investments in water infrastructure to either maintain or enhance service levels. The crux of the challenge lies in achieving a delicate balance between cost recovery and efficiency, against the backdrop of affordability and equitable access to water services. This balance is essential for sustainable development, as it ensures that water resources are used efficiently while remaining accessible to all segments of the population. When considering water stress (Indicator 6.4.2), as indicated by Ghana's freshwater withdrawal as a proportion of available freshwater resources—around 6% in 2020 with a slight increase to 6.3% in 2021—the country exhibits a relatively low level of water stress compared to many regions worldwide. Nonetheless, the decline in per capita renewable water resources from 1926 m^3/person in 2017–1725 m^3/person in 2021, nearing the water stress threshold of 1700 m^3 per person, signals an emerging challenge. This downward trend, though still within safe limits, underscores the imperative for sustainable water management practices. Such practices are crucial to avert future water stress, particularly considering population growth and the potential impacts of climate change, emphasizing the need for proactive measures to ensure water sustainability. The progress in integrated water resources management (IWRM), with its degree of implementation rising from 49% in 2017 to 57% in 2021 (Indicator 6.5.1), illustrates Ghana's commitment to a holistic approach to water management. This approach is indispensable for tackling the diverse challenges associated with water supply, sanitation, water quality, and environmental sustainability. The positive trend in IWRM implementation highlights efforts to enhance policy coordination, stakeholder engagement, and the adoption of best practices in water management across the country. Despite these advancements, achieving higher benchmarks for integrated management necessitates ongoing improvements in governance frameworks, data availability, and the strengthening of institutional capacities. Furthermore, the achievement of a 91% level in operational arrangements for water cooperation across Ghana's transboundary river, lake basins, and aquifers in 2020 stands out as a significant accomplishment (Indicator 6.5.2). This high level of cooperation indicates effective mechanisms in place for managing shared water resources, emphasizing the importance of collaborative water management and aquifer management within Ghana's water strategy. The distribution of this indicator, with 88% for river and lake basins and an impressive 96% for aquifers, underscores Ghana's dedication to regional stability and sustainable water use. Such collaborative efforts ensure that transboundary water resources are managed in a manner that benefits all involved parties, supporting both regional stability and the sustainability of ecosystems. The slight increase in the extent of water-related ecosystems, with surface water area expanding by 30% from 5564 km^2 during 2001–2005 to 7233 km^2 in the 2011–2015 period, is a testament to the dynamic nature of Ghana's environmental landscape (Indicator 6.6.1). This change, attributed to natural variability and the potential impacts of climate change, such as more pronounced flooding, underscores the necessity for adaptive management strategies. These strategies are crucial for responding to fluctuations in water availability, ensuring biodiversity conservation, maintaining water quality, and managing floods effectively. The expansion of water-related ecosystems illustrates the critical need for continuous monitoring and the development of robust management frameworks that can adapt to environmental changes, safeguarding the sustainability of these vital resources. On the financial front, the fluctuation in water- and sanitation-related ODA received by Ghana, which saw a peak of 159 million USD in 2015 before falling to 56 million USD in 2021, reflects shifts in international funding priorities and the global economic landscape (Indicator 6.a.1). The peak in 2015 indicates a strong international commitment to enhancing Ghana's water and sanitation infrastructure. However, the subsequent decline in ODA highlights the imperative for Ghana to diversify its funding strategies for water and sanitation projects. This includes increasing domestic investment and exploring innovative financing mechanisms to compensate for the reduction in international assistance. The efficiency and effectiveness in utilizing available funds have become more critical than ever to sustain the momentum of improvements in water and sanitation infrastructure, demonstrating the need for strategic financial planning and management to continue advancing towards the SDG 6 targets. Also, the level of community participation in water resources planning and management in Ghana represents a cornerstone of the country's approach to sustainable water and sanitation services (Indicator 6.b.1). Being assessed as high (level 3) in recent evaluations, this level of engagement signifies a robust framework for incorporating local communities into decision-making processes. High community participation ensures that the planning and management of water resources are reflective of and responsive to local needs and priorities. The transition from moderate to high levels of engagement not only demonstrates Ghana's success in promoting an inclusive approach to water and sanitation management but also emphasizes the effectiveness of participatory approaches in enhancing the sustainability of initiatives. This engagement promotes local ownership, ensures that interventions are culturally appropriate, and aligns with community needs, thereby bolstering the overall impact of water and sanitation projects. Ghana's water and sanitation sector has demonstrated significant progress, with access to basic drinking water sources rising from 71.5% in 2020 to 75% in 2021, aiming for 85.51% by 2024. This achievement is part of a broader effort to enhance water resource management and sanitation services, as outlined by the Ministry of Water Resources and Sanitation. Key initiatives have led to reductions in distribution losses and substantial advancements in liquid waste management, reflecting the sector's commitment to sustainable management and aligning with national development goals.

These achievements underscore the effectiveness of government policies and underscore the pivotal role of collaborations with NGOs and private entities, notably Zoomlion, in driving progress within the water and sanitation sector. Despite these strides, the sector faces persistent challenges, including funding shortages and the pressures exerted by rapid urbanization. The economic significance of the water and sewage sector is highlighted by its contribution of approximately 211 million GHS to Ghana's GDP in the second quarter of 2021, yet this Figure also points to the variability in its economic impact. The increase in the UNICEF-Ghana WASH program's budget to 558 million GHS in 2021, marking a 74% rise from the previous year, falls short of the estimated 1750 million GHS required to meet sector needs comprehensively. This funding gap, coupled with the threat of future budget reductions, jeopardizes the sustainability and expansion of water and sanitation services. Addressing these financial constraints is crucial for maintaining momentum towards achieving SDG 6 in Ghana. Collaborative efforts among government agencies, development partners, and various stakeholders are pivotal in this endeavor. Achieving universal access to clean water and sanitation services in Ghana by 2030 is not just a national goal but also aligns with the country's commitment to SDG 6.

5.1 The Way Forward: Boosting SDG 6 in Ghana

Ghana has made commendable progress toward improving sanitation conditions, but there is still much progress to ensure the all-inclusive availability of sanitation spaces for Ghanaians. UNICEF has played a pivotal role in Ghana by actively promoting behavior change, improving structures, and increasing access to basic sanitation facilities. These endeavors are crucial in minimizing water-related illnesses and enhancing the overall health and well-being of the population. To attain the goal of providing essential sanitation services to Ghanaians, the government needs to focus on establishing modern public toilet spaces and encouraging investments, especially by private sector players and NGOs, in constructing private toilets, particularly in slum areas. Additionally, Ghana has set its sights on eliminating open defecation by 2030, which requires a yearly minimization of 2% of the persons without access to toilets. To attain this, the government must review and amend some urban sanitation policies and the mandate that all newly built homes include water closets, sinks, baths, and other essential sanitation resources. Noncompliant households should face fines to ensure compliance (Ametepey, 2023; Asiamah, 2021). Unfortunately, the government's efforts in providing basic sanitation services have shown only minimal progress, as shown by the Voluntary National Review Report and the census report by the Ghana Statistical Service in 2021. However, there is hope for improvement through private and state investments in the sanitation sector and by giving it the necessary priority it merits. Achieving the goal of delivering vital sanitation services to every Ghanaian citizen is feasible, but it includes a significant commitment and dedication from all stakeholders, especially the government. The absence of the government's serious attention in the sanitation sector will prove challenging in meeting the targets showcased in SDG 6 (Ametepey, 2023; Asiamah, 2021). Several interventions can contribute to improving the situation. First, investing in private water production companies can extend their reach to remote and distant areas and supply clean drinking water. Expanding pipe-borne water networks to underserved regions and establishing water treatment plans that utilize local water resources within communities are also essential steps (Ametepey, 2023). The government should prioritize providing clean water to regions such as Oti, Northern, Northeast, and Savannah regions, which currently face significant challenges in accessing clean water throughout the country. A major obstacle in Ghana's water supplication undertakings is water pollution, primarily caused by illegal mining, locally known as “Galamsey”. This illegal mining activity has been attributed to the destruction of numerous water sources and continues to hinder advancement toward SDG 6. The VNR report indicates that despite minor enhancements to Ghana's water bodies, freshwater sources are still adversely impacted by illegal mining, uncontrolled pollution, and improper waste disposal into rivers. The government must tackle this issue by intensifying its assay to stop waste disposal, illegal mining, and indiscriminate pollution in river bodies. This strategy involves taking decisive action against illegal miners and their supporters and empowering key bodies such as the Minerals Commission to effectively track and regulate legal mining activities (Ametepey, 2023). Addressing water pollution is urgent, particularly considering the constrained availability of functional water filters. Intentions to substitute faulty filters have been considered by the government, but the estimated cost of USD 35 million poses an incredible challenge given the nation's current fiscal situation (Ametepey, 2023). Nevertheless, Ghana has made substantial strides in supplying water accessibility sources nationwide. However, the quality of the water supply still requires further improvement. Investments by state and private organizations in the water production sector, coupled with stringent monitoring of water basins, can help bridge the gap and enhance the overall water quality for the population.

6 Limitations

In conducting this research, certain assumptions were necessary to gauge Ghana's advancements towards achieving SDG 6. A fundamental assumption was that secondary data from government publications, international reports, and analyses from the private sector are accurate reflections of Ghana's situation. This research presumed consistency in the data collection methodologies across these diverse sources, which is essential for making valid temporal comparisons. Moreover, it was assumed that the Ghanaian government's publicly stated policies and commitments towards SDG 6 genuinely mirror their actual efforts and financial investments. A critical limitation identified in this study is the challenge of obtaining comprehensive data, particularly concerning Target 6.6, which focuses on the protection and restoration of water-related ecosystems. The last data collection for this target was in 2016, highlighting a significant gap and underscoring the need for another comprehensive review after a five-year period to assess current performance accurately. This gap underscores the broader issue of data accessibility, which is further compounded by a reliance on reports that may not always provide an unvarnished view due to potential government influence. The necessity for cross-verification with reports from international bodies becomes evident, suggesting a move towards providing real-time data for ongoing progress monitoring. Additionally, while Ghana shows potential in improving water and sanitation services, the achievement of safely managed services presents a considerable challenge, potentially jeopardizing the 2030 targets. The difficulty in obtaining actual data, hence the reliance solely on reports, poses a significant limitation. These reports, potentially influenced by governmental perspectives, highlight the importance of corroborating findings with international bodies. Such cross-checking is vital for a more balanced and less biased understanding of progress, advocating for the inclusion of live data mechanisms to track real-time advancements. Considering the importance of improved water and sanitation services, Ghana appears to be on a promising path toward achieving the targets of SDG 6. However, the goal of ensuring safely managed services by 2030 seems increasingly challenging, given the current trajectory. This research underscores the urgency for Ghana to prioritize comprehensive data collection and analysis, particularly for Target 6.6. Focused efforts on this target are essential, as they will significantly inform the status and health of water ecosystems within the country. Additionally, there is a pressing need for enhanced support and collaboration at both national and international levels, ensuring that adequate human and technological resources are dedicated to data collection for these critical areas.

7 Conclusions

Through the lens of the conducted research and its findings, it becomes clear that Ghana is making significant progress towards Sustainable Development Goal 6, aiming for universal access to clean water and sanitation. This progress is a result of concerted efforts from both the government and private sectors, with notable collaboration from organizations such as UNICEF and companies like Zoomlion. These partnerships have been instrumental in expanding clean water access and enhancing sanitation infrastructure across the nation. The adoption of advanced technologies and forward-thinking policies has led to a decrease in water distribution losses and improvements in liquid waste management, showcasing Ghana's proactive stance on sustainable water resource management. While there is potential for significant improvements in water and sanitation services, the goal of achieving universally safely managed services by 2030 is daunting. The challenges in acquiring accurate and impartial data suggest that, despite advancements in improving services, meeting the criteria for safely managed services throughout the country by the target year might be unattainable without substantial policy and operational shifts. Addressing these challenges is critical, and it requires Ghana to place a high priority on thorough data collection and analysis. Implementing systems for real-time data monitoring and fostering greater international cooperation could offer a clearer and more objective view of the nation's progress. Furthermore, the government and all stakeholders need to allocate resources and support judiciously, focusing particularly on underprivileged regions to ensure a balanced and inclusive approach to improving water and sanitation services. Ghana's journey towards achieving SDG 6 is a testament to the nation's commitment to enhancing the quality of life through improved water and sanitation services. However, to fully realize these goals, a concerted effort to address data collection challenges, ensure equitable service distribution, and enhance the sustainability of water resources is required. Through concentrated commitment and joint initiatives involving the government, academic institutions, and the private sector—spanning projects, research, and funding—Ghana is well-equipped to tackle the intricacies of this challenge. By harnessing these collaborative efforts, the nation stands on the brink of significant achievements towards a future where every citizen has access to clean water and sanitation, thus honoring its pledge to the Sustainable Development Goals.

References

Akoto-Bamfo, A. (2020). *An examination of the implementation of SDG 6: The case of Ghana* (Doctoral dissertation, University of Ghana, Legon).

Alba, R., Kooy, M., & Bruns, A. (2020). Conflicts, cooperation, and experimentation: Analysing the politics of urban water through Accra's heterogeneous water supply infrastructure. *Environment and Planning E: Nature and Space*, 2514848620975342.

Ametepey, S. O. (2023). *Access to basic sanitation services in Ghana: Are we making progress as a country*? Citi Newsroom. https://citinewsroom.com/2023/03/access-to-basic-sanitation-services-in-ghana-are-we-making-progress-as-a-country-article/

Appiah-Effah, E., Duku, G. A., Azangbego, N. Y., Aggrey, R. K., Gyapong-Korsah, B., & Nyarko, K. B. (2019). Ghana's post-mdgs sanitation situation: An overview. *Journal of Water, Sanitation and Hygiene for Development, 9*(3), 397–415. https://doi.org/10.2166/washdev.2019.031

Arku, R. E., Bennett, J. E., Castro, M. C., Agyeman-Duah, K., Mintah, S. E., Ware, J. H., et al. (2016). Geographical inequalities and social and environmental risk factors for under-five mortality in Ghana in 2000 and 2010: Bayesian spatial analysis of census data. *PLoS Medicine, 13*, e1002038.

Asiamah, I. K. (2021, December 1). *Report of the Committee on works and housing on the 2023 annual budget estimate of the Ministry of Sanitation and Water Resource*. Retrieved December 19, 2022, from http://ir.parliament.gh/handle/123456789/2124

Awuah, E., Nyarko, K. B., & Owusu, P. A. (2009). Water and sanitation in Ghana. *Desalination, 248*(1–3), 460–467. https://doi.org/10.1016/j.desal.2008.05.088

Bartels, L. M. (2016). *Unequal democracy*. Princeton University Press.

Basic sanitation information system. BaSIS. Retrieved December 19, 2022, from https://sanitationghana.org/about/#:~:text=Ghana%20has%20adopted%20the%20Community%20Led%2Total%20Sanitation,efforts%20to%20scale%20up%20CLTS%20to%20all%20districts

Bayu, T., Kim, H., & Oki, T. (2020). Water governance contribution to water and sanitation access equality in developing countries. *Water Resources Research*, 56, e2019WR025330. https://doi.org/10.1029/2019WR025330

Boamah, L. A. (2011). *The Environmental Sanitation Policy of Ghana (2010) and Stakeholder capacity: A case study of solid waste management in Accra and Koforidua* (pp. 1–2). Institutionen för Geovetenskaper.

Cha, S., Jin, Y., Elhag, M. S., Kim, Y., & Ismail, H. A. (2021). Unequal geographic distribution of waterand sanitation at the household and school level in Sudan. *PLoS ONE, 16*(10), e0258418.

Chopra, M., Singh, S. K., Gupta, A., Aggarwal, K., Gupta, B. B., & Colace, F. (2022). Analysis & prognosis of sustainable development goals using big data-based approach during COVID-19 pandemic. *Sustainable Technology and Entrepreneurship, 1*(2), 100012.

Cowell, F. (2015). *Measuring inequality* (3rd ed.). Oxford Scholarship Online.

Deshpande, A., Miller-Petrie, M. K., Lindstedt, P. A., Baumann, M. M., Johnson, K. B., Blacker, B. F., et al. (2020). Mapping geographical inequalities in access to drinking water and sanitation facilities in low-income and middle-income countries, 2000–17. *The Lancet Global Health, 8*(9), e1162–e1185. https://doi.org/10.1016/S2214-109X(20)30278-3

Dinar, A., & Saleth, R. M. (2005). Can water institutions be cured? A water institution's health index. *Water Science and Technology: Water Supply, 5*, 17–40.

Duran, D. C., Gogan, L. M., Artene, A., & Duran, V. (2015). The components of sustainable development-a possible approach. *Procedia Economics and Finance, 26*, 806–811.

Ezcurra, R., & Rodríguez-Pose, A. (2014). Government quality and spatial inequality: A cross-country analysis. *Environment and Planning A, 46*(7), 1732–1753.

FAO. (2022). Ghana. Country (or area) | SDG 6 Data. https://www.sdg6data.org/en/country-or-area/ghana#anchor_6.1.1

Fukuda, S., Noda, K., & Oki, T. (2019). How global targets on drinking water were developed and achieved. *Nature Sustainability, 2*(5), 429–434. https://doi.org/10.1038/s41893-019-0269-3

Garrick, D. E., Hall, J. W., Dobson, A., Richard, D., Quentin Grafton, R., Hope, R., Hepburn, C., Bark, R., Boltz, F., de Stefano, L., O'Donnell, E., Matthews, N., & Money, A. (2017). Valuing water for sustainable development. *Science, 358*(6366), 1003–1005. https://doi.org/10.1126/science.aao4942

Gelaye, F. A. (2018). *Converging intentions, diverging realities: Rights vs. growth-based approaches to safe sanitation provision in Addis Ababa, Ethiopia* (Doctoral dissertation, Massachusetts Institute of Technology).

Ghana Clean Water Access 2000–2022. MacroTrends. Retrieved December 23, 2022, from https://www.macrotrends.net/countries/GHA/ghana/clean-water-access-statistics#:~:text=Ghana%20clean%20water%20access%20for%202020%20was%2041.41%25%2C,for%202017%20was%2036. %5%2C%20a%201.64%25%20increasefrom%202016

Ghana Statistical Service. (2013). *2010 Population and Housing Census Report*. Accra, Ghana.

Ghana Statistical Service. (2022). *2021 Population and Housing Census Report*. Accra, Ghana.

Griggs, D., Stafford-Smith, M., Gaffney, O., Rockström, J., Öhman, M. C., Shyamsundar, P., Steffen, W., Glaser, G., Kanie, N., & Noble, I. (2013). Policy: Sustainable development goals for people and planet. *Nature, 495*(7441), 305–307. https://doi.org/10.1038/495305a

Meissner, R., & Jacobs, I. (2016). Theorising complex water governance in Africa: The case of the proposed Epupa Dam on the Kunene River. *International Environmental Agreements: Politics, Law and Economics, 16*, 21–48. https://doi.org/10.1007/s10784-014-9250-9

Ministry of Sanitation and Water Resources. (2022). *Medium Term Expenditure Framework (MTEF) for 2022–2025*. Retrieved December 17, 2022.

Mosello, B. (2017). *How to reduce inequalities in access to WASH: Urban water in Ghana*. Overseas Development Institute.

Moulds, S., Chan, A. C., Tetteh, J. D., Bixby, H., Owusu, G., Agyei-Mensah, S., ... & Templeton, M. R. (2022). Sachet water in Ghana: A spatiotemporal analysis of the recent upward trend in consumption and its relationship with changing household characteristics, 2010–2017. *Plos One, 17*(5), e0265167.

Mugagga, F., & Nabaasa, B. B. (2016). The centrality of water resources to the realization of Sustainable Development Goals (SDG). A review of potentials and constraints on the African continent. *International Soil and Water Conservation Research, 4*(3), 215–223. https://doi.org/10.1016/j.iswcr.2016.05.004

Mutono, N., Wright, J., Mutembei, H., & Thumbi, S. M. (2022). Spatio-temporal patterns of domestic water distribution, consumption and sufficiency: Neighbourhood inequalities in Nairobi, Kenya. *Habitat International, 119*, 102476.

National Development Planning Commission. (2022). (rep.). *Ghana Voluntary National Report* (pp. 61–65).

Neef, A., Lynne, G. D., Burkhardt, J., & Batchelor, C. (2009). Water governance literature assessment. *Water Alternatives, 24*(4), 1059–1078. https://doi.org/

Nyika, J., & Dinka, M. O. (2023). *Water challenges in rural and urban Sub-Saharan Africa and their management*. Springer Nature.

OECD. (2015a). *OECD principles on water governance*, no. June 1–22. https://doi.org/10.1017/CBO9781107415324.004

OECD. (2015b). The governance of water regulators. Results of the OECD survey on the governance of water regulators. The Governance of Water Regulators. https://doi.org/10.1787/9789264231092-en. Reports|JMP. Retrieved December 24, 2022, from https://washdata.org/reports-of-free-waste-bins/2021/

Rambla, X., & Langthaler, M. (2016). *The SDGs and inclusive education for all: From special education to addressing social inequalities*. http://hdl.handle.net/10419/142739

Roaf, V., de Albuquerque, C., & Heller, L. (2018). The human rights to water and sanitation: Challenges and Implications for future priorities. In *Equality in water and sanitation services* (pp. 26–43). Routledge.

Rogers, P., & Hall, A. W. (2003). *Effective water governance* (Vol. 7). Global Water Partnership.

Rusca, M., Boakye-Ansah, A. S., Loftus, A., Ferrero, G., & van der Zaag, P. (2017). An interdisciplinary political ecology of drinking water quality. Exploring socio-ecological inequalities in Lilongwe's water supply network. *Geoforum, 84*, 138–146.

Sallah-Phillips, A. (2006). *The institutional & legal framework for protecting poor communities' access to water supply and sanitation in the Senegal and Volta river basins*. http://hdl.handle.net/1826/1202

Salifu, L. Y., Nashiru, A., & Taylor, K. (2015). 31st WEDC international conference. In *Sanitation policy in Ghana—Assessing key elements and policy measures*. Asset Publishing Service.

Sanitation. (2022, November 30). Retrieved November 30, 2022, from https://www.unicef.org/ghana/sanitation

Sasu, D. D. (2022, March 2). *Ghana: Water and sewerage GDP value at current price*. Statista. Retrieved December 23, 2022, from https://www.statista.com/statistics/1272522/value-added-by-water-and-sewerage-to-GDP-at-current-prices-in-ghana/

SDG indicators—SDG indicators. (2021, December 20). Retrieved December 19, 2022, from https://unstats.un.org/sdgs/metadata/?Text=&Goal=6

Sun, S., & Han, J. (2021). Open defecation and squat toilets, an overlooked risk of fecal transmission of COVID-19 and other pathogens in developing communities. *Environmental Chemistry Letters, 19*, 787–795.

Tetteh, J. D., Templeton, M. R., Cavanaugh, A., et al. (2022). Spatial heterogeneity in drinking water sources in the Greater Accra Metropolitan Area (GAMA), Ghana. *Population and Environment, 44*, 46–76. https://doi.org/10.1007/s11111-022-00407-y

UNEP. (2022). *Ghana*. Country (or area) | SDG 6 Data. https://www.sdg6data.org/en/country-or-area/ghana#anchor_6.1.1

UNESCO, International Social Science Council, and Institute of Development Studies. (2016). *World Social Science Report 2016: Challenging inequalities: Pathways to a just world* (vol. 359). http://publishing.unesco.org/details.aspx?Code_Livre=5160

UNESCO/UNECE. (2022). *Ghana*. Country (or area)|SDG 6 Data. https://www.sdg6data.org/en/country-or-area/ghana#anchor_6.1.1

UNICEF Ghana. (2023, May 4). *Sanitation*. Related UNICEF Sites. https://www.unicef.org/ghana/sanitation:~:text=Major%20challenges%20affecting%20basic%20rural%20sanitation%20delivery%20includes-resources%20for%20capacity%20building%20and%20implementation%20of%20programmes

UNICEF Ghana. (2023, May 4). *Water*. Related UNICEF Sites. https://www.unicef.org/ghana/water

UNICEF. (2022, November 30). *Water*. Retrieved November 30, 2022, from https://www.unicef.org/ghana/water

United Nations. (2015). *Transforming our world: The 2030 agenda for sustainable development* (pp. 1–35).

UN-Water. (2015a). *A post-2015 global goal for water: Synthesis of key findings and recommendations from UN-Water*.

UN-Water. (2015b). *Water for a sustainable world*.

USAID. (2021). *Ghana water resources profile overview—globalwaters.org*. Ghana Water Resources Profile Overview. https://www.globalwaters.org/sites/default/files/ghana_country_profile_final.pdf

Wash Budget Brief 2021. (2021). Retrieved December 19, 2022, from https://www.unicef.org/ghana/media/4341/file/Budget%20Brief%20-%20Water%20Sanitation%20&%20Hygiene.pdf

Water Resources Commission. (2021). *Ghana Water Resource Report 2021*. Accra: Water Resources Commission. Retrieved from https://www.wrc-gh.org/publications/annual-reports/2021-water-resource-report

Water Resource Commission. (2013). *Riparian buffer zone policy for managing freshwater bodies in Ghana*.

WHO/OECD. (2022). *Ghana*. Country (or area) | SDG 6 Data. https://www.sdg6data.org/en/country-or-area/ghana#anchor_6.1.1

WHO/UNICEF. (2017). *Progress on drinking water, sanitation and hygiene: 2017 update and SDG baselines*. World Health Organization (WHO) and the United Nations Children's Fund (UNICEF). https://doi.org/10.1111/tmi.12329

WHO/UNICEF. (2015). *2015 Annual Report WHO/UNICEF joint monitoring programme for water supply and sanitation*. World Health Organization. https://doi.org/10.1007/s13398-014-0173-7.2

WHO/UNICEF. (2022). *Ghana*. Country (or area) | SDG 6 Data. https://www.sdg6data.org/en/country-or-area/ghana#anchor_6.1.1

Zoomlion Ghana Limited. (2021, April 26). *Zoomlion begins distribution of free waste bins*. Retrieved December 19, 2022, from https://zoomlionghana.com/zoomlion-begins-distribution

Sustainable Development and Community Engagement

Evaluating the Implementation of Sustainable Development Goal 11: A Focus on Sustainable Cities and Communities in Ghana

Simon Ofori Ametepey, Clinton Aigbavboa, Wellington Didibhuku Thwala, Prince Boateng, and Rexford Aboagye

Abstract

The United Nations Agenda for Sustainable Development (SDGs) is an ongoing aspiration to enhance places, lives, and the environment by 2030. The primary drivers of the sustainable development agenda are cities and communities. Therefore, this study aims to investigate Ghana's ongoing efforts to achieve SDG 11: Sustainable Cities and Communities, specifically in creating inclusive, safe, resilient, and sustainable cities and communities. The study adopted document analysis and secondary data from notable sources such as the United Nations, World Bank, Ghana Statistical Service, and other organizations to analyze and interpret the findings. Although some progress has been made to create sustainable cities, especially within two major cities in Ghana, Accra, and Kumasi, most of the targets are unlikely to be achieved by 2030. Specifically, target 11.1 assesses access to appropriate, safe, affordable housing, essential amenities, and upgraded slums. The metric for target 11.1 displayed the percentage of persons living in slums being 57.99%, a significant rise of about 3.24% since 2016. Improvements made to curb the development of slums have been stunted by rapid population growth and urban–rural migration. To attain all requirements and targets of SDG 11, all industry players and government have to collaborate and work expeditiously to reach a satisfactory level of achievement to prepare Ghana for further advancements in creating sustainable and environmentally friendly cities and communities, even after 2030.

Keywords

Resilient cities · SDG 11 · Slums · Sustainable cities and communities · Urbanisation · Waste management

1 Introduction

Sustainable Development has been described as the long-term continuation of development since there exist several interpretations of Sustainable Development due to its intricate nature (Gibson, 2006). The United Nations (UN) Conference on Sustainable Development held in 2012 emphasized that member states must create policies and programs for a set of Sustainable Development Goals (SDGs) to give due recognition to Sustainability and Sustainable Development. The SDGs were created to be ambitious and to include all areas of sustainable development, including economic, social, and environmental sustainability. The United Nations General Assembly established an Open Working Group to develop the SDG proposal (Bhattacharya et al., 2014). The endorsement of the SDGs in September 2015 was an important milestone that marked the end of a lengthy process that resulted in the establishment of the seventeen (17) SDGs. The 17 SDGs are sub-goals that will contribute to improving and maintaining

Simon Ofori Ametepey, Clinton Aigbavboa, Wellington Didibhuku Thwala, Prince Boateng and Rexford Aboagye these authors are contributed equally to this study.

S. O. Ametepey (✉) · R. Aboagye
Centre for Sustainable Development (CenSUD),
Koforidua Technical University, Koforidua, Ghana
e-mail: simon.ametepey@ktu.edu.gh

R. Aboagye
e-mail: harex97@gmail.com

S. O. Ametepey · C. Aigbavboa
Department of Construction Management and Quantity Surveying,
Faculty of Engineering and Built Environment,
University of Johannesburg, Johannesburg, South Africa

P. Boateng
Department of Building Technology, Faculty of Built
Environment, Koforidua Technical University, Koforidua, Ghana
e-mail: p.boatengh@yahoo.com

S. O. Ametepey · W. D. Thwala
Department of Civil Engineering, College of Science Engineering
and Technology, University of South Africa, Tshwane, South Africa

B. He et al. (eds.), *Urban Climate Change Adaptation*, Advances in Science, Technology & Innovation,
https://doi.org/10.1007/978-3-031-65088-8_14

the global development agenda (Nilsson & Costanza, 2015). Cities play a critical role in the global economy and have been described as the centers of employment, progress, and wealth generation. It is projected that most of the global population resides in cities. Cities worldwide serve as homes and workplaces for about 4.4 billion people representing 56.2% of the global population. These figures are expected to double by 2050 (UN DESA, 2020). Even though cities are the centers of progress and development globally, cities have been cited to contribute significantly to global greenhouse gas emissions. In that regard, cities have emerged as essential stakeholders in sustainability efforts and must be active participants in implementing the SDGs (Barrutia et al., 2015). Therefore, cities must be seen as potential solutions rather than contributors to global challenges. Throughout the Sustainable Development agenda, SDG 11 places a high value on cities and communities. SDG 11 emphasizes the significance of inclusive, resilient, and safe cities within the sustainable development framework. Recognizing the difficulty of conserving a habitable world and assuring resource availability for future generations, development, and consumption must follow a cyclical cycle that leaves no space for waste while ensuring safety and efficiency (Juraschek et al., 2017). Citizen engagement and the active involvement of local stakeholders have increasingly been recognized as valuable tools for sustainable urban development (Cottafava et al., 2019). Cities or urban areas have been seen to be active in driving development and innovation, given the higher rates of education and access to resources and information. Cities possess the potential to propel sustainable development. However, municipal governments must take action to guarantee sustainable urban livelihoods. This requires the proficient participation of locals and business interests in decision-making processes and the accountability of city authorities (Bulkeley et al., 2005; Pieterse et al., 2014). Incorporating urban big data into the urban development decision-making process can lead to the advanced efficiency of urban services and improved urban management.

Nonetheless, difficulties exist in emerging nations where urban shantytowns continue, underscoring the need for extensive urban planning and management (Debrah et al., 2020; Owusu-Manu et al., 2021). Against this backdrop, the aim of the study is to assess the progress of Ghana in achieving SDG 11—Sustainable Cities and Communities—by 2030. The objectives of the study are; firstly, identify the various targets and indicators of SDG 11 and secondly, measure the current state of progress made towards achieving the targets of SDG 11. The findings of this study provided valuable insights for policymakers and local authorities in their efforts to create a sustainable urban environment and to aid in achieving SDG 11 by 2030.

2 Literature Review

2.1 Urbanisation in Ghana

Urbanization is primarily characterized by the increase in population and the expansion of non-agricultural activities. Various countries utilize population thresholds to categorize communities or settlements as urban (Moreno, 2017). Doe et al., (2022) also propose that urban growth involves the enlargement of built-up areas within cities and communities. For instance, in Ghana, urban communities are designated as those with a population of 5000 or more, while in Denmark and Iceland, urban localities are defined as communities with a minimum of 200 people (Ghana Statistical Service, 2013; Moreno, 2017; Naab et al., 2013). Urbanization occurs as a result of economic, social, and physical processes that concentrate populations in towns and cities, resulting in transformations in land use, social lifestyles, and economic systems (Cobbina et al., 2020). Studies have revealed that urbanization has both positive and negative impacts on society, but the negative effects are more significant due to the loss of agricultural land and other related losses. The consequences of urbanization extend across all economic sectors and the entire social fabric. It has been identified as a central contributing factor to sprawling community layouts, the expansion of slums, increased urban poverty, widening income disparities, and heightened inequalities among populations (Ashiagbor et al., 2019; Horn & Van Eeden, 2018). Over the past six decades, the global urban population has witnessed a fourfold increase. In 1950, only about thirty percent (30%) of the world's population, approximately 0.8 billion people, lived in urban areas. Subsequently, the urban population reached one billion in 1959, two billion in 1985, three billion in 2002, and four billion in 2015 (UNPD, 2019).

Likewise, Ghana has witnessed a noticeable expansion in all regions, particularly the Greater Accra region, with the city's growth extending into the Awutu Senya East Municipality on the western side and the Ga North Municipality on the eastern side (Eminsang, 2011). Asabere et al., (2020) discovered that the overall pattern of growth in Ghana's metropolitan areas follows a comparable trend, where regions constituting the historical core undergo rapid land cover transformation into built-up zones, while peripheral and suburban areas continue to grow and merge with urban regions. These findings emphasize the intricate and dynamic relationship between urban open spaces, developed land, and vegetation. Land-use alteration has emerged as a crucial concern in Ghana, with Ntajal et al., (2022) indicating that the conversion of natural areas into built-up regions is a significant issue. This development has led to a decrease in urban vegetation cover and an increase in soil sealing, thereby elevating the risk of flooding. This underscores the complex linkages and feedback mechanisms between land use, flooding, water contamination, and waterborne illnesses. The upsurge of urbanization holds great significance in Ghana, Sub-Saharan Africa, and Asia, leading to increased pressure on infrastructure and the depletion of natural resources. Urbanization in Ghana has influenced the encroachment of wetlands, flood-prone areas, farmland conversion, and makeshift settlements. The depletion of urban natural resources has been attributed to the transformation of natural areas and improper waste disposal. Kumasi, known as the garden city, now has a mere seven percent (7%) of its land area occupied by natural environmental resources. Air pollution in several urban areas in Ghana has become a major health concern due to the presence of particulate matter. Air pollution prevails in industrial areas and roads, stemming from vehicle and machinery emissions that contain fumes, dust, smoke, and other chemicals (Mitchell & Sweeney, 2018).

Land transformation, also known as land conversion, refers to the process of altering land from its original state or previous use to different purposes, such as agricultural, residential, or industrial activities. In Ghana, like in numerous developing countries, the escalating land conversion has been linked to various factors, with urbanization being a primary driver. The rise in population directly correlates with an increased demand for housing and commercial infrastructures (Bonye et al., 2021). Consequently, population growth has led to the conversion of agricultural lands into urban areas, resulting in the depletion of fertile soils and a subsequent decline in agricultural productivity. Moreover, the continuous reduction of peri-urban agricultural regions poses a threat to both peri-urban farming and the long-term food security of the peri-urban underprivileged population. Land conversion for urban purposes stands as one of the most significant and irreversible impacts on the global biosphere. This conversion affects the local climate, causes habitat fragmentation, and poses a threat to biodiversity (Seto et al., 2011).

2.2 Sustainable Urbanisation - SDG 11 Framework

The structure for tracking success in reaching SDG 11 has been formed by integrating various goals and corresponding indicators or measurements (Allen et al., 2017). This structure functions as an instrument for decision-makers and interested parties to assess the progress towards fulfilling the designated goals of SDG 11. The encompassed indicators and measurements cover a broad spectrum of topics related to urban growth, the availability of cost-effective housing, sustainable transport systems, the development of secure and inclusive public areas, resilience to disasters, and ecological sustainability. Through the observation and examination of relevant data within the SDG 11 structure, governmental bodies, organizations, and local communities can obtain essential insights into the efficacy of policies, initiatives, and endeavours designed to cultivate safe, resilient, and inclusive urban areas and communities (Ramani & Hettiarachchi, 2022).

2.2.1 Target 11.1: Ensure Everyone Has Access to Appropriate, Safe, Affordable Housing and Essential Amenities and Upgrade Slums

The housing sector, including its related laws, institutions, and markets, has a profound influence on every facet of a country's economic and social growth. People living in adequate housing are more likely to enjoy better health, increased human capital, and greater access to urban opportunities (Ha, 2010). A well-operated housing system serves as a catalyst for the expansion of related industries, thereby enhancing overall economic development. Conversely, the severe repercussions of poor and insufficient housing pose substantial risks to economic and social advancement. For instance, the slum population in Sub-Saharan Africa grows at an estimated rate of 4.5% per year (Vandermeulen et al., 2011). As a result, it is vital for interested parties to establish conditions that safeguard slum inhabitants while also elevating their living conditions (Glaeser, 2014). The swift expansion of slums and informal settlements can be traced back to various causes, such as ineffective planning and inefficient local government handling of financial and administrative tasks (Collier & Venables, 2016). Living in slums is linked to financial limitations and non-monetary poverty regarding access to vital public services. Likewise, substandard housing, as characterized by a set of indicators from the American Housing Survey (AHS), includes deficiencies in physical aspects like plumbing, heating, electricity, and upkeep. The UN-Habitat City Prosperity Initiative states that poor housing negatively affects multiple facets of urban prosperity, leading to decreased equality and inclusion, restricted safety and livelihood opportunities, and inadequate connectivity and public space (Brown et al., 2020). Addressing substandard housing offers both obstacles and prospects for worldwide urban sustainability (Eloy et al., 2018).

However, the terms "adequate housing" and "slums" are frequently laden with political implications. More precise and nuanced definitions of these terms could promote a thorough and meaningful dialogue, spurring participation from all pertinent stakeholders and crafting customized solutions for particular situations and locations. The variation in substandard housing and slums among inhabitants is shaped by several factors, including the accessibility of affordable housing (Cuming, 2016). Offering affordable housing continues to be one of the most intricate challenges tied to rapid global urbanization. Studies show that individuals from all economic backgrounds in cities find it difficult to afford either homeownership or rent. In Africa, housing costs are 55% higher in relation to GDP per capita than in other regions, leaving numerous families strained by housing expenses and with restricted resources for other essential necessities like food, healthcare, and transport. The migration crisis has further intensified housing difficulties in the area (UN-Habitat, 2016). As per the 2030 Agenda, precise assessment of the percentage of people living in slums, informal settlements, and inadequate housing is essential for shaping suitable policies and initiatives to ensure access to proper housing for all and the upgrading of slums (Habitat, 2018). Informal settlements, often linked with slums, are devoid of official recognition for land, buildings, and services. They usually lack tenure security and formal basic services, and the housing may not adhere to existing planning and building regulations, often situated in areas prone to geographical and environmental hazards (ECF, 2020). Adequate housing is considered essential for everyone's right to a decent standard of living, as articulated in Article 25 of the Universal Declaration of Human Rights. Housing must fulfill minimum criteria, including legal tenure security to avert forced evictions and harassment, availability of services and facilities, and infrastructure such as safe drinking water and proper sanitation. Moreover, affordability is crucial, as housing cannot be deemed suitable if its cost hinders residents' ability to enjoy other basic rights (Russell et al., 2021).

2.2.2 Target 11.2 Ensuring Safe, Affordable Transportation for All, Emphasizing Road Safety and the Needs of Vulnerable Groups like Women, Children, the Disabled, and the Elderly

Building a community that embraces inclusivity, environmental consciousness, and long-term sustainability necessitates the provision of transportation that is secure, economical, and readily available. The urgent need to reshape the transportation industry has become more pronounced due to the challenges of climate change and the global goal of achieving intelligent and secure mobility for all (Budd & Ison, 2020). The transportation sector is integral to various facets of economic and social growth, contributing substantially to the worldwide consumption of oil, utilization of energy, and release of carbon dioxide. Therefore, a global effort to reform and update transportation infrastructures is urgently needed. At present, transportation activities around the world account for roughly 64% of total oil usage, 27% of overall energy consumption, and 23% of energy-associated carbon emissions (IISD, 2023). SDG 11 emphasizes the paramount importance of transportation in the realization of resilient and sustainable urban areas (Habitat, 2018). The introduction of novel policies and the embrace of emerging technologies, such as electric cars, buses, and carbon–neutral energy solutions, have proven vital in the fight against climate change and in guaranteeing fair access to transport services. The networks of sustainable transportation are essential in cultivating cities and communities that are both inviting and robust. The period following the Covid pandemic offers a chance for fundamental reforms in the global transportation systems, capitalizing on the improvements in emission control and air purity achieved during the worldwide response to the health crisis (Mead, 2021). The enlargement of public transportation infrastructures is a key component in the global pursuit of sustainable urban living (Moallemi et al., 2020). Enhancing public transport through the expansion of its reach, cost-effectiveness, and user-friendliness can lead to a decrease in dependence on private cars, a reduction in traffic bottlenecks, and a potential decrease in emissions of greenhouse gases. It's vital to tailor public transportation to meet the varied requirements of community members, such as the disabled, the elderly, and children, to guarantee their well-being, ease of use, and satisfaction (Fokdal et al., 2021).

2.2.3 Target 11.3 Fostering Inclusive, Sustainable Urbanization and Collaborative Planning in All Nations

The growth of urban areas has unquestionably fuelled swift social and economic progress, but it has simultaneously spawned numerous challenges such as overcrowding, traffic jams, housing deficits, limited resources, loss of biodiversity, "heat island" phenomena, and pollution of air and water (Liang et al., 2019). On a broader scale, urbanization has prompted the migration of individuals and their labour abilities from countryside to city areas, leading to an economic transition from agriculture-centric to industries dominated by manufacturing and services (Deng et al., 2018). The creation of the Sustainable Development Goals, especially SDG 11, targeting the development of cities that are secure, inclusive, resilient, and sustainable, has placed urbanization at the centre of global development strategies. This view extends beyond seeing urbanization simply as a population shift, acknowledging it as a transformative force affecting various aspects of worldwide growth. These objectives are interrelated, and the realization of one often necessitates tackling issues typically linked with another. Sustainable urban growth involves actions to diminish disparities, guaranteeing fair distribution of resources, amenities, and prospects for city inhabitants. Whether urbanization amplifies inequalities or encourages inclusive growth ultimately hinges on its administration (UN DESA, 2020). The advantages of carefully orchestrated and governed urbanization are considerable. Achieving sustainable urbanization through diverse programs and regulations results in enhanced living standards, economic expansion, judicious use of resources, and adaptability to climate alterations. Sustainable urban environments have been shown to encourage dense and interconnected urban growth, effective transport networks, energy-efficient structures, and other ecologically responsible behaviours (Butsch et al., 2017). Conversely, hasty and ill-conceived urban expansion has given rise to problematic scenarios that attract stakeholders' attention for adverse reasons. Such uncontrolled growth has been pinpointed as the primary root cause of issues like overcrowding, insufficient infrastructure and public facilities, contamination, and social marginalization (Habitat, 2013).

2.2.4 Target 11.4 Efforts to Maintain and Protect the World's Cultural and Natural Heritage

Historical sites of global significance are scattered across different countries and regions, each reflecting unique cultural and economic backgrounds (Jokilehto, 2017). These sites are categorized into four heritage types: cultural, natural, mixed, and those under the World Heritage Centre (Galland et al., 2016). Their preservation and maintenance are financed through two primary channels: public funds, encompassing national, regional, and local contributions, and private investments, including in-kind donations, non-profit support, and sponsorships. These financial resources are allocated to cover the operational costs and other expenses related to conservation activities (Meskell et al., 2014). The socio-economic advantages that cultural heritage offers to cities around the world are widely acknowledged. Local authorities act as vital facilitators in executing policies and strategies that benefit their communities, fostering inclusiveness and sustainable growth (UNESCO, 2016). UNESCO's designated locations, including World Heritage Sites, Biosphere Reserves, and Global Geoparks, are deemed essential for sustainable development across various ecological, social, and economic dimensions (Schaaf & Rodrigues, 2016). Well-maintained UNESCO-designated sites contribute both directly and indirectly to goals that foster a balance between human beings and nature. They offer essential resources, products, services, and insights into human evolutionary history, enriching our understanding of our relationship with the Earth (Luo et al., 2022).

2.2.5 Target 11.5 Reduce Fatalities, Affected Individuals, and Global GDP Losses from Disasters, Particularly Water-Related Ones, with a Focus on Protecting the Poor and Vulnerable

Disasters are manifesting with increasing frequency and severity around the world, a trend that has been noted by various studies (Leaning & Guha-Sapir, 2013). These catastrophic events are generally characterized as sudden incidents that interrupt regular activities, potentially leading to profound societal breakdown and loss of human life (Perry, 2007). Among the various types of disasters, river flooding is particularly notable for being one of the most financially burdensome natural tragedies (UNISDR, 2016). The toll of natural disasters is not only measured in human suffering but also in substantial economic losses. These events are inherently destructive and can cause significant disruptions. However, the body of research examining the relationship between disasters and economic growth is fragmented, and there is a lack of comprehensive data to draw definitive conclusions (Fiala, 2017). In the immediate aftermath of a disaster, the economic impact is typically negative, stemming from both direct and indirect damages Direct damages include the loss of human life, physical injuries, disabilities, and the destruction of tangible assets like homes, factories, and vital infrastructure. These immediate losses can further translate into a reduction in potential labour hours (and corresponding wages) and a decrease in anticipated production in sectors such as agriculture or industry. The wages that would have been earned, had the disaster not occurred, would have contributed to the nation's GDP, thereby indirectly affecting economic growth (Hochrainer, 2009). Given these considerations, Disaster Risk Reduction (DRR) has emerged as an essential component of both economic and social advancement, playing a key role in securing sustained progress. The intricate relationship between disaster risk management and sustainable development has been recognized in international accords and strategies, such as the Yokohama Strategy and Plan of Action for a Safer World (Tozier de la Poterie & Baudoin, 2015). These agreements underscore the importance of integrating disaster mitigation efforts into broader development goals, acknowledging the interconnected nature of these challenges.

2.2.6 Target 11.6 Minimizing Cities' Per Capita Environmental Impact, Focusing on Air Quality and Waste Management

The repercussions of swift urban growth on the environment are profound and demand serious attention. Such urban dilemmas present risks to both human communities and the fragile equilibrium of nature, with air contamination and waste handling being among the most critical challenges. Air pollution stands as a worldwide health crisis, leading to a rise in fatalities and illnesses related to respiratory, cardiovascular, and other health conditions. The effects of this pollution are particularly alarming in cities, where the expanding urban populace, notably in nations with lower and middle incomes, is more susceptible to harmful health consequences (Mir Alvarez et al., 2020). The influence of urbanization on a city's air quality is shaped by elements like industrial processes, energy utilization, and emissions from transportation. The degradation of air quality can result in severe public health crises, with a range of respiratory ailments and other medical problems associated with urban air contamination (WHO, 2021). The convergence of factors such as population concentration, economic growth, urban development, and industrialization has contributed to a surge in the global production of solid waste (UNEP, 2009). An estimated 1.9 billion tonnes of solid waste are generated by cities across the globe each year. In major urban centres in Ghana, the escalation in population and rapid urban development have led to a substantial increase in solid waste, surpassing the capabilities of local administrations to effectively manage and dispose of this waste. The country produces roughly 12,710 tonnes of solid waste daily, with a mere 10% being collected and processed at sanctioned locations (Miezah et al., 2015). The labour-intensive and frequently inefficient methods of waste collection and disposal present a formidable obstacle to solid waste management in Ghana's urban regions, creating complications for municipal authorities. The sheer volume of waste generated daily overwhelms both city officials and waste management enterprises (Fobil et al., 2017). Despite two decades of private sector participation in Ghana's waste management efforts, there has been no substantial improvement in collection efficiency or recycling practices. Municipal solid waste handling is an essential component of urban governance. Urban areas and communities produce significant quantities of waste, and improper disposal methods can lead to environmental harm and pose threats to public health. Sustainable waste management strategies must prioritize waste reduction, encourage recycling and composting, and implement suitable disposal methods to mitigate these risks (UNEP, 2021).

2.2.7 Target 11.7 Ensuring Access to Safe, Inclusive, and Accessible Green and Public Spaces, Especially for Women, Children, the Elderly, and People with Disabilities

Cities focusing on the welfare of their inhabitants often flourish, becoming more sustainable and prosperous. Such cities strive for social fairness and gender parity by enhancing access to communal spaces and resources, resisting undue privatization, and fostering an improved standard of living for all. Public areas, including streets, green spaces, parks, and leisure centres, are vital in forming communities, civic pride, and cultural identity. These spaces not only foster social connections and economic development but also rejuvenate neighbourhoods. The ongoing vibrancy and upkeep of public spaces lead to well-maintained, secure, and thriving urban surroundings, rendering the city a desirable place for living and working (Low, 2023). Public spaces should enrich various facets of city life, such as the overall quality of living, economic advancement, civic unity, safety, health, well-being, and transportation. In 2011, UN-Habitat was tasked by member states to unify efforts concerning public spaces, formulate and advocate policies on public spaces, coordinate knowledge sharing, and provide direct support to cities in their public space endeavours. Furthermore, the inherent right of individuals with disabilities to access and engage in higher education is widely recognized, and the presence of sufficient support structures is vital for the fulfillment of this right (Mensah et al., 2022).

3 Research Methodology

The evaluation of Ghana's progress towards attaining SDG 11 was based on the targets specified by the SDG 11 framework. A research design based on document analysis was employed to achieve this objective, incorporating data from various secondary sources. This approach ensured the cohesiveness, reliability, and depth of the research. Document analysis is a qualitative method of interpreting and extracting meaningful insights from electronic and physical documents. Through meticulous examination of published international and national documents, the study sought to gain a comprehensive understanding of the advancements made towards SDG 11. The study relied on secondary data collected over 15 years, emphasizing the data obtained within the last five years. The sources for this data included research papers, journals, books, reports, and other credible platforms. The Metrics and indicators were employed to assess the attainment of specific targets within the goal and to evaluate the contributions of various stakeholders, such as the government, UN, and private sector. To determine Ghana's commitment to achieving SDG 11, the study utilized secondary data from sources such as the Voluntary National Review (VNR) report 2021 population and housing census report, among others. Additionally, the study examined other national documents, such as the 2021–2022 Medium-Term Expenditure Framework (MTEF), to determine the government's financial contributions and other arrangements toward attaining SDG 11. The data obtained from these sources were analyzed to provide a clear depiction of Ghana's progress in pursuing SDG 11 and to compare Ghana's commitment to other developing countries and the standards set by the United Nations. Furthermore, the study further reviewed and discussed the contributions of major private sector players, international bodies, NGOs, and State-Owned Enterprises (SOEs) operating within Ghana, such as NADMO and UNICEF, towards SDG 11. In sum, this study seeks to provide an academic analysis of Ghana's progress toward attaining SDG 11, utilizing a diverse range of credible sources to evaluate the contributions of various stakeholders. By employing this methodology, the study aims to comprehensively assess Ghana's progress in achieving SDG 11—Sustainable Cities and Communities. This evaluation will offer valuable insights for policymakers, local authorities, and relevant stakeholders in their efforts to promote sustainable urban development in Ghana. The metrics or indicators used in the study are universally accepted systems of measurement by the United Nations for assessing the progress of SDG 11 in member states. However, it is worth noting that the availability of data for certain metrics may vary depending on a nation's development level. In Ghana's case, inadequate data for specific metrics can be attributed to factors such as a lack of facilities, financial resources, expertise (such as research personnel), and overall resources required for data collection. Additionally, the absence of development or progress towards the respective targets of SDG 11 may contribute to data gaps. To evaluate Ghana's progress towards SDG 11, this study relied on various sources of information, including the Voluntary National Review (VNR) for 2019 and 2022, Government official documents, Data from the Ghana Statistical Service, and Reports from local branches of the international development. By assessing the available data from these sources and aligning it with the metrics and targets outlined shown in Table 1, this study aims to evaluate Ghana's progress towards achieving SDG 11—Sustainable Cities and Communities.

4 Results

4.1 Ghana's Performance Towards Achieving SDG 11 - Sustainable Cities and Communities

4.1.1 Percentage of the Ghanaian Urban Population Residing in Slums, Informal Settlements, or Substandard Housing

The 2021 Population and Housing Census in Ghana revealed that 8.8 million individuals were residing in slums, making up a significant part of the total urban population, estimated at 18.4 million. This urban segment represents more than 57% of the country's entire population. Further data from the 2021 census showed that around 5.7% of urban dwellings in Ghana were makeshift accommodations, including metal containers, kiosks, wooden structures, unfinished buildings, and living spaces attached to commercial areas. The data below displayed in Table 2, shows that the proportion of the urban population living in slums, informal settlements, or inadequate housing cannot be eliminated anytime soon, considering the population growth, especially within urban centres.

Over recent years, there has been a consistent growth in the urban population's share of Ghana's total population. In 2016, urban residents made up 54.75% of the population, a figure that incrementally increased to 55.41% in 2017, 56.06% in 2018, 56.71% in 2019, and 57.35% in 2020. This upward trend persisted into 2021, with urban inhabitants constituting 57.99% of the overall population.

In 2010, the Kumasi Metropolitan Area in Ghana had a built density of 24% and experienced an annual growth rate of 6%. Meanwhile, the Accra Metropolitan Area had a built density of 22% and an annual growth rate of 4.5%. These figures provide a snapshot of the urban development level and the population growth rate in these metropolitan areas during that year as shown in Table 3.

4.1.2 Percentage of People with Easy Access to Public Transport, Categorized by Gender, Age, and Disability Status

The most recent information regarding the population's easy access to public transport was gathered in 2012 (Fig. 1), as reported by the Ministry of Roads and Highways. The findings revealed that an impressive 97.9% of people had convenient access to public transport. This aligns with the objectives of fostering cities and human settlements that are inclusive, safe, resilient, and sustainable. As per the Five-Year Strategic Plan for Gender Statistics report, this goal is intended to be evaluated over a five-year span, with data collection planned every two years through the Ministry of Transport and Communication (Ministry of Gender, Children and Social Protection and Ghana Statistical Services, 2018). Despite this plan, there have been no recent updates or progress reports on this specific target. This absence of current data generates concerns regarding the oversight and assessment of public transport accessibility and its ramifications for meeting the goals of sustainable and inclusive urban development. It underscores the necessity for consistent data gathering and scrutiny to guarantee that strides are being made to establish cities where all residents have fair and convenient access to public transportation.

In Ghana, the majority of the population, encompassing both genders and all age groups, enjoys convenient access to public transport. However, this accessibility does not extend to individuals living with disabilities. Most forms of

Table 1 SDG 11 Targets and indicators

Target	Indicators/ metrics
Target 11.1: Ensure access for all to adequate, safe, and affordable housing and basic services and upgrade slums	Indicator 11.1.1: Proportion of urban population living in slums, informal settlements, or inadequate housing
Target 11.2: Provide access to safe, affordable, accessible, and sustainable transport systems for all, improving road safety, notably by expanding public transport, with special attention to the needs of those in vulnerable situations, women, children, persons with disabilities, and older persons	Indicator 11.2.1: Proportion of the population that has convenient access to public transport by sex, age, and persons with disabilities
Target 11.3: Enhance inclusive and sustainable urbanization and capacity for participatory, integrated, and sustainable human settlement planning and management in all countries	Indicator 11.3.1: Ratio of land consumption rate to population growth rate Indicator 11.3.2: Proportion of cities with a direct participation structure of civil society in urban planning and management that operate regularly and democratically
Target 11.4: Strengthen efforts to protect and safeguard the world's cultural and natural heritage	Indicator 11.4.1: Total per capita expenditure on the preservation, protection, and conservation of all cultural and natural heritage by source of funding (public, private), type of heritage (cultural, natural), and level of government (national, regional, and local/municipal)
Target 11.5: Significantly reduce the number of deaths and the number of people affected and substantially decrease the direct economic losses relative to global gross domestic product caused by disasters, including water-related disasters, focusing on protecting the poor and people in vulnerable situations	Indicator 11.5.1: Number of deaths, missing persons and directly affected persons attributed to disasters per 100,000 population Indicator 11.5.2: Direct economic loss attributed to disasters in relation to global domestic product (GDP) Indicator 11.5.3: (*a*) Damage to critical infrastructure (*b*) number of disruptions to basic services attributed to disasters
Target 11.6: Reduce the adverse per capita environmental impact of cities, including by paying particular attention to air quality and municipal and other waste management	Indicator 11.6.1: Proportion of municipal solid waste collected and managed in controlled facilities out of total municipal waste generated by cities Indicator 11.6.2: Annual mean levels of fine particulate matter (e.g., PM2.5 and PM10) in cities (population weighted)
Target 11.7: Provide universal access to safe, inclusive, and accessible green and public spaces, in particular for women and children, older persons, and persons with disabilities	Indicator 11.7.1: Average share of the built-up area of cities that is open space for public use for all, by sex, age, and persons with disabilities Indicator 11.7.2: Proportion of persons victim of physical or sexual harassment, by sex, age, disability status, and place of occurrence, in the previous 12 months
Target 11.A: Support positive economic, social, and environmental links between urban, peri-urban, and rural areas by strengthening national and regional development planning	Indicator 11.a.1: Number of countries that have national urban policies or regional development plans that (a) respond to population dynamics; (b) ensure balanced territorial development; and (c) increase local fiscal space
Target 11.B: Substantially increase the number of cities and human settlements adopting and implementing integrated policies and plans towards inclusion, resource efficiency, mitigation and adaptation to climate change, resilience to disasters, and develop and implement, in line with the Sendai Framework for Disaster Risk Reduction 2015–2030, holistic disaster risk management at all levels	Indicator 11.b.1: Number of countries that adopt and implement national disaster risk reduction strategies in line with the Sendai Framework for Disaster Risk Reduction 2015–2030 Indicator 11.b.2: Proportion of local governments that adopt and implement local disaster risk reduction strategies in line with national disaster risk reduction strategies
Target 11.C: Support least developed countries, including through financial and technical assistance, in building sustainable and resilient buildings utilizing local materials	

Source UN | Department of Economic and Social Affairs

public transportation in the country, including smaller vehicles, buses, and motorcycles, are not designed to accommodate disabilities. As a result, they fail to provide convenient or accessible transportation options for those living with a disability, highlighting a significant area of concern and inequality within the nation's transportation system.

4.1.3 Preservation of Natural and Cultural Heritage in Ghana

The Ministry of Tourism, Arts, and Culture (MTAC) in Ghana has dedicated over GH₵ 88 million from 2010 to 2022 for the preservation and upkeep of the nation's cultural and natural heritage. The spending in this domain

Table 2 Urban population of Ghana

Year	Population	% of total population
2021	18,399,874	57.99
2020	17,820,023	57.35
2019	17,249,055	56.71
2018	16,687,441	56.06
2017	16,135,330	55.41
2016	15,593,581	54.75

Source Macrotrends.net

Table 3 Built Density and Annual Growth Rate of Metropolitan Areas in Ghana (2010)

Area	Built Density (%)	Annual Growth Rate
Kumasi Metropolitan Area	24%	6%
Accra Metropolitan Area	22%	4.5%

Source Toure et al., 2020

reached GH₵ 88 million between 2010 and 2021, accounting for 0.19% of the GDP as of 2021. This allocation for the targeted goal is glaringly insufficient. To achieve the desired objective by 2030, there is an urgent need for a substantial increase in funding, as the current amount falls woefully short of what is required for effective preservation and maintenance.

4.1.4 Number of Disaster-Related Deaths, Missing Persons, and Directly Affected Individuals Per 100,000 Population of Ghana

Figure 2 illustrates the varying numbers of deaths, missing persons, and individuals directly affected by disasters (referred to as DMPs) per 100,000 population across different years. In 2017, the figure stood at 997 DMPs for every 100,000 people, highlighting a considerable toll on communities. This number escalated to 1,063 per 100,000 in 2018, reflecting an even greater impact of disasters on the populace. Contrastingly, 2019 saw a reversal in this trend, with the DMPs dropping to 739 per 100,000, a sign of reduced disaster impact relative to preceding years. This decline might be a result of enhanced disaster management strategies implemented that year, leading to more effective responses and minimized adverse outcomes. Regrettably, the progress made in 2019 was short-lived, as 2020 experienced a substantial surge in DMPs, reaching 1184 per 100,000 individuals. This increase could be linked to particularly severe disaster occurrences or a confluence of elements that amplified vulnerability during that period.

On an encouraging note, 2021 witnessed a reduction in DMPs to 682 per 100,000 people, signifying a lessened disaster impact compared to the prior year. This decline in 2021 might be a reflection of successful disaster management interventions and reactions that aided in alleviating the detrimental effects of disasters on the community. The data underscores the importance of continuous monitoring and adaptive strategies in disaster management to mitigate the human toll.

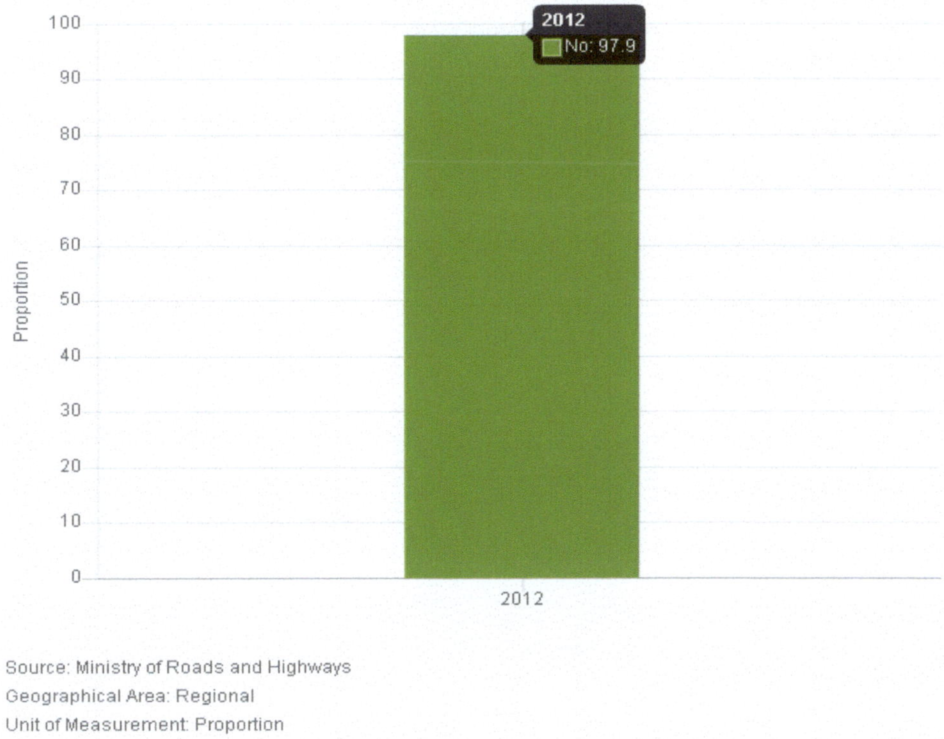

Fig. 1 Percentage of the population with convenient access to public transport (*Source* Ministry of Roads and Highways)

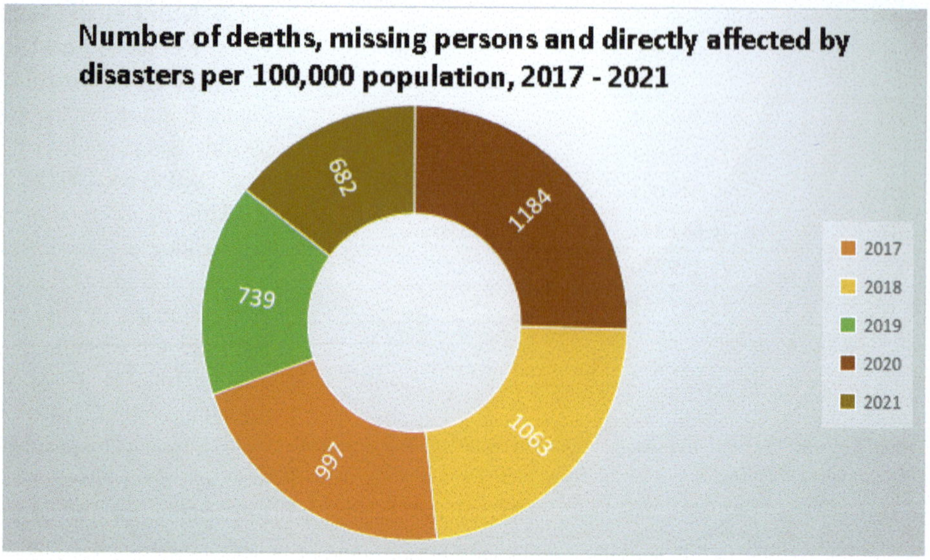

Fig. 2 Number of deaths, missing persons and persons directly affected by disasters. *Source* NADMO Annual Reports (2017–2021)

4.1.5 Percentage of Municipal Solid Waste Collected and Handled in Controlled Facilities Compared to Total City-Generated Waste

Figure 3 provides an in-depth analysis of the methods employed by households in Ghana to dispose of solid waste in 2010 and 2021. The data from 2010 shows that only 14.4% of the solid waste generated by households was handled through formal waste collection systems, reflecting inadequate waste management at that time. A concerning 61.5% of the waste was carelessly discarded in open areas, further emphasizing the lack of proper waste disposal practices during that period. Fast forward to 2021, the situation shows marked improvements in the handling of solid waste. The portion of solid waste collected through proper channels rose to 33.4%, signifying a substantial enhancement in waste collection efficiency and overall waste management. Concurrently, the percentage of waste thoughtlessly left in open spaces dropped to 37.5%, pointing to a decline in inappropriate waste disposal habits. However, the data also reveals a lingering challenge, as a considerable amount of waste remained uncollected in 2021. There was even a slight uptick in the percentage of uncollected solid waste, reaching 29.1%. This information underscores the progress made in waste management in Ghana, while also highlighting areas that still require attention and improvement.

Figure 4 offers a detailed examination of the approaches used for urban household solid waste disposal in 2021. According to the statistical breakdown, formal collection systems adeptly managed 51.4% of the waste produced by urban households. This figure represents a substantial part of the waste being effectively processed through suitable waste management practices. However, the data also

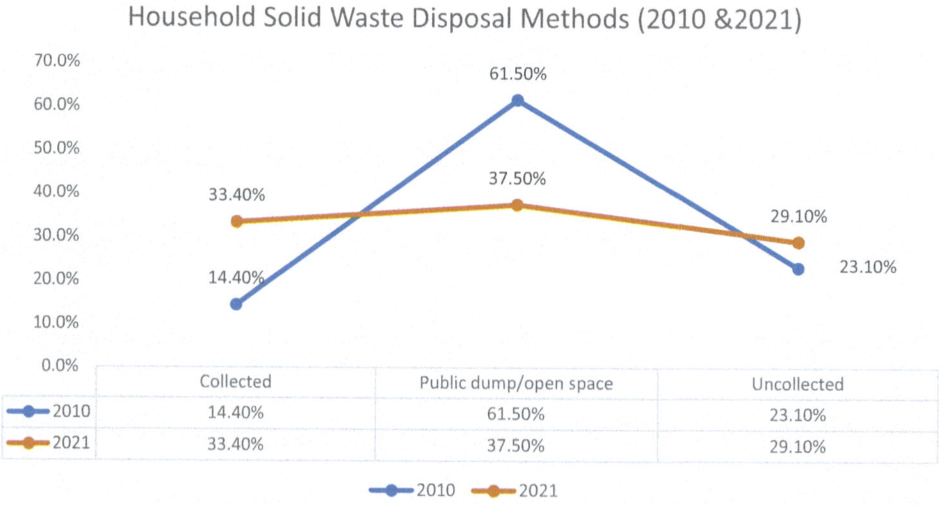

Fig. 3 Solid waste disposal of households in Ghana (2010 and 2021). *Source* PHC 2010 and 2021

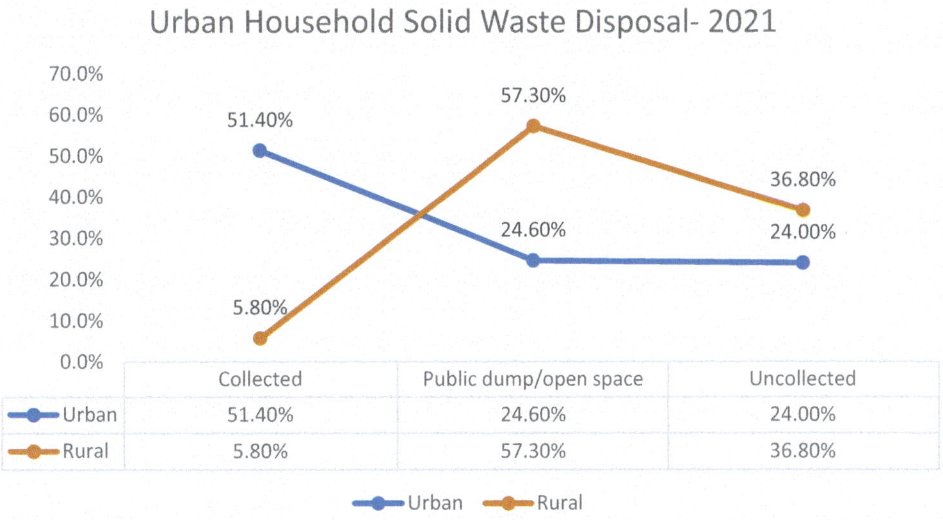

Fig. 4 Urban household solid waste disposal methods 2021. *Source* PHC 2010 and 2021

reveals areas of concern. A notable 24.6% of household waste was discarded in open spaces, a practice that can lead to environmental degradation and health risks. Additionally, 24.0% of the waste went uncollected, highlighting a significant portion of waste that was neither addressed nor properly managed. Overall, the statistics in Fig. 4 underscore the progress made in urban waste management, with over half of the waste being handled appropriately. Yet, they also point to ongoing challenges, particularly regarding waste left in open spaces or uncollected, emphasizing the need for continued efforts to enhance waste management practices in urban areas.

4.1.6 Percentage of Individuals Who Experienced Physical or Sexual Harassment in the Past year, Categorized by Gender, Age, Disability Status, and Location

The most recent information regarding the percentage of individuals in Ghana who have been subjected to physical or sexual abuse or violence dates back to 2015. As per the Ghana Family Life and Health Survey (Ghana Statistical Service, Ghana Health Service & ICF, 2023) of that year, approximately 27.7% of women in the country encountered at least one form of abuse or violence. This could encompass physical, social, sexual, psychological, or economic threats. Meanwhile, men represented 20% of the total proportion of individuals who experienced abuse.

5 Discussion

The Sustainable Cities and Communities Goal (SDG 11) of the Sustainable Development Goal is fundamentally concerned with promoting inclusive, safe, resilient, and sustainable cities and human settlements. Ensuring that all individuals have unfettered access to adequate, secure, and cost-effective housing and fundamental amenities is essential. Moreover, the objective of upgrading slums must also be pursued. In Ghana, over 47.8% of the urban population resides in slums, indicating that almost half of the urban population lives in such areas. Ghana's population in slums decreased from 39.3% in 2017 to 28.2% in 2020. Nevertheless, accomplishing the SDG objective by 2030 presents a daunting challenge, as insufficient efforts have been made to reverse the prevailing situation. To effectively address the issue of slums and attain the SDG target, a pressing demand exists for prompt and practical strategies to impede the escalating progression of slums in Ghana. For instance, through the Zongo Development Fund, the Government of Ghana has constructed 58 community water systems and 252 household toilet facilities to improve the living conditions in various slum-like settlements. Nonetheless, despite these interventions, much more needs to be done to achieve the Sustainable Development Goals outlined in the 2030 Agenda.

Furthermore, the issue of land use and urban expansion must be addressed in accordance with the Land Use and Urban Extent phases of target 11.1. In Ghana, land use and conversion rates in urban areas have increased, with high built density noted in regions like the Kumasi Metropolitan Area. These urbanized regions experience an average annual growth rate of more than 5%, indicating the need for proper land management and planning. Accra and Kumasi have witnessed significant additions of residential and non-residential built-up land use. Addressing the issue of slums in Ghana demands an expeditious and all-inclusive response. The government must direct supplementarily resources towards affordable housing programs, with a focus on land use planning and management. The attainment of SDG target 11.1 necessitates implementing efficacious measures to guarantee accessibility to satisfactory, secure, and reasonably priced housing for all, encompassing the enhancement of current slums. According to a study conducted by Doe et al., (2022), the rapid urbanization in Greater Accra has spread to neighboring municipalities, leading to a considerable increase in urban expanse. Specifically, the urbanized areas in Awutu Senya East Municipal, a nearby municipality, have amplified more than two-fold over the past six years, indicating a 152% alteration in urban expansion. The urbanized area of the municipality accounted for 49.37 km^2 in 2020, which constituted 51.45% of the municipality's total spatial extent. This analysis highlights the rapid growth observed in municipalities surrounding Ghana's metropolitan areas. Similar instances are recorded in Osei et al., (2015). Similarly, Sarfo et al., (2022) noted a significant decrease in Ghana's forested areas. Since 1970, Ghana has experienced a decline of 334.8 km^2 per year in forest cover. In contrast, built-up areas and farmlands have increased by 137.93 km^2 per year and 131.97 km^2 per year, respectively. The researchers identified increasing urbanization as a primary concern contributing to this trend. This indicates the detrimental impact of urbanization on Ghana's natural environment. Given the study's findings, it is crucial to consider the allocation of budgetary resources to preserve and protect Ghana's natural and cultural heritage. Data on budgetary allocation revealed that the government of Ghana allocated an average of GH 88million, which is less than one percent of the country's gross domestic product (GDP), for the preservation of natural and cultural heritage. However, Ghana's natural and cultural heritage significantly contributes to the country's GDP. In 2021, the tourism sector contributed about 3.1 billion United States dollars to GDP. As tourism initiatives such as "The Year of Return" and "Beyond the Return" have been launched to attract international tourists, it becomes necessary to allocate a portion of tourism's contribution to GDP to preserve and protect Ghana's natural and cultural heritage. This would help sustain the country's tourism potential.

Ghana's efforts towards significantly decreasing the number of fatalities, affected individuals, and direct economic repercussions resulting from disasters, particularly those related to water, with a focus on safeguarding impoverished and vulnerable communities, by 2030 have been inconsistent. The average number of fatalities, missing persons, and disaster-affected individuals per 100,000 people stands at 933. The National Disaster Management Organization (NADMO) attributes these figures to wind and rainstorms, floods, and domestic and bushfires. Timely and adequate disaster management is crucial to effectively reduce loss of life during natural disasters in Ghana. This necessitates allocating sufficient resources to NADMO and other security agencies, such as the Ghana National Fire Service, to enhance disaster response capabilities. Floods have become a significant challenge globally, resulting in social, economic, and environmental impacts. Urban flooding has emerged as a significant issue in many parts of the world, displacing millions of people and causing economic losses. Ghana, being susceptible to floods, faces similar challenges. Floods destroy developmental infrastructure and increase human casualties. In Pakistan, floods are widespread and uncontrollable, leading to significant economic losses. Ghana must address these challenges by implementing effective flood management strategies (Mensah & Ahadzie, 2020). Acknowledging that water-related disasters, such as floods, have substantial health impacts is imperative. Although there is limited epidemiological evidence on the health consequences of these disasters, global trends indicate the augmented frequency and intensity of floods and droughts. Addressing the health impacts of water-related disasters is crucial for ensuring the well-being of affected populations (Lee et al., 2020).

Moreover, Ghana must align its disaster management strategies with the Sustainable Development Goals (SDGs) established by the United Nations. SDG 11.5 mainly aimed to decrease the number of fatalities, affected individuals, and economic losses resulting from disasters. Ghana's commitment to achieving this goal necessitates a comprehensive approach that concentrates on bolstering resilience, enhancing early warning systems, and ensuring the availability of resources for disaster management. Particulate matter (PM) poses a noteworthy environmental concern in Ghana, aligning with Sustainable Development Goal 11 (SDG 11), which strives to establish cities and human settlements that are inclusive, safe, resilient, and sustainable. The harmful effects of PM on air quality and public health necessitate reducing its impact. Ghana's Revised Nationally Determined Contribution (NDC) under the Paris Agreement underscores the importance of addressing air pollution and municipal waste management. In Ghana, concerted efforts are being made to address the issue of particulate matter and its implications for SDG 11. The United Nations in Ghana is actively supporting the country in attaining the SDGs, including SDG 11, through collaborative partnerships and initiatives. Projects like IGES Ghana Eco-Transport aim to augment energy efficiency and curtail emissions, contributing to the transport sector's climate change mitigation and air quality enhancement. Moreover, studies on particulate matter pollution levels in traffic hotspots of Accra highlight the significance of comprehending exposure experiences and implementing measures to minimize the adverse effects on street traders and the general populace (WHO, 2021).

Indicator 11.6.1 mandates that the government and relevant stakeholders intensify their efforts toward achieving "universal solid waste control." There has been a significant decrease in municipal solid waste disposed of in public and open spaces. In 2021, over 37% of the municipal solid waste produced by households was discarded in open and public spaces. It is concerning that 29.1% of household solid waste is not collected. In urban areas, an astonishing 48.6% of household waste is not deposited in controlled sites, leading to dumping in public and open spaces. The management of waste in urban areas has proven to be challenging due to the rapidly increasing population and insufficient resources required for efficient waste management. The detrimental effects of improper household waste management include increased health risks and long-term degradation of natural landscapes and communities. It is crucial to address these challenges to ensure public health safety and environmental preservation. Evidence-based data is crucial for developing waste management strategies and making informed investment decisions in infrastructure and service expansion to combat these issues. To achieve proper waste collection and management in Ghana, it would be advisable to employ the UN Habitat's Ware Wise Cities Tool, which quantifies parameters that enable cities and countries to better manage resources, mitigate and prevent environmental pollution, create business and employment opportunities, and transition towards a circular economy.

6 Conclusion

The study analyzed various documents published by the government of Ghana and international bodies on the progress made towards achieving SDG 11, which seeks to put communities and cities in the development spotlight. Generally, the study results indicated that Ghana lacks data to measure the progress made in achieving all the targets of SDG 11. Despite the efforts of the government and its agencies towards SDG 11, Ghana faces significant challenges in efficiently monitoring, tracking, and evaluating the progress made. Notably, it was found that there has been an increase in the proportion of the urban population living in slums over the ten years. This is indicative of the persisting challenge in the provision of affordable, safe, and adequate housing for the urban population in Ghana. Again, the study highlighted the improvement in waste management in Ghana as the proportion of collected waste has increased. This calls for a more resilient and decisive approach to managing municipal solid waste in Ghana. The rapid urbanization in Ghana has become more than evident considering the increasing land consumption and land conversion rates in Ghana. Considering the study's findings, it is recommended that Ghana should prioritize adopting urban resilience strategies and principles in planning policies and regulations. This has the potential to address climate change challenges in urban areas, emerging environments, and health issues while ensuring that communities and cities in Ghana are much safer, more inclusive, and more sustainable. Investment in resilient infrastructure is crucial in ensuring safe urbanization. The expansion of essential services such as water, sanitation, and water supply can also potentially reduce the expansion of slum dwellings. The presence of these services in current slums can be a plausible way of reducing slums in urban areas. Again, the central and local governments should, as a matter of urgency, invest in creating robust data collection and monitoring mechanisms. This is crucial in making evidence-based decisions on developmental issues, policies, and programs. This will prove effective in assessing urban resilience, identifying vulnerabilities, and monitoring the progress of the country or local areas over some time. Finally, the government of Ghana should improve collaboration and partnerships among the stakeholder community to ensure that the views factored in decision-making are well-balanced and varied. The government can also capitalize on the expertise, innovations, and resources of the private sector, especially in financing green communities and cities development in Ghana.

References

Allen, C., Nejdawi, R., El-Baba, J., Hamati, K., Metternicht, G., & Wiedmann, T. (2017). Indicator-based assessments of progress towards the sustainable development goals (SDGs): A case study from the Arab region. *Sustainability Science, 12*(6), 975–989. https://doi.org/10.1007/s11625-017-0437-1

Asabere, S. B., Acheampong, R. A., Ashiagbor, G., Beckers, S. C., Keck, M., Erasmi, S., Schanze, J., & Sauer, D. (2020). Urbanization, land use transformation, and spatio-environmental impacts: Analyses of trends and implications in major metropolitan regions of Ghana. *Land Use Policy, 96*, 104707.

Ashiagbor, G., Amoako, C., Asabere, S. B., & Quaye-Ballard, J. A. (2019). Landscape transformations in rapidly developing peri-urban areas of Accra, Ghana: Results of 30 years. *Open Geosciences, 11*(1), 172–182.

Barrutia, J. M., Echebarria, C., Paredes, M. R., Hartmann, P., & Apaolaza, V. (2015). From Rio to Rio+20: Twenty years of participatory, long term oriented and monitored local planning?

Bhattacharya, D., Khan, T. I., & Salma, U. (2014). A commentary on the final outcome document of the Open Working Group on SDGs. *The SAIS Review of International Affairs, 34*(2), 165–177.

Bonye, S. Z., Yiridomoh, G. Y., & Bebelleh, F. D. (2021). Compliance with land use regulations in peri-urban areas in Ghana: a study of Bamahu and Danko residential areas in Wa municipality, Upper West Region. *GeoJournal, 86*(6), 2845–2859.

Brown, M., Tran, A. D., Cummings, C., Fay, L., Malone, D., Fyall, R., & Tsemberis, S. (2020). Attempting randomized housing first research in a community context: Reflections on failure. *American Journal of Community Psychology, 66*(1–2), 201–213.

Budd, L., & Ison, S. (2020). Responsible Transport: A post-COVID agenda for transport policy and practice. *Transportation Research Interdisciplinary Perspectives, 6*, 100151.

Bulkeley, H., Watson, M., Hudson, R., & Weaver, P. (2005). Governing municipal waste: Towards a new analytical framework. *Journal of Environmental Policy and Planning, 7*(1), 1–23.

Butsch, C., Kumar, S., Wagner, P. E., Kroll, M., Kantakumar, L. N., Bharucha, E., Schneider, K., & Kraas, F. (2017). Growing

'Smart'? Urbanization Processes in the Pune urban agglomeration. *Sustainability, 9*(12), 2335. https://doi.org/10.3390/su9122335
Cobbinah, P. B., Okyere, D. K., & Gaisie, E. (2020). Population growth and water supply: The future of Ghanaian cities. In M. KhosrowPour, S. Clarke, M. E.Jennex, & A. Anttiroko (Eds), Megacities and rapid urbanization: Breakthroughs in research and practice (pp. 96–117). Pennsylvania: IGI Global
Collier, P., & Venables, A. J. (2016). Urban infrastructure for development. *Oxford Review of Economic Policy, 32*(3), 391–409.
Cottafava, D., Cavaglià, G., & Corazza, L. (2019). Education of sustainable development goals through students' active engagement: A transformative learning experience. *Sustainability Accounting, Management and Policy Journal.*
Cuming, E. (2016). *Housing, Class and Gender in Modern British Writing, 1880–2012*. Cambridge University Press.
Debrah, C., Owusu-Manu, D. G., Kissi, E., Oduro-Ofori, E., & Edwards, D. J. (2020). Barriers to green cities development in developing countries: Evidence from Ghana. Smart Sustain. *Built Environ.*
Deng, Y., Zeng, Y., & Li, Z. R. (2018). Real estate prices and systemic banking crises. *Economic Modelling, 80*, 111–120.
Doe, B., Amoako, C., & Adamtey, R. (2022). Spatial expansion and patterns of land use/land cover changes around Accra, Ghana-Emerging insights from Awutu Senya East Municipal Area. *Land Use Policy, 112*, 105796.
ECF. (2020). ECF report. In *8th International Conference on Wireless Networks and Mobile Communications (WINCOM)*. https://doi.org/10.1109/wincom50532.2020.9272414
Eloy, S., Dias, M. Â., & Vermaas, P. E. (2018). User-centered shape grammars for housing transformations: towards post-handover grammars. *User-centered shape grammars for housing transformations: towards post-handover grammars.*
Eminsang, F. E. (2011). *Managing growth in Ghanaian cities-the role of peripheral district centres* (Doctoral dissertation).
Fiala, O. (2017). Natural disasters and individual behaviour in developing countries. Springer.
Fobil, J., Kolawole, O., Hogarh, J., Carboo, D., & Rodrigues, F. (2017). *Waste management financing in Ghana and Nigeria—How can the concept of Polluter-pays-principle (Ppp) work in both countries?*
Fokdal, J., Bina, O., Chiles, P., Ojamäe, L., & Paadam, K. (2021). *Enabling the City: Interdisciplinary and Transdisciplinary Encounters in Research and Practice*. Routledge.
Galland, P., Lisitzin, K., Oudaille-Diethardt, A., & Young, C. (2016). *World heritage in Europe today*. UNESCO Publishing.
Ghana Statistical Service, Ghana Health Service & ICF. (2023). *Ghana demographic and health survey 2022: Key indicators report*. Accra, Ghana and Rockville, Maryland, USA: GSS and ICF.
Gibson, R. B. (2006). Sustainability assessment: Basic components of a practical approach. *Impact Assessment and Project Appraisal, 24*(3), 170–182.
Ghana Statistical Service (GSS). (2013). *2010 National population and housing census: National analytical report*. GSS
Glaeser, E. L. (2014). A world of cities: The causes and consequences of urbanization in poorer countries. *Journal of the European Economic Association, 12*(5), 1154–1199.
Ha, S. (2010). Housing, social capital and community development in Seoul. *Cities, 27*, S35–S42. https://doi.org/10.1016/j.cities.2010.03.004
Habitat, U. N. (2013). *State of the world's cities 2012/2013: Prosperity of cities*. Routledge.
Habitat, U. N. (2018). Tracking progress towards inclusive, safe, resilient and sustainable cities and human settlements. SDG 11 synthesis report-high level political forum 2018.
Hochrainer, S. (2009). Assessing the macroeconomic impacts of natural disasters: Are there any?. *The World Bank Policy Research Working Paper Series.*
Horn, A., & Van Eeden, A. (2018). Measuring sprawl in the Western Cape Province, South Africa: An urban sprawl index for comparative purposes. *Regional Science Policy & Practice, 10*(1), 15–23.
IISD. (2023). IISD Annual Report 2022–2023. https://www.iisd.org/system/files/2023-10/iisd-annual-report-2022-2023.pdf
Jokilehto, J. (2017). *A history of architectural conservation*. Routledge.
Juraschek, S. P., Miller, E. R., Weaver, C. M., & Appel, L. J. (2017). Effects of sodium reduction and the DASH diet in relation to baseline blood pressure. *Journal of the American College of Cardiology, 70*(23), 2841–2848.
Leaning, J., & Guha-Sapir, D. (2013). Natural disasters, armed conflict, and public health. *New England Journal of Medicine, 369*(1836), 1842.
Lee, J., Perera, D., Glickman, T., & Taing, L. (2020). Water-related disasters and their health impacts: A global review. *Progress in Disaster Science, 8*, 100123.
Liang, L., Wang, Z., & Li, J. (2019). The effect of urbanization on environmental pollution in rapidly developing urban agglomerations. *Journal of Cleaner Production, 237*, 117649.
Low, S. M. (2023). *Why public space matters*. Oxford University Press.
Luo, L., Wang, X., & Guo, H. (2022). Contribution of UNESCO designated sites to the achievement of sustainable development goals. *The Innovation, 3*(3), 100227.
Mead, L. (2021). The Road to Sustainable Transport. *International Institute for Sustainable Development*. https://www.iisd.org/articles/deep-dive/road-sustainable-transport
Mensah, H., & Ahadzie, D. K. (2020). Causes, impacts and coping strategies of floods in Ghana: A systematic review. *SN Applied Sciences, 2*, 1–13.
Mensah, M. A., Campbell-Evans, G., & Main, S. (2022). Inclusion in Ghanaian public universities: A focus on student voices. *Disability & Society*, 1–23.
Meskell, L., Liuzza, C., Bertacchini, E., & Saccone, D. (2014). Multilateralism and UNESCO World Heritage: Decision-making, States Parties and political processes. *International Journal of Heritage Studies, 21*(5), 423–440. https://doi.org/10.1080/13527258.2014.945614
Miezah, K., Obiri-Danso, K., Kádár, Z., Fei-Baffoe, B., & Mensah, M. Y. (2015). Municipal solid waste characterization and quantification as a measure towards effective waste management in Ghana. *Waste Management, 46*, 15–27.
Ministry of Gender, Children and Social Protection, & Ghana Statistical Service. (2018). Five-year strategic plan for gender statistics (2018–2022). Ministry of Gender, Children and Social Protection in collaboration with the Ghana Statistical Services. Prepared by Marian A. Tackie, Private/Independent Consultant.
Mir Alvarez, C., Hourcade, R., Lefebvre, B., & Pilot, E. (2020). A scoping review on air quality monitoring, policy, and health in West African cities. *International Journal of Environmental Research and Public Health, 17*(23), 9151.
Mitchell, R., & Sweeney, F. (2018). Air Pollution. Scientific e-Resources.
Moallemi, E. A., Malekpour, S., Hadjikakou, M., Raven, R., Szetey, K., Ningrum, D., Dhiaulhaq, A., & Bryan, B. A. (2020). Achieving the sustainable development goals requires transdisciplinary innovation at the local scale. *One Earth, 3*(3), 300–313. https://doi.org/10.1016/j.oneear.2020.08.006
Moreno, E. L. (2017). Concepts, definitions and data sources for the study of urbanization. *The 2030 Agenda for Sustainable Development Produced by Experts Group on Cities, Human Mobility and International Migration*. New York: United Nations.

Naab, F. Z., Dinye, R. D., & Kasanga, R. K. (2013). Urbanization and its impact on agricultural lands in growing cities in developing countries: a case study of Tamale in Ghana. *Modern Social Science Journal, 2*(2), 256–287.

Nilsson, M., & Costanza, R. (2015). Overall framework for the sustainable development goals. *Review of targets for the sustainable development goals: The science perspective*, 7.

Ntajal, J., Höllermann, B., Falkenberg, T., Kistemann, T., & Evers, M. (2022). Water and health nexus—land use dynamics, flooding, and water-borne diseases in the Odaw River Basin, Ghana. *Water, 14*(3), 461.

Osei, F. E., Balogun, B. O., & Afrifa, C. G. (2015). Identifying and quantifying urban sprawl in the Greater Accra region of Ghana from 1985 to 2014. International Journal of Science and Research (IJSR), *4*(1).

Owusu-Manu, D. G., Debrah, C., Oduro-Ofori, E., Edwards, D. J., & Antwi-Afari, P. (2021). Attributable indicators for measuring the level of greenness of cities in developing countries: Lessons from Ghana. *Journal of Engineering, Design and Technology, 19*(3), 625–646.

Perry R. W. (2007). Chapter 1: What is a disaster? In H. Rodríguez, E. L., Quarantelli, & R. R. Dynes (Eds.*), Handbook of Disaster Research. Handbooks of Sociology and Social Research* (pp. 1–15). Springer Science + Business Media, LLC

Pieterse, E., Smit, W., Kayizzi-Mugerwa, S., Shimeles, A., & Yeméogo, D. (2014). Institutions, decentralisationand urban development. *Urbanization and socioeconomic development in Africa: Challenges and opportunities* (pp. 42–80).

Ramani, S. V., & Hettiarachchi, H. (2022). *SDG11*. Taylor & Francis.

Russell, H., Privalko, I., McGinnity, F., & Enright, S. (2021). *Monitoring Adequate Housing in Ireland.*

Sarfo, I., Shuoben, B., Otchwemah, H. B., Darko, G., Kedjanyi, E. A. G., Oduro, C., Folorunso, E. A., Alriah, M. A. A., Amankwah, S. O. Y., & Ndafira, G. C.(2022). Validating local drivers influencing land use cover change in Southwestern Ghana: A mixed-method approach. *Environmental Earth Sciences, 81*(14), 367.

Schaaf, T., & Rodrigues, D. C. (2016). *Managing MIDAs: Harmonising the Management of Multi internationnaly Designated Areas: Ramsar Sites, World Heritage Sites, Biosphere Reserves and UNESCO Global Geoparks*. Gland, Switzerland: IUCN International Union for Conservation of Nature and Natural Resources.

Seto, K. C., Fragkias, M., Güneralp, B., & Reilly, M. K. (2011). A meta-analysis of global urban land expansion. *PLoS ONE, 6*(8), e23777.

Toure, S. I., Stow, D. A., Clarke, K., & Weeks, J. (2020). Patterns of land cover and land use change within the two major metropolitan areas of Ghana. *Geocarto International, 35*(2), 209–223.

Tozier de la Poterie, A., & Baudoin, M. A. (2015). From Yokohama to Sendai: Approaches to participation in international disaster risk reduction frameworks. *International Journal of Disaster Risk Science, 6*, 128–139.

UN DESA. (2020). United Nations Department of Economic and social affairs. https://www.un.org/development/desa/en/news/population/2018-revision-of-world-urbanization-prospects.html

UNEP (2021). *Turning off the tap: How the world can end plastic pollution and create a circular economy*. Available at: https://wedocs.unep.org/bitstream/handle/20.500.11822/42277/Plastic_pollution.pdf?sequence=3

UNEP. 2009. *Converting waste plastics into a resource—Assessment guidelines*.

UNESCO. (2016). *Culture: urban future: Global report on culture for sustainable urban development*. UNESCO Publishing.

Un-Habitat. (2016). *Planning sustainable cities*. https://doi.org/10.4324/9781315541389

UNISDR. (2016). UNISDR Annual Report 2015: 2014–15 Biennium Work Programme Final Report. United Nations.

UNPD. (2019). World population prospects 2019 highlights. United Nations

Vandermeulen, V., Verspecht, A., Vermeire, B., Van Huylenbroeck, G., & Gellynck, X. (2011). The use of economic valuation to create public support for green infrastructure investments in urban areas. *Landscape and Urban Planning, 103*(2), 198–206.

WHO. (2021). www.who.int. https://www.who.int/publications/i/item/WHO-HEP-ECH-AQH-2021.8

Energy Performance Improvement in Lebanese Family House

Fatima Belok and Yathreb Sabsaby

Abstract

Given the depletion of fossil fuel resources, energy efficiency is becoming significant. As energy efficiency controls emissions, which affect the average world temperature, it is also critically vital for addressing climate change. Also, according to the generalized technique of moments estimate results, there is a significant rise in real and ideal energy consumption as a result of urbanization, but less efficiency in energy use. Due to these reasons energy efficiency is becoming increasingly important in this century. Improving the energy efficiency of existing residential buildings in Lebanon has not yet been sufficiently addressed in the latest guidelines and studies. Since the techniques of energy saving differ from one building to another, depending on the type of the existing building and the climate zone, saving strategies are difficult to generalize. This study first examines a case study in Nicosia, the capital of Cyprus, which experiences an energy crisis comparable to that of Lebanon. The study investigates into various energy-efficient improvements for Cyprus's existing buildings, which are intended to serve as a model for the Lebanese situation. Subsequently as an example of an actual residential building type in Lebanon, this study focuses on Terrace Family Houses. The effectiveness of the improvement is then assessed by comparing the new state with the existing condition state following amelioration and the implementation of energy-efficient techniques in Lebanon. This paper aims to investigate different techniques and strategies to improve the energy efficiency of existing Lebanese family houses by emphasizing the significance of thermal transmittance of building materials in exterior wall construction. These techniques focused on applying Low-e double glazing and wall insulation to test the U-values' impact on energy saving in the different case study apartments. In order to measure the energy performance improvement value, various tools are used: Teslo 635, which measures U-values, FLIR Thermal imaging tool, and Design Builder as energy modeling software. The paper's outcome measures the effectiveness of the applied techniques in decreasing the consumption of energy in Lebanese Terrace Family homes. Thus, research findings could serve as a starting point to legislate rules for a National Thermal Standard and Energy Code, which could be applied to existing and new residential building designs.

Keywords

Energy performance · Energy efficiency · U-value · Terrace family house · Thermal standards

1 Introduction

Lebanon has faced several financial crises starting in October 2019. Many accidents have accelerated this financial collapse, such as the Port Beirut Explosion in 2020, and the COVID-19 pandemic. These consecutive problems put Lebanon on the list of the highest inflation rates, placing it in second place globally in 2022 (see Fig. 1 in Derhally, 2022). Also, the inflation for gas, electricity, and water increased consecutively. Every culture needs energy, and the manner in which it is consumed affects social, economic, and environmental aspects of life. High energy use has a number of disadvantages including increased expenses, air pollution, damage to the environment, and

F. Belok (✉) · Y. Sabsaby
Faculty of Architecture-Design and Build Environment, Beirut Arab Unibersity, Beirut, Lebanon
e-mail: F.belok@bau.edu.lb

Y. Sabsaby
e-mail: y.sabsaby@bau.edu.lb

B. He et al. (eds.), *Urban Climate Change Adaptation*, Advances in Science, Technology & Innovation,
https://doi.org/10.1007/978-3-031-65088-8_15

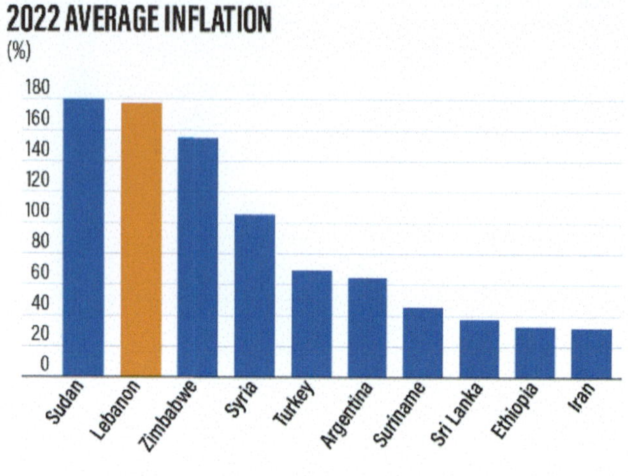

Fig. 1 Shows the average of inflation in different countries, and how Lebanon occupied second highest place in 2022. *Source* Derhally (2022)

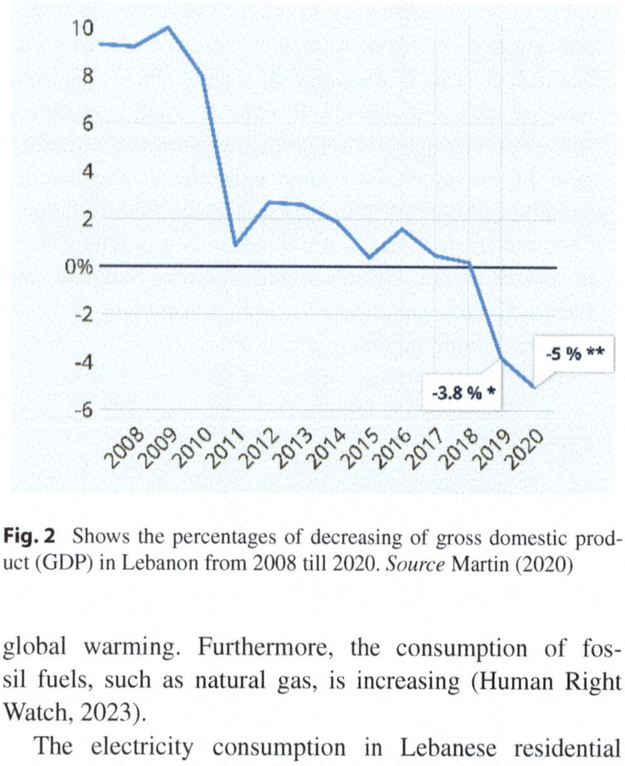

Fig. 2 Shows the percentages of decreasing of gross domestic product (GDP) in Lebanon from 2008 till 2020. *Source* Martin (2020)

global warming. Furthermore, the consumption of fossil fuels, such as natural gas, is increasing (Human Right Watch, 2023).

The electricity consumption in Lebanese residential buildings has mainly increased due to the use of air-conditioning systems, which are not energy-efficient. A survey on 1200 households in Lebanon was conducted by the Human Right Watch in 2022 to demonstrate the existence of an electricity crisis that forces people into poverty and delays the accessibility of their essential needs, such as water and food. Thus, almost 90% of the surveyed households were unable to pay for their essential services (see Fig. 2 in Martin, 2020), because of the electricity cost and the decreasing of gross domestic product (GDP) (El Charif, 2022).

Globally, the planet's future is threatened by increasing energy consumption, whereby 10% of the world's population abuses 90% of energy resources. Building activities are responsible for 30% of global final energy consumption and 26% of worldwide energy-related greenhouse gases (with 8% of emissions coming directly from buildings and 18% from the generation of heat and electricity used in buildings) (Iea, 2023). Compared to the other building sectors such as commercial and educational, 2/3 of the energy is consumed by residential buildings. Energy efficiency, which reduces the energy used and improves the building's performance simultaneously, could be the solution to the previous problems. Efficient systems have numerous benefits, such as improving comfort, increasing life-cycle cost savings, adding market value to the building, and reducing CO_2 emissions and consumption of fuels (Iea, 2011). Reducing energy use is essential to addressing the climate emergency. Enhancing efficiency have been shown to be the most economical strategy for addressing climate change with lowering energy waste, making financial savings, and extending the use of renewable energy resources at a reasonable price. Energy efficiency has many positive effects on the environment. It significantly decreases emissions of carbon dioxide, both direct emissions from burning fossil fuels and indirect emissions from the generation of electricity. In addition, urbanization leads to significant increases in both actual and ideal energy consumption, but it also causes a drop in energy usage efficiency, according to the generalized method of moments estimate results (Sheng et al., 2017). These factors all demonstrate how essential it is to improve energy efficiency strategies. For that, different researcher discussed the energy efficiency from different point of view such as the following, some studies focus on the effect of the energy sector on the Lebanese economy, and finally evaluates the viability of introducing appropriate energy-saving measures in the building sector. In addition, other studies evaluate the long-term benefits of implementing energy-efficient choices in the construction sector (Chedid & Ghajar, 2004). Additional studies examine whether online surveys can be used to gather information about general views about energy efficiency (EE) and zero energy housing (ZEH) in the US and Lebanon. Also, the connection between energy efficiency and climate change in Lebanon was the topic of another research (Jabbour, 2021).

1.1 Energy Approach in Cyprus and Lebanon

Petroleum and oil products are essential sources to produce electricity in Lebanon and Cyprus. Both countries depend on the imports of such products, which account for, e.g., 84% of the total electricity generation in Cyprus in 2021

(Statista Research Department, 2023). In Lebanon, the residential and commercial sectors have been reported to consume about 80% of the country's electricity, also Cyprus's analysis of the final energy consumption in 2012 have been demonstrated that the bigger amount of energy is consumed by the transportation sector, then followed by the residential sector (Konstantinou & Knaack, 2013), also. Since 2009, the total supply of primary energy in Lebanon has increased each year. The analysis shows that sectors in Lebanon and Cyprus suffer almost from the same problems regarding energy supply and efficiency, which means that the retrofitting applied in Cyprus can be helpful if applied in Lebanon.

1.2 Energy Performance Improvement Techniques in Buildings

This research focuses only on the residential typology. In the residential sector, energy efficiency improvement is related to five points: (i) Ventilation, (ii) Hot water, (iii) Heating, (iv) Cooling, and (v) Lighting (Rosenfeld et al., 1998). The principles for retrofitting and improving energy performance in residential buildings, which are related to the mentioned five points, are the following (Rosenfeld et al., 1998):

- Solar energy and heating sources

Active solar energy frameworks can produce power, hot water, and space conditioning. There are techniques for cooling spaces and distributing heat without using refrigeration cycles, such as using earth pipe cooling.

- Insulation

It is very important to select a fitting type of insulation since the choice between the various existing types can affect the building's thermal performance. In addition, the thermal bridges should be taken into consideration and avoided because it reduces the effectiveness of insulation, increases the energy spent, the risk of condensation, mold and rot.

- Passive solar heating

For passive solar heating, different techniques can be chosen, such as glazing type, roof-mounted solar air collectors, wind current windows, and pipes for pre-heating or precooling.

- Cooling load reduction

Various factors and principles affect the building's cooling load. These factors are, e.g., building orientation, selected materials, and building form, which are all related to the building climate, causing the architect to play a very important role (Rosenfeld et al., 1998). The building's orientation is an important factor in reducing cooling load by minimizing the amount of wall facing the sun. In Lebanon, the sun faces south for the most of the day. In addition, the cooling load could be reduced by inserting high reflective shading materials in the opening area or by using a specific type of glazing with low heat gain. Also, inside the building, new efficient lighting spots could be used to save energy and, at the same time, produce enough lux (Asghar, 2023).

1.3 Research Need, Objectives and Methods

A key factor in thermal applications and energy efficiency studies are building envelopes and their materials, including walls and roofs. However, the lack of a thermal standard for the design of new and existing buildings is the main problem of the Lebanese construction industry. Energy efficient building envelopes are thus not effectively utilised in Lebanon. Consequently, the thermal insulation factor is the focus of this study rather than the other three previously mentioned principles for increasing energy efficiency because it can be taken consideration from the initial stages of construction, saving a significant amount of time, money, and energy if it is respected. Therefore, this study aims to fill this gap by focusing on the thermal characteristics in residential building envelopes of the Lebanese building stock and the U-value that describes the ability of materials to pass thermal energy.

A residential apartment building located in Cyprus serves as a case study reference for energy efficiency improvement. The founded solutions in the Cyprus example are applied in the existing local apartment building in Koura, North of Lebanon, to measure the efficiency of these techniques in Lebanon.

Nicosia, the Capital of Cyprus, is 241.6 km away from Beirut. Both cities have the same time zone. Subtropical steppe is the climate classification of Nicosia, while Mediterranean, hot summer is the climate classification of Beirut, Lebanon. Nicosia and Beirut almost have the same average temperature, consecutively 21.85 and 21.21 °C, creating a difference of less than 1 °C. In addition, the average annual relative humidity in Nicosia, Cyprus, and Beirut, Lebanon is similar, which is consecutively 61.7%

and 69.3%. Furthermore, the temperature difference between Nicosia and Lebanon in the hottest month is nearly 3 °C, while in the coldest month it is equal to 1 °C. From this analysis, it is deduced that Lebanon and Cyprus have similar climates, which means that the energy improvement techniques applied in Cyprus, could also be applied in Lebanon. Thus, two case studies are conducted, one in Cyprus and one in Lebanon, with the aim of applying similar retrofitting and energy efficiency techniques in Lebanon as those used in Cyprus for the reasons mentioned above. Afterwards, the data of the existing case study in Lebanon called the Jaara Terrace Family House is gathered using the Design Builder software, a thermal imaging tool, and a Teslo 635 thermometer that measures U-values. This study compares the current scenario with the improved one by focusing on the thermal transmittance of materials (U-value). Consequently, the U-value help to indicate the energy efficiency performance of the building. In addition, it determines different energy efficiency strategies and parameters that decrease energy use and consumption in Lebanese Terrace Family homes. For example, these strategies include the effect of the external wall's area, the external walls' U-values for thermal transmittance that may vary depending on the materials selected, as well as how the walls are oriented and exposed to sunlight. Subsequently, an analysis is conducted to compare the efficiency of the enhancement strategies between the before and after improvement scenarios.

2 Cyprus Terrace Family House Study

The following paragraph discusses a selected case study in Lebanon, where solutions founded in Cyprus example are applied, to see their effectiveness on it. There are seven blocks in the Terrace Family House complex. The

Fig. 3 A realistic image for The terrace family housing complex under study. *Source* Serghides and Katafygiotou (2013)

starting, middle, and end blocks are the blocks that were chosen for the evaluation (see Figs. 3 and 4 in Serghides & Katafygiotou, 2013). This decision examined the two peripheric units with the most external wall area exposed to the environment, and the middle with the least exposure. These three dwellings have the same characteristics and electromechanical systems. A measurement of their energy consumption is made, using Design Builder software, a thermal imaging tool, and a Teslo 635 thermometer tool. Each unit is then equipped with retrofitting techniques, using Design Builder software, which is able to do various simulation to visualize the outcomes of these techniques' implementations. Following that, a comparison is done between those three units to see how their performances differed from one another, taking into account the various locations of the units. Finally, a conclusion related to the consumption of energy in the dwellings is deduced. In this study, the values of the heating and cooling framework are documented.

2.1 Existing Condition of Buildings

The Terrace Family Houses lie in the capital city of Nicosia. The blocks have the following characteristics: pitched roof covered with tiles but non-protected, no roof insulation, the ground contacts directly all the floor slabs, block walls, and single-glazed windows. Each block contains the following zones: kitchen, WC, living room, three bedrooms and a bathroom. These components are divided into two floors (Serghides & Katafygiotou, 2013). The cooling systems are composed of split units with various functions. The U-values and thermal attributes are presented in a table (see Table 1 in Serghides & Katafygiotou, 2013).

North South West is the orientation for (Block 1) in the terrace family house, including three windows on the south façade. 525.52 kWh/(m^{2}a), 7 kWh/(m^{2}a) is the total primary energy consumption concerning (Block 1). This energy is delivered from a solar panel located on the top of the roof. This consumption is equal to 404.42, 7 kWh/(m^{2}a) in (Block 2), which is orientated toward the Northeast and Southwest. In addition, (Block 3) which is oriented northeast and northwest has a total primary energy consumption equal to 601.16, 7 kWh/(m^{2}a). Concerning (Block 1), the total consumption of energy reaches 248, and 210.31 kWh/(m^{2}a) in (Block 2), while 307.63 kWh/(m^{2}a) in (Block 3). Concerning the consumption of energy for heating and cooling in (Block 1) they are consecutively 90.22 kWh/(m^{2}a), and 136.14 kWh/(m^{2}a), concerning (Block 2) they are 102.14 and 85.45 kWh/(m^{2}a), while for the ended (Block 3) they are 143.40 and 139.43 kWh/(m^{2}a).

A building's energy efficiency is evaluated by an Energy Performance Certificate (EPC), which ranges from A (the most efficient) to G (the least efficient). The government's

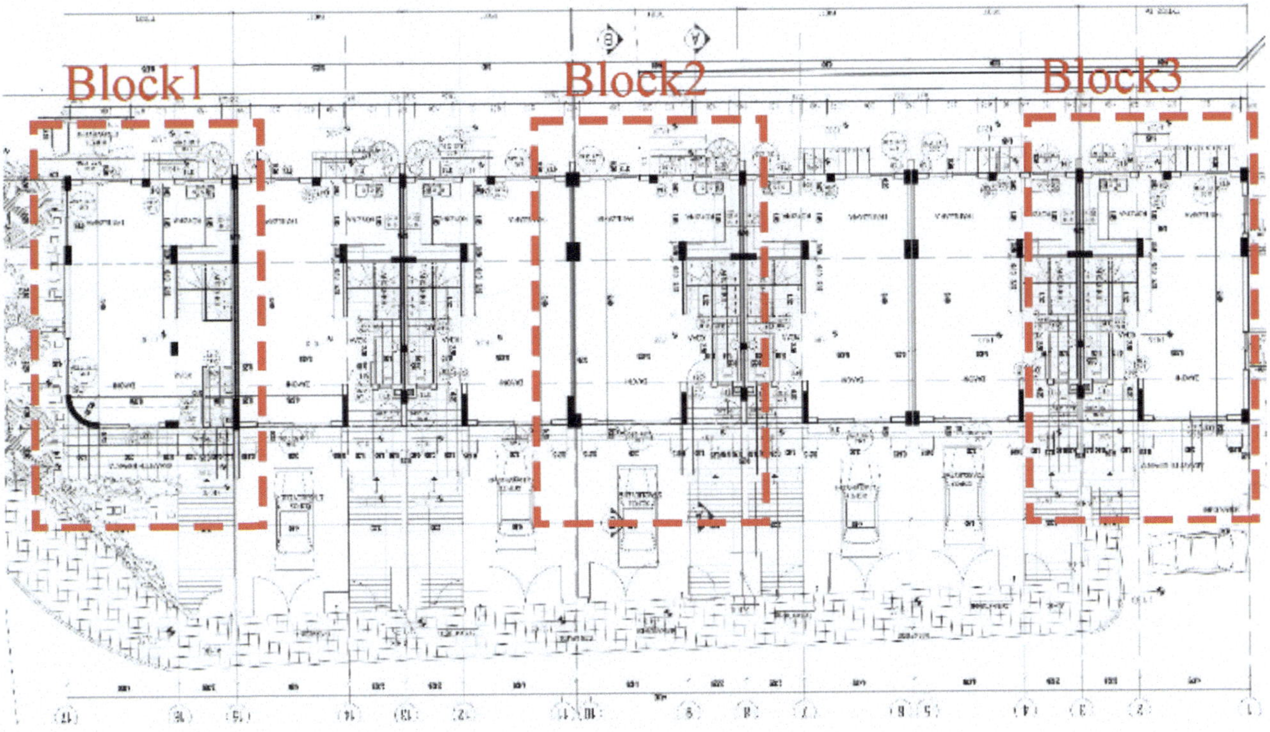

Fig. 4 The architectural plans of the complex under study. *Source* Serghides and Katafygiotou (2013)

Table 1 Thermal attributes of the building envelope for terrace family housing complex. *Source* Serghides and Katafygiotou (2013)

Construction element	U-value W/(m^2k)
Pitched roof with horizontal ceiling	3.42
External Walls	1.42
Floor in contact with the ground	0.91
Double glazed windows	3.80

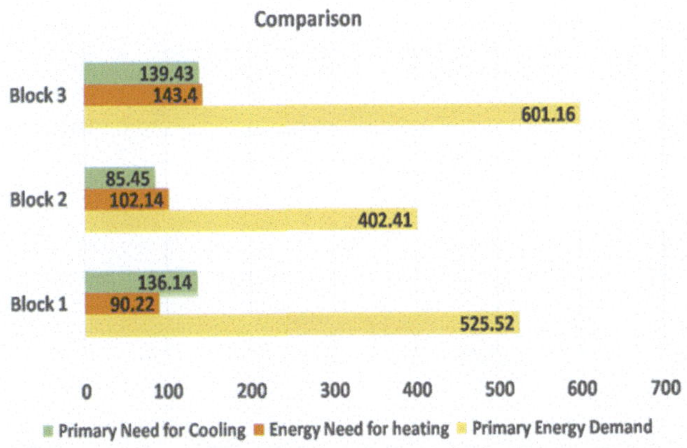

Fig. 5 Shows a comparison of the primary energy consumption, energy need for heating and cooling for the existing blocks 1, 2 and 3. *Source* Serghides and Katafygiotou (2013)

Standard Assessment Procedure (SAP) assigns a numerical score, broken down into bands as follow:

- Green (92–100 SAP points) is the EPC rating A.
- Green (81–91 SAP) points is the EPC rating B
- Green (69–80 SAP points) is the EPC grading C
- Yellow (55–68 SAP points) is the EPC rating D.
- Amber (39–54 SAP points) is the EPC rating E.
- Orange (21–38 SAP points) is the EPC rating F
- Red (1–20 SAP points) is the EPC rating G

The three blocks received a certificate for energy performance (EPC), which are (Block 1) and (Block 3) F, while (Block 2) E. Through similar investigations of the above results, the energy needs for (Block 3) are the highest followed by the House (Block 1) and then by (Block 2). A comparison of the three blocks is done, (see Fig. 5 in Serghides & Katafygiotou, 2013).

2.2 Performance Improvement Scenarios

The following paragraph describes techniques for improving energy performance for Blocks 1, 2, and 3. These techniques are just applied to the building envelopes, keeping the same methods for cooling and heating. This enhancement is done using energy conservation measures (ECMs)

(Serghides & Georgakis, 2012). Various improvements are done to achieve essential needs for building envelopes for EPBD Directive. These improvements are: (i) adding heating insulation equal to 30 mm, called extended polystyrene, to the exterior wall to achieve better insulation, (ii) adding heat insulation inside the slabs with a thickness equal to 50 mm, and (iii) replacing the single-glazed windows with double-glazed. After this enhancement, the new U-values are recorded to show the difference compared to the original vsoalues. There is a remarkable enhancement (see Tables 2 and 3 in Serghides & Katafygiotou, 2013) concerning the classification, energy, cooling and heating consumption.

The three Blocks moved from one classification to another as follows: Block 1 moved from class F to class D, Block 2 moved from class E to class C, and Block 3 moved from class F to class D after improving their energy performance. Table 3 illustrates the total primary energy consumption, which is saved as 39.23% in (Block 1), 47.50% in (Block 2), and 36.64% in (Block 3) (see Figs. 6 and 7 in Serghides & Katafygiotou, 2013). Furthermore, with regard to (Block 1), there is a drop in overall energy consumption, which is currently equivalent to 149.08 kWh/(m^{2}a). This decrease was also observed in (Blocks 2 and 3), where it became 112.44 and 190.47 kWh/(m^{2}a) in that order. Consequently, after the improvement, there is a noticeable difference and a decline

As shown in the previous figures, the most energy saving was scored in the middle Block 2, which has the lowest area of external walls. Thus, it can be argued that the smaller external wall area will benefit from the upgrade more than the larger one. Furthermore, it is noticeable that the more the surface area of the external walls is reduced, the higher is the energy efficiency of buildings, as contact between external weather influences and the interior spaces is minimized.

Table 2 Shows thermal characteristics of the building envelope for the complex. *Source* Serghides and Katafygiotou (2013)

Construction element	U-value W/(m^{2}k)
Pitched roof with horizontal ceiling	0.60
External walls	0.69
Floor in contact with the ground	0.91
Double glazed windows	3.23

3 Application in Lebanese Terrace Family House

In Lebanon, the selected case for application is called Jaara Terrace Family House located in North Lebanon, Koura, as shown in Fig. 8. Three example building units were chosen from the Terrace Family House complex, having different positions: the starting dwelling called Block A (west-oriented), the middle one called Block B, and the ending one called Block C (east-oriented), (see Fig. 8 in author, 2023). These units have the same characteristics, structure, and building envelope. Each unit has 121 m^2 in area (see Fig. 9 in author, 2023). U-values and energy consumption, including fuel, lighting, heating, and cooling, are measured in two cases: the first involves comparing the energy consumption

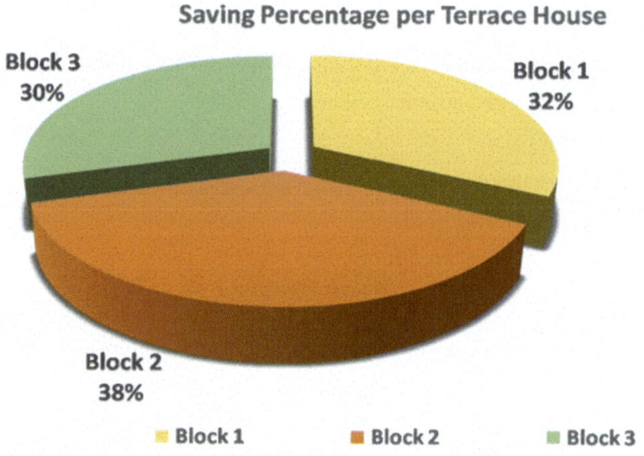

Fig. 6 Shows the saving percentage after improvement for blocks 1, 2 and 3. *Source* Serghides and Katafygiotou (2013)

Table 3 Shows a comparison between energy consumption and classification for Block 1, 2 & 3 in Cyprus Terrace family house, before and after improvement (Author, 2024)

Simulation type	Block 1		Block 2		Block 3	
	Existing conditions	EE improvement	Existing conditions	EE improvement	Existing conditions	EE improvement
Classification	Class F	Class D	Class E	Class C	Class F	Class D
Saving energy consumption (%)	–	39.23%	–	47.50%	–	36.64%
Cooling Kwh/(m^{2}a)	136.14	75.36	85.45	32.58	139.40	82.30
Heating kWh/(m^{2}a)	90.22	51.98	102.14	51.13	143.40	83.30

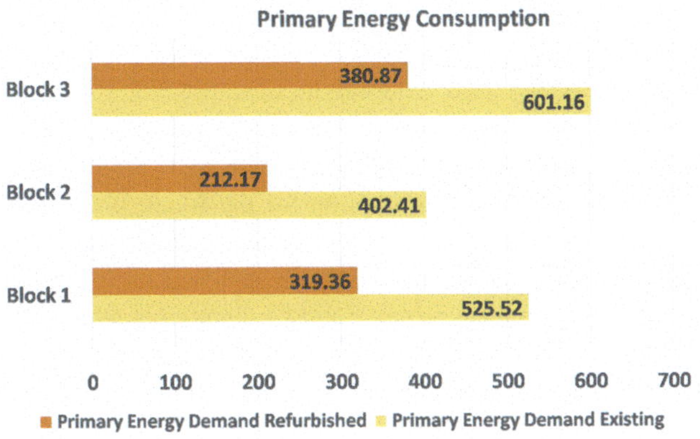

Fig. 7 Shows the calculated primary energy consumption for the existing and the refurbishment for block 1, block 2, and block 3. *Source* Serghides and Katafygiotou (2013)

results between the three units under the current conditions using measurement tools as explained in (1.3. Research need,objectives and metods). The second scenario involves applying performance improvement scenarios and using design builder software as a simulation tool to compare the energy consumption results between the units.

The main complex facades oriented to South-East orientation have an external surface temperature equal to 28.1 °C (see Fig. 10 in author, 2023), while the second façade North-West orientation has an external surface temperature equal to 26.6 °C (see Fig. 11 in author, 2023). These thermal measurements were taken using FLIR as a thermal imaging tool on 5/6/2023 at 12:00 AM, which is considered the most solar-radiated hour in the middle of the summer season. Thermal imaging is used to determine the most external wall solar exposure that needs to be considered in the improvement strategies.

3.1 Existing Condition of Buildings

The Jaara Terrace Family House dwellings are composed of an apartment unit of 121 m^2 area. They consist of a living room, kitchen, WC, two bedrooms, and a bathroom, (see Fig. 12 in author, 2023). The model of the dwellings is done in the design builder software as energy model software.

The U-values of external walls and the glazing are measured by using the Testo 635 tool as a surface temperature and humidity measurement instrument (see Fig. 13 in

Fig. 8 Shows the location on google earth as zoom in and zoom out for the Jaara Terrace Family House. *Source* Author

Fig. 9 Shows autoCAD plan for each dwelling in block A, B and C in Jarra Terrace Family House. *Source* Author (2023)

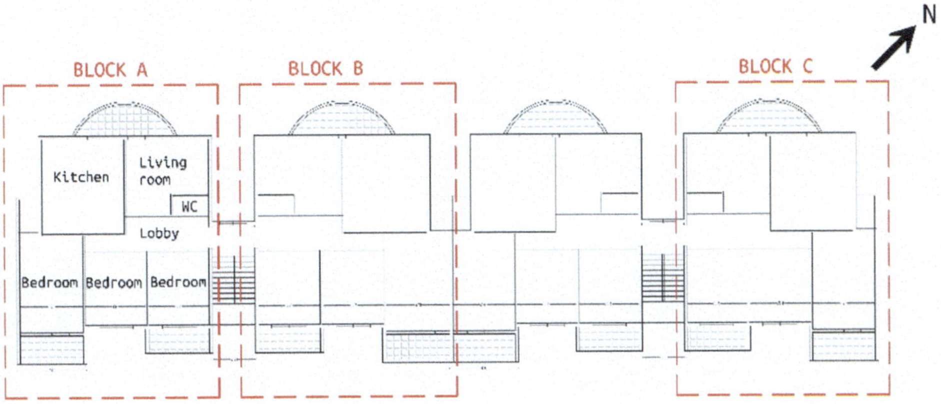

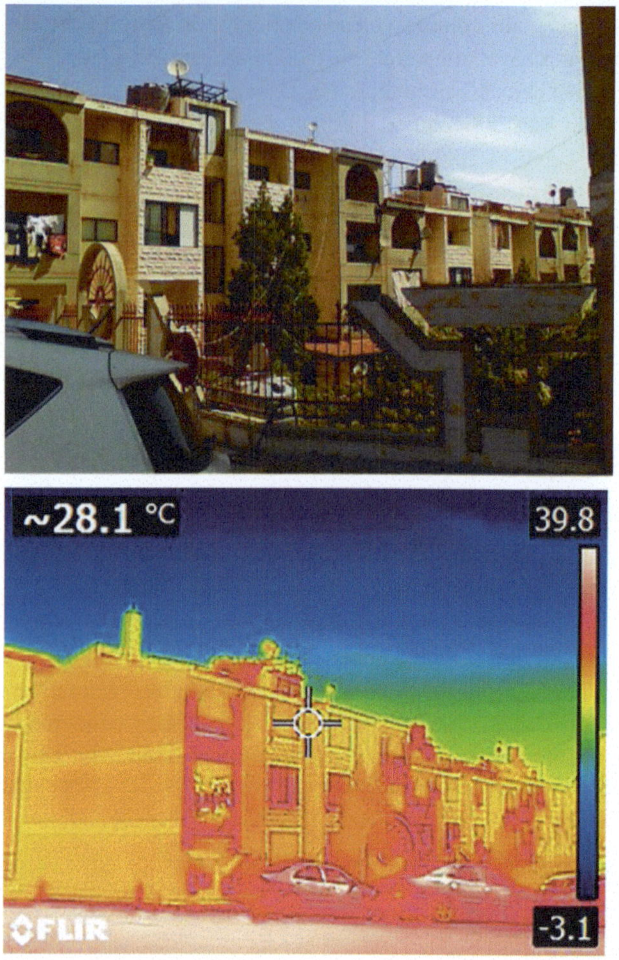

Fig. 10 South-East façade with thermal photo. *Source* Author (2023)

author, 2023). The external wall assembly consists of a single masonry concrete block wall, with a thickness equal to 20 cm without any insulation, the floor slab in contact with the ground, the pitched roof in non-protected, and single glazed windows. The U-value of external wall scored 2.86 w/m^2-K. The external glazing consists of single clear glass with a U-value of 5.6 w/m^2-K. The lighting and HVAC systems were entered as the conventional system used in conventional Lebanese houses. Lighting fixture is suspended incandescent lamps with a dual control system. The HVAC system used is the split units for each room.

Based on the previous input data, the simulation is generated to specify the energy consumption in the different units (see Table 4 in author, 2023) as follows:

- Block A has three external façades (105 m^2 external walls) with five openings covering 26 m^2 confronting north-west (13.2 m^2) and south-east. WWR (Window to Wall Ratio) = 0.24. The total energy consumption for BlockA is 215.65 kWh/(m^{2}a), the energy consumption for heating is 4 kWh/(m^{2}a), for cooling is 44.39 kWh/(m^{2}a), for hot water, is 106.4 kWh/(m^{2}a), and for lighting is 61 kWh/(m^{2}a).
- Block B has two external façades (64.5 m^2 external walls) with five openings covering 26 m^2, confronting north-west (13.2 m^2) and south-east. WWR = 4.8. The total energy consumption for the Block B achieves 201.7 kWh/(m^{2}a), the energy consumption for heating is 3.2 kWh/(m^{2}a), for cooling is 33.59 kWh/(m^{2}a), for hot water, is 106.4 kWh/(m^{2}a), and for lighting is 59.8 kWh/(m^{2}a).
- Block C has three external façades (105 m^2 as external wall) with five openings covering 26 m^2, facing north-west (13.2 m^2) and south-east. WWR = 0.24. Similar to BlockA with opposite orientation. The total energy consumption for Block C is 206.6 kWh/(m^{2}a), the energy consumption for heating is 4 kWh/(m^{2}a), for cooling is 36.78 kWh/(m^{2}a), for hot water is 106.4 kWh/(m^{2}a), and for lighting is 60.1 kWh/(m^{2}a).

3.2 Performance Improvement Scenarios

Based on the previous existing model simulation, the cooling and heating load in the peripheric apartments (Block A and C) was higher than the middle one (Block B). Therefore, the proposed energy-efficient improvement is concentrated on increasing the external wall's thermal performance by adding expanded polystyrene insulation and replacing the clear glazing with double Low-E glazing.

In addition, all three apartments (Block A, B, and C) had significant lighting fuel usage. In order to reduce and modify the lighting operating schedules, tiny fluorescent bulbs with stepped control systems were installed in place of all the lighting fixtures. All the improvement strategies are summarized (see Fig. 14, in author, 2023).

A comparison between the existing condition, before, and after improvement of energy efficiency, in the three Blocks A, B, and C for Jaara Terrace Family House, is done (see Table 5 in author, 2023). Remarkably, the annual fuel consumption is reduced after improvement, especially in Blocks B and C, which were consecutively 2,4409.29 Kwh and 2,5003.4 Kwh and reduced to 1,8739.3 and 1,9200 Kwh. Thus both blocks save almost 23% of the energy after energy effeceincy improvment. Additionally, there has been a noticeable improvement in the amount of energy saved by the lighting following the change. For example, the lighting in Block A used to consume 7354.05 Kwh, but after upgrades and modifications, the energy used to light the building now only uses 1104.74 Kwh. Furthermore, following improvements and adjustments, the cooling consumption is nearly dropped in half. For example, in Block B, it was equal to 4064.58 Kwh in the pre-improvement state, but after improvements, it was reduced to 2846.56 Kwh.

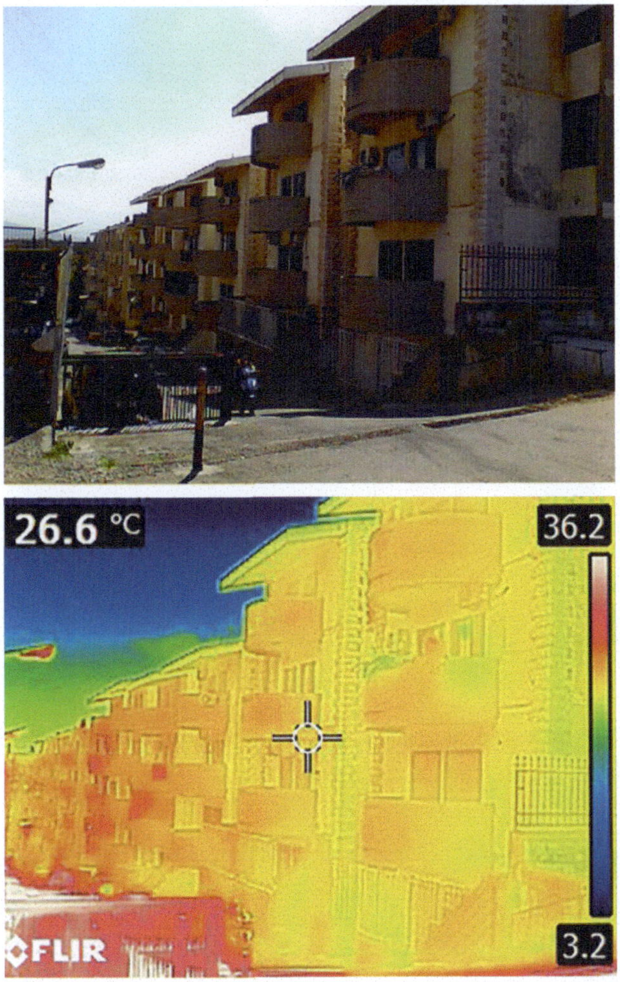

Fig. 11 North-West façade with thermal photo. *Source* Author (2023)

The U-values have been reduced by almost half; for example, in the external wall, after adding the expanded polystyrene, the U-value has been reduced from 2.864 to 1.329 w/m^2-k. Consequently, it has decreased by almost 50%, which is a significant outcome. Thus, after making some changes to the blocks, such as replacing the clear windows with double Low-E glazing and adding expanded polystyrene insulation thermal performance insulations, an improvement in energy efficiency is noticed.

4 Conclusion

The main problem in the Lebanese building sector is the absence of a thermal standard for new and existing building design. These standards are considered a milestone in achieving energy efficiency in buildings in different countries. Therefore, this study is considered a first step in evaluating the proposed thermal characteristics of construction materials in the building's external wall envelope. The case studies chosen in Lebanon and Cyprus are a prototype of Terrace Family houses. Cyprus's case study is considered a benchmark for the Lebanese case due to the similarity in climate, energy sources, and economic crises. Thus, as explained in the application section, the enhancements in the selected Lebanese project were made by utilizing Low-e windows, adding external polystyrene insulation to the external walls, and making small adjustments to the lighting systems. The impact of these improvements was measured using an energy modeling software Design Builder, comparing the existing and the proposed scenario based on the energy demand in both. Different parameters directly

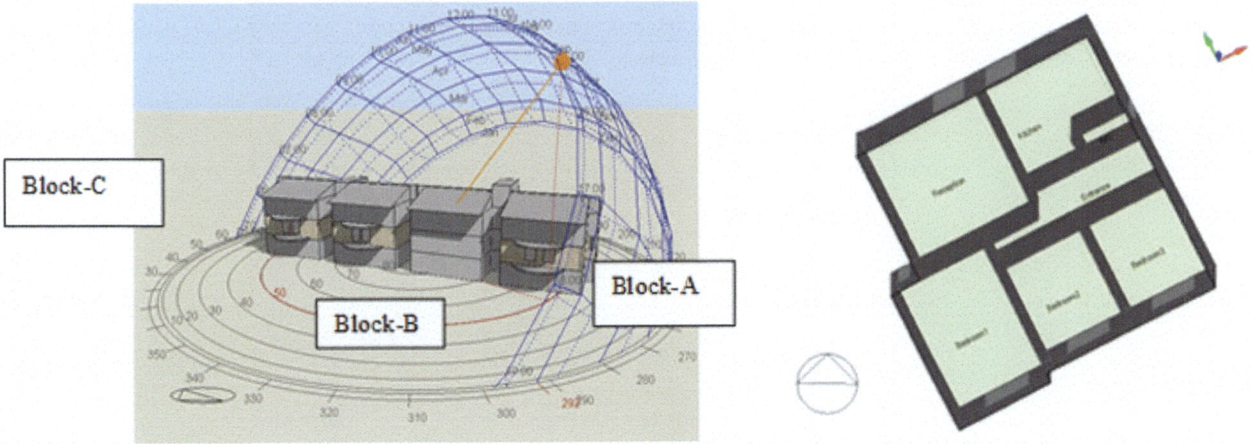

Fig. 12 Shows the Jaara Terrace Family House model using design builder software. *Source* Author (2023)

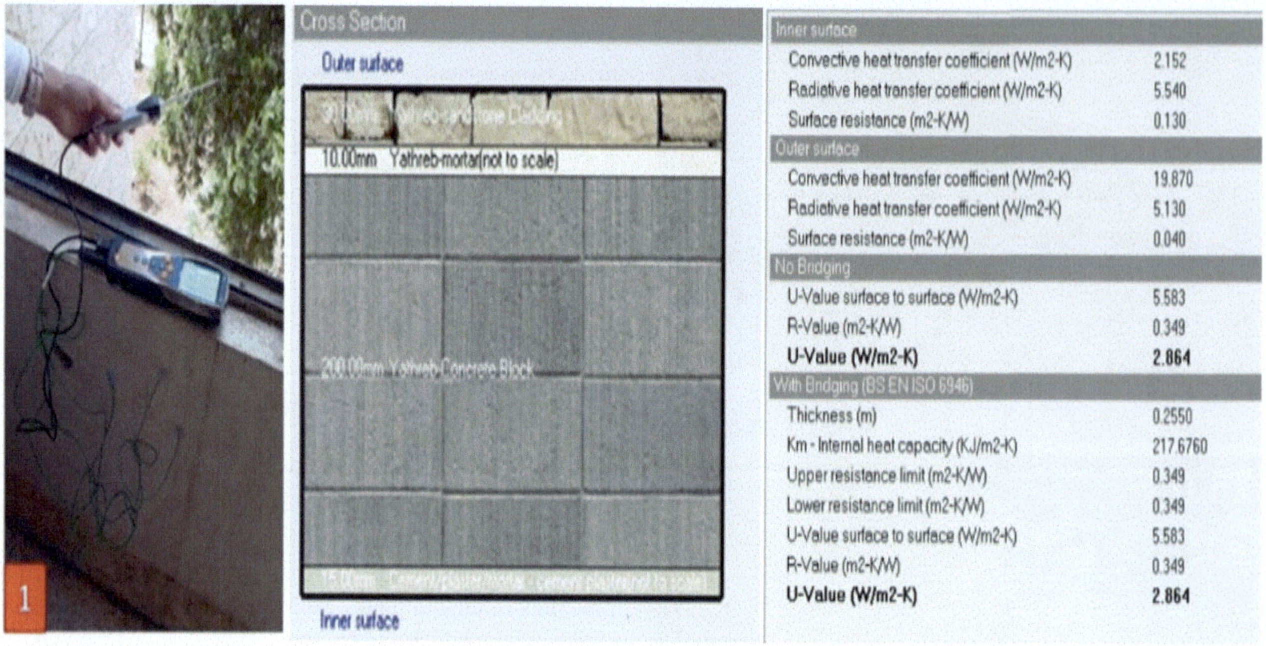

Fig. 13 Shows the tools used to measure and compute U-Value including Teslo 635 and design builder, which also provide an illustration of the exterior wall layers in their current state. *Source* Author (2023)

Table 4 Shows the annual fuel consumption values, total fuel, and fuel breakdown graphs, lighting, heating, cooling and DHW values for the existing condition of Jaara Terrace Family House. *Source* Author, 2023

	Block A	Block B	Block C
Fuel total	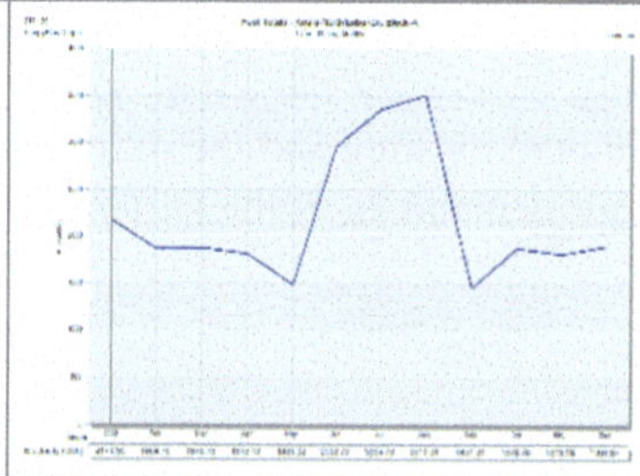	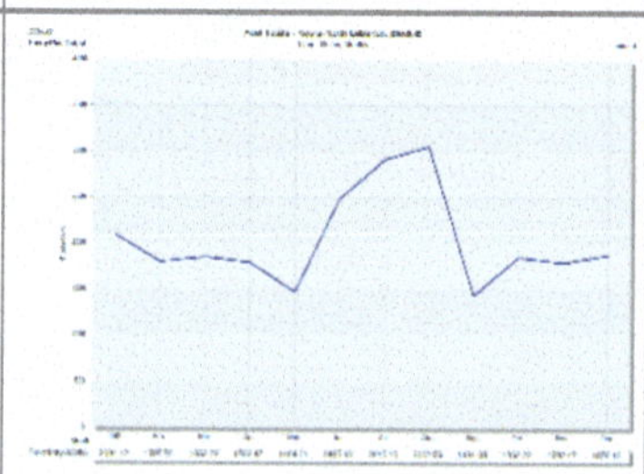	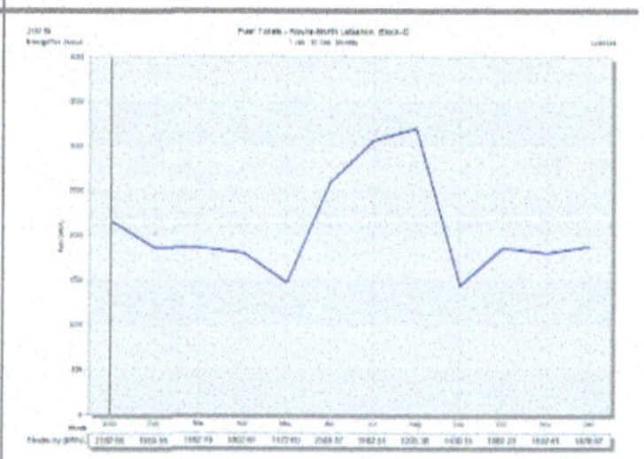
Annual fuel consumption (Kwh)	2,6094.14	2,4409.29	2,5003.4
Fuel breakdown graphs	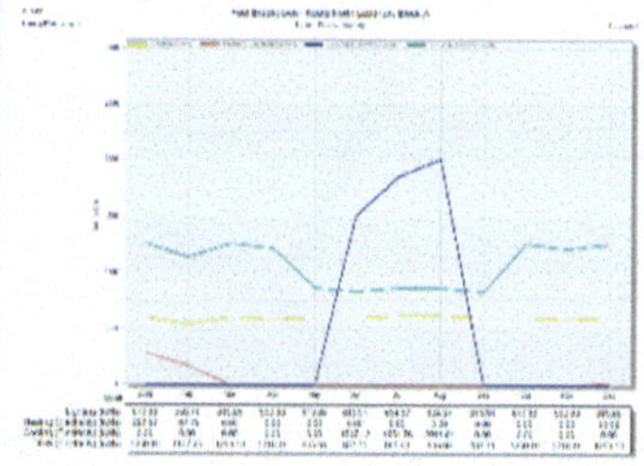	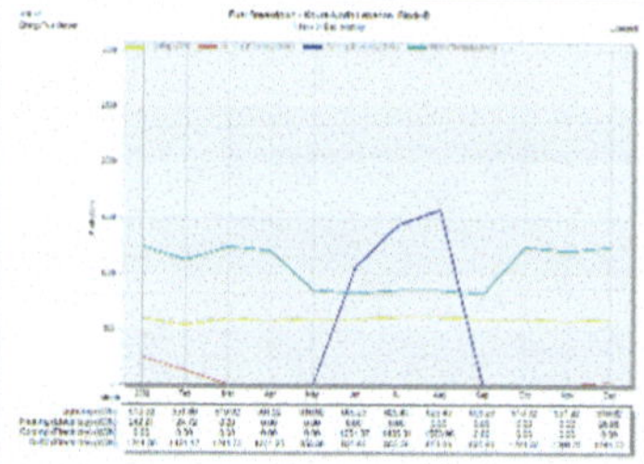	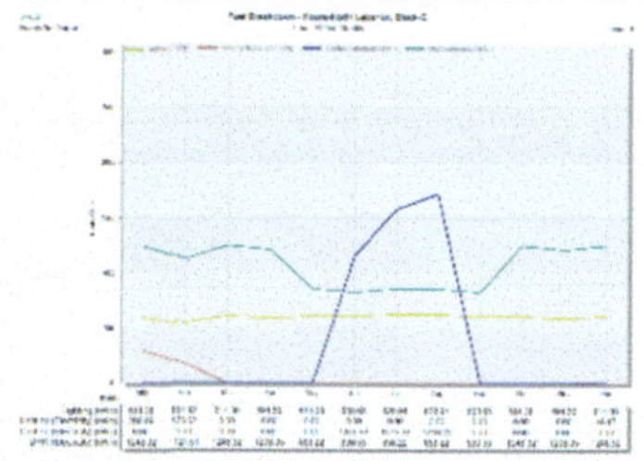
Lighting (Kwh)	7354.05	7243.67	7284.87
Heating (Kwh)	484.26	393.58	492.78
Cooling (Kwh)	5373.5	4064.58	4451.1
DHW (Kwh)	1,2878	1,2878	1,2878

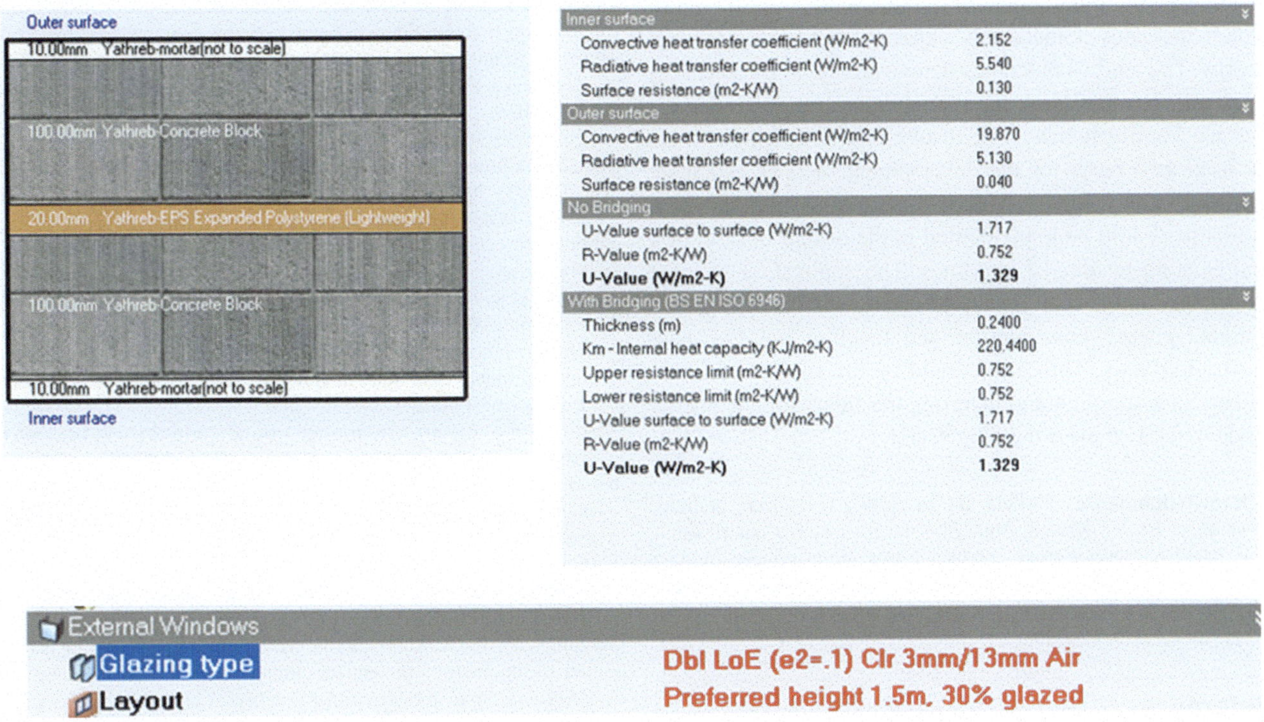

Fig. 14 Shows thermal properties for the wall and glazing after modification to improve energy efficiency. *Source* Author (2023)

Table 5 Shows a comparison between energy and fuel consumption for blocks A, B, and C in Jaara Terrace family house, before and after improvements. *Source* Author 2023

Simulation type	Block A			Block B			Block C		
	Existing conditions	EE improv	Savings (%)	Existing conditions	EE improv	Savings (%)	Existing conditions	EE improv	Savings (%)
Annual fuel consumption (Kwh)	2,6094.14	2,0784	30	2,4409.29	1,8739.3	23	2,5003.4	1,9200	23
Lighting (Kwh)	7354.05	1104.74	85	7243.67	1633.55	77	7284.87	1617.33	77
Cooling (Kwh)	5373.5	3975.31	26	4064.58	2846.56	30	4451.1	3080.4	31

affected the energy consumption, as highlighted by the comparison. These parameters can be summarized as follows:

- The size of the area of the external wall
- The thermal transmittance U-values of the external walls
- The orientation and solar radiation exposure on the external walls

In Lebanon, the Terrace Family House is well-known because so many buildings exist of this type. This study demonstrates that the middle residential unit in the Terrace Family Home in the two case studies (Lebanon and Cyprus) achieved the most energy savings, with up to 30% less energy used for cooling. Thus, decreasing the external wall area will decrease the contact with exterior weather changes and the energy needed for cooling. Therefore, the clustered residential complex could be more energy efficient than detached residential buildings. Another fact proved in this study is the importance of building orientation in energy efficiency, as the Block-A (west-oriented) score showed high energy consumption, where energy efficiency improvement is essential. Furthermore, the added insulation also greatly impacted the cooling load in other-oriented buildings. This study was focused on the architectural factors that could be used in new or existing building. Therefore, domestic hot water (DHW) systems are excluded from the improvement techniques like integrated solar water heater panels.

Building envelopes and their components, such as the walls and roof, are recognized as being particularly important in thermal applications and studies (Zhang et al., 2023). Thus, this study focused on the thermal characteristics of residential building envelopes, particularly the U-value, which indicates a material's capacity to transfer thermal energy. The study's outcomes demonstrate that these energy efficiency strategies reduce energy usage in Lebanese Terrace Family homes. The results of this research might be used as a basis for the development of laws establishing an Energy Code and National Thermal Standard. These guidelines could be implemented in the design of both new and existing residential buildings. The application of these standards could be the first step toward decreasing the total national energy consumption and tackling the Lebanese electricity crisis. This study could be the subject of future research on applying and inverting the International Energy Code to the Lebanese building sector.

Acknowledgements I would like to express my sincere gratitude to architect Fatima Khaled, from who we got inspired our research. It is based on a prior study conducted as a group project during a postgraduate program. I would additionally like to express gratitude to Mireille Hussein for reviewing the grammar and English.

References

Asghar, W. (2023). *Lighting and energy efficiency in PE power exam.*

Chedid, R. B., & Ghajar, R. F. (2004). Assessment of energy efficiency options in the building sector of Lebanon. *Energy Policy, 32*(5), 647–655. https://doi.org/10.1016/s0301-4215(02)00328-2

Derhally, M. (2022). https://www.thenationalnews.com/business/economy/2022/08/17/lebanon-expected-to-record-second-highest-inflation-rate-globally-in-2022/

El Charif, K. (2022). *Lebanon from August 2015 to October 2019: Attempts to Rehabilitate a Nation.*

Human Rights Watch. (2023). *'Cut off from life itself': Lebanon's failure on the right to electricity [en/ar]—lebanon, ReliefWeb.* Available at: https://reliefweb.int/report/lebanon/cut-life-itself-lebanons-failure-right-electricity-enar

Iea. (2011). https://www.iea.org/reports/world-energy-outlook-2011

Iea. (2023). *Buildings—energy system.* IEA. https://www.iea.org/energy-system/buildings

Jabbour, N. (2021). *Zero energy and energy efficiency perceptions, a comparative survey analysis from the US & Lebanon [Preprint].* https://doi.org/10.13140/RG.2.2.24597.24805.

Konstantinou, T., & Knaack, U. (2013). An approach to integrate energy efficiency upgrade into refurbishment design process, applied in two case study buildings in Northern European Climate. *Energy and Buildings, 59*, 301–309.

Martin, N. (2020). Retrieved from https://www.dw.com/en/lebanon-china-debt-default-economy-imf/a-52706863

Rosenfeld, A. H., Akbari, H., Romm, J. J., & Pomerantz, M. (1998). Cool communities: Strategies for heat island mitigation and smog reduction. *Energy and Buildings, 28*, 51–62.

Serghides, D. K., & Georgakis C., (2012). The building envelope of mediterranean houses—optimization of mass and insulation. *Journal of Building Physics, 36*(1), 83–98.

Serghides, D. K., & Katafygiotou, M. C. (2013). The role of materials in the energy efficient retrofitting of traditional buildings. In: *Materials and processes for energy: Communicating current research and technological developments*, pp. 978–984.

Sheng, P., He, Y., & Guo, X. (2017). The impact of urbanization on energy consumption and efficiency. *Energy & Environment, 28*(7), 673–686 (14 pages). https://www.jstor.org/stable/90014163.

Statista Research Department. (2023). Retrieved from https://www.statista.com/statistics/1235188/cyprus-distribution-of-electricity-production-by-source/#:~:text=Oil%20and%20petroleum%20products%20are,total%20electricity%20generation%20in%202021

Planning and Designing a Sustainable Mobility System in Rural Areas

Muhammad Junaid and Maddalena Ferretti

Abstract

Sustainable transportation planning and design play a vibrant role in the transformation of the transport sector to become more environmentally friendly. Sustainable mobility in cities is a widely debated topic, but there is a lack of research and perspectives on sustainable mobility in rural areas. The aim of this paper is to provide an overview and critical analysis of recent studies on sustainable mobility in rural areas in Europe. From the selected studies, we highlight possible solutions for a more sustainable transport system in marginal territories, which could support decision-making processes. The paper consists of two parts. The first part is a preliminary analysis of the concept of sustainable mobility, and the second part is a critical analysis of best practices. The analysis highlights relevant aspects to consider when planning and designing transport solutions in rural areas. Particularly, we notice a growing attention to citizens' participation and engagement, which often becomes a decisive factor in achieving shared and efficient solutions. Also, the analyzed sustainable mobility systems show that the fragile population, especially elderly people, can considerably benefit from shared mobility solutions in rural areas, as they substantially increase their possibility of accessing services and infrastructure. The best practices are further divided into four categories: micromobility, mobility as a service (MaaS), transport on demand, and shared mobility. From the critical comparison of the analyzed best practices, we conclude that transport problems in rural areas can be more efficiently addressed with alternative solutions other than traditional public transport modes. This aspect will be further evaluated in the follow-up of the research, funded within the PON Program of the Italian Ministry of University and Research, in the Marche Region of Italy, as an assessment of the design and planning of a sustainable mobility strategy in rural areas.

Keywords

Rural areas · Sustainable rural mobility · Transport planning · Planning and design

1 Introduction

Rural areas, mostly dependent on private transport, have suffered from transport poverty for decades due to various reasons. The very low local population, long travel routes, and uncertain demands at specific times of the year (seasonality) make it challenging to fulfil travel needs with an efficient public transport service as a sustainable substitute for private vehicles (Cotrill et al., 2020; Naess et al., 2019). In this paper, we adapt the definitions for rural areas from the US Department of Agriculture and EUROSTAT: "Rural areas are defined as open countryside where population densities are equal to or less than 500 people per square mile or 300 persons per square km" (US Dept. of Agriculture, 2019); according to EUROSTAT, "rural areas are all areas outside urban clusters. 'Urban clusters' are clusters of contiguous grid cells of 1 square km with a density of at least 300 inhabitants per square km and a minimum population of 5000" (Eurostat, 2021). Traditionally, private transport is the governing solution to accessibility in rural areas, especially for older people and households with children (Graham et al., 2018; Mattioli, 2014). The mobility necessities and access requirements differ among groups in rural areas. Permanent residents' mobility needs are mainly

M. Junaid (✉) · M. Ferretti
DICEA Department of Construction, Civil Engineering and Architecture, UNIVPM, Ancona, Italy
e-mail: m.junaid@pm.univpm.it

M. Ferretti
e-mail: m.ferretti@staff.univpm.it

B. He et al. (eds.), *Urban Climate Change Adaptation*, Advances in Science, Technology & Innovation,
https://doi.org/10.1007/978-3-031-65088-8_16

for education, work, healthcare, socialization, and leisure activities. At the same time, other users, like younger and older citizens who cannot travel freely, must be provided with additional services and ad hoc solutions (Hiltunen & Rehunen, 2014). Lately, there has been an increasing demand for demand-responsive transport (DRT) as well as shared mobility solutions in rural areas worldwide (Alonso-Gonzalez et al., 2018; Mounce et al., 2020; Wright, 2013). Whereas in urban and metropolitan areas, the share is even higher and more noticeable (Cohen, 2019).

Implementing "novel mobility solutions (other than fixed route public transportation)", especially in the countryside, has been rather slow, so transport systems have developed more effectively only in the last decade (Liyanage et al., 2019). With "novel mobility solutions," researchers refer to demand-responsive systems and shared mobility solutions. These innovative solutions offer green and ecological alternatives to polluting private transportation and improve accessibility in rural areas. However, good internet access and know-how in communication technology are required for novel mobility solutions (Velaga et al., 2012).

Internet access for mobile devices and good digital skills to use technological tools for mobility are often preliminary requirements to set up new mobility services (Pot et al., 2020). However, rural areas often suffer from inefficient mobile networks and internet access due to low and uneven population distribution compared to urban regions (Bauchinger et al., 2021; Berg & Ihlstrom, 2019). Also, the digital literacy of older people is a subject of concern in rural areas. Thus, in these contexts, there are boundaries for using transport services utilizing digital technologies.

In the era of climate change, finding and implementing sustainable mobility solutions for rural areas is urgent, considering the profound impacts of the transportation sector on the environment and human health. According to Swinburn et al. (2019) and Tranter (2010), access to food, employment, healthcare, education, and other amenities is made possible by transportation, but car-dominated transportation systems are a major cause of global health issues like obesity, air pollution, climate change, and traffic accidents. Transportation also contributes to health issues, including noise-related problems, environmental degradation, stress, and a range of diseases linked to physical inactivity. These diseases encompass type 2 diabetes, cardiovascular disease, and dementia, as well as bowel and breast cancers (Woodcock et al., 2009).

"Transport is responsible for about a quarter of the EU's total greenhouse gas (GHG) emissions and causes air pollution, noise pollution, and habitat fragmentation" (EU Environment Agency, 2023). Furthermore, the transportation sector is also the largest contributor to nitrogen oxide emissions, having grown further since 1990 (EU Environment Agency, 2023). Therefore, the UN aims to cut CO_2 emissions from transportation by 15% by 2050.

Nowadays, it is necessary to shift how we use and even think about mobility. Good practices of multimodal mobility for rural areas have to be envisaged to provide public administrations with correct solutions to be implemented in their territories. Connectivity issues will have to be addressed to compensate for the rural–urban divide while fostering innovation both in the digital and ecological fields, facilitating the introduction of new tools, enhancing the necessary skills, and trying to progress towards a fair and just balance between city and countryside (Anon, 2022).

1.1 Problem Statement and Study Aim

This study is a component of a broader research project centered on sustainable mobility in rural areas. The project is supported by a grant dedicated to innovative Ph.D. studies provided by the Italian Ministry for University and Research. The broader research of this project is based in Italy. Italian and European rural areas boast rich cultural and natural resources, but they have encountered challenging conditions and negative trends in the past decades, affecting both demography and socio-economic structures. These areas are threatened by a progressively aging population and abandonment of built heritage, as well as by reduced service provision and inhibited accessibility. In Italy, the Department for Cohesion set up an innovative and progressive policy for the so-called 'inner areas' in 2014: "Italy's National Strategy for 'Inner Areas' addresses 72 inner areas, overall, encompassing 1077 municipalities and about 2,072,718 inhabitants, stretches over 60% of the national surface, and hosting 52% of Italian municipalities and 22% of its population" (A strategy for inner areas in Italy, 2014; Barca et al., 2014). Selected inner areas of the Marche region have an average population of 29,000 people, spread across multiple municipalities. Poor accessibility is a major cause of their depopulation.

The aim of this paper is to analyze environment-friendly, socially inclusive, and economically viable alternative solutions for mobility in rural areas in Europe as a means to enhance accessibility to peripheral territories and counteract progressive depopulation and abandonment. The study analyzes fourteen best practices that fulfil the pre-defined criteria of sustainability outlined in the following section. From the selected cases, we extract potential solutions for more sustainable transport systems in rural territories. These solutions are useful to assess the Infrastructure Plan developed and approved for the Marche Region (Italy) in 2023, helping decision-makers to address possible issues.

1.2 What Is Sustainable Mobility?

There is no agreed-upon definition of sustainable transportation. However, the majority of definitions continue to emphasize the significance of social, environmental, and economic goals and the importance to integrate them all (Bojković et al., 2010). Gillis et al. (2016) created the well-known concept of sustainable transportation. Other pertinent writers contributed knowledge to his concept afterwards, which can be summarized as follows:

- The promotion of a modal shift (Banister, 2008), i.e., the reduction of automobile travel and the strengthening of more ecologically friendly modes of transportation like bicycles and public transportation (Molin et al., 2016; Sdoukopoulos et al., 2019).
- A decrease in the necessity for travel aims to minimize both the frequency of trips and the distance covered per journey (Banister, 2008). Strategies to alleviate traffic congestion are intricately connected with urban planning initiatives (Hickman and Banister, 2019).
- Enhanced efficiency in transportation systems can be achieved by implementing shared vehicle ownership and incorporating low-emission vehicles (Hickman and Banister, 2019). This leads to an improvement in overall performance (Banister, 2008).
- Digitalization, exemplified by smart applications or mobility-as-a-service solutions (European Commission, 2019), introduces diverse intermodal options and multi-modalities for passenger transportation. This entails providing customers with a variety of mobility options and encouraging a shift in travel preferences away from the use of automobiles (Alyavina et al., 2020a, 2020b).

1.2.1 Criteria for Sustainable Mobility

The criteria for sustainable mobility must be designed around its definition. Summarizing the above definition, the key elements of sustainable mobility are as follows:

The promotion of low-emission, environmentally friendly, economically viable, and socially inclusive modes of transportation, such as walking, cycling, public transit, and electric vehicles. It also involves the development of infrastructure that supports these modes, urban planning that prioritizes accessibility and reduces the need for extensive travel, and the integration of technological solutions for more efficient and eco-friendly transportation systems.

The most crucial aspect of these key elements is the promotion of "low-emission, energy-efficient, and environmentally friendly modes of transportation." This statement implies a focus on public transport modes that genuinely emit minimal CO_2 emissions (Jiangping, 2012). Moreover, strategies to pursue the objective of sustainable mobility cannot be limited to producing or using less polluting transport systems, although this is of fundamental importance (Gallo & Marinelli, 2020). In this study, we will assess the sustainability of a mode of transport based on their CO_2 emissions, while also considering safety, social aspects, and economic viability.

2 Methodology

The methodology of this research serves to understand the environmental impacts of transportation, as well as the strategies for sustainable transportation already in place, and to critically analyze sustainable mobility cases.

The overall methodological structure of this paper is shown in Fig. 1. In the beginning, a comprehensive review of the literature demonstrates the effects of transportation on different environmental aspects, including environmental pollution, air pollution, and soil consumption due to impermeable streets. Additional literature was gathered concerning European and Italian strategies to mitigate the environmental effects of transportation. One approach to alleviate these effects is by adopting sustainable mobility. To identify sustainable, implemented solutions, we used grey literature (materials and research produced by organizations outside academic publishing) and academic published literature. These solutions are tested on the ground and identify innovative mobility ideas for European rural and peripheral areas. The best practices identification is based on the following criteria: the cases must be implemented solutions in rural and peripheral areas; the population of the area must not exceed the one that defines a rural area according to EUROSTAT; "contiguous grid cells of 1 square km with a density of at least 300 inhabitants per square km and a minimum population of 5000" (Eurostat, 2021); they must be based in the European context; they might be related to the Mediterranean context to investigate the possibility of a specificity of this geographical area; they must be sustainable according to the predefined criteria for sustainability in Sect. 1.2.1. There are many cases that fulfill the above criteria, but we have collected fourteen examples from various parts of Europe that address mobility issues more effectively based on their provision of service to each and every citizen, irrespective of their age, gender, and impairment.

The next methodological step (categorization) is the organization of the best practices into four categories representing the most frequent, effective, and innovative modes of sustainable transport in the European context (Poltimae et al., 2022). The categories were acknowledged during the preliminary analysis of several European projects, such as the SMARTA project, MAMBA project, and Inclusion

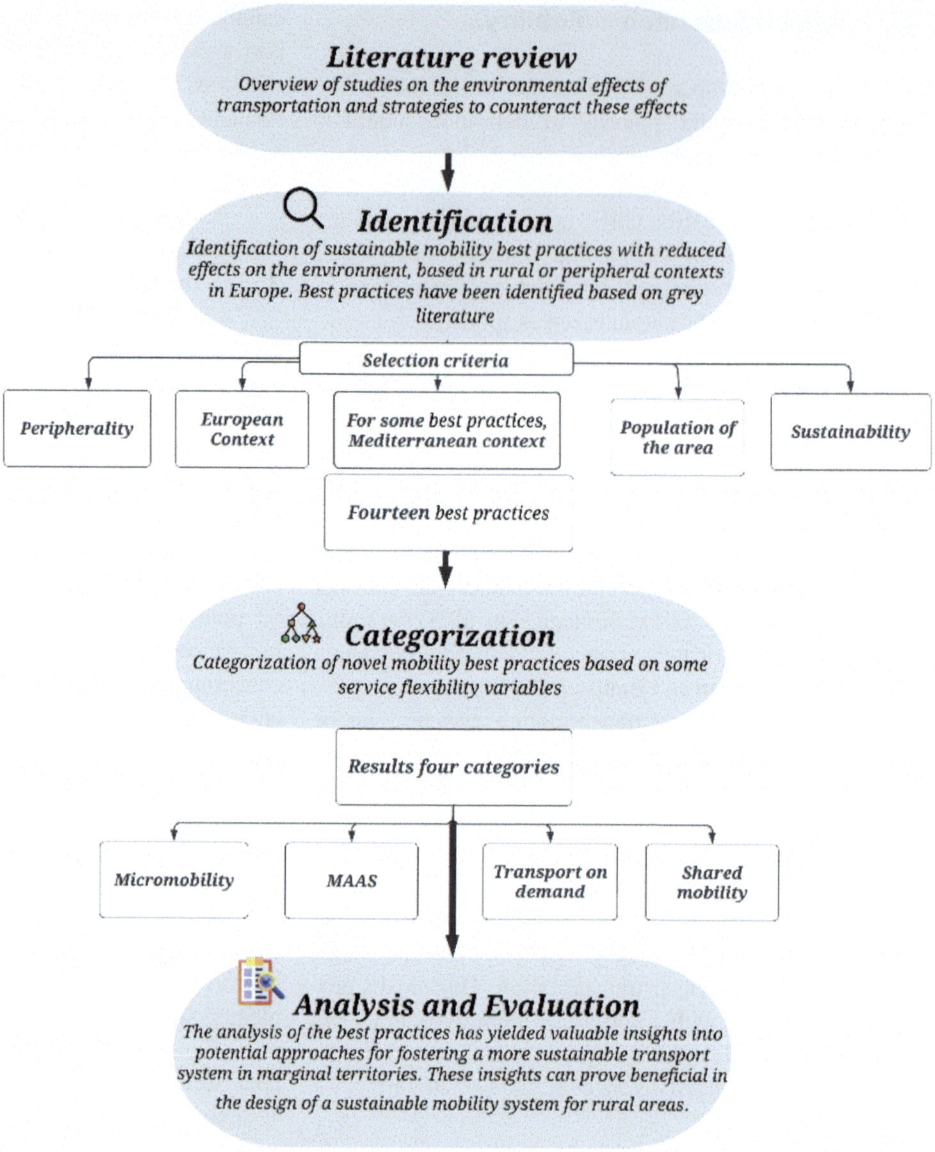

Fig. 1 Methodological framework of this article. by Junaid (2024) (Author own elaboration)

h2020. (SMARTA, 2021; MAMBA, 2020; Inclusion h2020, 2020): (1) Micromobility (low-speed mobility, light-weight vehicles, etc.); (2) Mobility as a Service (integration of different transport systems into one in order to facilitate booking processes and intermodality); (3) Transport-on-Demand (transport service that operates only according to the specific demand of people and does not have any fixed timetable or route); (4) Shared Mobility (mobility services that can be shared among citizens, including a van, car sharing, etc.). These categories are dominant mobility solutions for rural areas (Bronsvoort, 2019; Shahid et al., 2023).

Furthermore, the categorization is based on five transport service flexibility variables and specificities (Poltimäe et al., 2022): (1) type of route (e.g., flexible, fixed route, etc.), (2) number of stops (e.g., fixed stops, flexible stops, etc.), (3) flexibility of the mode (e.g., choosing routes, travel times, forms of transport service providers, etc.), (4) booking methods (user-friendly or technical know-how needed), and (5) sharing of vehicles.

Due to the limited scope of our work, we selected four of the fourteen best practices, one from each category, to explain them in detail in the next chapter. This in-depth analysis aims to understand the specific transport solutions adopted for these rural contexts, the reasons for their selection, how the mobility system is planned, and how the implementation process works.

3 Literature Review

The first objective of this article is to provide an overview of sustainable mobility modes in rural areas that counteract the negative environmental impacts of current transportation modes and use more efficient and performative technological solutions. As an initial step, we have examined the impacts of climate change and explored how sustainable mobility solutions can mitigate them, considering aspects of safety and long-term budget efficiency. Furthermore, we reviewed the ongoing strategies trying to address sustainable solutions for rural areas in the European and Italian contexts. The following sections synthesize this analysis.

3.1 Impact of Transport Systems on the Environment

According to Taibat et al. (2018), the transportation sector accounts for 23% of the overall GHG emissions, making it the second-largest contributor to emissions worldwide. As income levels rise and infrastructure expands globally, transportation-related emissions are rising at a faster rate than emissions from any other sector (such as agriculture or commerce). Experts predict that by 2050, annual transportation emissions will have doubled due to the rapidly growing demand for transportation (Taiebat et al., 2018). Particulate matter (PM) and gaseous pollutants exceed standard air quality limits in many major European cities. Road traffic emissions are often cited as the primary source of these exceedances (European Environment Agency, 2010). In addition, urbanization has led to increased soil consumption, e,g. transportation infrastructure creating more impermeable surfaces like roads, parking lots, walkways, and rooftops. This reduction in permeable surfaces that can naturally absorb and clean stormwater has caused a decline in the quality of stormwater runoff (Brabec et al., 2002).

Transportation can also degrade soil quality, with soil erosion and contamination being the two primary environmental effects on soil health. Ports and other coastal transportation infrastructure have a big influence on soil erosion. The scale and scope of wave actions are changing due to shipping activities, which are causing damage in narrow channels like riverbanks. A significant amount of fertile land has been lost as a result of building highways or lowering surface impermeability grades for port and airport developments. Noise pollution is another way in which transportation can directly affect human health. Vehicle noise and port, airport, and railyard operations can also impact people's health, especially regarding the capacity for sleep and consequent problems (Rodrigue, 2020).

3.1.1 Best Practice

To better understand the effects of different transportation modes in urban and rural areas, we analyze a study on the environmental CO_2 impacts of different mobility service options, in particular on-demand services. The study was conducted by Dang et al. (2021) quantified the effects of on-demand mobility services on sustainability in terms of emissions and traffic (Dang et al., 2021). Taking into account the spatial characteristics (presented in Table 1) of the two locations in Switzerland, Glarus South is used as an example of a rural area, Basel-Saint-Louis as an urban area, as shown in Fig. 2.

In the rural area of Glarus South, motorized transportation makes the highest contribution, accounting for 71% of total trips; public transportation makes up 21% of the total trips in the area, and slow (non-motorized) transportation comes in last with 6%. Each resident in Glarus makes an average of 3.1 daily trips. It was estimated that the combined number of daily trips made by all residents and workers in Glarus South was 32,000 (of which 6700 via public transportation).

In the urban area of Basel-St. Johann, each person makes three trips a day. It was estimated that all residents and workers make about 146,000 trips a day based on the territorial principle. 30% of these passengers travel via public transportation a day, 11% via slow transportation, and 58% via motorized private transportation.

Six different types of mobility options are considered in this study: (1) Motorized private transportation (MTP); (2) Public transportation (PT) Suburban Railway; (3) Public Transportation—Bus traffic; (4) Slow traffic (pedestrian and bicycles); (5) Taxi; and (6) On-demand services. The on-demand services are further divided into four categories: (1) On-demand line operation, which combines the benefits of on-demand collective transportation with public transportation. The system is based on an analysis of current public transportation routes and their deficiencies. Where there is a lack of transportation networks, minibuses are implemented to transport passengers to significant intersections or virtual stops close to their addresses, where they can switch to public transport. Such virtual stops are thought to be roughly 250 m from the front door; (2) On-demand public transportation supplement, which is a transportation service provided as an adjunct to public transportation during off-peak and marginal hours; (3) Replacement of traditional public transportation with on-demand services; (4) Commercial on demand, which is a transport service with independent rates. The service is run by a commercial provider, setting their own prices with no restrictions from the government. Thus, the offer is not integrated into the public transport system and is in competition with the existing public

Table 1 Author's adaptation of existing data sourced from Dang et al. (2021), showing spatial characteristics of the best practice area

Type of space	Example localities	Area in km^2	Population density (inhabitants/km^2)	Workplaces per km^2	Existing transport offer	Number of passengers' trips per day	Modal split (in relation to trips)
Rural peripheral municipality	Glarus South	430	22	1.87	One train line, several bus lines (30 to 60 min intervals)	32,000	MPT: 71% Public transport: 21% Non-motorized: 6% Other: 2%
Urban municipality of a large agglomeration	Basel-St. Johann	2.24	8341	10,241	Several train, suburban railway, and bus lines (frequency intervals)	146,000	MPT: 58% Public transport: 30% Non-motorized: 11% Other: 2%

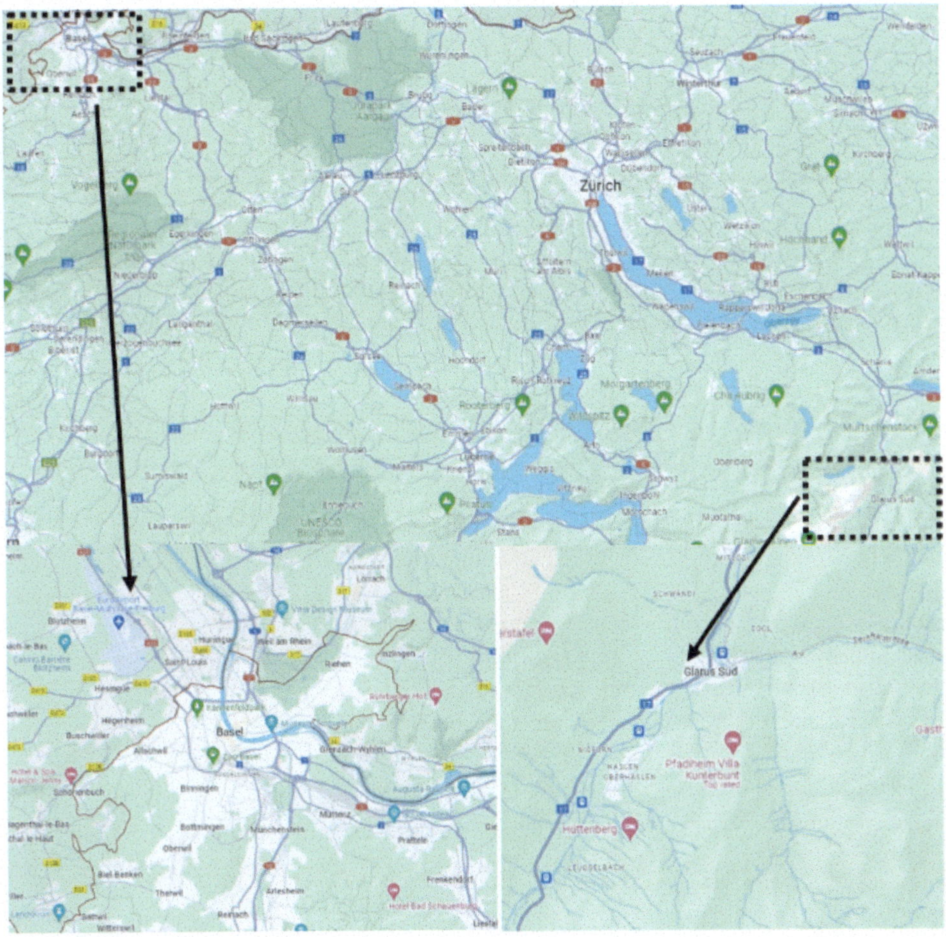

Fig. 2 Best practice area on the map. *Source* Google Maps

transport offer. (Federal Office for Special Development Bern, Switzerland, 2015; Federal Office for Special Development Dichte und Mobilitätsverhalten, 2018).

Figure 3 displays two charts, 3a and 3b with a comparison of the grams of CO_2 equivalent per passenger kilometer for the four on-demand service options and the five conventional modes of transportation in both regions. The calculations from the above charts show that motorized private transportation and taxis produce the highest percentage of CO_2 emissions. Finally, on-demand service options are more pollutant than slow or public transportation but less pollutant than taxis and motorized private transportation in both the urban and rural areas considered in the study.

In summary, transportation significantly influences the environment and living conditions, amplifying the climate crisis and adversely affecting various aspects of people's wellbeing. This results in a multifaceted, complex, and problematic long-term situation. Consequently, studies and strategies have recently been implemented focusing on innovative and sustainable mobility concepts at both the European and national levels to confront this challenge.

3.2 Sustainable Strategies for Mobility in Rural Areas

3.2.1 A European Vision for Rural Areas

The European Commission has developed a common vision for 2040 as the long-term vision for the EU's rural areas. While recognizing the diversity of the involved contexts, the vision also identifies shared challenges and opportunities. Using a collaborative approach based on the engagement of different stakeholders, such as rural businesses and groups of interest, the commission produced a broad vision and an all-inclusive plan for rural development to assist the local actors in realizing their full potential in the near future (European Union, 2020). The four complementary areas of action determined for the long-term vision for stronger, connected, and resilient rural areas are shown in Fig. 4.

The general methodology is based on the detection of best practices that should serve as flagship projects to assist rural municipalities in implementing sustainable

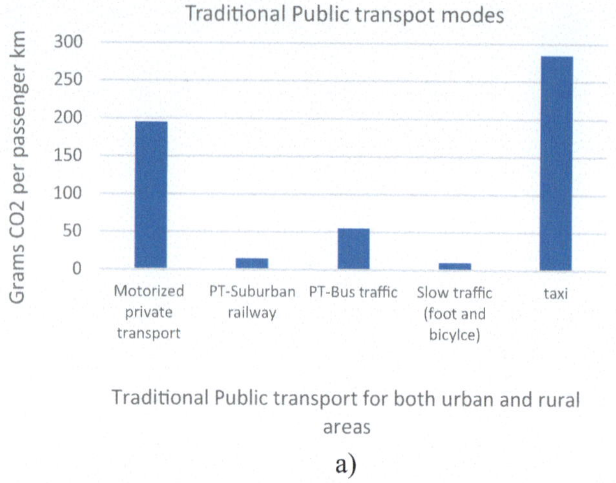

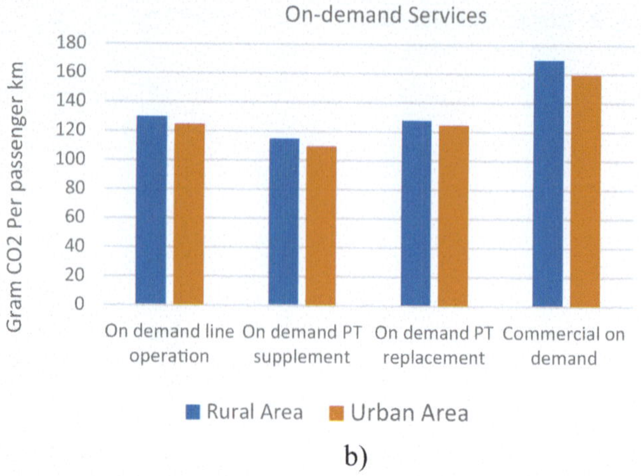

Fig. 3 Author's adaptation of an existing image sourced from Dang et al. (2021), **a** showing the CO_2 emission from traditional public transpport modes and **b** showing the CO_2 emission from different on-demand services in rural and in an urban

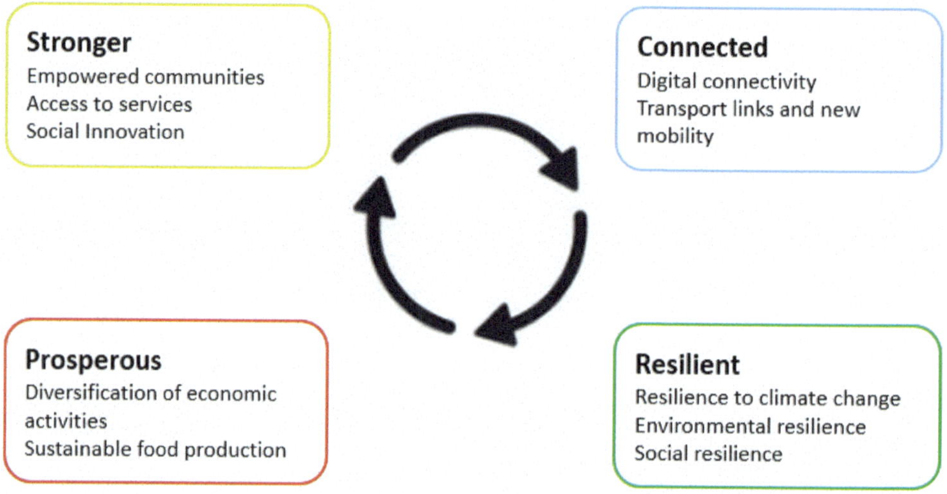

Fig. 4 The Author;s adaptation of an existing image sourced from the EU Rural Vision 2040, showing four complementary areas of action of the EU vision

solutions within their borders. Reducing the digital divide is a major priority. Indeed, the EU vision aims to bridge the gap between rural and urban areas by promoting digital innovation and the use of new technologies. This includes improving transportation modes, as highlighted in a 2021 survey of EU residents. The survey revealed that 79% of European citizens prioritize the improvement of rural areas. According to a majority of 65%, the province or local area should be empowered to determine the allocation of funds for rural areas and specifically for public services. 44% of respondents identify transportation infrastructure and connections as a crucial necessity and requirement for these contexts (EU Commission Eurobarometer, 2021).

3.2.2 Climate Neutral EU Mobility

Transportation is responsible for 25% of greenhouse gas emissions in Europe (Marche region, 2023). Consequently, significant efforts have been made in recent years to decrease this percentage. The European Commission released the European Strategy for Low-Emission Mobility in July 2016. Its objective is to achieve a reduction of at least 60% in transportation emissions by 2050 compared to 1990 levels. This aligns with the commitments made in the Paris Climate Agreement and the 2030 Agenda for Sustainable Development. The strategy focuses on three key areas: improving transportation system efficiency through digital technologies and transitioning to low-emission modes, promoting alternative energy sources like biofuels and electricity, and accelerating the adoption of low- and zero-emission vehicles (EU Commission low emission mobility, 2016).

Another European strategy to mitigate transportation environmental impact is "Climate-Neutral and Smart Cities." Its objective is to establish 100 smart and climate-neutral cities by 2030, serving as hubs for innovation and experimentation. The goal is for all European cities to follow their example by 2050 (European Commission, 2020). Similarly, "clean and sustainable mobility" seeks to reduce the transportation sector's contribution to climate change. This strategy aims to achieve the EU's climate targets by decarbonizing the transportation sector in the future (European Council, 2021).

3.2.3 Italian Strategy for Sustainable Mobility

In 2014, Italy introduced a significant strategy for rural areas known as the "National Strategy for Inner Areas" (Barca et al., 2014). While the strategy faced criticism in subsequent years, particularly regarding the effectiveness of its indicators for identifying rural areas, it successfully drew attention to the challenges faced by these regions compared to more urbanized areas. The strategy aimed to address these challenges through dedicated funding. Initially implemented for the 2014–2020 programming period, it has since been extended until 2027. The strategy applies to all regions and macro-areas in Italy that meet specific criteria, such as distance from railways and a lack of education and health services. Its objectives include supporting economic and social recovery, generating employment, promoting social inclusion, and halting the demographic decline in inner areas in terms of population size and age profile (Agenzia per la Coesione Territoriale, 2013).

In the context of infrastructure and mobility in the Marche Region, a new plan called the Marche Infrastructural Plan 2032 was approved in October 2023. The plan is guided by four principles: sustainability, safety, multimodality, and interconnection. It aims to transition from the current comb structure of territorial connections to a more interconnected mesh structure. This transition offers benefits such as reduced CO_2 emissions and travel times, as well as promoting balanced development across the territories. The plan encompasses four key objectives. Firstly, it aims to address the deficiency in port connections, which significantelty impacted the downgrade experienced in 2018. Secondly, it seeks to connect the region to the Ten-T Corridor in Europe, connecting water infrastructure with motorized transportation and railway infrastructure through various intermodal modes. The third objective involves a paradigm shift in the regional infrastructure network to become more interconnected. Lastly, the plan prioritizes environmental, social, and economic sustainability by focusing on modern infrastructure that promotes environmental preservation and favorable business conditions, while also strengthening the social structure. The realization of the plan is designed in three phases: from now until 2027, from 2027 to 2032, and beyond 2032 (Marche Region, 2023).

3.3 Safety in the Transportation System

Safety is one of the most crucial aspects for society in the field of transportation. In this context, safety refers to the strategies and precautions used to keep drivers from getting hurt or killed (Statista, 2023). The media and transportation industry often portray mass transit as riskier than driving, but experts agree that driving is the single most dangerous activity, with 31.9% of total accident fatalities attributed to transportation-related incidents in 2010 (Global road safety, 2023).

A recent study estimates that the risks involved in driving a car or truck are about 60 times higher than those involved in taking public transportation, such as a bus. (Herrman & Herman, 2023).

Similarly, the safest way to travel in Europe is by bus or coach (European Commission, 2015). While bus-related traffic accidents in Europe annually kill around 500 people, car drivers and passengers account for nearly half of the fatalities (19, 917) in 2021 (Eurostate Statistics, 2021).

To increase road safety and reduce death tolls on roads, there is an EU strategy called "Strategic Action Plan on Road Safety and EU Road Safety Policy Framework 2021–2030." The strategy aims for a 50% reduction target for road deaths and serious injuries by 2030. The further aim of the strategy is to reach zero road deaths by 2050 through innovative road safety solutions ('Vision Zero') (European Commission, 2022).

3.4 Budget Friendly Mobility Modes

Numerous factors, including the destination, the number of passengers, and the type of goods being transported, are likely to affect which mode of transportation is most cost-effective per kilometer. However, the following forms of transportation are thought to be the most economical:

Bicycles are the most economical form of transportation. It only needs the rider's own energy for propulsion. Bike maintenance costs are also comparatively low, which makes them an extremely economical choice for short trips. Another cheap form of transportation is walking. Like riding a bicycle, walking is a free form of transportation that requires no fuel, electricity, or upkeep. Furthermore, public transportation like buses and trains can accommodate large numbers of people at once; they are usually more expensive to operate per kilometer than bicycles but less expensive than private vehicles or taxis. The minimum amount required for a public transportation service is around 32,500 US dollars per passenger bus, although it depends on the type of bus, its passenger capacity, etc. [ITS Department of Liaison Evaluation, 2020]. Due to the lower maintenance and fuel expenses of electric vehicles, they may be less expensive to operate than conventional gasoline or diesel vehicles. Carpooling enables several people to share the expense of gas and auto maintenance and can also be a financially advantageous choice.

It is crucial to remember that the cost-effectiveness of a mode of transportation may vary depending on the laws, taxes, and infrastructure of a particular area. Furthermore, for a particular person or journey, the most economical form of transportation in terms of cost per kilometer might not always be the best or most convenient choice (Quora Q&A, 2024).

4 Analysis and Results

As explained in the methodology chapter, we identified fourteen best practices and organized them into four categories. In this article, for synthesis, we have selected one best practice for each detected category. They are described in their main aspects and contents in the following text and discussed in the next chapter.

4.1 Micromobility (MM), Cairngorm National Park (CNP), United Kingdom

The project is categorized in the micromobility category and is one of the pilot labs of the H2020 project Inclusion (Inclusion, 2020). The core purpose is to increase accessibility options in the National Park of Cairngorm, located in the Scottish Highlands (UK). The final goal is that the inhabitants purchase their own e-bike and experience the journey through the park with their own vehicles. The primary users of this project are elderly people, people with problems using public transport, teenagers, young adults, other locals facing the high fuel costs in rural regions, and tourists.

Prior to implementation, the planning and research group studied documents and literature such as the Aviemore work report (written by AECOM and led by HITRANS in collaboration with The Highland Council and CNPA), Halfords (a UK blogging site that revealed that 62% of the users who buy an e-bike in the UK are on average 55 years old), and Wheels for Wellbeing (a study that highlighted that 18% of disabled cyclists bought an e-bike).

Some useful features of the parks include various cycling routes such as the NCN7 cycle network, as shown in Fig. 5, which is 550 miles from Sunderland to Inverness. Before arriving in Inverness in the Scottish Highlands, two national parks are accessible from it. National Cycle Route 7 covers the west part of the park, which runs from Glasgow to Inverness. Cycle Route 7 links up the communities of Badenoch and Strathspe. Another cycle route is the Deeside Way, linking Dinnet and Ballater in the Royal Deeside area of the CNP. Sections of the Speyside Way (cycling track in Aviemore) and Dava Way (cycling track in Grantown-on-Spey) are also important cycling tracks. The planning of the pilot case involves the analysis of the area (i.e., demographic data, transport services, visitors and tourists per year, vulnerable groups). On the basis of available bicycle infrastructure and people's needs, the pilot project of an e-bike scheme is tested. The e-bike service benefited more able-bodied citizens to enjoy their journeys for health and leisure purposes (of the total population, 15% of the inhabitants were users over 65, and 20% were 56–65 years old). More than 65% of the citizens had never experienced an e-bike before, but all of them stated they would try one. Four of the e-bike users in the community have purchased an e-bike inspired by the project. The user reviews about the project are stated in Fig. 6.

Furthermore, this case is a good example for the implementation of a micromobility project in rural areas, demonstrating a viable and affordable method of implementing e-bike sharing programs in rural areas through public–private partnerships with neighborhood bike shops and the resources that are readily available."

4.1.1 Sustainability of Micromobility

The best practice is a sustainable means of transport implemented in a rural territory. Studies have shown that e-bike use positively contributes to health in sedentary lifestyles (Gojanovic et al., 2011). Research indicates that active travel modes like cycling permit high levels of interaction with the outside world, which may result in higher levels of mental well-being compared to car travel (Van Wee & Ettema, 2016). Lower population densities may pose challenges to the accessibility of bus stops and train stations. For instance, in the Netherlands, more than 80% of the population resides within 7.5 km of a rail station (Kager et al., 2016). Expanding the cycling range emerges as a viable solution to enhance access to transportation hubs, thereby broadening the range of available transport options.

Additionally, cycling presents potential drawbacks in terms of comfort and ease of use compared to car travel. E-biking, in particular, may necessitate specific clothing and protection against varying weather conditions (Popovich et al., 2014). Slower average speeds and diminished carrying capacity could pose challenges in integrating e-bike commutes with other activities, such as transporting children to school or attending meetings in different locations (Jeekel, 2014).

The environmental impact of this mode of transport in terms of CO_2 is far less than that of other motorized

Fig. 5 NCN7 cycle network in Cairngorm National Park, UK *Source* OS Maps

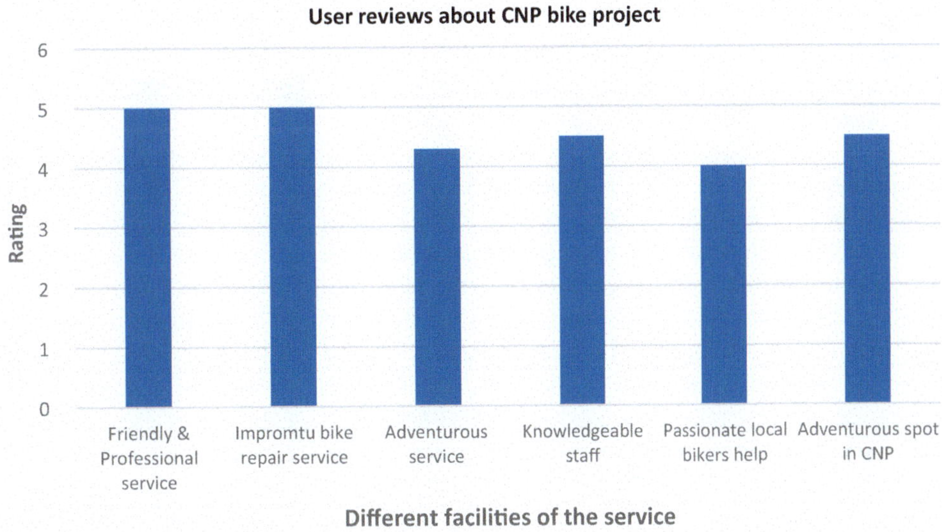

Fig. 6 Author's adaptation of existing data sourced from (Cairngorm guided mountain bike descent, 2021), showing user reviews about the CNP pilot project

vehicles. Lifecycle emissions vary for conventional bicycles and e-bikes, with estimates in China showing 4.7 g of CO_2 per person-kilometer for conventional bicycles and 22 g for e-bikes (Hollingsworth et al., 2019; Kou et al., 2020) while European sources suggest emissions of 21 g for conventional bicycles and 22 g for e-bikes (Blondel et al., 2011), and another European study indicates 5 g for conventional bicycles and 25 g for e-bikes per person-kilometer (Weiss et al., 2015).

Implementing a bicycle-sharing system costs between 3000 and 5000 US dollars, according to the US Transportation Department, while building bike network infrastructure depends on factors like topography and urban density. In CNP, bike rental services range from 7.14 to 20 pounds per hour, with non-electric bikes available at 40 pounds for 2 days (ITS Department of Liaison Evaluation, 2020; Aviemore Bikes, 2024).

Additionally, there are many safety concerns related to micromobility. According to (Derriks & Mak, 2007; Beck et al., 2007) in contrast to motor vehicles, bicyclists encounter a 1.8-times higher risk of injuries related to accidents, a 1.3-times higher risk of death, and a 10-times higher accident rate (Wegman et al., 2012). The majority of bicycling fatalities take place on urban roads and typically involve a motor vehicle, with head injuries ranking as a leading cause of death (Gaudet et al., 2015). The best practice also addresses various social aspects, targeting active elderly individuals, those with reduced mobility, young adults, teenagers, local residents facing fuel poverty due to high rural fuel costs, and tourists (Inclusion, 2020).

4.2 MaaS, Budapest, Hungary

"MaaS, or mobility as a service, combines different types of transportation with transportation-related services to create a single, all-inclusive mobility service" (Maas Alliance, 2022). By eliminating the need for numerous ticketing and payment processes, MaaS provides end users with the added benefit of enabling mobility through a single application and payment channel. For the implementation of the best practice in Budapest, the developers of the pilot case combine different modes of transportation in the Budapest "go" application and generate different scenarios for the first mile, last mile, and main journey, i.e., "Scenario 1: Private car for first mile, then park and ride (P&R), then public transport for the main journey, and then taxi for the last mile. Scenario 2: private car for first mile, then park and ride (P&R), then public transport for the main journey, and then bike sharing for the last mile. Scenario 3: taxi for the first mile, then public transport for main journey, and then bike sharing for the last mile. Scenario 4: taxi for the first mile, then car sharing for main journey, and then public transport for the last mile journey" (Georgakis et al., 2020). To provide the best scenario for the users according to the situation and the history of the users, algorithmic steps were performed, e.g., if the user does not have a driving license, then the application will not suggest the option of using their own private car. This best practice is used as a guide to understand the MaaS system.

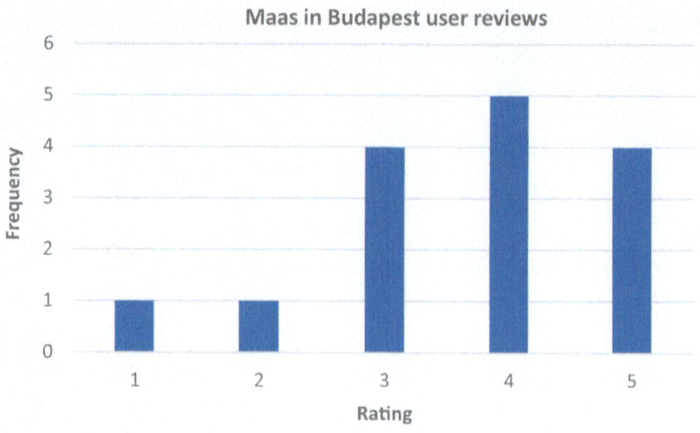

Fig. 7 Author's adaptation of existing data sourced from the (BudapestGo app Efficiency, 2023), about user reviews of Mobility as a Service in Budapest, 1 being the worst and 5 the best rating

The user reviews, as shown in Fig. 7, state positive and negative results. The low rating, such as 1 showing the negative results while 5 showing the positive results. The negative ratings are mostly complaining about the complexity of the application. The positive reviewers commented about the high level of in-depth information and the integration of many services.

4.2.1 Sustainability of MaaS

Assessing the sustainability of a Mobility as a Service (MaaS) system directly is challenging since MaaS is an innovative concept, not just an individual mode of transport. It has the potential to encourage a shift in individuals' travel behavior away from dependence on private cars (Elena et al., 2020). MaaS is anticipated to enhance sustainability by promoting increased sharing of mobility resources, thereby reducing the overall number of vehicles required and streamlining multimodality through seamless integration (Strömberg et al., 2018). To extend the accessibility of public transport, MaaS relies on shared-use alternatives to private cars, encompassing options like taxis, car-sharing, ride-sharing, ride-hailing, and bike-sharing (Jittrapirom et al., 2017). The MaaS system contributes to sustainability because it promotes public transport and shared mobility through its integration, providing a wide range of options for its users. Moreover, the primary users of MaaS in Budapest are students, full-time employees (both male and female), and individuals aged between 30 and 50, highlighting the social inclusivity of the service (Georgakis et al., 2020).

Furthermore, the implementation of a MaaS concept does not require a higher cost, as its implementation costs involve the development of an algorithm for service integration, app development, and service management (Mitropoulos et al., 2023). The price for the users is 9500 HUF, equal to 24.43 Euro for a monthly pass, which enables you to travel anywhere in Budapest the whole month (BudapestGo MaaS App, 2024).

The MaaS is a crucial concept for rural areas, because it enhances the quality of life for rural inhabitants, plays a significant role in the development of sparsely populated regions, and elevates their competitiveness in human resources, aiming to alleviate transport deprivation and reduce social exclusion (Misso et al., 2017). One of the major issue associated with MaaS for its users is data protection. As MaaS heavily relies on user data, the concern for data privacy and security becomes paramount. Users must have confidence that their personal information, location data, and payment details are secure. Yet, in light of prominent instances of data breaches and misuse of user information, ensuring that MaaS platforms can uphold privacy and security standards is becoming increasingly challenging (Levinson, 2023).

4.3 Transportation on Demand (ToD), Vidzeme, Latvia

This project is a demand-responsive transport system implemented in Vidzeme, Latvia. The municipality is facing significant challenges related to declining population trends caused by migrations towards major cities in search of better services and facilities. One of the main drivers behind these migrations is the inadequate regional transport network, which has a detrimental impact on the residents' quality of life. It hinders their daily activities, such as work, and access to essential services like community services, healthcare, and education. Additionally, it affects their social and cultural lives (Maximising mobility and access to services (MARA), 2020).

The design of the mobility system involved several key activities. Firstly, the needs of users, as well as the interests, limitations, and potentials of municipalities and service providers, were identified through interviews, surveys, and meetings with various target groups in potential focus areas. This process helped clarify the mobility concept, including how the service would operate and how users would book rides. The emphasis was not only on finding answers but also on actively involving people in proposing new ideas to enhance the overall strategy. This approach increased curiosity and awareness among residents, motivating them to try out the new services. The mobility service was initially tested in two pilot cities from 2019 to 2020. Users registered independently with the municipality and could book a small van, bus, or car via phone 24 h in advance. In cases of high demand, trips were prioritized based on necessity, such as giving priority to those needing transportation for medical visits. Information about the transport-on-demand service was disseminated through local media, posters, and registration forms.

The pilot case was launched at the end of 2019 and concluded in April 2020, when 92 single riders and 493 additional passengers were reported to have used the service in one of the two pilot cities. The service was used mostly to reach community services. In the second pilot city, the service totaled 605 trips, of which 160 were single riders and 1,154 were additional passengers. Similarly, in this case, the trip's purpose was mostly related to visiting community services, and the users were mostly women and the elderly (Maximizing mobility and access to services (MARA), 2020). The user reviews for ToD are depicted in Fig. 8. In this instance, the reviews focus not on the overall service but rather on its efficiency in various areas and for different demographics. For instance, one user assigned a 4.5 rating to the service's importance in rural areas, while another gave it a 4 rating for effectiveness among older individuals. The highest rating, a 5, pertains to the service's effectiveness for disabled persons. However, there are concerns regarding trust issues associated with this service. Consequently, in gaining people's trust, its performance is subpar, receiving only a 2 rating.

4.3.1 Sustainability of ToD

The best practice, as depicted in Fig. 8, is an implemented solution in a rural region that has been embraced by the residents, as evidenced by their feedback. The best practice is closely related to the aim of this research, as low-demand areas can be served with transportation on demand (Campisi et al., 2021; Franco et al., 2012).

The sustainability of ToD services can be measured by their CO_2 emissions, economic viability, and social inclusivity i.e., according to a study carried out by Dang et al. (2021) in Switzerland, the average CO_2 emission from on-demand transport services in rural areas is 135.5 g per person-kilometer and in urban areas, the emission rate is 129.75 g per person per kilometer.

The budget required for on-demand services is less compared to other public transport services like buses, rail, etc. In this case, mobility can be provided by a car, van, or bus. This service does not require a large fleet of buses or vans for small rural areas (Maximizing mobility and access to services (MARA), 2020). The ToD service is even free of charge in Alūksne and Mazsalaca, Vidzeme, Latvia, as the service is financed by the European Union (Maximizing mobility and access to services (MARA), 2020).

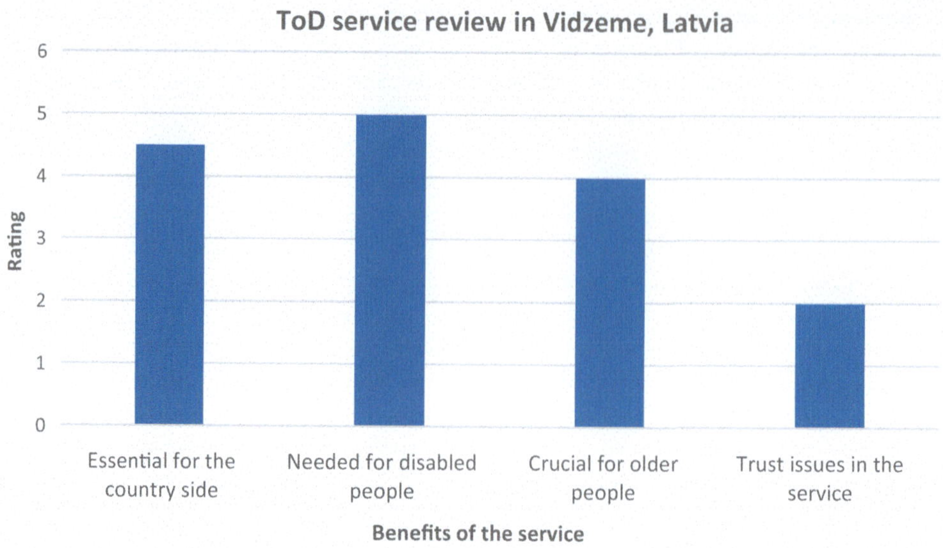

Fig. 8 Author's adaptation of existing data sourced from the Mamba project 2020, about user reviews of the ToD service in Vidzeme, 1 being the worst and 5 the best rating

On-demand transport services are considered safe. Safety is increased by providing a seat belt for each individual passenger in the vehicle and safety seats for children. Additionally, there are surveillance cameras to monitor the vehicle from inside, and after a specific time in the evening, commuters can request the vehicle at their door (Edmonton, 2024). The service is socially inclusive as it enables older citizens, citizens living on the countryside and citizens with disabilities to benefit from it and fulfill their daily needs (Maximizing mobility and access to services (MARA), 2020).

In summary, on-demand solutions have their advantages and disadvantages. One advantage is that, unlike fixed-route transit, where passengers have to wait at designated bus stops and walk long distances, on-demand transit offers reliable services right at their doorstep. The shuttle arrives within minutes after passengers walk to a nearby virtual stop, ensuring swift transportation to their final destinations with guaranteed arrival times. On-demand services provide consistency and convenience for commuters. As a result, people living in rural areas have easier access to affordable transportation thanks to dynamic, demand-responsive transport. This positive effect outweighs the decreased reliability of on-demand services compared to public transport (RideCo, 2023; Liftango, 2022).

4.4 Shared Mobility, Hesse, Germany

The "Mobilfalt" pilot case in Hesse, Germany, is categorized within the shared mobility category. It is a car pooling project aiming to increase mobility for rural residents and some specific users (e.g., children and young people without driving a license or car; people with an ecological and green approach to life; residents of rural villages and regions; and older people who need to reach out to social activities, e.g., to meet other people). The project was initiated by the North Hessian Public Transport Company (NVV) in 2013 in Hesse, one of the German Federal States. The main problem is the migration of citizens to big cities. Limited transport services reduce the capacity of people to reach important places by public transport. Therefore, car numbers are high. The aim of the project is to facilitate ride sharing of rides for people who must travel along the same route (Mobilfalt, 2018).

The service can work by finding available rides on the NVV platform published by drivers. Users can book trips online, and drivers receive ride requests. If confirmed, users can utilize the trip; otherwise, Mobilfalt arranges a taxi to still guarantee mobility. Drivers receive fixed compensation from the application, while users pay a low flat rate for each trip. Due to the initially low price/trip of Mobilfalt, the users had the misconception that the service was of low quality. So, the prices of the service have been raised to align with regular public transport prices to counter the misconception among users. The user's transportation needs are important to assess in rural areas. In this case, the identification of their needs and their transportation habits were evaluated through interviews and feedback (about which type of solution is better for the region) and thus taken into consideration in the development of the project. The first online service was built on the premises of previous software called "dial-a-bus," also developed by NVV. The application is then used to book the trip, while the bus stops are used as meeting points for riders.

4.4.1 Sustainability of Shared Mobility

According to this case, shared mobility is another dominant means of transport in rural areas. It is useful in rural areas where public transport is not viable. Shared mobility has so many advantages, including: (1) reductions in energy consumption and emissions, (2) congestion mitigation, and (3) reduced parking infrastructure demand (Shaheen et al., 2018). Shared mobility, especially carpooling, has some drawbacks as it doubles the travel time compared to solo driving, particularly for long distances (Monchambert, 2020). Loss of personal independence to travel is another major limitation of this service (Gheorghiu & Delhounme, 2018).

One factor in the sustainability of a carpooling system can be measured by its impact on reducing carbon emissions. A study related to the famous carpooling service "BlaBlaCar" was conducted by the French research institute Le BIPE, based on real usage data from BlaBlaCar's community of 70 million carpoolers. Le BIPE analyzed actual usage data spanning 12 months and conducted surveys with 6,884 members across eight countries. The results show that the carpooling initiative doubles the number of people traveling while operating a carbon-saving network. In total, 1.6 million tons of CO_2 were saved by BlaBlaCar carpoolers in 2018, with 3.9 passengers occupancy per vehicle (Intelligent BlaBlaCar, 2021; Transport, 2019). In examining the environmental impact of a carpooling system, it is pertinent to determine the grams of CO_2 emitted per person-kilometer. Diverse studies indicate varying car emission rates: 306g CO_2/person-km in China (Hollingsworth et al., 2019; Kou et al., 2020), 271g by the European Cyclist's Federation [Blondel et al., 2011], and 240g from European lifecycle assessments (Weiss et al., 2015). By amalgamating these findings, the average estimated CO_2 emission for a car in a carpooling system stands at 272 g per person-kilometer. Combining these three studies, the average CO_2 emission for a car is estimated to be 272 g per person-kilometer. The average global car occupancy rate is 1.5 passengers per vehicle (Wolfram et al., 2020). According to this rate, a car can emit a total of 408 g of CO_2 per kilometer. In the scenario where a vehicle emits 408 g of CO_2 per kilometer and accommodates 3.9 passengers, the emission rate translates to 104.6 g of CO_2 per passenger per kilometer.

For most ride-sharing companies, the budget is needed to develop an app and manage it, as drivers on this service usually use their own cars and drivers and passengers use their own devices to access to the platform, e.g., NVV and BlaBlaCar (BlaBlaCar, 2006; Mobilfalt, 2018). For app development like BlaBlaCar, the estimated cost is around $1,8000–2,0000 (Nurture, 2019). In comparison, the average taxi booking app development cost varies between $1,5000 and $2,5000 (Pillai, 2023). Mobilfalt is also budget-friendly for its users, charging a cost of only 1 Euro for every ride, regardless of its duration or the distance covered, and the driver gets a fixed compensation of 0.30 Euro per km (Mobilfalt, 2018).

Nevertheless, safety in carpooling is a significant concern for its users. Sharing a ride with another person is often viewed as a risky practice, and this perception, whether associated with concerns about physical safety or the safeguarding of privacy, serves as a significant obstacle to carpooling (Correia & Viegas, 2011; Créno and Cahour, 2015). When you carpool, you agree to give someone else responsibility for your travel, which entails certain risks. Research indicates that there is a significant decrease in interest in carpooling due to these risks (Ciari & Axhausen, 2013). The social inclusivity of this service is clear from its objectives. The objectives of the service are to enable kids and teenagers without a driving license, green lifestyle pursuers, villagers with hobies that require a ride, retirees, who have a desire to help/drive others and meet people in their cars, and villagers longing for more socializing and meeting places. The user satisfaction with the service is illustrated in Fig. 9.

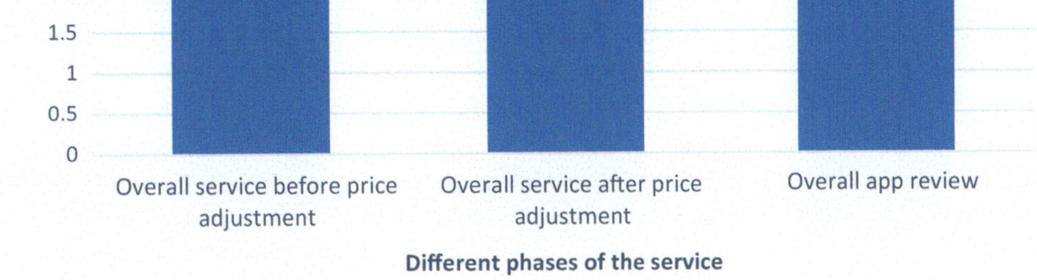

Fig. 9 Author's adaptation of existing data sourced from the NVV mobile and Google apps, showing user reviews about the Mobilfalt service in Germany, 1 being the worst and 5 the best rating

4.5 Synthesis of the Best Practices Analysis

The best practices highlight the primary, sustainable, and innovative transportation modes that are currently prevalent in European rural areas. The key advantage of these transportation modes is their flexibility, on-demand availability, allowing users to hire them at any time without adhering to a fixed schedule or route. Additionally, these modes of transport offer cost-effectiveness, as the fare is divided among all the passengers. The overall details of the best practice analysis, including the sustainability of the four categories, are summarized in Tables 2, 3 and 4. They provide useful insights about planning and designing sustainable mobility solutions for rural areas. According to this analysis, the main aspects to be considered are the following:

- Operational efficiency and cost effectiveness:
  - E-bike sharing schemes can achieve cost-effectiveness through partnerships with local bike shops.
  - Collaborating with local shops can offer additional services, reducing operating costs, including maintenance, and fostering community involvement.
- User-centric approach:
  - Analyzing the needs and preferences of target populations is crucial for designing effective services.
  - Involving motivated individuals on-site is recommended to encourage participation and utilization of the transport system once implemented.
  - Motivating individuals to try new services and shift away from private vehicle usage is essential.
- Technology and user experience:
  - Emphasizing the importance of a user-friendly booking system in ToD or shared mobility is crucial.
- Role of public transport operators:
  Existing public transport operators can play a significant role in addressing transportation challenges, especially in rural areas.
- Social inclusion for the elderly:
  - Recognizing the need for more opportunities, such as shared mobility, to facilitate social interactions and participation in activities for the elderly.

Table 2 indicates that MM emits the least CO_2 compared to ToD and shared mobility. While MM is budget-friendly, drawbacks include its limited suitability for long distances, with an average trip of 11.4 km for e-bikes (Helms & Lienhop, 2015). Studies show varying results, such as an average distance of 14.5 km in Flanders (Mobiel, 2014). Micromobility is weather-sensitive and poses safety concerns, as highlighted in several studies (Beck et al., 2007; Derriks & Mak, 2007; Robert et al., 2021; Wegman et al., 2012).

Shared mobility has a lower CO_2 emission per person-kilometer than ToD but higher than MM. It offers cost savings, environmental benefits, and all-weather usability (Arbeláez, 2023), promoting social inclusivity. However, safety concerns, especially sharing rides with strangers, pose challenges and can significantly decrease user adoption (Correia & Viegas, 2011; Créno & Cahour, 2015; Ciari & Axhausen, 2013). Additionally, the mode increases travel time by picking up other passengers, impacting user independence (Monchambert, 2020; Gheorghiu & Delhounme, 2018).

ToD, or Demand-Responsive Transport, enhances safety with features like surveillance cameras and built-in seats. It offers weather-resistant and inclusive services, catering to residents regardless of age, gender, or disability. Affordability and comfort are additional advantages. However, user concerns include longer waiting times for pickups, inconsistent pricing, and lower dependability compared to public transportation (RideCo, 2023; Liftango, 2022).

MaaS, or Mobility as a Service, is a conceptual framework rather than a specific transport mode, aimed at integrating various services into a unified platform. It proves efficient not only in urban but also in rural areas (Elena et al. 2020; Barreto et al., 2018).

In summary, each transportation category examined demonstrates social inclusivity with unique advantages and disadvantages. MM exhibits the lowest CO_2 emissions, followed by shared mobility. ToD places a premium on safety, while MM demands the least budget. Furthermore, ToD's subscription charges are favorable, and in this case, the service is entirely free. It is important to note that determining a singular best practice can vary based on location, rendering the identification of an overall superior approach challenging.

5 Discussion and Final Considerations

The sustainability of the transport system is a crucial element that needs specific attention because the transportation industry has many negative environmental effects. The transport sector is responsible for about a quarter of the EU's total greenhouse gas (GHG) emissions (EU Environment Agency, 2023). To neutralize these environmental effects, there are different types of strategies and policies at the European level. The objective of these approaches is to mitigate greenhouse gas (GHG) emissions in the transportation sector, working towards a specific future target (Council-Clean and sustainable mobility for a climate-neutral EU, 2021), however these strategies are not enough to completely counteract the negative effects of transportation. Furthermore, the utilization of private vehicles has a more pronounced impact on the environment (Shen et al., 2022). Rural societies are

Table 2 Sustainability comparison of the four categories in rural areas, Author own elaboration based on various scientific literature cited in Chap. 4

Category	Environmental effect (CO_2 emission) grams/person-kilometer	Budget required	Subscription charges	Safety	Social inclusivity
Micromobility (MM)	23 g (average of three studies, see Sect. 4.1.1)	Less budget required compared to PT	Subscription charges for users is 7.14 to 20 pounds/h, from a bike service in CNP see Sect. 4.1.1	Less safe compared to other public transport (PT) modes	Needs people to be able to be active
MaaS	MaaS can reduce CO_2 because it promotes public transport modes	Less budget, as its implementation only involves development of an algorithm and app development	Monthly subscription charges are 24.43 euros, from the BudapestGo app, see Sect. 4.2.1	Safety of personal information, location data, and payment information is a concern for users	Needs people to be able to understand the online service
Transport on demand ToD	129.75 (average value of four types of on-demand services, see Sect. 4.3.1)	Requires a smaller budget compared to traditional public transport and more compared to shared mobility and MM	The service is totally free in Alūksne and Mazsalaca districts of Vidzeme region and financed by EU	Safe compared to MM and shared mobility	Empower older and disabled people and people living on the country side to use mobility
Shared mobility	104.6 (author's own calculation based on other studies, see Sect. 4.4.1)	In case of driver's own vehicle, it required less budget (required budget is only for app development and its management, ranges from 15–25,000 US dollars, see Sect. 4.4.1)	Subscription charges is only 1 euro for each trip cited from the Mobilfalt service, see Sect. 4.4.1	Safety is lower compared to ToD	Enable people with reduced mobility to use the service and fulfill their needs, require people to travel with strangers,

Table 3 Identified best practices across Europe

Category	Type of route	No. of stops	Booking methods/subscription	Ride-sharing	Vehicle type	Example cases
Micromobility (MM)	Flexible	No stops	Required unless ownership of, e.g., a bike is given	Shared or sometimes as an "individual means of transport" (A specific category of shared mobility)	Bikes, electric scooters, skate-boards etc	(1) Trani city micromobility system (Italy)5 (2) Cairngorm national park micromobility system (UK) (3) Hajnowka e-bike scheme (Poland)
Mobility as a Service (MaaS)	Different for each transport mode	Fixed	Required	Shared	Different types	(1) MaaS scheme for Budapest (Hungary) (2) MaaS in rural Scotland (3) MaaS in rural Finland
Transport on Demand (ToD)	Flexible	Flexible	Required	Shared	Bus, minibus or car	(1) On-demand services in Vidzeme region (Latvia) (2) On-demand bus service in Catalonia (Spain) (3) Shared and on-demand bus service in Beilsko (Poland) (4) Transport on demand system "ALFA" (Germany)
Shared mobility	Flexible	Semi- flexible	Required	Shared	Mostly cars and e-bikes	(1) Mobilfalt (Hesse Germany) (2) Shared mini bus in Beilsko, (Poland) (3) Community shared car system Cuxhaven (Germany) (4) Carpooling in Vejle (Denmark)

Table 4 Author's own observation from the best practices, showing similarities and differences between the four categories

Category	Similarities	Differences
Micromobility (MM)	Micromobility and shared mobility are mostly related to private transport (commuters can share a person's private vehicle and the community center will provide a private means of transport according to the requirement)	Micromobility needs an investment to be established Micromobility is not subscribed by people frequently (sometimes); users are mostly tourists and residents
Shared mobility		Any private owner of a vehicle can establish shared mobility Subscriptions are not frequent Even a single person can make use of shared mobility
Mobility as a service (MaaS)	MaaS and transport on demand are similar to public transport (MaaS is the integration of public transport modes). ToD commuters need to register the trip, and when the required number of people is reached, then ToD service will operate	MaaS needs an intermodal hub to switch from one means of transport to the other MaaS needs only a single subscription People frequently subscribe to MaaS
Transport on demand (ToD)		ToD does not need any additional space for exchange For ToD, you need to register every time you need a trip Subscriptions are not frequent On-demand service is requested only if the required number of people is reached

more car-centric due to the inefficiency or absence of public transport. In the UK, rural residents hold 0.3 more cars than urban households (JM Dargay, 2002). Moreover, there is still a growing focus on transport strategies that primarily focus on urban contexts (Nemtanu et al., 2016). Therefore, it is necessary to shift attention towards rural transport policies and strategies that offer sustainable, efficient, and effective solutions to address their unique challenges. In rural areas, public transport is not an optimal solution due to low demand and a sparse population. Additionally, European rural regions boast a wealth of built heritage and tourist sites, attracting millions of visitors each year (UNWTO, 2020). With increasing tourist flows and higher seasonality, it often becomes economically inefficient to maintain a public transport system in these areas.

In rural areas, sustainable and innovative modes of transportation can replace conventional public transport. The public transport system is not economically viable in these low density areas due to the low demand. For example, in Catlunya, a region of Spain, a conventional bus line was re-designed into on-demand flexible transport, because the fixed schedule did not fully respond to the needs of the people (SMARTA, 2021). These alternative solutions to conventional public transport have been successfully tested, meeting the needs of both residents and tourists while remaining economically viable and sustainable. These solutions include micromobility, mobility as a service, transport-on-demand services, and shared mobility. User response analyses have shown that these solutions largely fulfill the users' requirements. This study thoroughly examined these four alternatives, collecting best practices from various parts of Europe. A total of fourteen cases have been gathered for this research, with one case from each category discussed in detail in this paper. The study concludes that when designing solutions for rural areas, there are relevant features to consider, specifically referring to mobility in these regions, listed in Sect. 4.5 of this paper.

User engagement is a crucial consideration when designing effective solutions. Shared mobility can particularly benefit the elderly, who often lack driving skills and seek social interaction. Another finding from this research is the significant role that existing public transport operators can play in addressing transportation challenges in rural areas. However, providing affordable alternative transportation in these regions is challenging, as many residents in rural and peripheral areas face internet connectivity issues. Therefore, an offline and user-friendly booking system for on-demand or shared mobility services would be highly valued in these areas. Lastly, addressing the needs of rural areas is best achieved through alternative mobility solutions discussed in this paper rather than relying solely on fixed-route public transportation.

Micro-mobility, Shared mobility and on-demand solutions offer a more cost-effective alternative to traditional fixed-route or dial-a-ride transit by combining the efficiency of mass transit with the convenience of private transportation (Liftango, 2022). These services stand out for their ability to optimize vehicle utilization and service coverage while providing passengers with unparalleled convenience compared to earlier transit models. These solutions are typically tailored to meet the needs of users and effectively respond to them.

Based on the preliminary considerations and available data, it is conceivable that micro-mobility, shared mobility, and on-demand transportation have the potential to solve transportation problems in rural areas, particularly for vulnerable groups such as the elderly population and young people, as explored in some of the best practices in this article. However, it is crucial to acknowledge that these hypotheses are still in their early stages and require further development and analysis. Therefore, additional research is necessary to outline effective solutions for the rural areas.

One future step of this research is to focus on the Marche Region in Italy, as an assessment of the design and planning of a sustainable mobility strategy in rural areas. In this regard, the authors started research on the statistical data of the region, showing that the Marche Region exhibits an average income status among its residents. The total income reported by a considerable number of taxpayers in 2022 is noteworthy. Although the average falls below the national and Central Region averages, there has been a positive increase compared to the previous year. The Marche region holds the eleventh position in economic status among Italian regions. Notably, Ancona boasts the highest average income, followed by Pesaro Urbino, Macerata, Ascoli Piceno, and Fermo (Agenzia Nazionale Stampa Associata (ANSA), 2023). Furthermore, the region is in a stage of development, and many projects need to be realized. Regarding mobility, a major development was announced in October 2023, which is the "Marche Infrastructural Plan 2032."

The primary guiding principles that require attention are sustainability, safety, inter-multimodality, and interconnection, which necessitate a shift from the territory's current "comb" connection structure to a "mesh" structure. The implementation of the plan is divided into three phases: until 2027, 2027–2032, and beyond 2032. The plan is designed to achieve several objectives. One of the key goals is to establish strong connections between the sub-Apennine region, the surrounding region, Italy, and other European countries. Additionally, the infrastructure network will be strategically developed to address territorial disparities, slow population loss, and create new employment opportunities, specifically emphasising on the sub-Apennine region. In summary, the region's low population density, limited public transportation, last-mile connectivity challenges, geographical complexity, and community engagement, as well as the recent approval of the afformentioned planning framework for the regional infrastructure system, make the Marche a possible candidate to verify innovative mobility solutions in rural areas.

Competing Interest The research work is part of PhD research funded by the Italian Ministry of University and Research under the "PON Research and Innovation 2014–20: Doctoral Studies on Green Solutions" cycle. A.Y. (2021–22) under the code specific code "DOT1301159", CUP: I31B21005710007. The authors declare that they have no known competing financial interests or personal relationships that could have appeared to influence the work reported in this paper.

Acknowledgements The authors acknowledge support from the Department of Construction, Civil Engineering, and Architecture (DICEA) of the Università Politecnica delle Marche and from the Italian Ministry of University and Research for the funding of the PhD grant. Indeed, the research work is part of PhD research funded by the Italian Ministry of University and Research under the "PON Research and Innovation 2014-20: Doctoral Studies on Green Solutions" cycle. A.Y. (2021-22) under the code specific code "DOT1301159", CUP: I31B21005710007.

References

A strategy for inner areas in Italy—Agenzia per la ... Available at: https://www.agenziacoesione.gov.it/wpcontent/uploads/2020/07/MUVAL_31_Aree_interne_ENG.pdf

Agenzia per la Coesione Territoriale (2013) National Strategy for "Inner Areas" SNAI (no date). Available at: https://www.agenziacoesione.gov.it/strategia-nazionale-aree-interne/?lang=en

Alonso-Gonzalez, M. J., Liu, T., Cats, O., Van Oort, N., & Hoogendoorn, S. (2018). The potential of demand- responsive transport as a complement to public transport: An assessment framework and an empirical *evaluation. Transportation Research Record, 2672*, 879–889.

Alyavina, E., Nikitas, A., & Tchouamou Njoya, E. (2020a). Mobility as a service and sustainable travel behaviour: A thematic analysis study. *Transportation Research Part f: Traffic Psychology and Behaviour, 73*, 362–381. https://doi.org/10.1016/j.trf.2020.07.004

Anon, (2022). A long-term vision for the EU's rural areas. European Commission—European Commission. Available at: https://ec.europa.eu/info/strategy/priorities-2019-2024/new-push-european-democracy/long-term-vision-rural.

Ansa, R. (2023). Nelle marche il 74,5% dichiara Redditi fino a 26mila Euro—Notizie, Agenzia ANSA. Available at: https://www.ansa.it/marche/notizie/2023/06/07/nelle-marche-il-745-dichiara-redditi-fino-a-26mila-euro_20db6ede-235f-4afc-9a26-0b7a8b-d9eb0a.html#:~:text=Le%20Marche%20si%20collocano%20all,reddito%20inferiore%20a%2015mila%20euro

Arbeláez Vélez, A. M. (2023) 'Environmental impacts of shared mobility: A systematic literature review of life-cycle assessments focusing on car sharing, carpooling, bike sharing, scooters and moped sharing. *Transport Reviews*, 1–25. https://doi.org/10.1080/01441647.2023.2259104

Aviemore Bikes, Electric Bike Hire. (2024). Available at: https://www.aviemorebikes.co.uk/bike-hire-demo-guiding/electric-bike-hire/

Barca F., Casavola P., & Lucatelli S. (eds., 2014) Strategia nazionale per le Aree interne: Definizione, Obiettivi, Strumenti e Governance. Materiali Uval. 2014; Volume 31. Available online: https://www.agenziacoesione.gov.it/wp-content/uploads/2020/07/MUVAL_31_Aree_interne.pdf

Barreto, L., Amaral, A., & Baltazar, B. (2018) Mobility as a Service (MaaS) in rural regions: An overview. In: *2018 International conference on intelligent systems (IS), Funchal, Portugal*, pp. 856–860. https://doi.org/10.1109/IS.2018.8710455

Banister, D. (2008). The sustainable mobility paradigm. *Transport Policy, 15*, 73–80.

Bauchinger, L., Reichenberger, A., Goodwin-Hawkins, B., Kobal, J., Hrabar, M., & Oedl-Wieser, T. (2021). Developing sustainable and flexible rural urban connectivity through complementary mobility services. *Sustainability, 13*, 1280.

Beck, L. F., Dellinger, A. M., & O'Neil, M. E. (2007). Motor vehicle crash injury rates by mode of travel, United States: Using exposure-based methods to quantify differences. *American Journal of Epidemiology, 166*(2), 212–218.

Berg, J., & Ihlstrom, J. (2019). The importance of public transport for mobility and everyday activities among rural residents. *Social Sciences, 8*, 58.

BlaBlaCar. (2021). Available at: https://blog.blablacar.com/newsroom/news-list/zeroemptyseats#:~:text=Le%20BIPE%2C%20a%20member%20of,operating%20a%20carbon%2Dsaving%20network

BlaBlaCar. (2006). Available at: https://blog.blablacar.com/about-us

Blondel, B., Mispelon, C., & Ferguson, J. (2011). Cycle more often 2 cooldown the Planet! Quantifying CO_2 savings of cycling. Available at: https://repository.difu.de/jspui/handle/difu/186307.

Bojković, N., Anić, I., & Pejčić-Tarle, S. (2010). One solution for cross-country transport-sustainability evaluation using a modified ELECTRE method. *Ecological Economics, 69*, 1176–1186.

Brabec, E., Schulte, S., & Richards, P. L. (2002). Impervious surface and water quality: A review of current literature and its implications for watershed planning. *Journal Planning Literature, 16*, 499–514.

Bronsvoort, K. (2019). Exploring alternative public transport in rural areas (Master thesis). Retrieved from: https://repository.tudelft.nl/islandora/object/uuid%3Ab64b3429-c521-430b-93bf-148944260281?collection=education

Budapest go App Efficiency—Budapest Forum, Tripadvisor. Available at: https://www.tripadvisor.com/ShowTopic-g274887-i263-k14250141-Budapest_go_app_efficiency-Budapest_Central_Hungary.html

BudapestGO Maas App (2024) *Intelligent Transport*. Available at: https://www.intelligenttransport.com/transport-news/133138/bkk-budapestgo-maas-app/

Cairngorm guided mountain bike descent—Review of Cairngorm Mountain, Aviemore, Scotland, Tripadvisor. Available at: https://www.tripadvisor.com/ShowUserReviews-g186537-d555324-r138186949-Cairngorm_Mountain-Aviemore_Aviemore_and_the_Cairngorms_Scottish_Highlands_Scotla.html

Campisi, T. et al. (2021). Innovative Solutions for sustainable mobility in areas of weak demand. some factors influencing the implementation of the DRT system in Enna (Italy). In: *International conference of computational methods in sciences and engineering* ICCMSE 2020 [Preprint]. https://doi.org/10.1063/5.0047765.

Ciari, F., & Axhausen K.W. (2013). Proceedings of transportation research board 92nd Annual Meeting. 2013. In: Carpooling in Switzerland: Public attitudes and growth strategies.

Cohen, K. (2019). Human behavior and new mobility trends in the United States, Europe and China. Working Paper, No. 024.2019. Fondazione Eni Enrico Mattei (FEEM), Milano.

Correia, G., & Viegas, J. M. (2011). Carpooling and carpool clubs: Clarifying concepts and assessing value enhancement possibilities through a Stated Preference web survey in Lisbon, Portugal. *Transportation Residential Part A: Policy Practice, 45*(2), 81–90. https://doi.org/10.1016/j.tra.2010.11.001.[CrossRef][GoogleScholar]

Cottrill, C. D., Brooke, S., Mulley, C., Nelson, J. D., & Wright, S. (2020). Can multi-modal integration provide enhanced public transport service provision to address the needs of vulnerable populations? *Research in Transport Economics, 83*, 100954. https://doi.org/10.1016/j.retrec.2020.100954

Créno, L., & Cahour B. Proceedings of the 22nd ITS World Congress (paper ITS-2466) 2015. Perceived risks and trust experience in a service of carpooling.

Dang, L., von Arx, W., & Frölicher, J. (2021). The impact of on-demand collective transport services on sustainability: A comparison of various service options in a rural and an urban area of Switzerland. *Sustainability, 13*, 3091. https://doi.org/10.3390/su13063091

Dargay, J. M. (2002). Determinants of car ownership in rural and urban areas: a pseudo-panel analysis. *Transportation Research Part E: Logistics and Transportation Review, 38*(5):351–366, ISSN 1366–5545. https://doi.org/10.1016/S1366-5545(01)00019-9

Derriks, H. M, Mak, P. M. (2007). IRTAD special report. Underreporting of Road Traffic.

Edmonton. (2024). On demand transit safety and accessibility, City of Edmonton. Available at: https://www.edmonton.ca/ets/on-demand-transit-safety-security

Elena, A., Alexandros, N., & Eric, T. N. (2020) Mobility as a service and sustainable travel behaviour: a thematic analysis study. *Transportation Research Part F: Traffic Psychology and Behaviour, 73*, 362–381, ISSN 1369–8478, https://doi.org/10.1016/j.trf.2020.07.004.

European commission. (2015). Mobility & Transport—Road Safety. Available at: https://road-safety.transport.ec.europa.eu/eu-road-safety-policy/priorities/safe-vehicles/archive/buses-andcoaches_en#:~:text=Transport%20by%20bus%20and%20coach%20is%20the%20safest%20mode%20of%20road%20travel

Council-Clean and sustainable mobility for a climate-neutral EU. (2021). Consilium. Available at: https://www.consilium.europa.eu/en/policies/clean-and-sustainable-mobility/

EU Environment Agency. (2023). *Transport and mobility*. Available at: https://www.eea.europa.eu/en/topics/in-depth/transport-and-mobility

EU Rural vision (no date) Rural Vision. Available at: https://rural-vision.europa.eu/rural-vision_en#:~:text=A%20vision%20for%20rural%20areas%20towards%202040,The%20long%2Dterm&text=It%20recognises%20the%20diversity%20of,consultations%20and%20stakeholder%2Dled%20events

Eueopan commission, EU Road Safety: Towards 'Vision Zero' (2022) *European climate, infrastructure and environment executive agency*. Available at: https://cinea.ec.europa.eu/publications/eu-road-safety-towards-vision-zero_en

EU commission low emission mobility. (2016). Available at: https://ec.europa.eu/commission/presscorner/detail/en/MEMO_16_2497

European Commission. (2019). https://ec.europa.eu/info/law/better-regulation/have-your-say/initiatives/12438-Sustainable-and-Smart-Mobility-Strategy_en

European Commission Eurobarometer. (2021). Available at: https://europa.eu/eurobarometer/surveys/detail/2229

European Environment Agency. (2011). *Annual report 2010 and Environmental statement*. http://www.eea.europa.eu/publications/annual-report-2010

Eurostat. (2021). Definition of rural area. Available at: https://ec.europa.eu/eurostat/web/rural-development/methodology

Eurostate Statistics explained. (2021). *Statistics explained*. Available at: https://ec.europa.eu/eurostat/statistics-explained/index.php?title=Road_accident_fatalities_-_statistics_by_type_of_vehicle&oldid=583880#:~:text=In%202011%2C%20the%20EU%20recorded,to%2019%20917%20in%202021

EU Commission. (2020). Research and Innovation. Available at: https://research-and-innovation.ec.europa.eu/funding/funding-opportunities/funding-programmes-and-open-calls/horizon-europe/eu-missions-horizon-europe/climate-neutral-and-smart-cities_en

Federal Office for Special Development Bern, Switzerland. (2015). Available online: https://www.are.admin.ch/dam/are/de/dokumente/verkehr/dokumente/mikrozensus/modalsplit_agglomerationen.pdf.download.pdf/Modalsplit_Agglomerationen_2015.pdf

Federal Office for Special Development Dichte und Mobilitätsverhalten. (2018). Available online: https://www.are.admin.ch/are/de/home/medien-und-publikationen/publikationen/grundlagen/dichte-und-mobilitatsverhalten.html

Franco, P., Bell, M., & Galatioto, F. (2012). The role of electronic vehicle and fare data for carbon accounting in the delivery of sustainable bus operation. In: 19th intelligent transport systems world congress, ITS 2012, pp. EU-00613

Gallo, M., & Marinelli, M. (2020). Sustainable mobility: A review of possible actions and policies. *Sustainability, 12*(18), 7499. https://doi.org/10.3390/su12187499

Gaudet, L., Romanow, N. T. R., Nettel-Aguirre, A., Voaklander, D., Hagel, B. E., & Rowe, B. H. (2015). The epidemiology of fatal cyclist crashes over a 14-year period in Alberta Canada. *BMC Public Health, 15*, 1142.

Georgakis, P., Almohammad, A., Bothos, E., Magoutas, B., Arnaoutaki, K., & Mentzas, G. (2020). Heuristic-based journey planner for mobility as a service (MaaS). *Sustainability, 12*, 10140. https://doi.org/10.3390/su122310140

Gheorghiu, A., & Delhomme, P. (2018). For which types of trips do French drivers carpool? Motivations underlying carpooling for different types of trips. *Transportation Reserve Part A Policy Practice, 113*, 460–475. https://doi.org/10.1016/j.tra.2018.05.002

Gillis, D., Semanjski, I., & Lauwers, D. (2016). How to monitor sustainable mobility in cities? Literature review in the frame of creating a set of sustainable mobility indicators. *Sustainability, 8*, 29.

Global Road Safety. (2023). *Centers for disease control and prevention*. Available at: https://www.cdc.gov/injury/features/global-road-safety/index.html

Gojanovic, B., Welker, J., Iglesias, K., Daucourt, C., & Gremion, G. (2011). Electric bicycles as a new active transportation modality to promote health. *Medicine and Science in Sports and Exercise, 43*, 2204–2210. https://doi.org/10.1249/MSS.0b013e31821cbdc8

Graham, H., de Bell, S., Flemming, K., Sowden, A., White, P., & Wright, K. (2018). The experiences of everyday travel for older people in rural areas: A systematic review of UK qualitative studies. *Journal of Transport and Health, 11*, 141–152.

Helms, K., & Lienhop, M. (2015). Pedelction—Mobilitätsmuster, Nutzungsmotive und Verlagerungseffekte. In: Conference Paper At NationalerRadverkehrskongress, Porsdam, 18–19 May 2015.

Herrman & Herrman, P.L.L.C. (2023a). *Public transportation safety: Is mass transit safer than driving?, Hermann & Hermann*. Available at: https://www.herrmanandherrman.com/blog/public-transportation-safety-mass-transit-actually-safer-driving/

Hickman, R., & Banister, D. (2019). Transport and the environment. In: J. Stanley, & D. A. Hensher (Eds.), *A research agenda for transport policy* (pp. 25–33). Cheltenham, UK: Edward Elgar Publishing.

Hiltunen, M. J., & Rehunen, A. (2014). Second home mobility in Finland: Patterns, practices and relations of leisure oriented mobile lifestyle. *Fennia, 192*(1), 1–22.

Hollingsworth, J., Copeland, B., & Johnson, J. X. (2019). Are e scooters polluters? The environmental impacts of shared dockless electric scooters. *Environment Research Letters, 14*, 084031. https://doi.org/10.1088/1748-9326/ab2da8

Inclusion. (2020). *INCLUSION European Project—Pilot Labs*. Available at: http://www.h2020-inclusion.eu/inclusion-pilot-labs/cairngorms-national-park-uk/index.html

Intelligent Transport. (2019). Available at: https://www.intelligenttransport.com/transport-news/77856/car-pooling-social-environmental-impacts/

ITS Department of Liaison Evaluation. (2020). *ITS deployment evaluation*. Available at: https://www.itskrs.its.dot.gov/costs

Jeekel, H. (2014). Social exclusion, vulnerable groups and driving forces: Towards a social research based policy on car mobility. *Case Studies Transportation Pollution, 2*, 96–106. https://doi.org/10.1016/j.cstp.2014.06.005

Jiangping, Z. (2012). Sustainable Transportation in the US: A Review of Proposals, Policies, and Programs since 2000

Jittrapirom, P., Caiati, V., Feneri, A. M., Ebrahimigharehbaghi, S., Alonso-González, M. J., & Narayan, J. (2017). Mobility as a Service: A critical review of definitions, assessments of schemes, and key challenges. *Urban Planning, 2*(2), 13–25.

Kager, R., Bertolini, L., & Te Brömmelstroet, M. (2016). Characterization of and reflections on the synergy of bicycles and public transport. *Transportation Research A Policy Practice, 85*, 208–219. https://doi.org/10.1016/j.tra.2016.01.015

Kou, Z., Wang, X., Chiu, S.F., Anthony, D., & Cai, H. (2020). Quantifying greenhouse gas emissions reduction from bike share systems: a model considering real-worldtrips and transportation mode choice patterns. *Resources Conservation Recycle, 153*, 104534.https://doi.org/10.1016/j.resconrec.2019.104534

Levinson, D. M. (2023) *Transportist: Maas or Saam: Mobility-as-a-service pro and con, transportist.net*. https://www.transportist.net/p/maas-or-saam-mobility-as-a-service

Liftango. (2022). *The benefits of demand-responsive transport in rural areas, RSS*. Available at: https://www.liftango.com/resources/benefits-of-demand-responsive-transport-in-rural-areas#:~:text=Because%20rural%20transport%20systems%20are,dial%2Da%2Dride%20transit

Liyanage, S., Dia, H., Abduljabbar, R., & Bagloee, S. A. (2019). Flexible mobility on-demand: An environmental scan. *Sustainability, 11*, 1262.

Maas-Alliance. (2022). MAASAlliance. Available at: https://maas-alliance.eu/homepage/what-is-maas/

Marche region. (2023). *Regional infrastructure plan—regione.marche.it*. Available at: https://www.regione.marche.it/portals/0/Infrastrutture_Trasporti/Marche2032/All_1_Interventi.pdf

Mattioli, G. (2014). Where sustainable transport and social exclusion meet: Households without cars and car dependence in Great Britain. *Journal of Environmental Policy & Planning, 16*(3), 379–400.

Maximizing mobility and access to services (MARA). (2020). Nordregio.org. Available at: https://nordregio.org/wp-content/uploads/2022/10/Maximising-mobility-and-access-to-services-in-rural-areas-.pdf

Misso, F. E., Pani, F. E., Porru, S., Repetto, C., & Sansone, U. (2017). Opportunities and boundaries of transport network telematics. In: Interreg- Central.Eu, 2017.

Mitropoulos, L., Kortsari, A., Mizaras, V., & Ayfantopoulou, G. (2023). Mobility as a service (MaaS) planning and implementation: challenges and lessons learned. *Future Transportation, 3*(2), 498–518. https://doi.org/10.3390/futuretransp3020029

Mobiel 21, (2014). *Only Data Source Website Text*: http://www.mobiel21.be/en/content/97-pedelec-commuters-love-their-bikes

Mobilfalt. (2018). Car pooling as a complement to public. Available at: https://www.mambaproject.eu/wp-content/uploads/2018/07/WP-2_good-practice-cases-_Mobilfalt-FINAL.pdf

Mobility and accessibility in rural areas-Regional action plan (2020). Mara-mobility.eu. Available at: https://www.mara-mobility.eu/images/RAPs/Action_Plan_Vidzeme_Latvia_VPR.pdf

Molin, E., Mokhtarian, P., & Kroesen, M. (2016). Multimodal travel groups and attitudes: A latent class cluster analysis of Dutch travelers. *Transportation Research Part A Policy Practice, 83*, 14–29.

Monchambert, G. (2020). Why do (or don't) people carpool for long distance trips? A discrete choice experiment in France. *Transportation Research Part A Policy Practice, 132*, 911–931. https://doi.org/10.1016/j.tra.2019.12.033

Mounce, R., Beecroft, M., & Nelson, J. D. (2020). On the role of frameworks and smart mobility in addressing the rural mobility problem. *Research in Transport Economics, 83*, 100956.

Næss, P., Xue, J., Stefansdottir, H., Steffansen, R., & Richardson, T. (2019). Second home mobility, climate impacts and travel modes: Can sustainability obstacles be overcome? *Journal of Transport Geography, 79*, 102468. https://doi.org/10.1016/j.jtrangeo.2019.102468

Nemtanu, F., Schligensiepen, J., Buretea, D., & Iordache, V. (2016). Mobility as a service in smart cities. In *Proceeding of the 9th international conference responsible entrepreneurship vision development and ethics, Bucharest, Romania*, pp. 13–27.

Nurture, L. (2019). *How much would it cost to develop an carpooling app platform like Blablacar?, Lets Nurture—An IT Company Nurturing Ideas into Reality*. Available at: https://www.letsnurture.com/how-much-carpooling-app-platform-like-blablacar.html

Pillai, S. K. (2023). How much does it cost to develop a carpooling app?, Medium. Available at: https://medium.com/mqos-technologies/how-much-does-it-cost-to-develop-a-carpooling-app-682c129fac9

Poltimäe, H., Rehema, M., Raun, J., & Poom, A. (2022). In search of sustainable and inclusive mobility solutions for rural areas. *European transport research review, 14*(1), 13. https://doi.org/10.1186/s12544-022-00536-3

Popovich, N., Gordon, E., Shao, Z., Xing, Y., Wang, Y., & Handy, S. (2014). Experiences of electric bicycle users in the Sacramento California area. *Travel Behaviour Society, 1*, 37–44. https://doi.org/10.1016/j.tbs.2013.10.006

Pot, F. J., Koster, S., Tillema, T., & Jorritsma, P. (2020). Linking experienced barriers during daily travel and transport poverty in peripheral rural areas: The case of Zeeland, the Netherlands. *European Journal of Transport and Infrastructure Research, 20*(3), 29.

Quora. Available at: https://www.quora.com/Which-mode-of-transport-is-cheapest-to-operate-on-a-per-km-basis

RideCo. (2023). Available at: https://www.rideco.com/post/on-demand-transit-benefits#:~:text=On%2Ddemand%20transit%20provides%20passengers,for%20their%20shuttle%20to%20arrive

Robert, N. B. (2021). Scootin' in the rain: Does weather affect micromobility?, *Transportation Research Part A: Policy and Practice, 149*, 114–123, ISSN 0965–8564. https://doi.org/10.1016/j.tra.2021.05.003

Rodrigue, J.-P. (Ed.). (2020). *The geography of transport systems*. Routledge.

Sdoukopoulos, A., Pitsiava-Latinopoulou, M., Basbas, S., & Papaioannou, P. (2019). Measuring progress towards transport sustainability through indicators: Analysis and metrics of the main indicator initiatives. *Transportation Research Part d: Transport and Environment, 67*, 316–333.

Shaheen, S., Cohen, A., & Bayen, A. (2018). *The benefits of carpooling*. UC Berkeley: Transportation Sustainability Research Center. https://escholarship.org/uc/item/7jx6z631.

Shahid, H., Valtteri, A., Taha, K., & Pekka, L. (2023). Sustainability of smart rural mobility and tourism: A key performance indicators-based approach. *Technology in Society, 74*:102287, ISSN 0160–791X, https://doi.org/10.1016/j.techsoc.2023.102287.es that transport has a networthy effect on wellbeing.

Shen, Y. S., Lin, Y. C., Cui, S. H., Li, Y. M., & Zhai, X. X. (2022). Crucial factors of the built environment for mitigating carbon emissions. *Science of the Total Environment, 806*, 150864.

SMARTA. (2021). *Good Practices*. Available at: https://ruralshared-mobility.eu/good-practices/

Statista Portal. (2023). Numbers of Car-Sharing Users Around the World. Available at: https://www.statista.com/statistics/415636/car-sharing-number-of-users-worldwide.

Strömberg, H., Karlsson, I. C., & Sochor, J. (2018). Inviting travelers to the Smorgasbord of sustainable urban transport: Evidence from a maas field trial. *Transportation, 45*(6), 1655–1670. https://doi.org/10.1007/s11116-018-9946-8

Swinburn, B. A., Vivica, I., Kraak, S. A., Vincent, J., Atkins, P. I., Baker, J. R., Bogard, H. B., Calvillo, A., De Schutter, O., Devarajan, R., et al. (2019). The global syndemic of obesity, undernutrition, and climate change: The lancet commission report. *Lancet, 393*, 791–846. https://doi.org/10.1016/S0140-6736(18)32822-8

Taiebat, M., Brown, A. L., Safford, H. R., Qu, S., & Xu, M. (2018). A review on energy, environmental, and sustainability implications of connected and automated vehicles. *Environmental Science and Technology, 52*, 11449–11465.

Tranter, P. J. (2010). Speed kills: the complex links between transport, lack of time and urban health. *Journal Urban Health Bulletin N. Y. Academy Medicine, 87*:155–166. https://doi.org/10.1007/s11524-009-9433-9.

UNWTO. (2020). *World tourism organization technical note*. Available at: https://www.unwto.org/world-tourism-day-2020/tourism-and-ruraldevelopment-technical-note

US Department of Agriculture. (2019). ERS—What is Rural? Available at: https://www.ers.usda.gov/topics/rural-economy-population/rural-classifications/what-isrural/#:~:text=This%20delineation%20of%20built%2Dup,with%20fewer%20than%202%2C500%20people

van Wee, B., & Ettema, D. (2016). Travel behaviour and health: A conceptual model and research agenda. *Journal of Transport and Health, 3*, 240–248. https://doi.org/10.1016/j.jth.2016.07.003

Velaga, N. R., Beecroft, M., Nelson, J. D., Corsar, D., & Edwards, P. (2012). Transport poverty meets the digital divide: Accessibility and connectivity in rural communities. *Journal of Transport Geography*.

Wegman, F., Zhang, F., & Dijkstra, A. (2012). How to make more cycling good for road safety? *Accident Analysis and Prevention, 44*(1), 19–29.

Weiss, P., Dekker, A., Moro, H., Scholz, M., & Patel, K. (2015). On the electrification of road transportation—A review of the environmental, economic, and social performance of electric two-wheelers *TransportationResearch Part D: Transport and Environment, 41*, 348–366. https://doi.org/10.1016/j.trd.2015.09.007

Wolfram, P., Tu, Q., Hertwich, E., & Pauliuk, S. (2020). Documentation of the transport-sector model within the RECC model framework v1.0. https://doi.org/10.5281/zenodo.3631938.

Woodcock, J., Edwards, P., Tonne, C., Armstrong, B. G., Ashiru, O., Banister, D., Beevers, S., Chalabi, Z., Chowdhury, Z., Cohen, A., et al. (2009). Public health benefits of strategies to reduce greenhouse-gas emissions: Urban land transport. *Lancet, 374*, 1930–1943. https://doi.org/10.1016/S0140-6736(09)61714-1

Wright, S. (2013). Designing flexible transport services: Guidelines for choosing the vehicle type. *Transportation Planning and Technology, 36*(1), 76–92.

Emirati Local Communities as Balanced Ecosystems: Pilot Study for Individualistic Residents-Led Urban Agriculture Within Housing Clusters

Khaled Galal Ahmed, Noor Abdelwahed, Sohila Abdelrehim, Amal Al Askari, Wedad Alyamahi, Raghad Younes, and Hadia Al Haddad

Abstract

Among the 17 Sustainable Development Goals, SDG15-Life on Land is a vital goal that calls for preserving terrestrial ecosystems in a way that contributes to maintaining the productivity of agricultural systems and greatly supports the mitigation of the harmful impacts of the climate change. Urban Agriculture (UA) seems a very promising, albeit still understudied, approach that could significantly contribute to achieving this goal. In the cities of the United Arab Emirates (UAE), such as Al Ain, widely underdeveloped semi-public open spaces have originated from the sprawling urban forms adopted in the designs of Emirati urban communities. It has been noticed almost in every Emirati urban community that the residents are individualistically embarking on UA activities, that vary in size and locations, in these vast shared semi-public spaces around which their housing plots are clustered. Through their UA activities the residents have managed to partially convert most of the semi-public spaces of their local communities into urban farms/gardens. This pilot study aims to better understand the magnitude and aspects of these individualistic residents-led UA activities through an exploratory case study method in the semi-public spaces of five typical housing clusters utilizing some relevant qualitative tools of map analysis, field observations, and semi-structured interviews with the residents. An urban community in Al Ain city was selected as a representative case study to undertake the research investigations. The results of this pilot exploratory research have led to better understanding of the processes behind the residents-led UA activities and indicated a considerable contribution to achieving SDG 15. This is believed to open the door for further research about UA as a potentially more productive community-led UA practice.

Keywords

Urban agriculture · Urban community · SDG 15 · Ecosystems · Residents · Housing · UAE

1 Introduction

The Sustainable Development Goal (SDG) 15: Life on Land calls for preserving terrestrial ecosystems in a way that contributes to maintaining the productivity of agricultural systems and help mitigate climate change damaging impacts. Terrestrial ecosystems provide a series of goods, raw materials for construction and energy, food, and a series of ecosystem services including the capture of carbon, maintenance of soil quality, and provision of natural habitat that caters for biodiversity. Accordingly, these terrestrial ecosystems would contribute to reducing the risks of natural disasters, regulating climate, and maintaining the productivity of agricultural systems. It is claimed that preserving balanced ecosystems would also support climate change mitigation efforts and thus help preserve biodiversity for

K. Galal Ahmed (✉) · N. Abdelwahed · S. Abdelrehim · A. Al Askari · W. Alyamahi · R. Younes · H. Al Haddad
United Arab Emirates University, Al Ain, UAE
e-mail: kgahmed@uaeu.ac.ae

N. Abdelwahed
e-mail: 201950025@uaeu.ac.ae

S. Abdelrehim
e-mail: 201950139@uaeu.ac.ae

A. Al Askari
e-mail: 201850021@uaeu.ac.ae

W. Alyamahi
e-mail: 201904178@uaeu.ac.ae

R. Younes
e-mail: 201950106@uaeu.ac.ae

H. Al Haddad
e-mail: 201470236@uaeu.ac.ae

B. He et al. (eds.), *Urban Climate Change Adaptation*, Advances in Science, Technology & Innovation,
https://doi.org/10.1007/978-3-031-65088-8_17

future generations (UNEP, 2022). Locally in the United Arab Emirates (UAE) and in line with SDG 15, the government pledged in its SDGs 2030 Plan to preserve the environment while pursuing its economic growth. This entails the preservation of biodiversity in the interest of the future generations through the adoption of plans, strategies, and policies that promote the sustainable use of vital resources in the majorly sandy desert (The Official Portal of the SDGs in the UAE, 2022).

On the other hand, while balanced ecosystems allow for multiple species to live and interact in the long term (Guneralp et al., 2015), it is acknowledged that there is desperate need to meet the current challenges of the biodiversity loss and habitat degradation associated with the unsustainable urban expansion. An ecosystem that has a high negative environmental impact happens when it has unbalanced components and therefore would have lower resilience affecting the capacity of these components to stand stresses. So, when a pandemic or a conflict crisis happens, it may be difficult for people in cities to find food (Orsini & D'Ostuni, 2022). Today, about half of Earth population live in urban areas. By 2050, two thirds of the global population will live in cities. So, while urban areas represent mainly human habitats, they also include areas for synanthropic species. In line with this notion, recently urban Green Infrastructure (GI) has arisen as a means for redefining urban ecosystems as socio-ecological systems. Urban GI is basically considered as a multi-functional network of urban green spaces situated within the boundary of the urban ecosystem. As such, urban ecosystems would include both physical and biological components that interact with each other. Both components could be considered in synergy when the living environment is appropriate for people and for urban biodiversity alike. With the increasing tendency towards urbanization, the importance of GI in urban and peri-urban areas has significantly grown in the last few years (Interreg Europe, 2020).

Defined as the activity of food production with its associated processes within urban areas, Urban Agriculture (UA) is considered a vital component of GI. The growing world-wide interest in UA is principally attributed to the increasing pressures on urban environments by global climate change and the need to create sustainable food systems. Even with the noticeable rising numbers of studies about GI, UA, and their possible contribution to achieving SDG 15, there is still a significant need for more (Evans et al., 2022; Steele, 2017). It is evident that UA has essential benefits on different environmental, social, and economic sustainability levels. Environmentally, UA contributes to the reduction of urban heat island effect that causes temperature increase in cities (about 5.6 °C higher than rural areas) due to the presence of large concentration of built-up areas where huge parking lots and rooftops of buildings absorb and re-emit heat (Center for Science Education, 2023). Through increasing areas of green spaces that cool down their surroundings, UA helps reduce this effect. Secondly, UA can considerably reduce greenhouse gas emissions by encouraging more regional production of food. This would avoid importing food through aviation and shipping transportation which are among the most contributing sectors in producing greenhouse gases emissions, especially in the UAE where over 80% of food is imported (Mohammed bin Rashid School of Government, 2018). In addition, through UA, grown plants consume Carbon Dioxide (CO_2) as one of the well-known greenhouse gases. Thirdly, UA helps preserve biodiversity through supporting the variety of living things in the urban environment by providing places for them to eat and reside (Ritchie et al., 2020). For instance, UA produces numerous species of fruits, vegetables, flowers, etc., that all foster the maintenance of life-supporting insects and critters (Martin, 2016; Rabin, 2022).

As per its social benefits, UA promotes healthy lifestyles in various ways. First, with the concerns of obesity spreading in urban areas and the increasing awareness on healthy diets, UA initiatives keep people mindful to their food and encourage them to increase portions of vegetables and fruits in their meals. Second, UA helps increase community bonding as it contributes to social capital especially when it is being utilized as a community shared activity that brings together distinct groups in local communities (Armstrong, 2000; Wakefield et al., 2007).

Furthermore, UA has its economic benefits from different angles. First, it supports food security. The standard definition of food security by the Food And Agriculture Organization (FAO) is when sufficient food is available and easily accessible by people. UA contributes to the efforts of the provision of self-sufficient and sustainable food systems especially when it creates local value chains through which the small surplus produced from UA activities is sold to local traders. By producing local food from UA activities, a shorter value chain is endorsed. This not only lowers the cost but also adds a meaningful environmental value. Second, UA enhances livelihood and employability especially for the urban poor in many developing countries. For example, in Senegal it was estimated that the residents' contributions in urban and sub-urban agriculture have improved the economic status by providing jobs for over 14,000 persons by 2013 (Urban Development & Resilience Unit—World Bank, 2013). Clinton et al. (2018) estimate that when UA is sufficient, its outcomes of food production, nitrogen fixation, energy savings, pollination, climate regulation, soil formation, and biological control of pests could be worth as much as $80–160 billion annually. In an ideal situation, UA could lead to healthy local food production and vivid ecosystems in which sustainable agriculture, local economy, and local communities are synergically integrated (The URBES Project, 2014).

2 Individualistic Residents-Led UA Activities

Individualistic residents-led UA is the least studied form of UA activities, as the global and local focus is directed towards the 'institutionalized' and largely 'top-down' community collective UA activities. Individualistic Residents-led UA is an activity usually undertaken by households with no official permissions (Express & Star, 2012). In literature, the individualistic residents-led UA activities have been categorized into three main types, namely, guerrilla gardening, backyard/on-plot agriculture, and commercial farming. First, the guerrilla gardening is an 'unauthorized' gardening practice that usually occurs on an abandoned site, no-man land, or sometimes within a private property (Binder, 2014). An example of this practice is the guerrilla gardening that an Eastern European Immigrant started in Arundel Street, close to the city of Birmingham, UK. As claimed by the Immigrant, his initiated activity has successfully managed to gather the residents of the street together despite their perceptible social and cultural disparities. These guerrilla gardening activities were undertaken in the neglected public spaces by volunteers from the local community who ultimately managed to transform these areas into thriving community gardens (Erwin, 2022; Hardman et al., 2018).

The second type of the individualistic residents-led UA is the backyard/on-plot agriculture. In this activity, residents cultivate crops within and sometimes around their housing plots. An example for this type is in Accra, the capital of Ghana, where almost 60% of the households managed to cultivate a huge area of 50 to 70 hectares through backyard agriculture. The residents secured the needed water for irrigation mainly from rain (Weltbank & World Bank, 2013). Another example is in Harare, Zimbabwe, where the relatively wet climate and the large housing plots have encouraged the backyard/on-plot agriculture activities. The residents managed to produce maize in the wet season and vegetables throughout the whole year. In the dry seasons, residents usually rely on tap water and borehole water for irrigation. A third case is in Nairobi, the capital of Kenya, where the urban poor families intensify their UA activities not only in the backyards of their houses but also on the 'no-man' land along roadsides, riverbanks, and railway lines. Most residents in Nairobi rely on themselves in undertaking the UA activities and do not hire helpers. While rainwater is the main source for irrigation, wastewater has been used as a fertilizer (Freeman, 1991; Weltbank & World Bank, 2013).

The third type of the individualistic residents-led UA activity is the commercial farming where residents generate profit out of their UA activities. Unlike the conventional commercial UA activities, in this residents-led commercial farming there is no local authority intervention (Tractorgyan, 2022). An example of this type is found in the city of Abidjan, the capital of Ivory Coast, where the residents were allowed to develop their individual commercial farms within the city's projected airport perimeter. Due to the high level of groundwater, the muddy soil of this land helped the residents to grow vegetable easily (Ittner, 2022).

Locally in the UAE, the existing urban form of the Emirati local communities were designed based on car-oriented urban neighborhood strategies where the density of the built-up area was extremely low with a Floor-Area-Ratio (FAR) equals 0.11 or even less. In these urban sprawl communities, vast areas of public and semi-public open spaces, which were originally designed to allocate local amenities and recreational facilities, have become practically no-man and underused land (Galal Ahmed & Alipour, 2021). This has inevitably led to deteriorated ecosystems in these urban communities. But on the other hand, it has been noticed in almost every Emirati urban community, especially in Al Ain desert and oasis city, that the individualistic residents-led UA activities have converted some or all of the semi-public spaces, especially those juxtaposed to the housing plots, into urban farms or gardens that vary in size and magnitude.

3 Research Objectives, Questions, and Method

The main aim of this study is twofold. First is to explore the community self-initiated UA processes of the individualistic residents-led activities in a selected case study representing the Emirati urban communities. Second, is to investigate if such individualistic residents-led UA activities could help realize SDG15 through the contribution to balanced ecosystems and biodiversity in the studied urban community. Accordingly, the research poses two questions. First is "How individualistic residents-led UA is currently practiced in the selected Emirati urban community?" And second is "Could these practices help achieve SDG15?" To answer these two questions, 'Al Dhaher New' urban community in Al Ain city was selected as a representative case study for the existing urban communities in the UAE. Al Ain city was chosen because it is a favorite location for thousands of Emirati families who live in public housing urban communities. Meanwhile, Al Dhaher New urban community was selected because it was developed more than 20 years ago following an urban design pattern that was frequently repeated in many Emirati cities. During this long period, many residents of this urban community have individually and informally embarked on UA activities, especially in the semi-public open spaces around which their houses are clustered. This makes this urban community an ideal case study to investigate the status quo of the individualistic residents-led UA activities. Five typical patterns of the semi-public spaces in the neighborhood were defined based on their spatial sizes and forms. Such significant variations made it important to explore the UA practices in each one of them (Fig. 1).

Google Earth Pro (2023) engine was utilized to explore the gradual development of the individualistic residents-led UA activities over time in the semi-public open spaces in Al Dhaher New urban community, starting from 2008, when the pioneering individualistic residents-led UA activities started to emerge, until the year 2023. Through both the satellite map analysis and the field visits of these spaces, the location, size, and shape of the residents-developed urban farms and gardens were defined. One semi-public open space of each of the defined five housing cluster categories was carefully selected to represent the most intensive residents-led UA activities. These semi-public spaces were frequently visited by the researchers to record the individualistic residents-led UA activities in as much detail as possible. The aim was to accurately define the claimed areas of the UA activities within each of the five defined semi-public spaces. Also, the analysis aimed to explore the boundary-definition methods, the types of cultivated vegetation, and the sources of irrigation water. Figure 2 illustrates the five selected semi-public spaces within the housing clusters in Al Dhaher New urban community (boundaries are highlighted in a yellow color).

Another utilized exploratory tool was the interviews with the residents who have been involved in the observed UA activities in each of the five selected urban spaces. While it was planned to have the interviews with 25 families residing in these housing clusters to reasonably represent the local community, but only 20 families accepted to undertake the interviews. Still, those 20 families represent about 10% of the total number of families in Al Dhaher New urban community, which is considered as a good number of interviews given the nature of the interviews as face-to-face and in-depth interviews with open-ended questions.

Appendix 1 shows the interview script including the list of the 8 asked questions in both Arabic, as the native

Fig. 1 The five typical patterns of the semi-public spaces forming the housing clusters of 'Al Dhaher New' urban community in Al Ain city. *Source* Google Earth Pro, 2023 and The Authors

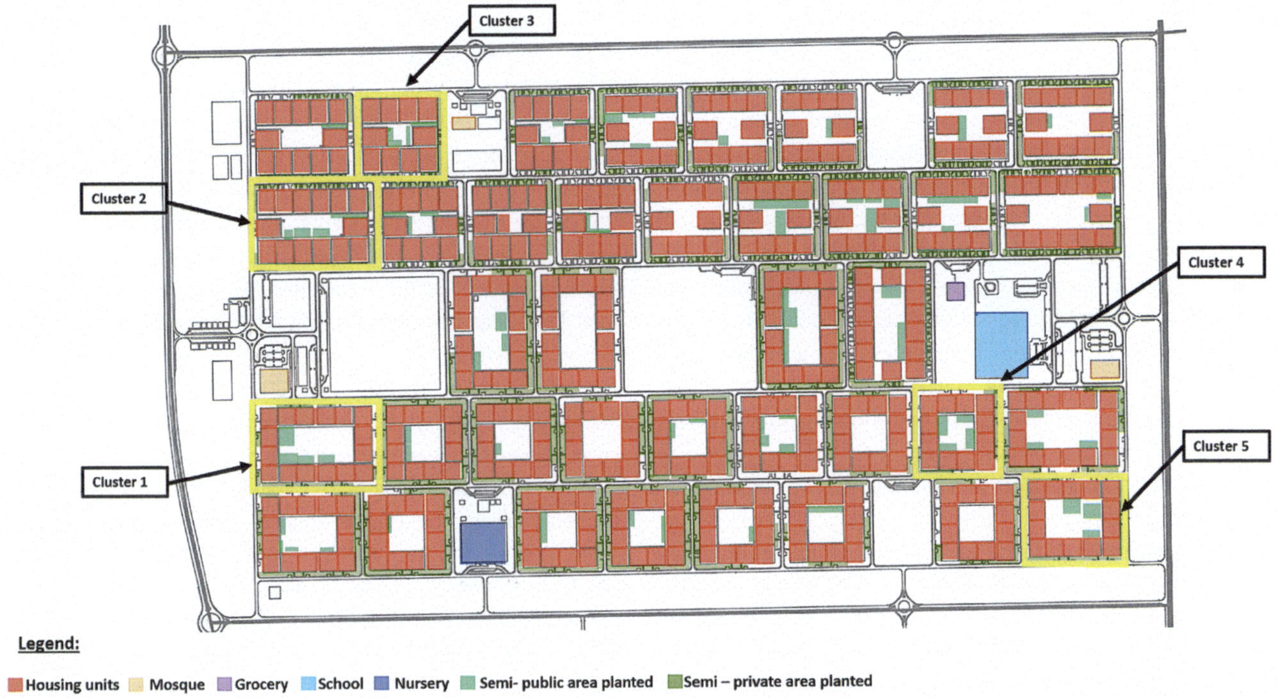

Fig. 2 The locations of the selected five studied housing clusters for investigations in Al Dhaher New urban community (Boundaries are highlighted in a yellow color). *Source* The Authors

language of the interviewed residents, and English. The asked questions covered the different aspects of the current individualistic residents-led UA activities as follows:

- Q1: What is the purpose of your UA activities in the shared semi-public spaces of your housing cluster?
- Q2: How did you define the location and size of your developed UA area?
- Q3: Do you usually carry out the UA activities on your own (as a family), or do you usually hire professional farmers to help you?
- Q4: What type of vegetation do you usually grow? and why?
- Q5: From where do you get the water for irrigation?
- Q6: Have you ever had a dispute with a neighbor over the practiced UA activities within the shared semi-public area?
- Q7: What are the UA related problems that you usually encounter, if any?
- Q8: In your opinion, how these problems could be solved?

And to give them the chance to add more information, the interviewed families were asked by the end of the interview to add anything they wish regarding the addressed topics in the 8 asked questions.

4 Results

4.1 Results of the Map Analysis and Field Observations of the Residents-Led UA Activities

The map analysis of Al Dhaher New urban community showed that the neighborhood lacks any professionally developed recreational and/or green areas in the public or the semi-public open spaces despite being originally planned as such (Fig. 2). The land-use map shown in Fig. 2 indicated the presence of limited numbers of local amenities mainly including mosques, a grocery store, a school, and a nursery. As mentioned above, the map analysis was chronologically conducted each 5 years starting form 2008, then 2013, followed by 2018, and concluded by 2023. This helped the researchers to depict the gradual development of the individualistic residents-led UA activities in the semi-public spaces in each of the five selected housing clusters. In general, such activities were limited in scale in 2008 but along the following years, the residents managed to individually develop, as households, significant areas of these underused and underdeveloped semi-public spaces as urban farms and/or gardens. The residents-led UA activities have been mostly close to the back or side fences of the housing plots that overlook the shared semi-public space in each of the housing clusters. The sizes of the cultivated areas varied considerably among houses but also kept moderately and cautiously extending towards the core of the shared space.

The field observations in the five selected housing clusters proved that the residents managed to individually cultivate a variety of crops (like cabbage, red pepper, eggplant, etc.), native trees especially palm trees, and various other plants using simple and conventional irrigation systems. Some residents have also grown flowers and ornamental plants. Through field observations, it was noticed that the residents have defined the boundaries of their developed UA areas using light-weight materials such as wood sticks, steel posts, and mesh fabrics. In the following sub-sections, the results of the analysis of the residents-led UA activities in each of the five selected housing clusters are briefly discussed, one by one.

Housing cluster 1: From the chronological satellite map analysis along the years from 2008 to 2023, it is revealed that the residents of 5 out of the 12 houses overlooking the semi-public shared urban space have embarked on UA activities since 2013 (Fig. 3a). The UA activities have sensibly expanded as indicated by the increased size of the cultivated areas by almost each house since 2013. Meanwhile, through field observations it was evident that palm trees were intensively planted especially near the fences of the houses to also cater for privacy for the households. Some residents have planted lavender and other types of flowers (Fig. 3b).

Housing cluster 2: In 2008 the residents of only 4 out of the 10 houses overlooking the semi-public shared space embarked on UA activities. This number increased to 6 houses since 2013 (Fig. 4a). Like housing cluster 1, the individualistic residents-led UA activities have significantly increased over the years. By 2023, and due to the overall limited size of the semi-public shared urban space in this cluster, especially when compared to the open space in housing cluster 1, the expansion of the residents' developed UA areas has been confined by the presence of active alleyways (*sikakk*) as clearly shown in the ground traces in satellite images of 2018 and 2023 (Fig. 4a). In addition, during the field visits to this housing cluster, palm trees were observed in almost all cultivated areas where they have been intensively planted outside the back fences of the houses (Fig. 4b). Figure 4b also shows the used conventional irrigation system through domestic water taps and the simple fence made out from dried trunks of small tree to surround and protect the plants in the cultivated area.

Housing cluster 3: Since 2008, the residents of 4 out of the 6 houses overlooking the 'smallest' semi-public shared space among the five housing clusters have embarked on UA activities (Fig. 5a). Seemingly, the active alleyways hindered the ability of two of the houses overlooking the semi-private space to embark on UA activities as this would block the passageway through the space. Noticeably, still one of the two houses has planted a line of palm trees by its fence, while the two houses directly overlooking the

Fig. 3 **a** Chronological development of the residents-led UA activities in housing cluster 1 from 2008 until 2023 (*Source* Google Earth Pro, 2023). **b** Observed palm trees and lavender flowers planted by residents in the semi-public spaces in housing cluster 1 (*Source* The Authors)

Fig. 4 **a** Chronological development of the residents-led UA activities in housing cluster 2 from 2008 until 2023 (*Source* Google Earth Pro, 2023), **b** Observed palm trees and other ornamental plants, conventional irrigation system through water taps, and wooden fences in the semi-public spaces of housing cluster 2 *Source*: The Author

alleyways have partially encroached into the alleyways leaving just a limited space for pedestrian (Fig. 5a). The field visits revealed intensively planted palm trees in the developed areas within the shared semi-public space of this housing cluster. Also, various types of light-materials fences defining the boundaries of the cultivated areas were observed including shrubs, steel supports and meshes, and fabrics. The steel mesh of one of the fences was covered by climbing plants apparently for privacy (Fig. 5b).

Housing cluster 4: In 2008 the residents of only one house out of the 8 houses overlooking the semi-public shared space limitedly started some UA activities. This number jumped to 7 houses by 2013 and continued as such until 2023 (Fig. 6a). The 7 individually cultivated areas have increased in size over the years but the areas of them were obviously affected by the presence of the alleyways especially in 2023. This limitation and the moderate overall size of the semi-public space in this housing cluster are expected to restrict future expansion of the cultivated areas. Besides palm trees plantation, as in all visited housing clusters, other types of trees and various crops were also observed. Again, light-materials such as wooden sticks, were used to define the individually cultivated areas. In addition, the conventional irrigation system via domestic water taps was also used in the visited areas (Fig. 6b).

Housing cluster 5: In 2008 only 1 of the 10 houses sharing the semi-public open space of this housing cluster was strongly involved in individualistic UA activities. But, by 2013 five houses were intensely embarked on UA activities as well with obvious conventional agreement among neighbors about the boundaries of the individual UA area allocated for each house (Fig. 7a). For one house, the UA activities have significantly decreased in 2018 and 2023, most likely due to the reuse of the alleyways which were previously fully concealed by a cultivated area in 2013 (Fig. 7a). Like the other housing clusters, the field visits revealed intensive palm trees plantation in this housing cluster, along with some other crops and decorative flowers (Fig. 7b).

4.2 Interviews Analysis

The interviews conducted with the 20 families who have been directly involved in the observed UA activities in the five selected housing clusters in Al Dhaher New urban community have helped clarify the issues that would not be possible to reveal through the conducted map analysis or field observations, as mentioned above. The qualitative responses of the interviewed residents to the 8 asked questions in the semi-structured interviews were briefly concluded and their various responses were categorized into shared themes and then quantitatively presented in a form of pie-charts for ease of visualization. In their responses to the first question about "What is the purpose of your UA activities in the shared semi-public open spaces of your housing cluster?", more than half of the interviewees (12/20) indicated that the

Fig. 5 **a** Chronological development of the residents-led UA activities in housing cluster 3 from 2008 until 2023 (*Source* Google Earth Pro, 2023), **b**. Observed palm trees intensively planted within the housing cluster with fences developed for the cultivated areas made of diverse materials including shrubs, fabric, and steel mesh covered with plants. *Source* The Authors

purpose of their UA activities is to simply grow crops. Six out of the 20 interviewed residents mentioned that they cultivated ornamentation flowers for the sake of beautification and decoration of these open spaces to be used as house gardens. In addition, three other purposes were mentioned, each by only one interviewee as follows: to control dust, to purify air, and to take part in 'saving the environment' (Fig. 8).

The second question was "How did you define the location and size of your developed UA area?". Eight of the 20 interviewed residents responded that they followed no 'obvious' rules as the semi-public open space for them is a mere wide 'empty area'. Meanwhile, 5 out of the 20 interviewed residents planed for their needed UA area to be suitable for the subdivisions of different crops they intend to grow. A few of them selected the areas of their UA activities according to the availability of needed agriculture recourses such as water resources (3/20 interviewees), presence of fertile soil (1/20 interviewees), and being less exposed to direct sun (1/20 interviewees). Two out of the 20 interviewed residents answered that Al Ain Municipality inspectors guided them about the 'conventional' (rather than official) restrictions that control the size and location of the individual UA areas for each house, mainly to be reasonable and to match the local practice of other houses in the area (Fig. 9).

In their responses to the third question about if they have carried out the UA activities on their own as a family or needed to hire 'professional' farmers to help them, 7 out of the 20 interviewees mentioned that they carried the UA activities on their own and did not ask for professional help. Still, half of the respondents mentioned that they got 'partial' help from hired farmers. Meanwhile, 3 out of the 20 interviewees mentioned that they hired a farmer to help them in the UA activities. One of the interviewed residents was a horticulture student and he put his learnt knowledge in practice to sufficiently accomplish the UA activities (Fig. 10).

The fourth question was "What type of vegetation do you usually grow? and why?". The interviewed families stated that they have been growing various types of vegetation in their developed urban farms. Palm trees are the most common plants that residents grow as mentioned by 14 out of the 20 interviewees. This was followed by leafy vegetables such as watercress, mint, etc., as indicated by 10 interviewees. Fruits was grown by 6 interviewees, including

Fig. 6 **a** Chronological development of the residents-led UA activities in housing cluster 4 from 2008 until 2023. *Source* Google Earth Pro, 2023, **b** Observed various types of trees and crops, and the domestic water tap irrigation system used in the developed agriculture areas in housing cluster 4 (*Source* The Authors)

mangoes, figs, and pomegranates. Different types of flowers especially jasmine was mentioned by 3 interviewees. While basil was mentioned by only 1 interviewee (Fig. 11a). For the second part of the question, more than half of the interviewees mentioned that they select the types of vegetation based on their houses use needs. Three of the respondents mentioned that they also consider sharing harvested crops with their neighbors when they decide about the vegetation type. Only one of the respondents mentioned that his family think about beautification of the outdoor space so that they decide to grow only flowers. Another one stated that his family usually considers selling some of the produced crops when they decide about the type of vegetation they will grow (Fig. 11b).

The fifth question was about the source of irrigation water. Most of the respondents (18 out of 20) stated that they rely entirely on the domestic water provided by the water supply company in Al Ain. Only one interviewee mentioned that his family also uses house-recycled water, and another one mentioned that his family uses ground water from a water well they dug to irrigate their cultivated area. The sixth question was "Have you ever had a dispute with a neighbor over the practiced UA activities within the shared semi-public area?". None of the interviewed residents mentioned that they had disagreement with neighbors about their conducted UA activities. They refereed the absence of disputes to the mutual respect of the firmly adhered-to local customs and conventions regarding the 'right of precedence' in the UA activities and recognizing the boundaries of each other's cultivated area. The seventh question was "What are the UA related problems that you usually encounter, if any?". Eight of the 20 interviewees complained about the damaging impacts of the harsh weather conditions in Al Ain and the irrigation water scarcity on their grown plants. Another important problem as mentioned by 8 respondents is the agricultural pests. This was followed by a third problem of the inappropriate soil for agriculture as mentioned by 3 respondents. Less significant reported issues are that some stray animals occasionally damage the plants, residents' poor knowledge about UA, and high cost of the domestic water consumption when it is used for irrigation. Meanwhile, 4 out of the

Fig. 7 **a** Chronological development of the residents-led UA activities in housing cluster 5 from 2008 until 2023 (*Source* Google Earth Pro, 2023), **b** Observed intensively planted palm trees in housing cluster 5 along with some decorative flowers and other crops (*Source* The Authors)

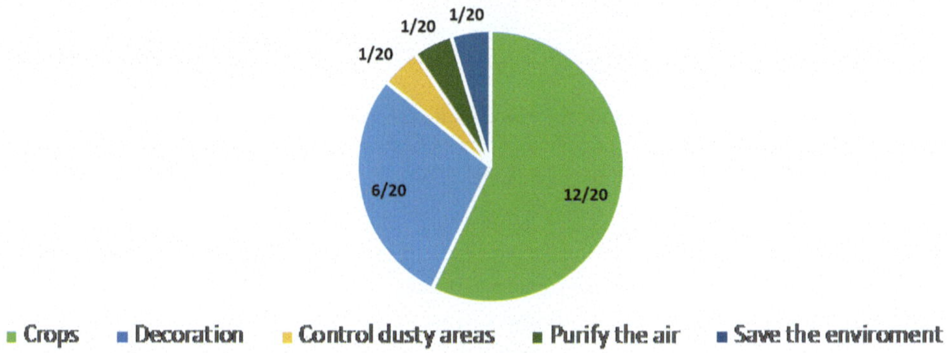

Fig. 8 Purpose of the individualistic residents-led UA activities (*Source* The Authors)

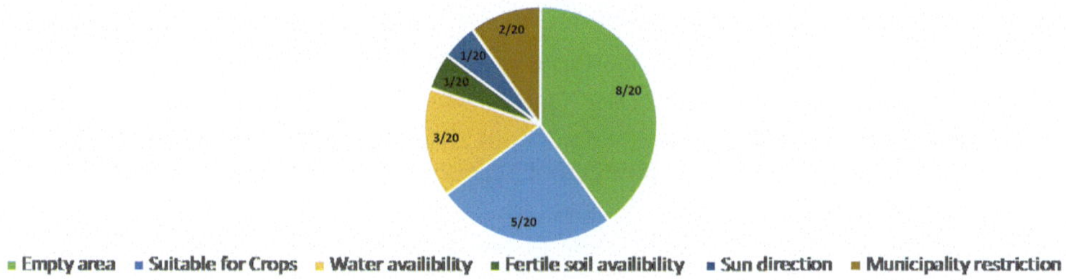

Fig. 9 Defining the location and size of the individually developed UA areas in the semi-public spaces. *Source* The Authors

20 interviewees reported that they did not face any issues (Fig. 12).

In relation to the previous question, the interviewed residents were lastly asked about how in their opinion these problems could be solved. For the harsh weather problem and irrigation water scarcity, one interviewee recommended watering the plants in the early morning or late evening to avoid losing water by evaporation due to the very high temperature, especially in summer. Another interviewee suggested improving the practiced conventional irrigation methods and that the municipality distributes seeds for the types of crops that can grow under such harsh weather conditions. For fighting the devastating impact of the agricultural pest, using insecticides was cautiously suggested by 3 interviewees. They added that the lack of knowledge about UA could be solved by holding professional workshops for

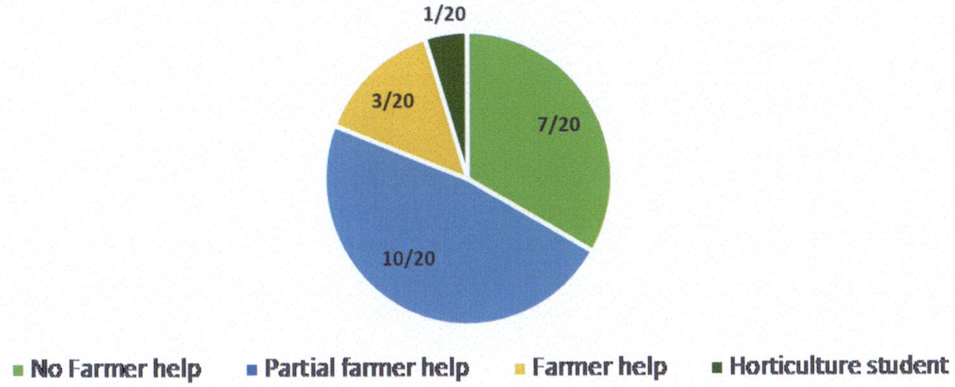

Fig. 10 Utilized professional agriculture help in the UA activities. *Source* The Authors

Fig. 11 **a**, **b** Preferred types of vegetation grown by residents. *Source* The Authors

Fig. 12 The interviewees' defined problems related to their individualistic UA activities. *Source* The Authors

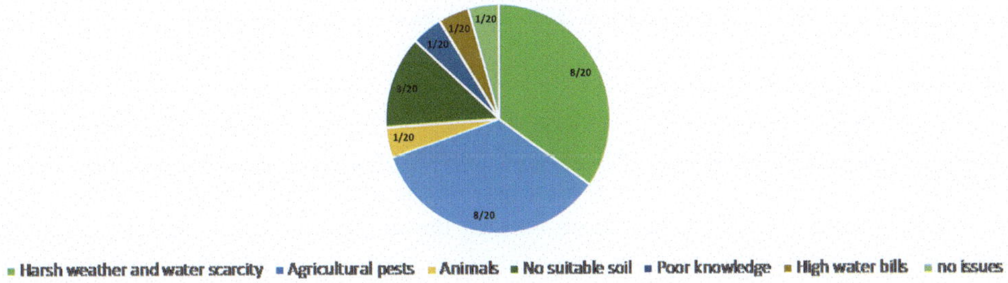

interested residents to raise their awareness and knowledge about UA. As for the high cost of domestic water consumption bills, 3 interviewed residents recommended choosing the right type of plants and trees that do not need much water.

5 Discussion and Conclusions

This pilot study has investigated the individualistic residents-led UA activities in the underdeveloped semi-public open spaces forming housing clusters in Al Dhaher New neighborhood as a representative case study for existing Emirati urban communities in Al Ain city. The aim is to better understand these activities and hence, assess their possible contribution in achieving SDG 15 in the UAE. Map analysis, field observations, and semi-structured interviews with a sample of Al Dhaher New urban community residents, who have been intensively embarked on UA activities, were utilized as the main research investigation tools. The research outcomes might be concluded as follows. First, the individualistic residents-led UA activities were mainly driven by the residents' desire to grow eatable crops especially vegetables and fruits for their own use and to share with their neighbors. The desire to have gardens for beautification of the front and back spaces came second in significance. Second, many residents have shown the ability to undertake the UA activities on their own, still some families sought partial help from professional farmers.

Third, the vast underused and undeveloped semi-public open spaces in the housing clusters resulting from the adopted sprawling urban form designs, have given the chance to the residents to secure land for their UA activities and for many of them to also expand the individually cultivated areas over the years. The residents have usually cultivated the semi-public areas attached to the back and side fences of their housing plots. The size and shape of the cultivated areas, as defined by each house through light-materials, have been affected by the availability of suitable soil for agriculture and the presence of access to essential agriculture resources especially irrigation water. Fourth, while the semi-public open spaces in the studied urban community vary in sizes, the shape and size of the individually developed UA areas were seemingly not much affected by the sizes of these semi-public spaces themselves where most of the houses have developed reasonable areas as urban farms. But on the other hand, the presence of alleyways (*sikkak*), that are linking the semi-public spaces with the main streets, has significantly affected the shape and size of the cultivated areas, and even sometimes eliminated them entirely especially for the houses that are overlooking the alleyways directly.

Fifth, as for the challenges facing the individualistic residents-led UA activities, the first problem is the lack of agricultural knowledge. Indigenous agricultural knowledge is gained over generations by trial and error. Sadly, many of the modern city residents, including those of Al Ain, do not have this knowledge. The second main challenge is the scarcity of the irrigation water that is actually facing all agriculture sectors in the UAE. The widely used conventional irrigation methods in cultivation are extremely inefficient in a country as arid as the UAE, where 60% of the water evaporates before being absorbed by plants (Dakkak, 2022).

When comparing the findings of this pilot study with the individualistic residents-led AU experiences in the developing countries, as discussed in Sect. 2, one can easily find some commonalities as well as substantial differences. Similarities include some close challenges specifically the lack of professional/local authorities help that might lead to the loss of crops and resources. The main difference is that residents-led UA activities in developing countries are mostly considered as means of survival and securing resources for the urban poor (Ittner, 2022; Weltbank & World Bank, 2013), whereas in the UAE residents have usually much easier accessibility to needed resources for undertaking UA including available irrigation water through domestic water supply companies (albeit costly), reasonable financial capability to hire a professional farmer, if needed, and so on.

As an answer to the second question, the research outcomes clearly indicate that the individualistic residents-led UA activities have compensated for the absence of the originally designed, but never developed, green and recreational facilities in the public and semi-public open spaces of their urban community. The on-going growth of the magnitude of the diversified plants and crops produced through the residents' individualistic UA activities have contributed to the increasing green area coverage. Accordingly, these activities have considerably enhanced the urban ecosystem and biodiversity of the urban area. This is supported by the variety of crops, trees, and flowers grown by the residents that all cater for nurturing the natural habitat of the urban community. This certainly would contribute to the UAE's effort for achieving SDG 15.

Based on these findings, some actions are recommended as follows:

- Residents in local urban communities in the UAE should be encouraged to keep undertaking the lead for UA activities in the semi-public open spaces of their housing clusters. So, to facilitate that in the design of the urban form of the local urban communities, the location and size of the areas allocated for the individualistic residents-led UA activities, especially in the semi-public spaces of the housing clusters, should be carefully defined in the design stage to mainly avoid the impeding impact of the alleyways on the ability of the residents of each household to develop their own UA area.
- The problem of the lack of agricultural knowledge of the residents practicing UA in the UAE could be overcome by raising their awareness about UA. Such awareness should not be only directed to local communities, but it should also include the related public sector, the municipalities, and other concerned local authorities. Fortunately, the Abu Dhabi Farmers Services Center (ADFSC) provides training on agricultural best practices. The ADFSC plans the appropriate crops cultivation cycles for each season and allows farmers to submit expressions of interest for which crops they would prefer to grow (Mohammed bin Rashid School of Government, 2018). Such practices need to be extended to UA activities as well.
- As an efficient alternative irrigation system to the conventional inefficient ones, irrigation drip system is considered one of the most effective irrigation systems as it provides drop-by-drop watering directly on roots of the plants through its pipelines and valves. This method could save about 46% of conventionally consumed irrigation water. Still, capital cost of this system compared to the conventional one is high. Also, the drip system

must use only clean and filtered water to avoid clogging (Kader, 2018). So, other sources of irrigation water might be taken into consideration as well. For example, reusing condensed water discharged from air conditioners, especially with their long operational periods due to the very hot temperature in the UAE, could be a partial source for irrigation water with an estimated amount of about 20 to 75 Liter of water per house per day according to the outside weather temperature (Grant, 2021). Currently this condensed water is wasted as it is usually removed by a drip line or hose outside the house.

While the outcomes of this pilot study have explored the individualistic residents-led UA activities that could help achieve SDG 15: Life on Land, but these UA activities should be studied on a larger scale beyond the limited semi-public open spaces within housing clusters to include all types of urban open spaces in the local urban communities in the UAE. Also, there is a need to investigate the attitudes of the residents to embark on wider community-led activities in the public and semi-public open spaces in their urban communities, side-by-side with their current individualistic UA activities. This would make UA activities more rewarding for both the urban ecosystem and the local economy alike. Finally, exploring the facilitating rules that other stakeholders could play in supporting the residents-led UA activities should also be included in the agenda for future UA studies, especially the enabling roles of municipalities and local authorities.

Acknowledgements The authors would like to thank the Office of Research and Sponsored Projects at the United Arab Emirates University for funding this research under the SDG Fund Program in 2022-23 Academic Year—Grant Number: G00004090.

Appendix 1:Interview Script

Thank you for giving me your time. My name is …., Research Assistant at the United Arab Emirates University, Department of Architectural Engineering. This study is part of a research project that aims to investigate the residents-led Urban Agriculture activities and its potential in contributing to enhancing SDG 15 in 'Al Daher New' urban community. Your responses to the posed questions about your own experience, as a family, in Al Dhaher New urban community will help us better understand the mechanisms of these UA activities. Kindly note that the information you provide in this interview will remain confidential and will be exclusively used for the purpose of this research. The interview is expected to last between 10–15 min.

No	Questions	الاسئلة
1	What is the purpose of your UA activities in the shared semi-public spaces of your housing cluster?	ما الهدف من قيامك بالزراعة في المساحة المشتركة بين المنازل؟
2	How did you define the location and size of your developed UA area?	كيف قمت بتحديد المنطقة والمساحة المناسبة وللزراعة؟
3	Do you usually carry out the UA activities on your own (as a family), or do you usually hire professional farmers to help you?	هل تقوم (أنت وعائلتك) عادة بأعمال الزراعة أم تستعين بمزارعين محترفين للمساعدة في هذه الأعمال؟
4	What type of vegetation do you usually grow? and why?	ما نوع المحاصيل التي تقوم عادة بزراعتها؟ ولماذا تقوم بزراعة هذه المحاصيل تحديدا؟ (لاستخدامها بالمنزل \ لبيعها\...)
5	From where do you get the water for irrigation?	هل هنالك مياه مخصصة لأنشطة الري؟ من اين تحصل عليها؟
6	Have you ever had a dispute with a neighbor over the practiced UA activities within the shared semi-public area?	هل قد حصلت مشكلة بينك وبين الجيران بسبب الزراعة بالأماكن العامة المشتركة بينكم؟
7	What are the UA related problems that you usually encounter, if any?	ما هي المشاكل التي تواجهها عادة اثناء عملية الزراعة؟
8	In your opinion, how these problems could be solved?	برأيك كيف يمكن حل هذه المشاكل؟
Finally, would you like to add anything regarding the addressed topics in the previous 8 questions?		وأخيرا هل تود إضافة أي شيء متعلق الموضوع ولم أسأله لك في المقابلة؟

References

Armstrong, D. (2000). A survey of community gardens in upstate New York: Implications for health promotion and community development. *Health and Place, 6*, 319–327.

Binder, A. (2014). *Exploring perceptions of guerrilla practices. Guerrilla Gardening in Manchester*. University of Salzburg. https://www.urbanallotments.eu/fileadmin/uag/media/STSM/Binder_shortreportSTSM_final.pdf

Center for Science Education. (2023, June 23). *Urban heat Islands*. SCIED. https://scied.ucar.edu/learning-zone/climate-change-impacts/urban-heat-islands#:~:text=An%20urban%20heat%20island%20(UHI,the%20surrounding%20natural%20land%20cover

Clinton, N., Stuhlmacher, M., Miles, A., Aragon, N., Wagner, M., Georgescu, M., Herwig, C., & Gong; P. (2018). A global geospatial ecosystem services estimate of urban agriculture, *Earth's Future*, 6(1), 40–60. https://doi.org/10.1002/2017EF000536

Dakkak, A. (2022, June 23). *Irrigation systems in the United Arab Emirates*. EcoMENA. https://www.ecomena.org/irrigation-systems-in-uae/

Erwin, D. (2022). *Urban and peri-UA case studies—Overview, conclusions and recommendations. An annex to Urban and peri-UA—From production to food systems*. FAO.

Interreg Europe. (2020). *Urban ecosystems: The importance of green infrastructure and nature-based solutions for the development of sustainable cities. A policy brief from the policy learning platform on environment and resource efficiency*. European Regional Development Fund.

Evans, D., Falag´an, N., Hardman, C., Kourmpetli, S., Liu d, L., Mead, B., & Davies, J. (2022). Ecosystem service delivery by urban agriculture and green infrastructure—A systematic review. *Ecosystem Services, 54*, 101405. https://doi.org/10.1016/j.ecoser.2022.101405

Express and Star. (2012, April 2022). *Guerilla gardeners clean up at plots*. Express & Star.Com. https://www.expressandstar.com/news/2012/08/20/guerilla-gardeners-clean-up-at-plots/

Freeman, D. (1991). *City of farmers: Informal urban agriculture in the open spaces of Nairobi, Kenya*. McGill-Queen's Press.

Galal Ahmed, K., & Alipour, S. (2021). More dense but less walkable: The impact of macroscale walkability indicators on recent designs of Emirati urban communities. *City, Territory and Architecture, 8*(12). https://doi.org/10.1186/s40410-021-00140-3.

Google Earth Pro (2023) *Al Dhaher New urban community*. Google Earth Pro https://earth.google.com/web/@24.0793068,55.86494646,1933.52541715a,0d,35y,89.9534h,5.3356t,0.0002r?utm_source=earth7&utm_campaign=vine&hl=en

Grant, B. (2021, April 22). *AC condensation for plants: Is irrigating with AC water safe*. Gardening Know How. https://www.gardeningknowhow.com/garden-how-to/watering/is-irrigating-with-ac-water-safe.htm.

Guneralp, B., Perlstein, A., & Seto, K. (2015). Balancing urban growth and ecological conservation: A challenge for planning and governance in China. *Ambio, 44*, 532–543. https://doi.org/10.1007/s13280-015-0625-0

Hardman, M., St Clair, R., Adams, D., Armitage, R., Larkham, P., & Sherriff, G. (2018). UA: Evaluating informal and formal practices. *Northwest Geography, 18*(1), 1–11.

Ittner, I. (2022, June 23). *Informal farming in the city*. D+C. https://www.dandc.eu/en/article/abidjan-cote-divoire-many-farms-face-uncertain-future

Kader, B. (2018, April 22). New Abu Dhabi irrigation system saves 46% water. *Gulf News*. https://gulfnews.com/uae/environment/new-abu-dhabi-irrigation-system-saves-46-water-1.1338984.

Martin, G. (2016, June 21). *Food, nature and community: Urban agriculture's synergies with ecological and social sustainability*. IAFOR. https://think.iafor.org/food-nature-community-urban-agricultures-synergies-ecological-social-sustainability/

Mohammed bin Rashid School of Government. (2018, Jan 23). *Advancing food security in the UAE*. MBRSG. https://mbrsgcdn.azureedge.net/cmsstorage/mbrsg/files/85/859ddec7-f5ed-48dd-99dd-4e1b8f326112.pdf

Orsini, F., & D'Ostuni, M. (2022). The important roles of urban agriculture. *Frontiers Young Minds, 10*, 701688. https://doi.org/10.3389/frym.2022.701688

Rabin, J. (2022, June 23). *7 Benefits of urban agriculture*. Microhabitat. https://microhabitat.ca/7-benefits-of-urban-agriculture/

Ritchie, H., Roser, M., & Rosado, P. (2020, April 22). *CO_2 and greenhouse gas emissions*. Our World in Data. https://ourworldindata.org/co2-and-other-greenhouse-gas-emissions.

Steele, K. (2017). *Urban farming workbook*. Elemental Group.

The Official Portal of the SDGs in the UAE. (2022, June 23). Goal 15: Sustainably manage forests, combat desertification, halt and reverse land degradation, halt biodiversity loss. UAESDGs. https://uaesdgs.ae/en/goals/life-on-land

The URBES Project (2014). *Urban agriculture: landscapes connecting people, food and biodiversity—Factsheet 7*. Stockholm Resilience Centre: Stockholm.

Tractorgyan. (2022, June 23). *What is commercial farming? Features, forms and advantages*. Tractorgyan. https://tractorgyan.com/tractor-industry-news-blogs/898/features-advantages-and-forms-of-commercial-farming

UNEP (2022, June 23). GOAL 15: Life on land. UNEP. https://www.unep.org/explore-topics/sustainable-development-goals/why-do-sustainable-development-goals-matter/goal-15.

Urban Development and Resilience Unit—World Bank. (2013, April 22). Urban Agriculture: findings from four city case studies. World Bank. https://documents1.worldbank.org/curated/en/434431468331834592/pdf/807590NWP0UDS00Box0379817B00PUBLIC0.pdf

Wakefield, S., Yeudall, F., Taron, C., Reynolds, J., & Skinner, A. (2007). Growing urban health: Community gardening in South-East Toronto. *Health Promotion International, 22*(2), 92–101.

Weltbank and World Bank. (2013). *UA: Findings from four city case studies*. World Bank.

Environmental Solutions and Innovations

Using Phytoremediation to Reduce Algal Blooms in Campus Green–Blue Space: A Research-Through-Design Study

Vipavee Surinseng, Yupa Chromkaew, Chulalux Wanitchayapaisit, Ekachai Yaipimol, Nadchawan Charoenlertthanakit, and Pongsakorn Suppakittpaisarn

Abstract

This research investigated the use of phytoremediation to mitigate algal blooms within campus green–blue spaces, taking Ang-Kaew University, Chiang Mai, Thailand, as a case study. In warmer climates, algal bloom can occur due to non-point source pollutants and stagnating water, thus decreasing the health and ecological benefits these spaces provide. Through a research-through-design methodology, which is a landscape architectural approach, this study explores the implementation of phytoremediation strategies to reduce water pollution. The processes of planting site selection, plant selection, and design implementation are systematically recorded, providing insights for future studies. Preliminary tests for plant survivability and potential water treatment efficacy were explored. However, long-term studies are required to conclude the efficacy of the design. This study pioneers the application of phytoremediation in green–blue spaces in subtropical campuses through research-through-design, encouraging further research and longitudinal studies in the field.

Keywords

Phytoremediation · Research-through-design · Green-blue spaces · Water sensitive urban design · Green stormwater infrastructure

V. Surinseng · Y. Chromkaew · C. Wanitchayapaisit · E. Yaipimol · N. Charoenlertthanakit · P. Suppakittpaisarn (✉)
Department of Plant and Soil Sciences, Faculty of Agriculture, Chiang Mai University, Chiang Mai, Thailand
e-mail: pongsakorn.sup@cmu.ac.th

V. Surinseng
e-mail: vipavee.s@cmu.ac.th

Y. Chromkaew
e-mail: yupa.c@cmu.ac.th

C. Wanitchayapaisit
e-mail: chulalux.w@cmu.ac.th

E. Yaipimol
e-mail: ekachai.y@cmu.ac.th

N. Charoenlertthanakit
e-mail: mrsnadchawan.c@cmu.ac.th

C. Wanitchayapaisit · N. Charoenlertthanakit
Faculty of Engineering, Chiang Mai University, Chiang Mai, Thailand

1 Introduction

1.1 Eutrophication and Phytoremediation

Eutrophication is a phenomenon in which excessive nutrients dissolve into water bodies. It is a widespread environmental issue that often leads to harmful algal blooms and the degradation of aquatic ecosystems (Anderson et al., 2002). Non-point source pollutants, such as agricultural runoff and urban stormwater, contribute to eutrophication by introducing high levels of nitrogen and phosphorus into water bodies (Jiang et al., 2007). Eutrophication can be particularly problematic in warmer climates where stagnant water exacerbates the growth of algal blooms (Erell et al., 2012).

Phytoremediation, a sustainable and cost-effective method, can combat eutrophication in aquatic environments (Zhang et al., 2007). Phytoremediation involves removing, stabilizing, and transforming pollutants from water and soil using plants (Kennen & Kirkwood, 2015). Specifically, aquatic plants play a crucial role in absorbing excess nutrients and contaminants. The potential of phytoremediation to mitigate eutrophication and algal blooms has gained growing significant attention due to its environmental benefits and adaptability (Ansari et al., 2014; Kennen & Kirkwood, 2015; Rugh, 2001; Zhang et al., 2007).

In the current trend of phytoremediation and eutrophication, most recent studies tend to explore plant species that

B. He et al. (eds.), *Urban Climate Change Adaptation*, Advances in Science, Technology & Innovation,
https://doi.org/10.1007/978-3-031-65088-8_18

performed best in nutrient removal from the ecosystem in various time scales (Dean et al., 2022; El-Sheekh et al., 2021; Wang et al., 2022). However, there is a lack of consideration in the synchronistic process of biodiverse phytoremediation systems, which are more akin to the natural process. Most experiments only focus on which plants or planting media perform best in different matrices (Dietz & Clausen, 2005; Venvik & Boogaard, 2020). However, the actual natural ecosystems consist of different levels and species of plants and soil organisms working together to achieve a series of well-rounded regulating performance goals regarding stormwater management (Jaber et al., 2012; Wanitchayapaisit et al., 2022).

1.2 Landscape Design and Phytoremediation

Landscape design plays a pivotal role in the successful implementation of phytoremediation strategies. Integrating phytoremediation into green–blue spaces requires a balance between site characteristics, plant selection, and aesthetic aspects (Banjerdkij et al., 2013; Davis et al., 2009; Hunt et al., 2015). Landscape architects and designers have a vital role in bridging the gap between scientific research and practical implementation, ensuring that phytoremediation projects are not only effective in pollutant removal but also visually appealing and compatible with their surroundings (Davis et al., 2009; Lord et al., 2015).

Studies have demonstrated that well-designed green–blue spaces can enhance ecosystem services and human well-being (Shanahan et al., 2015; Suppakittpaisarn et al., 2017). By incorporating phytoremediation into landscape design, green–blue spaces can act as multifunctional systems, providing recreational spaces while concurrently serving as ecological infrastructure for water quality improvement (see Fig. 1) (Suppakittpaisarn et al., 2019, 2020). Landscape architects play a critical role in selecting appropriate plant species, determining suitable locations, and creating aesthetically pleasing designs that optimize the efficiency and effectiveness of phytoremediation projects (Duzenli et al., 2017; Suppakittpaisarn, 2021).

Landscape architects, biologists, engineers, and other related fields have started to develop knowledge sets, guides, and examples of how biodiverse phytoremediation landscapes could be designed (Davar et al., 2021; Nissim & Labrecque, 2021; Podhajska et al., 2023; Shen et al., 2022). However, comparing these guides is difficult due to the differences in site-level factors, such as area sizes, stormwater quantity, and usage, and regional characteristics, such as climate, rainfall, and national policies. Moreover, there is a lack of case studies towards the global south, where countries may experience issues of flooding, eutrophication, and non-point source pollution in stronger ways than the developed world due to global climate change and limited resources (Haque, 2021).

1.3 The Need for Links Between Researchers and Designers

The successful implementation of phytoremediation in green–blue spaces necessitates strong links between

Fig. 1 Examples of different urban landscapes that can address phytoremediation by slowing down water and absorbing excess nutrients (image produced by authors)

researchers and designers. Collaboration between these two disciplines is crucial to address the complex challenges of water quality improvement and landscape design (Lenzholzer et al., 2013; Milburn et al., 2003; Nijhuis & Bobbink, 2012). While researchers provide scientific knowledge on pollutant removal efficiency and plant selection, landscape architects offer expertise in site analysis, spatial planning, and visual integration (Suppakitpaisarn, 2021).

Research-through-design, as an interdisciplinary approach, facilitates effective communication and cooperation between researchers and designers. This methodology allows landscape architects to actively engage with scientific findings and incorporate them into design decisions (Deming & Swaffield, 2011). By involving researchers in the design process, landscape architects can make evidence-based choices, maximizing the ecological and aesthetic benefits of phytoremediation projects (Nijhuis & Bobbink, 2012). Moreover, the collaboration encourages continuous learning and adaptation, fostering innovation and sustainability in green–blue space design (Brink et al., 2017; Nijhuis & Bobbink, 2012).

Research and design connections are still a new and developing trend. Recently, designers and researchers have worked to develop guides and toolkits to work together, such as the collaborations between Sasaki Associates, USA, and the local universities (Lagro et al., 2023). In addition, researchers in Thailand worked with designers in developing phytoremediation strategies for Tha Phae Gate, Chiang Mai, Thailand (Wanitchayapaisit et al., 2023). However, researchers, planners, designers, and decision-makers still need more cases to build a consistent work of discipline and examples to establish a systematic way of going forward (International Federation of Landscape Architects, 2022; Lagro, 2023; Lagro et al., 2023; Suppakitpaisarn, 2021; Wanitchayapaisit et al., 2023).

1.4 Climate Change and Environmental Sustainability

With the recent trends of global climate change, there are possibilities that eutrophication can become worse. Unpredictable global climate change patterns and prolonged drought can extend the period of the algal bloom, and local species might be less resilient due to the changing conditions in their surrounding (Anderson et al., 2002; Nazari-Sharabian et al., 2018). For landscape designers and researchers, the knowledge accumulated regarding landscape design and environmental management may change based on the new climate patterns and flux in rainfall, introduced species, and changes in human behaviors due to the changes in seasonal patterns (Suppakitpaisarn & Charoenlertthanakit, 2017). Eutrophication, especially if it persists in large parts of urban landscapes, is an issues that must be tackled in the interests of ecological sustainability. Advanced solutions in biodiverse phytoremediation to mitigate eutrophication can contribute to the following UN's Sustainable Development Goals items (Hák et al., 2016).

- Goal 11: Sustainable Cities and Community—well-designed urban landscapes that address ecological issues can lead towards more sustainable cities and communities.
- Goal 15: Life of Land—preserving and re-introducing biodiversity into stormwater management research can lead to higher biodiversity in the terrestrial environment.
- Goal 17: Global Partnership—establishing case studies in different parts of the world can inspire partnerships between researchers and designers across the globe.

1.5 Research Objectives

Building on the research gap of missing examples of biodiverse phytoremediation landscapes that address eutrophication in tropical climates, our main research objective is to explore the design and implementation processes of a biodiverse phytoremediation system on a field study in Thailand. It aimed to create a collaboration between researchers and designers through research-through-design and field monitoring process to advance the understanding of how effective the landscape might be.

This implementation could serve as a case study to build a consistent guide for phytoremediation system design towards Thailand and places with similar climates.

2 Methodology

2.1 Site Selection

Ang Kaew is one of the two reservoirs used for the main campus of Chiang Mai University. The water is used to produce drinking and utility water for the student dormitory and Chiang Mai campus. It is a large reservoir, with 89,524 sqm of water surface which holds up to 288,000 cubic meter of water at its full capacity. The landscape around Ang Kaew also serves as a recreational green space for campus students. During the summers, Ang Kaew is facing issues with eutrophication, which may harm the students and pets using the spaces, as well as increase the water treatment costs for usage (see Fig. 2). A preliminary investigation suggests that the issue may stem from non-point

Fig. 2 Eutrophication of Ang Kaew water (image produced by authors)

source pollution and nutrients from the campus and nearby communities. Thus, there is a need for a phytoremediation landscape to help with the water quality issue.

2.2 Research-Through-Design Approach

This study used the research-through-design approach. It is a widely used approach in landscape architecture research that applies adductive reasoning to answer the research questions. While this approach is still developing and has some limitations, including generalizability, it is an adduction research that could spark ideas towards solutions as well as generate the process of thinking towards future solutions (Deming & Swaffield, 2011; Wanitchayapaisit et al., 2022). The research method involves three stages: data analysis, design, monitoring and evaluation.

- Data collection and analysis: the researchers conducted the data analysis of the site, tested the water, and joined the stakeholder meetings, including CMU administrative officers, campus engineers, and representatives from the Faculty of Science, to understand the requirements of the site. Water volume, depth, and quality were collected from seven spots of the reservoir, two from the outlets and five from the inner parts, to understand the morphology of the reservoir. Soil assessment by landscape experts, as well as a typical landscape scale site analysis, were also performed to identify the suitable locations for the installation.
- Design: Once the suitable phytoremediation sites were identified, the landscape architects and designers selected suitable plants and an installation plan based on the nutrient and carbon sequestration capacity obtained from the knowledge base in Thailand. Two landscape architects and a planting design expert worked together to develop a design by creating a series of schematics, testing the ideas, discussing the implementation, and forming a collective agreement. The design also underwent consideration for aesthetics, land use, and biodiversity to ensure other facets of sustainability.
- Monitoring and evaluation: The water samples were collected twice a month from February to August to test for Nitrogen, Phosphorus, and Biochemical Oxygen Demand (BOD). However, due to a weather event on one of the scheduled collection dates, the researchers did not

collect one point of data, leaving 11 times of data collection. The researchers observed and recorded species survivability and coverage using photographs and visual estimation.

3 Results

The results are discussed in four sections following the structure of the research-through-design methods. First, the analysis guides the selection of the site of implementation. Second, the design section explaines the design from conceptual drawing to planting design. Third, the monitoring and evaluation section describes the results of the implementation after six months. Finally, we showed the workflow between the researchers and the designers for future replications and modifications.

3.1 Analysis and Implementation Locations:

First, the site analysis identified the bank conditions around the reservoir into four types as follows (see Fig. 3).

- Type A: Steep and eroded bank covered with mowed grasses.
- Type B: Steep bank covered with mowed grasses.
- Type C: Shallow bank covered with sparse vegetation.
- Type D: Shallow bank covered with dense spontaneous vegetation.

Second, the designers decided on four different design locations from Type C. The locations around the reservoir were selected due to the depth of the water appropriate for the riparian landscape, the existing vegetation, and soil suitability. The planting spaces were approximately 712 square meters in total (see Fig. 4).

3.2 Design

- Design plan: The research designers created the design under the concept 'Connection between Land and Water'. Because the design of Ang Kaw's recreational spaces has already been implemented, the designers focused on filling the available spaces and enhancing the experiences with riparian planting designs at the linkage between the water from the reservoir and the landscape around it. The illustration of the design can be seen in Fig. 5.

- Planting design: While the designers and researchers originally aimed for native species, the construction team could not find sufficient native plants in the market. Instead, we aimed for plants that are well-adapted to the environment, have phytoremediative qualities, and are

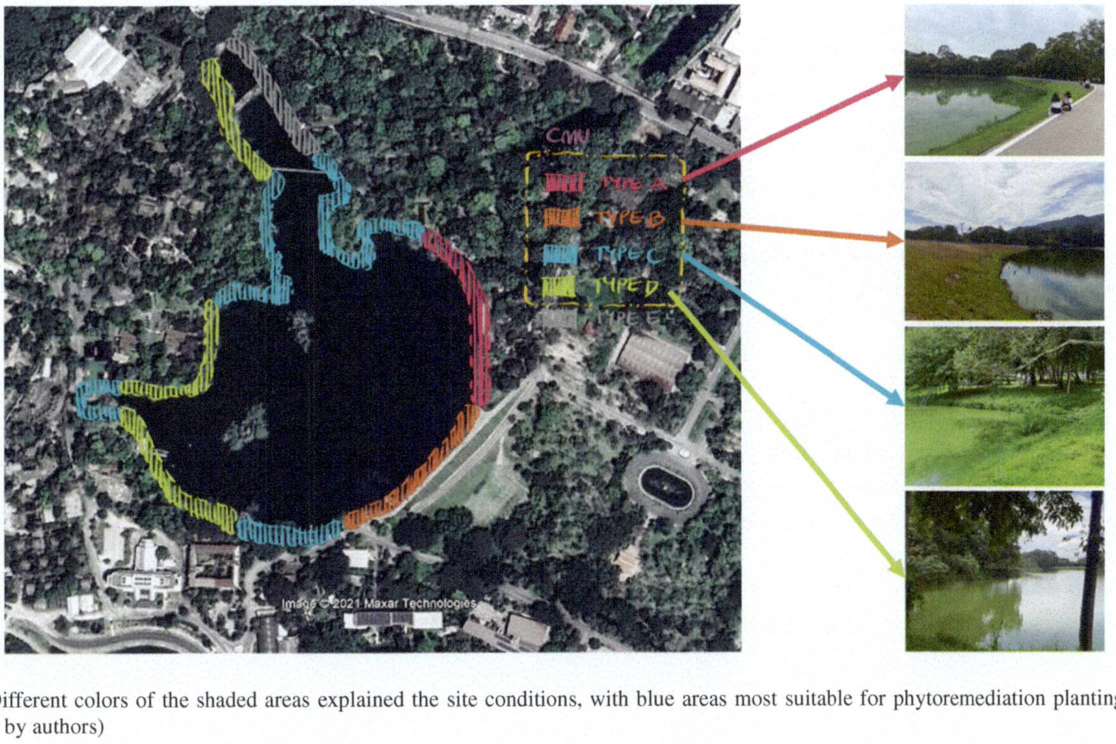

Fig. 3 Different colors of the shaded areas explained the site conditions, with blue areas most suitable for phytoremediation planting (image produced by authors)

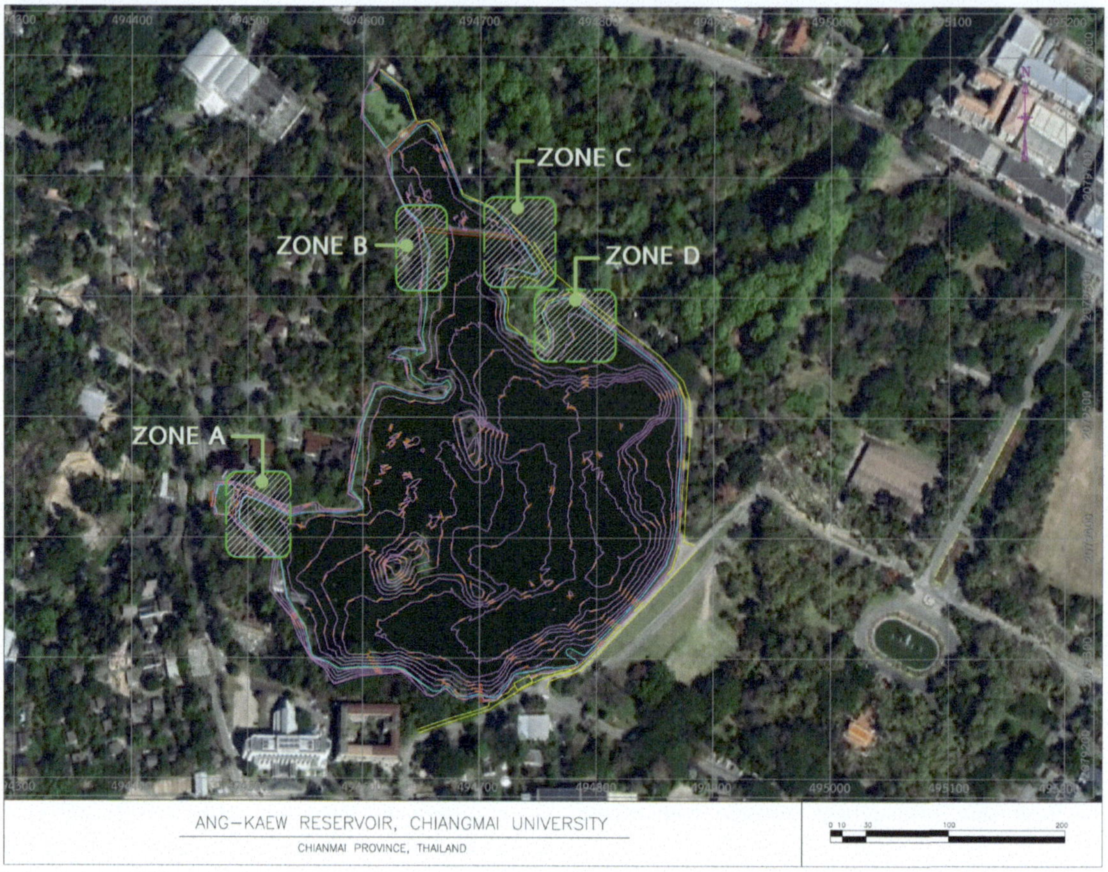

Fig. 4 Implementation locations, which are done in four different zones (image produced by authors)

available in the planting market. The designers focused on plants with high stem porosity for effectiveness in nutrient absorption and plants that have deep roots and can stabilize riparian landscapes, reducing soil erosion into the water. The plants were also tolerant of both wet and dry conditions to ensure their survival across the seasons. The plans of the design can be seen in Figs. 6, 7, 8 and 9. The species and reasoning were placed in Table 1. The installation and the site after installation can be seen in Figs. 10, 11 and 12.

- Design implementation: The design implementation was conducted after the approval of the Chiang Mai University in early February, which was the middle of the dry season. The researchers and designers installed the work with groups of construction workers and horticulturalists, with the lead designer overseeing the installation to ensure it went according to the design and calculation.

3.3 Monitoring and Evaluation

- Plant survivability: All 712 square meters of the planting spaces installed at the beginning of the study were still covered with the introduced vegetation. However, the population within the system have greatly shifted. Only 10 of 20 species survived but expanded to cover the entire the designed spaces. The species included the following: *E. cordifolius, C. esculenta var. aquatilis, T. geniculata L., Heliconia psittacorum L.f., Canna spp., P. amaryllifolious, C. fragrans, Dieffenbachia spp., and M. hastata.*

The researchers suggested that these plants survived because they were designed to be placed deeper in the banks, protected from wind and water currents; thus, they established faster and expanded until they filled the space and succeeded over other species.

Fig. 5 Design illustration of one of the planting locations, showing the plant species selection in its context (image produced by authors)

- Water quality changes: The mean water quality was compared before and after the implementation for descriptive statistics. The sampled water had a lower biochemical oxygen demand (BOD) value than before planting. Furthermore, the samples had lower Phosphorus values. However, they had slightly higher Nitrogen values. The results are presented in Table 2. This means that while overall water quality may have improved, some nutrients may not be reduced by the planting in this case.

3.4 Researcher-Designer Workflow

In this section, the designer-researcher workflow within this study has been summarized. At the initiation phase, they visit the site and decide what they aim to achieve. During the investigation, researchers look into the water's chemical and biological properties, while designers examine the land around the water. In planning and design phases, they used their expertise to figure out where to place things and how to visually represent their ideas, choosing plants that clean

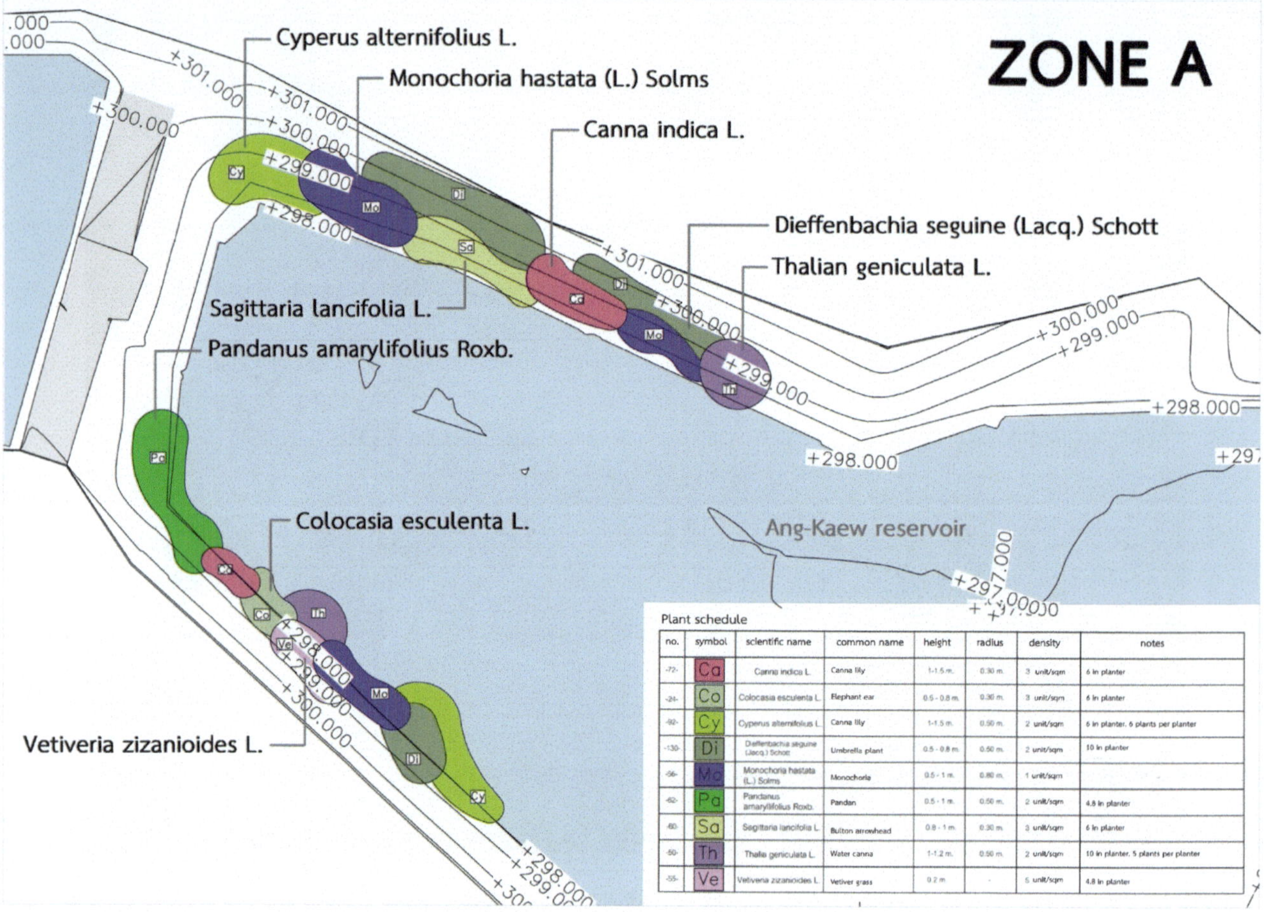

Fig. 6 Planting design in Zone A (image produced by authors)

the water and deciding where they should go. As the project progresses, they install the solutions, monitor how well they are working, and make changes based on what they learn to keep improving the water quality. The detailed workflow between the researchers and designers are represented in Table 3.

4 Discussion

The findings of this research demonstrate the promising potential of phytoremediation as an effective strategy for mitigating algal blooms and improving water quality in green–blue spaces. The research-through-design

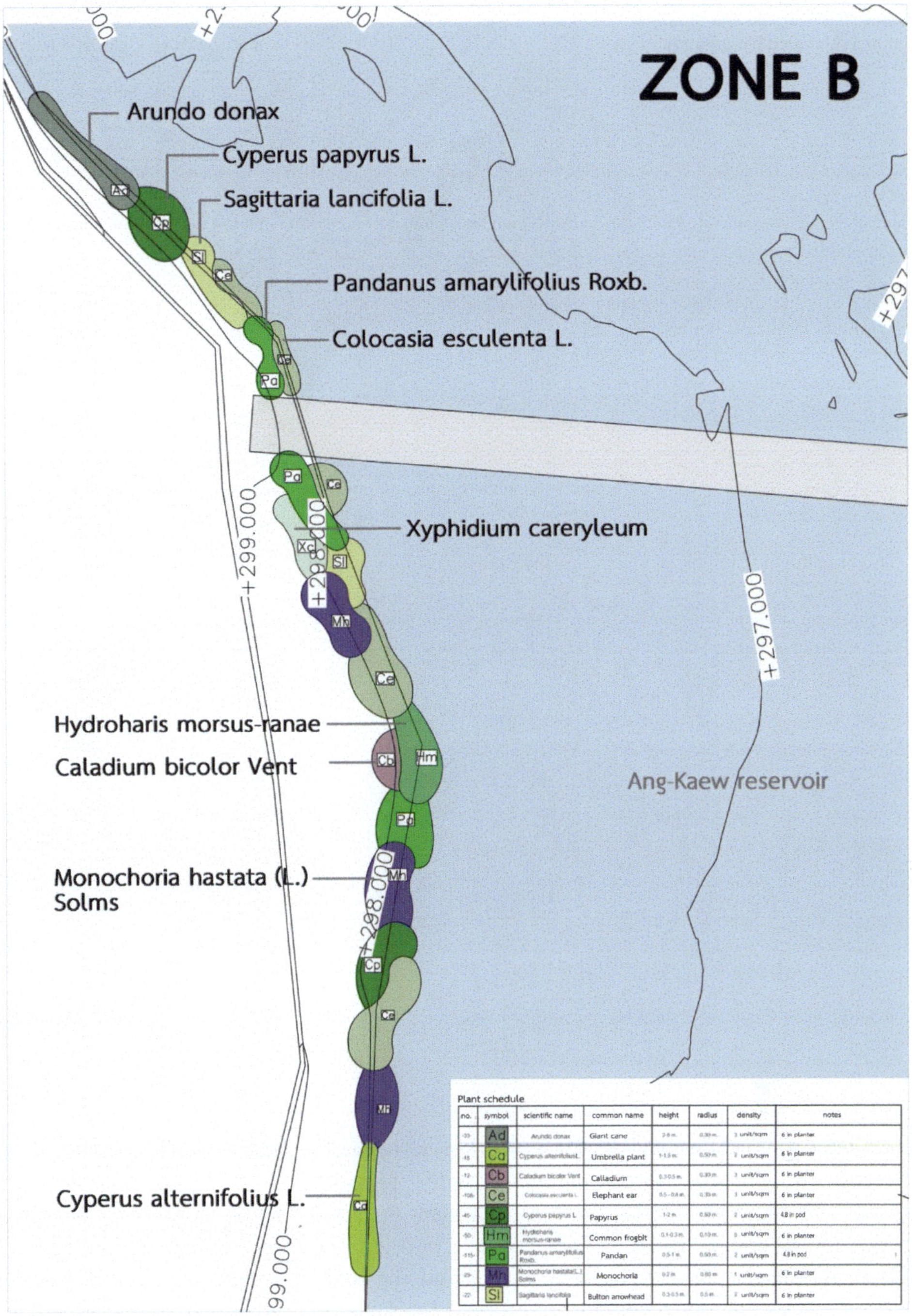

Fig. 7 Planting design in Zone B (image produced by authors)

methodology proved to be a valuable approach, providing a systematic way to analyze the site, select appropriate plant species, and design the implementation of phytoremediation strategies.

The observed improvements in water quality after the phytoremediation efforts included reduced levels of biochemical oxygen demand (BOD) and phosphorus, which agreed with the previous studies regarding

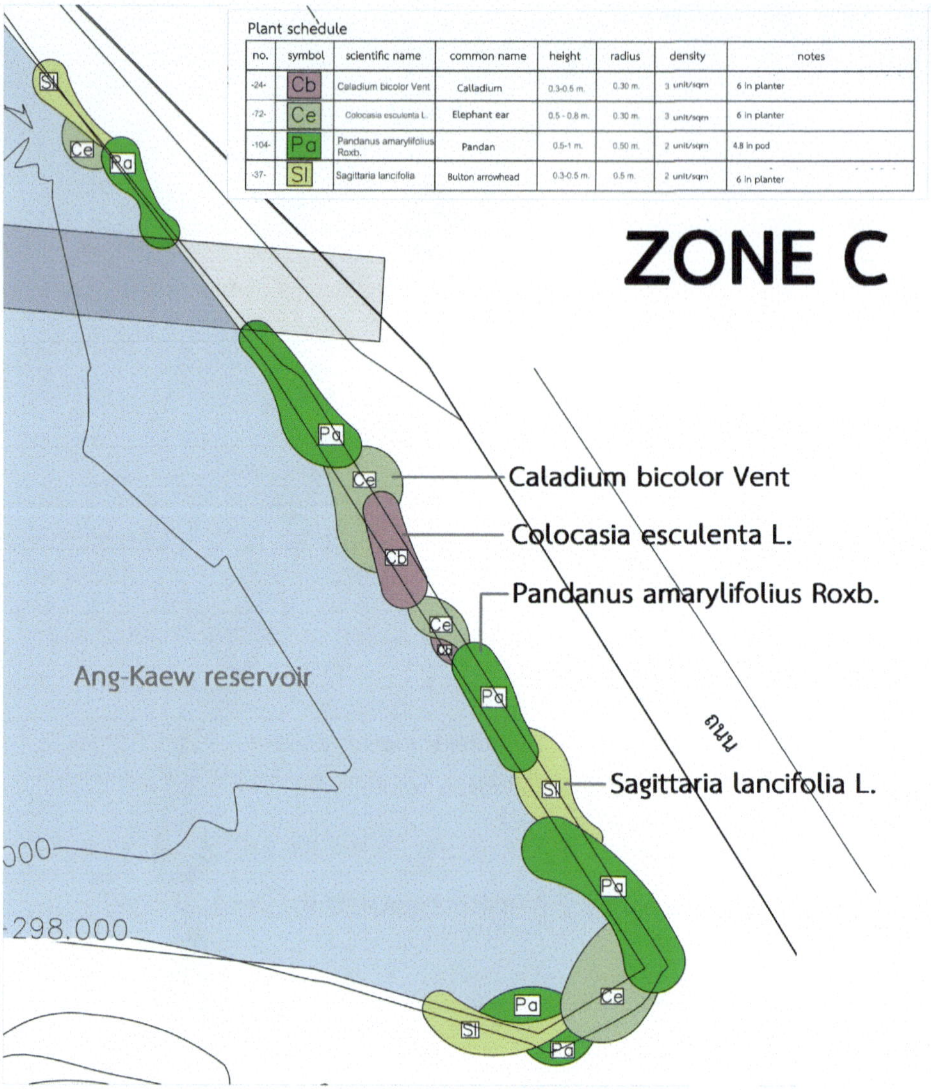

Fig. 8 Planting design in Zone C (image produced by authors)

phytoremediation (Zhang et al., 2007). However, the slight increase in nitrogen levels disagreed with previous research. It might be due to residues from plant material, which the laboratory research does not take into account (Dietz & Clausen, 2005). To fully understand the long-term efficacy of phytoremediation in different environmental conditions, it is crucial to conduct extended monitoring and evaluation studies. Future studies might have to monitor the water at least one year before and after the installation to ensure the comparisons between the previous and designed conditions. Furthermore, the water quality around the nearby inlets and drainage system should be monitored to understand possible fluxuation of the external pollutants.

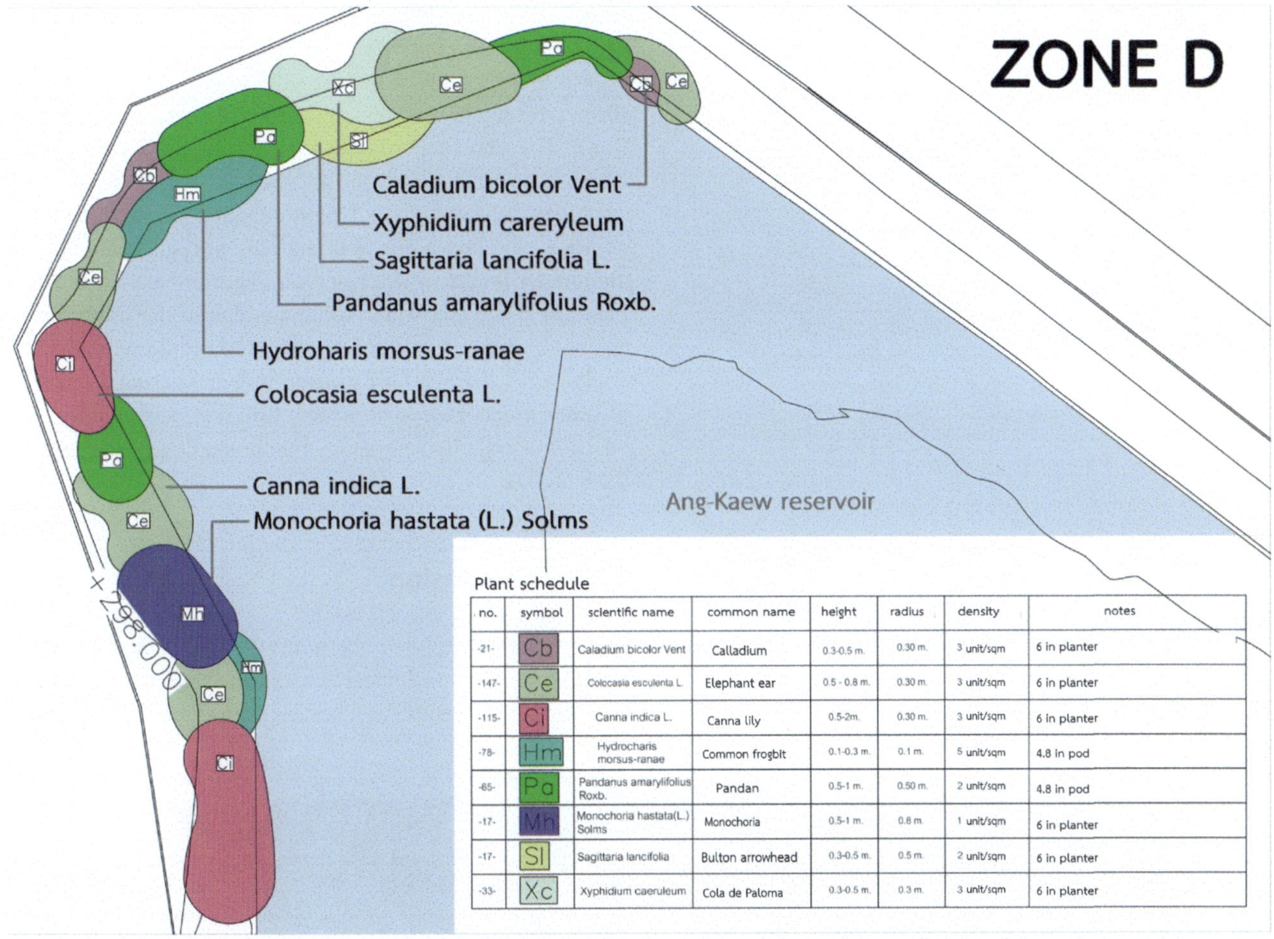

no.	symbol	scientific name	common name	height	radius	density	notes
-21-	Cb	Caladium bicolor Vent	Calladium	0.3-0.5 m.	0.30 m.	3 unit/sqm	6 in planter
-147-	Ce	Colocasia esculenta L.	Elephant ear	0.5 - 0.8 m.	0.30 m.	3 unit/sqm	6 in planter
-115-	Ci	Canna indica L.	Canna lily	0.5-2m.	0.30 m.	3 unit/sqm	6 in planter
-78-	Hm	Hydrocharis morsus-ranae	Common frogbit	0.1-0.3 m.	0.1 m.	5 unit/sqm	4.8 in pod
-65-	Pa	Pandanus amarylifolius Roxb.	Pandan	0.5-1 m.	0.50 m.	2 unit/sqm	4.8 in pod
-17-	Mh	Monochoria hastata(L.) Solms	Monochoria	0.5-1 m.	0.8 m.	1 unit/sqm	6 in planter
-17-	Sl	Sagittaria lancifolia	Bulton arrowhead	0.3-0.5 m.	0.5 m.	2 unit/sqm	6 in planter
-33-	Xc	Xyphidium caeruleum	Cola de Paloma	0.3-0.5 m.	0.3 m.	3 unit/sqm	6 in planter

Fig. 9 Planting design in Zone D (image produced by authors)

While it has not been proven that the current design is sufficient in addressing the eutrophication issues in the future, it is a great example of how landscape architects may work with researchers to address this challenge and such design outcomes. Thus, the outcomes of the study have met the objective set at the beginning of this study. The flow of analysis, synthesis, implementation, evaluation, and monitoring was recorded and could be repeated in future studies and works.

This research provides a novel contribution because it is one of the first few attempts to combine research-through-design with a field monitoring process to advance the understanding of the landscapes' effectiveness. It is also one of the first few implementations of phytoremediation of recreational campus landscapes in Thailand. Furthermore, the concept and process of this study can be adapted into a larger international context, such as similar corporate landscapes, housing, and as a part of green infrastructure in a

Table 1 Plant species selection and reasoning

Scientific name	Reasoning
Monochoria hastata (L.) Solms	Porous and spongy stems
Cannaceae	
Cyperus papyrus	
Cyperus haspan L	
Colocasia esculenta var. aquatilis	
Colocasia esculenta 'Black Coral' (Taro)	
Alocasia indica Schott	
Thalia geniculata L	
Cyperus alternifolius	
Echinodorus cordifolius	
Echinodorus cordifolius	
Heliconia	
Arundo donax	
Heliconia psittacorum L.f	
Pandanus amaryllifolius	Deep roots for bank stabilization
Dieffenbachia	
Callisia fragrans	
Hymenocallis littoralis	
Xyphidium caeruleum	
Vetiveria zizanioides Nash	
Vetiveria nemoralis A.Camus	

larger city in similar climates to Thailand, such as Laos, Malaysia, Southern China, and Singapore.

While the study's findings offer valuable insights for further research, it is essential to acknowledge its limitations. The short-term monitoring period limits a comprehensive understanding of the seasonal variations and the sustained effectiveness of phytoremediation over time. In addition, the space for planting was not sufficient for the amount of water in the reservoir. This imbalance between planting space and water volume is due to the limitation of the bank space. Additionally, the site-specific nature of the study raises the need for caution in generalizing the results to other green–blue spaces with different geographical and environmental characteristics. Future studies should address these issues.

5 Conclusion

This research explores the potential of phytoremediation to enhance the ecological and health benefits of green–blue spaces in campus recreational spaces in Chiang Mai,

Fig. 10 Installation of the site (image produced by authors)

Fig. 11 Site after installation (Zone B) (image produced by authors)

Thailand. By addressing the study's limitations and continuing to explore the optimization of plant selection and implementation techniques, institutions such as the Chiang Mai University can take proactive steps towards creating more sustainable and resilient environments. As the world faces increasing environmental challenges such as climate changes and water pollution, the application of phytoremediation

Fig. 12 Site after installation (Zone D) (image produced by authors)

Table 2 Changes in water quality after installation

	Before planting (mg/L)	After six months (mg/L)
BOD	9.60	0.818
Total phosphorus	2.87	0.27
Total nitrogen	0.00167 mg/:L	0.00228 mg/L

in green–blue spaces holds promise as a practical and eco-friendly solution to preserve the health and ecological integrity of these valuable campus resources. Further interdisciplinary research, collaboration, and long-term monitoring are essential to unlock the full potential of phytoremediation in enhancing the sustainability of green–blue spaces.

Table 3 Researcher-designer workflow

Stage of the project	Researchers	Designers
Initiation	Site visit, stakeholder briefing, developing goals and objectives for the project	
Investigation	Investigate chemical and biological property of the water and possible causes	Investigate bank conditions and possible implementation locations
Planning	Identify scope, placement, and overall installation and monitoring plans of the project	
Conceptual design		Develop conceptual design and visual representation of the project
Detail design	Identify possible phytoremediation strategies and the quality of plants needed for the project	Decide on plant species, placements, and interactions with the existing infrastructure
Installation		Manage installation to make sure it goes according to plan
Primary monitoring	Collect data and considering the results. Inform designers of the results	
Continuation of the project (future part)	Continue to monitor the changes	Systematically adapt from the monitor results

Acknowledgements This research project was supported by Chiang Mai University under the Project: Landscape improvement towards Vipavee Surinseng. We also thank the Faculty of Sciences for their support.

References

Anderson, D. M., Glibert, P. M., & Burkholder, J. M. (2002). Harmful algal blooms and eutrophication: Nutrient sources, composition, and consequences. *Estuaries, 25*(4), 704–726.

Ansari, A. A., Gill, S. S., Khan, F. A., & Naeem, M. (2014). Phytoremediation systems for the recovery of nutrients from eutrophic waters. In *Eutrophication: Causes, consequences and control*, (Vol. 2, pp. 239–248).

Banjerdkij, P., Danwandee, D., & Thippitak, S. (2013). *Development of bioremediation-bioretention application to reduce global warming effect*. http://librae.mju.ac.th/goverment/20111119104834_librae/11410.pdf

Brink, v. d. A., Bruns, D., Tobi, H., & Bell, S. (2017). Research in landscape architecture: methods and methodology.

Davar, S. A., Rostami, P., & Moradi, M. (2021). Planting design utilizing phytoremediation of Garden-Rangeland ecotypes in urban green space (case study of the new city of Pardis). *PalArch's Journal of Archaeology of Egypt/egyptology, 18*(18), 547–558.

Davis, A. P., Hunt, W. F., Traver, R. G., & Clar, M. (2009). Bioretention technology: Overview of current practice and future needs. *Journal of Environmental Engineering, 135*(3), 109–117.

Dean, S., Akhtar, M. S., Ditta, A., Valipour, M., & Aslam, S. (2022). Microcosm study on the potential of aquatic macrophytes for phytoremediation of phosphorus-induced eutrophication. *Sustainability, 14*(24), 16415.

Deming, M. E., & Swaffield, S. (2011). *Landscape architectural research: Inquiry, strategy, design*. John Wiley & Sons.

Dietz, M. E., & Clausen, J. C. (2005). A field evaluation of rain garden flow and pollutant treatment. *Water, Air, and Soil Pollution, 167*(1–4), 16.

Duzenli, T., Yilmaz, S., & Alpak, E. M. (2017). The effects of model making on design and learning in landscape architecture education. *Eurasian Journal of Educational Research, 17*(70), 121–134.

El-Sheekh, M., Abdel-Daim, M. M., Okba, M., Gharib, S., Soliman, A., & El-Kassas, H. (2021). Green technology for bioremediation of the eutrophication phenomenon in aquatic ecosystems: A review. *African Journal of Aquatic Science, 46*(3), 274–292.

Erell, E., Pearlmutter, D., & Williamson, T. (2012). *Urban microclimate: Designing the spaces between buildings*. Routledge.

Hák, T., Janoušková, S., & Moldan, B. (2016). Sustainable development goals: A need for relevant indicators. *Ecological Indicators, 60*, 565–573.

Haque, A. N. (2021). Climate risk responses and the urban poor in the global South: The case of Dhaka's flood risk in the low-income settlements. *International Journal of Disaster Risk Reduction, 64*, 102534.

Hunt, W. F., Lord, B., Loh, B., & Sia, A. (2015). *Plant selection for bioretention systems and stormwater treatment practices*. Springer Nature.

International Federation of Landscape Architects. (2022). *The Profession*. International Federation of Landscape Architects,. Retrieved May 27th from https://www.iflaworld.com/the-profession

Jaber, F., Woodson, D., LaChance, C., & York, C. (2012). Stormwater management: Rain gardens. *The Department of Soil and Crop Sciences and Texas A&M AgriLife Communications, The Texas A&M System, USA* (pp. 20).

Jiang, C., Fan, X., Cui, G., & Zhang, Y. (2007). Removal of agricultural non-point source pollutants by ditch wetlands: Implications for lake eutrophication control. *Eutrophication of Shallow Lakes With Special Reference to Lake Taihu, China*, 319–327.

Kennen, K., & Kirkwood, N. (2015). *Phyto: Principles and resources for site remediation and landscape design*. Routledge.

Lagro, J. (2023). Landscape journal workshop. Council of Educators in Landscape Architecture.

Lagro, J., Ryan, R. L., Mendenahll, A., & Popowsky, R. (2023). Align realign: Building the disciplines. Council of Educators in Landscape Architecture.

Lenzholzer, S., Duchhart, I., & Koh, J. (2013). 'Research through designing'in landscape architecture. *Landscape and Urban Planning, 113*, 120–127.

Lord, B., Sia, A., Hunt, W. F., & Loh, B. (2015). *Plant selection for bioretention systems and stormwater treatment practices*. Springer.

Milburn, L. -A. S., Brown, R. D. J. L., & planning, u. (2003). The relationship between research and design in landscape architecture. *64*(1–2), 47–66.

Nazari-Sharabian, M., Ahmad, S., & Karakouzian, M. (2018). Climate change and eutrophication: A short review. *Engineering, Technology and Applied Science Research, 8*(6), 3668.

Nijhuis, S., & Bobbink, I. (2012). Design-related research in landscape architecture. *Journal of Design Research, 10*(4), 239–257.

Nissim, W. G., & Labrecque, M. (2021). Reclamation of urban brownfields through phytoremediation: Implications for building sustainable and resilient towns. *Urban Forestry and Urban Greening, 65*, 127364.

Podhajska, E., Drzeniecka-Osiadacz, A., Halarewicz, A., Grech, D., Podhajski, B., Zienowicz, M., Bąbelewski, P., & Liszewski, M. (2023). Phytoremediation as an urban paradigm in promoting the health-potential of small green areas. *Sustainable Cities and Society*, 104684.

Rugh, C. L. (2001). Mercury detoxification with transgenic plants and other biotechnological breakthroughs for phytoremediation. *In Vitro Cellular and Developmental Biology-Plant, 37*(3), 321–325.

Shanahan, D. F., Fuller, R. A., Bush, R., Lin, B. B., & Gaston, K. J. (2015). The health benefits of urban nature: How much do we need? *BioScience, 65*(5), 476–485.

Shen, X., Handel, S. N., Kirkwood, N. G., Huang, Y., & Padua, M. G. (2022). Locating the responsive plants for landscape recovery: A toolkit for designers and planners. *Ecological Restoration, 40*(1), 33–35.

Suppakitpaisarn, P. (2021). *Research and design in landscape architecture (การวิจัยและออกแบบในงานภูมิสถาปัตยกรรม)*. O.S. Printing House.

Suppakitpaisarn, P., & Charoenlertthanakit, N. (2017). Current views of landscape architects' roles in environmental conservation: A discourse analysis. The National Kasetsart University Kamphaeng Saen Conference, Nakhon pathom, Thailand.

Suppakittpaisarn, P. (2021). *Research and design in landscape architecture*. O.S. Printing House.

Suppakittpaisarn, P., Chang, C.-Y., Deal, B., Larsen, L., & Sullivan, W. C. (2020). Does vegetation density and perceptions predict green stormwater infrastructure preference? *Urban Forestry & Urban Greening*, 55. https://doi.org/10.1016/j.ufug.2020.126842

Suppakittpaisarn, P., Jiang, X., & Sullivan, W. C. (2017). Green infrastructure, green stormwater infrastructure, and human health: A review. *Current Landscape Ecology Reports, 2*(4), 96–110.

Suppakittpaisarn, P., Larsen, L., & Sullivan, W. C. (2019). Preferences for green infrastructure and green stormwater infrastructure in urban landscapes: Differences between designers and laypeople. *Urban Forestry and Urban Greening, 43*, 126378.

Venvik, G., & Boogaard, C. F. (2020). Infiltration capacity of rain gardens using full-scale test method: Effect of infiltration system on groundwater levels in Bergen, Norway. *Land, 9* (12), 520.

Wang, X., Jain, A., Chen, B., Wang, Y., Jin, Q., Yugandhar, P., Xu, Y., Sun, S., & Hu, F. (2022). Differential efficacy of water lily cultivars in phytoremediation of eutrophic water contaminated with phosphorus and nitrogen. *Plant Physiology and Biochemistry, 171*, 139–146.

Wanitchayapaisit, C., Charoenlertthanakit, N., Surinseng, V., Yaipimol, E., Rinchumphu, D., & Suppakittpaisarn, P. (2023). Enhancing water-sensitive urban design in Chiang Mai through a research-design collaboration. *Sustainability, 15*(22), 16127.

Wanitchayapaisit, C., Suppakittpaisarn, P., Charoenlertthanakit, N., Surinseng, V., Yaipimol, E., & Rinchumphu, D. (2022). Rain garden design for stormwater management in Chiang Mai, Thailand: A research-through-Design Study. Nakhara: Journal of Environmental Design and Planning, *21*(3), 222–222.

Zhang, X.-B., Peng, L., Yang, Y.-S., & Chen, W.-R. (2007). Phytoremediation of urban wastewater by model wetlands with ornamental hydrophytes. *Journal of Environmental Sciences, 19*(8), 902–909.

A Study on Participatory Attitudes and Behaviors of Rural Residents in Sanitation Facilities

Simei Wu, Yujun Huang, and Yong Zhou

Abstract

Improving sanitation is an important step towards achieving the United Nations Sustainable Development Goals. Lack of sanitation can lead to the emergence of infectious diseases, which can further lead to serious consequences such as death, poverty and uneven development. Compared with urban areas, rural sanitation facilities are more dispersed, sanitations and toilets are relatively independent, and more dependent on users to manage and maintain them. Therefore, the attitude and willingness of village residents to improve sanitation facilities is crucial to the improvement of sanitation facilities. In order to further explore the attitudes and behavioral mechanisms of village residents towards sanitation improvement. This paper compares and analyzes research on the attitudes of village residents toward improving sanitations and toilets, and combines the results of existing research to give recommendations for enhancing the participation of village residents in sanitation improvements. This review can provide a new idea for future research on the participation attitudes and behaviors of village residents to improve sanitation.

Keywords

Village · Sanitations · Toilets · Attitude · Behavior

S. Wu (✉) · Y. Huang · Y. Zhou
School of Management, Xi'an University of Architecture and Technology, Yanta Road 13, Xi'an 710055, China
e-mail: simeiwu@xauat.edu.com

Y. Zhou
e-mail: zhouyong@xauat.edu.cn

1 Introduction

Improvement of sanitation facilities is an important step towards achieving the United Nations Sustainable Development Goals. The United Nations aims to achieve adequate and equitable sanitation and hygiene for all and to eliminate open defecation by 2030, particularly paying attention to the needs of women, girls and vulnerable groups.

Lack of sanitation can lead to the spread of epidemics and increased mortality rates, and can lead to a range of serious consequences such as poverty and uneven development. And improved sanitation will greatly reduce the likelihood of disease emergence. Surveys of low-income countries found that improved sanitation was effective in reducing the risk of diarrhoea in children under 5 years of age (Merid et al., 2023). Through the study of the impact of rural sanitation facilities on migrant workers, it was found that the implementation of sanitary latrines and non-hazardous latrines had a positive impact on the overall health of migrant workers (Zhang et al., 2020). Therefore, It is necessary to improve sanitation facilities in rural areas.

But according to a new report released on the eve of the first major United Nations conference on water resources in 2023, 46 percent of the world's population lacks access to basic sanitation. Due to the weaker overall economic situation in rural areas than in cities, and the complex topography of rural areas, the living is scattered. Therefore, rural sanitation facilities have many problems compared with cities in many aspects such as layout and management. An analysis of the status of sanitation reforms in India and Ghana over the past two decades reveals that sanitation coverage in rural areas remains low (Kanyagui & Viswanathan, 2022).The gap between rich and poor in terms of access to and use of toilet facilities is narrowing, but a quarter of rural villages still have poor-quality toilets (Budhathoki, 2019). By analyzing the amount and changes of rural domestic sewage and the treatment of rural domestic

B. He et al. (eds.), *Urban Climate Change Adaptation*, Advances in Science, Technology & Innovation,
https://doi.org/10.1007/978-3-031-65088-8_19

sewage, it is pointed out that there are problems such as insufficient investment in infrastructure construction, unreasonable system planning, and difficulty in sewage collection and treatment in rural sewage treatment (Guo et al., 2022). By analyzing the data on rural sanitations in recent years, it is found that the scale of construction of rural sanitations has generally increased in recent years, and the gap between villages and towns has narrowed, but the degree of improvement of the facilities is still unbalanced between regions (Wang et al., 2021).

At the same time, rural sanitation facilities are more decentralized and sanitations are relatively independent of latrines. There is a greater reliance on users to manage and maintain them. Moreover, in rural sewage treatment, farmers are the beneficiaries and responsible persons. Farmers are also the mainstay of rural environmental management, and there are serious drawbacks to rural environmental management that deviates from farmers. Therefore, the improvement of rural sanitation facilities is more dependent on the support of farmers. Therefore, this paper compares and analyzes the research on rural residents' attitudes towards improving sanitations and toilets, and combines the existing research results to give suggestions for improving rural residents' participation in sanitation facilities. This review can provide a new idea for future research on the participation attitudes and behaviors of rural residents in improving sanitation facilities.

2 Attitudes of Rural Residents Towards Improved Sanitation

2.1 Attitudes of Rural Residents Towards Improved Toilet Facilities

In previous research studies it was found that rural residents have very different attitudes towards improved toilet facilities. And there are a number of factors that influence the attitudes of the population towards improved toilet facilities. Through a survey of 300 randomly selected households, it was found that demographic, socio-economic and institutional factors had a significant impact on households' willingness to pay (Makwinja et al., 2019). At the same time, the advantages and disadvantages of the toilet facilities will also affect the willingness of farmers to participate and pay. Therefore, the following section will summarize and analyze the previous article in terms of each of these four factors. Relevant studies can be found in Table 1.

Demographic factors, including age and education, affect the willingness of rural residents to participate in latrine conversion. The level of education of the farmers is an important influencing factor. Most of the residents are willing to set up eco-latrines in their homes. Educational level affects their willingness to pay for eco-latrines (Eom et al., 2021). Surveys have found that rural households with lower levels of education sometimes fail to comply with sanitation guidelines and refuse to use improved toilet facilities (Guo et al., 2021). Household income also influences rural residents' willingness to participate. Household income has a significant negative effect on the improvement of toilet facilities for rural residents (Dare, 2014). In addition to household income, age is also an important factor influencing rural residents' willingness to participate in toilet renovation. Through a survey of 657 households in six villages in Shandong Province, it was found that with age, villagers' willingness to change toilets showed a significant downward trend (Liu & Fan, 2017).

Socioeconomic factors such as finance and technology are also important factors influencing the willingness of rural residents to participate in toilet renovation. Taking Yanghe Town, Xinyang City, Henan Province as an example, it is found that due to insufficient funds, rural villagers are not very willing to change toilets in the "toilet revolution" (Li, 2021).Through an experiment in Nakuru, Kenya, it was found that the willingness to pay for high-quality pour-flush latrines was low. Less than 5% of families are willing to pay the full cost because of lack of funds (Peletz et al., 2021).A survey of rural areas with high toilet penetration rate found that households were willing to bear the cost of emptying and transporting sludge (Balasubramanya et al., 2017).

Institutional factors also affect the willingness of rural residents to participate in toilet conversion. Government subsidies significantly adjusted the relationship between ecological cognition and toilet renovation, but there were differences in the moderating effect of different remediation actions (Huang & Yao, 2021).The simultaneous implementation of subsidies and hygiene marketing programmes can effectively influence the attitudes of poor households towards improving latrines (Hoo et al., 2022).In addition to government subsidies, the institutional environment is also an important institutional factor. The institutional environment will directly affect the willingness of rural residents to participate in toilet renovation (Hu et al., 2021).Some key policies and business strategies influence households' willingness to improve pit latrines (Burt et al., 2019).

The condition of the toilet facilities is also an important factor influencing the willingness of rural residents to participate in toilet renovation. Satisfaction with pre-renovation toilet facilities has a direct impact on rural residents' willingness to participate. A survey of rural Senegal found that men who were dissatisfied with existing facilities were more willing to pay for improvements to ventilated improved pit (VIP) latrine (Hall et al., 2015).Families' satisfaction with toilets affects their willingness to improve them (Mamo et al., 2023).Moreover, the benefits to the

Table 1 Attitudes of rural residents towards improved toilet facilities

Types of influencing factor	Author	Region	Study results	Influencing factor
Demographic factor	Young Sook Eom et al.	Korea	Most residents are willing to build ecological toilets in their homes. Education affects their willingness to pay for eco-toilets	Educational attainment
Demographic factor	Shaomin Guo	Rural Western China	Rural households sometimes refuse toilet facilities by not following hygiene guidelines	Educational attainment
Demographic factor	Alaba Modupe Dare	Nigeria	Household income has a significant negative impact on rural residents' improvement of toilet facilities	Household income
Demographic factor	Liu Yantao & Jin Yinqi	Shandong Province	The willingness to change toilets decreases with age	Age
Socioeconomic factor	Li Mengting et al.	Henan Province	Farmers' willingness to change toilets is not strong	Lack of funding
Socioeconomic factor	Peletz Rachel et al.	Kenya	The willingness to pay for high-quality Pour-Flush Latrines is well below market price	Lack of funding
Socioeconomic factor	Balasubramanya Soumya et al.	Rural areas with a high toilet penetration rate	Households are willing to bear the cost of emptying and transporting sludge	Toilet penetration rate
Institutional factor	Huang Hua & Yao Shunbo	Shaanxi Province	Government subsidies significantly adjusted the relationship between ecological cognition and toilet renovation, but there were differences in the moderating effect of different remediation actions	Government subsidies
Institutional factor	Yi Rong Hoo et al.	Cambodia	The simultaneous implementation of subsidies and hygiene marketing programmes can effectively influence the attitude of poor households towards latrine	Subsidy and marketing programs
Institutional factor	Hu Desheng et al.	Shaanxi Province	The institutional environment directly affects the willingness of rural residents to participate	Institutional environment
Institutional factor	Zachary Burt et al.	Rwanda	Some key policies and business strategies influence households' willingness to improve pit latrines	Key policies and business strategies
Facility factor	Ralph P. Hall et al.	Senegal	Men who are dissatisfied with existing facilities are more willing to pay for improved ventilated pit (VIP) latrine	Satis Satisfaction with existing facilities
Facility factor	Biruk Getachew Mamo	Ethiopia	The level of satisfaction of rural households with toilets affects their willingness to improve them	Satisfaction with the toilet
Facility factor	Liu Qijun et al.	Ningxia Province	There is a significant discrepancy between farmers' toilet conversion and their use behavior	The degree of perception of the benefits after toilet improvement

renovated toilet will also affect the willingness to change the toilet. Among the farmers who have completed the toilet renovation, because the farmers believe that the role of the toilet after the renovation is not large, 39.69% of the sanitary toilets have been idle for a long time, and the change of toilets by farmers is significantly different from their use behavior (Liu et al., 2023).

2.2 Attitudes of Rural Residents Towards Improved Sanitations

Like toilet facilities, rural residents' attitudes towards improving sanitations are also different, and the influencing factors can be roughly divided into demographic factors, socioeconomic factors and institutional factors. Relevant studies can be found in Table 2.

Previous research has found that demographic factors have a serious impact on rural residents' attitudes toward improved sanitations. The educational level of rural residents has a positive impact on residents' attitudes towards improving sanitations. It was found that rural residents who are still deficient in wastewater and stormwater management (WSM) and are highly educated are willing to pay for improvements (Rashid & Pandit, 2020).A survey of rural northern India found that educated people are willing to pay more for improved drainage (Chauhan et al., 2019).Rural residents have the willingness to participate in the improvement of rural living environment, but the consistency between participation willingness and behavior is poor, and the education level of rural households is conducive to the transformation of farmers' willingness to participate in behavior (Sun et al., 2020).In addition to educational attainment, the income situation of rural households also affects their attitudes. Through a study of Yunnan Province, it was found that most rural residents who are willing and able to earn are willing to pay for rural domestic sewage treatment (Wu et al., 2023).The household income of farming households is an important factor influencing their willingness to pay for improved sanitations.

Socioeconomic factors are also important factors influencing rural residents' attitudes towards improved sanitations. Among the socioeconomic factors, the most far-reaching influence on the attitudes of rural residents is the financial factor. A survey of four regions in Italy and France's northwestern Mediterranean found that fiscal and financial incentives stimulate interest in water-related green infrastructures (WrGIs) (Piacentini & Rossetto, 2020).Rural residents lack understanding and support for rural domestic sewage treatment due to lack of funds (Yang et al., 2016). Providing loan funds to poor households increases their willingness to pay for sanitations (Mulatya et al., 2021). Moreover, freshwater prices can also affect the participation attitudes of rural residents. If freshwater prices rise high enough for consumers, residents may be willing to fund a decentralized wastewater treatment plan (Chopra & Das, 2019).

Institutional factors can also influence rural residents' attitudes towards improved sanitations. Policy expectations and support are a key institutional factor. Farmers in pure farming areas are more willing to accept improvements in sanitations as having higher expectations for policies, believing that the environment is safer (Zhang et al., 2018). By strengthening policy support and improving relevant laws and regulations, rural residents can be more willing to participate in investment in improving sanitations (Yan et al., 2016). Government subsidies also have an important impact. Experimental data collected on the outskirts of Lusaka, Zambia, found that tenants with government subsidies were willing to pay extra in exchange for improved sanitations (Tidwell et al., 2019).Cross-subsidies help support low-income households to pay for improved sanitations (Acey et al., 2019). Using multivariate logistic regression through interviews in 13 villages in three western provinces of China, Shaomin Guo et al. found that social acceptability significantly influenced attitudes toward improved sanitation use (Guo et al., 2021).

The condition of the sanitations is also an important factor influencing the willingness of rural residents to participate. Understanding the benefits of standardized sanitation makes it easier to increase people's willingness to pay for improved sanitations (Yasin et al., 2020). At the same time, the sewage situation after renovation is also one of the factors affecting the attitude of rural residents. If rural residents understand the sewage situation after renovation, they will tend to implement convenient and efficient sewage disposal plans and pursue a clean and tidy rural living environment (Xu & Zhang, 2019). Using logistic regression modeling based on a survey of 211 rural households in Kenya, Diana Mutuku Mulatya et al. analyzed and found that the proximity of rural households to public toilets affects their willingness to improve their toilets (Mulatya et al., 2021).

3 Discussion

This paper comprehensively reviews the previous articles, and sorts out the different attitudes of rural residents towards improving toilets and sanitations, and the factors influencing rural residents' attitudes towards improving facilities. The study found,

(1) Among demographic factors, educational attainment has an impact on attitudes towards improving toilet facilities and sanitations. The more educated rural

Table 2 Attitudes of rural residents towards improved sanitations

Types of influencing factor	Author	Region	Study results	Influencing factor
Demographic factor	Mohammad Rashid & Debapratim Pandit	India	Rural residents believe that wastewater and stormwater management (WSM) is still flawed and are willing to provide some costs for improvement	Educational attainment
Demographic factor	A. Deepak Chauha et al.	India	Educated people are willing to pay more for drainage improvements	Educational attainment
Demographic factor	Sun Qianlu	Tibet	Farmers' participation attitude and behavior consistency are poor	Educational attainment
Demographic factor	Wu jian et al.	Yunnan Province	Most farmers are willing and able to pay for rural domestic sewage treatment	Household income
Demographic factor	Chen Tao et al.	Henan Province	The household income status of farmers is an important factor affecting their willingness to pay	Household income
Socioeconomic factor	Simone Maria Piacentini et al.	Italy and France	There is a lot of interest in Water-related green infrastructures (WrGIs)	Fiscal and financial incentives
Socioeconomic factor	Diana Mutuku Mulatya et al.	Kenya	Loan funds to poor families will increase the willingness to pay for sanitations	Loan funds are provided to poor families
Socioeconomic factor	Yang Xiaoying et al.	Rural areas where sewage treatment facilities have been built	Farmers lack of understanding and support for rural domestic sewage treatment	Fundraising and operational management
Socioeconomic factor	Vasudha Chopra and Sukanya Das	India	If prices for fresh water rise high enough for consumers now, residents may be willing to fund a decentralized sewage treatment plant	Fresh water prices
Institutional factor	Zhang Zhengfeng et al.	Pure agricultural area	In pure agricultural areas, the higher the expectation of the policy, the safer the environment, the more willing to accept remediation and pay more attention to the implementation and guarantee of compensation	Expectations for policy
Institutional factor	Yan Tingwu et al.	Hubei Province	By strengthening policy support and improving relevant laws and regulations, we are more willing to participate in the improvement	Policy support
Institutional factor	James B. Tidwell et al.	Zambia	Tenants were willing to pay extra for improved drainage	Government subsidies
Institutional factor	Charisma Acey et a.l	Kenya	Cross-subsidies will help support low-income families to pay for improved sanitations	Cross-subsidization
Facility factor	Mohammad Zeqi Yasin et al.	Indonesia	Farmers found that understanding the benefits of standardized health facilities made it easier to improve people's willingness to pay	Learn about the benefits of standardized sanitation
Facility factor	Xu Jinhong and Zhang Xintong	Hebei Province	Farmers tend to implement convenient and efficient sewage discharge schemes, and pursue a clean and tidy rural living environment	Effluent condition after renovation

people are, the more they are aware of toilet facilities and sanitations, and the higher their willingness to improve toilet facilities and sanitations. However, educational attainment also influences rural residents' attitudes towards the use of renovated sanitations. The lower the level of education, the more inconsistent the willingness and use behaviour of rural residents to participate in improvement. At the same time, household income is also an important factor. Household income has a significant negative impact on rural residents' participation in improving toilet facilities. The lower the household income, the lower the willingness of rural residents to participate. Conversely, households with adequate incomes are more willing to participate in improving sanitations. This may be because households with lower household incomes cannot afford to upgrade their sanitations, while households with sufficient household incomes may be more willing to retrofit their sanitations in pursuit of a higher quality of life. In addition, age can affect the willingness of rural residents to participate in toilet facilities to varying degrees. Studies have found that as they age, rural residents are less willing to change toilets. This may be because older residents use old toilets longer, making them reluctant to accept improved toilet facilities.

(2) Among socioeconomic factors, lack of funding has a significant impact on attitudes towards participation in toilet facilities and sanitations. A lack of social funding may mean that rural residents will need to pay more household income, so rural residents will be less willing to participate in improving toilet facilities. The lack of social funding can also reduce the willingness of rural residents to participate in improving sanitations. However, research has found that rural residents will be willing to participate in the improvement of sanitations if they can provide funds for households, or if there is a better way to raise funds. The prevalence of rural toilets is an important factor influencing the willingness of rural residents to improve toilet facilities. In areas with high toilet penetration rate, rural residents are more willing to improve toilet facilities when they are fully aware of the benefits of toilet facilities. However, no studies have found that drainage coverage affects rural people's attitudes towards participating in improving drainage. The price of fresh water affects the attitude of rural residents towards improved sanitations. If the price of fresh water is too high, rural residents will be willing to pay for improved drainage.

(3) Among the institutional factors, policy and government subsidies are the two main influencing factors. Both of these factors affect the willingness of rural residents to participate in improving toilet facilities and sanitations. First, the better the policy environment or the higher expectations for key policies, the greater the willingness of rural residents to participate in improving toilet facilities. As with toilet facilities, rural residents are more willing to participate in improving sanitations if they have higher expectations and support for policies. This may be because better policy support may reward rural residents who actively participate in improving sanitations, which in turn influences their willingness to participate. Secondly, if government organizations can provide subsidies or have better marketing plans for rural residents who participate in improving toilet facilities, it will increase the motivation of rural residents to participate in improving toilet facilities. At the same time, cross-subsidies provided by the government can influence families who are hesitant to participate in drainage renovation due to lack of funds, which in turn increases the willingness of rural residents to participate in drainage improvement.

(4) Among the factors of the facilities, the satisfaction of the facilities before the renovation and the benefits of the facilities after the renovation are the two main factors affecting the willingness of rural residents to participate. First, satisfaction with the toilet facilities before the renovation will affect the willingness of rural residents to participate in the renovation of the toilet facilities. If rural households are dissatisfied with the existing toilet facilities in their homes, they may seek a better experience. Then they will be more willing to improve. However, the presence of sanitations was not found to affect the attitude of rural residents towards improved sanitations. Studies have found that rural dwellers are more willing to participate in sanitation improvement if they understand the benefits of modern sanitation. The benefits of the renovated facilities have an impact on the willingness of rural residents to improve toilet facilities and sanitations. Rural residents who are aware of the benefits of improved toilet facilities are more willing to participate in improving them. At the same time, rural residents who understand the status of the renovated sewage will be more willing to participate in the improvement of sanitations.

4 Conclusion

From the above comparison, it is easy to find that rural residents' attitudes towards improving sanitation and improving toilets have both similarities and differences. Rural residents' age, education level, income and other factors affect their attitudes to different degrees. Since toilets are closely related to rural residents' lives, rural residents are more willing to improve toilets, especially when the original toilets are not in good condition. However, the impact of sanitation on rural residents' lives is relatively small, and rural residents often find it difficult to perceive the effect of sanitation, and therefore refuse to pay for improving sanitation. In addition, some rural areas are still dominated by public toilets, so the distance to public toilets affects their attitudes towards improving sanitation. Whereas sanitation is mostly located far away from the villagers' residence, rural residents are unable to understand the effect of the investment and tend to refuse to improve sanitation.

Therefore, when improving toilets, attention should be paid to the actual needs of rural residents and the rural operating environment. While improving sanitation, publicity and education campaigns on sanitation should be carried out to make rural residents fully aware of the benefits of sanitation, so as to ensure that the improved sanitation can operate effectively.

Acknowledgements This work is supported by the Shaanxi Provincial Education Department Key Research Program Project (project no.22JZ039); the Nature Science Foundation of Shaanxi Province (project no. 2023-CX-RKX-152); the Social Science Research Special Youth Project of Shaanxi Province (project no. 2023qn0088); the Social Science Research Project of Shaanxi Province (project no. 2023R192).

References

Acey, C., Kisiangani, J., Ronoh, P., Delaire, C., Makena, E., Norman, G., Levine, D., Khush, R., & Peletz, R. (2019). Cross-subsidies for improved sanitation in low income settlements: Assessing the willingness to pay of water utility customers in Kenyan cities. *World Development, 115*, 160–177. https://doi.org/10.1016/j.worlddev.2018.11.006

Balasubramanya, S., Evans, B., Ahmed, R., Habib, A., Asad, N., Rahman, M., Hasan, M., Dey, D., Camargo-Valero, M., & Rao, K. C. (2017). Take it away: The need for designing fecal sludge disposal services for single-pit latrines. *Journal of Water, Sanitation and Hygiene for Development, 7*(1), 121–128.

Budhathoki, C. B. (2019). Water supply, sanitation and hygiene situation in Nepal: A review. *Journal of Health Promotion, 7*, 65–76.

Burt, Z., Sklar, R., & Murray, A. (2019). Costs and willingness to pay for pit latrine emptying services in Kigali, Rwanda. *International Journal of Environmental Research and Public Health, 16*(23), 4738.

Chauhan, A. D., Rai, B. D., Khan, C., & Patel, D. N. (2019). A statistical analysis of consumer's willingness to pay for improved water supply system: A study of some villages of Uttar Pradesh India. *International Journal of Research in Advent Technology, 10*(6), 15–23.

Chopra, V., & Das, S. (2019). Estimating willingness to pay for wastewater treatment in New Delhi: Contingent valuation approach. *Ecology, Economy and Society-the INSEE Journal, 2*(2354–2020–1322), 75–108.

Dare, A. M. (2014). *Rural households' access, willingness to pay (WTP) and factors influencing WTP for safe water and sanitation in Southwest Nigeria*.

Eom, Y. S., Oh, H., Cho, J., & Kim, J. (2021). Social acceptance and willingness to pay for a smart Eco-toilet system producing a community-based bioenergy in Korea. *Sustainable Energy Technologies and Assessments, 47*, 101400.

Guo, F., Chen, Y., Wang, G., & Zhang, W. (2022). Present situation, problems and development suggestions of ruraldomestic sewage treatment in China. *Water & Wastewater Engineering, 58*(S1), 68–72.

Guo, S., Zhou, X., Simha, P., Mercado, L. F. P., Lv, Y., & Li, Z. (2021). Poor awareness and attitudes to sanitation servicing can impede China's Rural Toilet Revolution: Evidence from Western China. *Science of the Total Environment, 794*, 148660.

Hall, R. P., Vance, E. A., Van Houweling, E., & Huang, W. (2015). Willingness to pay for VIP latrines in rural Senegal. *Journal of Water, Sanitation and Hygiene for Development, 5*(4), 586–593.

Hoo, Y. R., Joseph, G., Rivera, R., Smets, S., Nguyen, H., Ljung, P., Um, S., Davis, G., & Albert, J. (2022). Strategic complements: Poverty-targeted subsidy programs show additive benefits on household toilet purchases in rural Cambodia when coupled with sanitation marketing. *PLoS ONE, 17*(6), e0269980.

Hu, D., Wang, Y., Wang, F., Kong, X., & Wang, B. (2021). Farmer's cognition, institutional environment and willingness to participate in the improvement of living environment—The mediating effect of information trust. *Journal of Arid Land Resources and Environment, 35*(06), 15–23.

Huang, H., & Yao, S. (2021). Ecological cognition, government subsidy and farmers' willingness to participate in the renovation of rural human settlement environment. *Journal of Statistics and Information, 36*(12), 80–91.

Kanyagui, M. K., & Viswanathan, P. (2022). Water and sanitation services in India and Ghana: An assessment of implications for rural health and related SDGs. *Water Policy, 24*(6), 1073–1094.

Li, M. (2021). Study on Farmers' Willingness in the rural "toilet revolution"—Taking Yanghe Town, Xinyang City, Henan Province as an Example. *Rural Economy and Science Technology, 32*(11), 242–244.

Liu, Q., Liu, S., Li, Y., Wang, X., & Li, Z. (2023). Study on the paradox between farmers' changing toilets and their using behaviors: Based on a micro-survey data in Ningxia. *Journal of Chinese Agricultural Mechanization, 44*(04), 230–238.

Liu, Y., & Fan, Y. (2017). An analysis of the factors influencing rural residents' willingness to change toilets—A survey of Town C, Wudi County, Shandong Province, China. *China Rural Health, 9*, 10–13.

Makwinja, R., Kosamu, I. B. M., & Kaonga, C. C. (2019). Determinants and values of willingness to pay for water quality improvement: Insights from Chia Lagoon, Malawi. *Sustainability, 11*(17), 4690.

Mamo, B. G., Novotný, J., & Ficek, F. (2023). Barriers for upgrading of latrines in rural Ethiopia: Disentangling a sanitation socio-technical lock-in. *Local Environment*, 1–19.

Merid, M. W., Alem, A. Z., Chilot, D., Belay, D. G., Kibret, A. A., Asratie, M. H., Shibabaw, Y. Y., & Aragaw, F. M. (2023, Jun 15). Impact of access to improved water and sanitation on diarrhea reduction among rural under-five children in low and middle-income countries: A propensity score matched analysis. *Tropical Medicine and Health, 51*(1), 36. https://doi.org/10.1186/s41182-023-00525-9

Mulatya, D. M., Were, V., Olewe, J., & Mbuvi, J. (2021). Willingness to pay for improvements in rural sanitation: Evidence from a cross-sectional survey of three rural counties in Kenya. *PLoS ONE, 16*(5), e0248223.

Peletz, R., Delaire, C., Kones, J., MacLeod, C., Samuel, E., Easthope-Frazer, A., & Khush, R. (2021). Will households invest in safe sanitation? Results from an experimental demand trial in Nakuru, Kenya. *International Journal of Environmental Research and Public Health, 18*(9), 4462.

Piacentini, S. M., & Rossetto, R. (2020). Attitude and actual behaviour towards water-related green infrastructures and sustainable drainage systems in four North-Western Mediterranean Regions of Italy and France. *Water, 12*(5). https://doi.org/10.3390/w12051474

Rashid, M., & Pandit, D. (2020). Rural communities' perception of and willingness to pay for wastewater and stormwater management infrastructure in Bihar, India. *Journal of Water, Sanitation and Hygiene for Development, 10*(1), 36–47.

Sun, Q., Fang, K., & Liu, T. (2020). Impact of social norms and public supervision on the willingness and behavior of farming households to participate in rural living environment improvement: Empirical analysis based on generalized continuous ratio model. *Resources Science, 42*(12), 2354–2369.

Tidwell, J. B., Terris-Prestholt, F., Quaife, M., & Aunger, R. (2019). Understanding demand for higher quality sanitation in peri-urban Lusaka, Zambia through stated and revealed preference analysis. *Social Science & Medicine, 232*, 139–147.

Wang, X., Wen, H., Zhang, Y., Guo, F., Chen, Y., & Wang, G. (2021). Current situation, problems and countermeasures of water supply and drainage facilities construction in villages and towns in China. *Environmental Protection Science, 47*(03), 47–53.

Wu, J., Xu, Y., Xu, S., & Zhang, H. (2023). Research on willingness to pay for rural domestic sewage treatment and its influencing factors: Evidence from Lashihai Watershed in Yunnan Province. *Resources and Environment in the Yangtze Basin, 32*(04), 842–854.

Xu, J., & Zhang, X. (2019). Analysis of the willingness to implement the "toilet revolution" in rural areas and the influencing—Factors based on the research of Shijiazhuang farmers. *Statistics and Management, 2*, 99–101.

Yan, T., He, K., & Zhang, J. (2016). Analysis of social capital influencing farmers 'willingness of environmental protection investment: Evidence from empirical study on reusing agricultural wastes in Hubei Rural Areas. *China Population, Resources and Environment, 26*(01), 158–164.

Yang, X., Yuan, J., Yao, M., Huang, W., & Wang, D. (2016). Statuesque of rural sewage treatment in China and its development strategies—A case study of rural area in South of Jiangsu Province. *Journal of Fudan University (Natural Science), 55*(02), 183–188+189.

Yasin, M. Z., Shahab, H. A., Nuryitmawan, T. R., Arini, H. R. B., Rumayya, R., Rohman, M. S., & Sari, F. R. (2020). *Willingness to pay for improved sanitation in Indonesia: Cross-sectional difference-in-differences*.

Zhang, S., Li, Y., Zhang, Y., Lu, Z.-N., & Hao, Y. (2020). Does sanitation infrastructure in rural areas affect migrant workers' health? Empirical evidence from China. *Environmental Geochemistry and Health, 42*, 625–646.

Zhang, Z., Wen, Y., & Wang, R. (2018). Impact factors of farmers' willingness in rural residential land readjustment: Comparing Jiangshan City in Zhejiang Province and Panshan County in Liaoning Province. *China Land Science, 32*(03), 28–34.

Research on Zhe-Style Dwelling Houses from the Perspective of Architecture Culture—Taking the Traditional Village in Xianju County as an Example

Gaochuan Zhang, Jiajia Lv, Qiaoyuan Lin, and Min Fang

Abstract

Rural architecture serves as a significant carrier in the revitalization of rural culture, together with the local natural scenery and customs, forming the distinctive rural landscape. This paper explores and extracts nine distinctive cultural elements of Zhe-style dwelling houses, including house layout, facade composition, gable form, street scale, wind and rain corridors, roof design, building materials, overall color scheme, and decorative carvings, through the identification of landscape genes using four modes, Based on this, an exploration of Zhe-style vernacular dwelling construction in Wengsen Street, Shuangmiao Township, Xianju County, in the Zhejiang central region is conducted, leading to the discovery of a new traditionalist design approach that combines theoretical application and Zhe-style culture practices. In addition, by promoting the integration of Zhe-style residential buildings with the natural environment, it is beneficial to reduce environmental damage and improve the adaptability and sustainability of residential buildings.

Keywords

Zhe-style dwelling houses · Landscape genes · Traditional village · Culture · Sustainable development

1 Introduction

The 19th National Congress of the Communist Party of China pointed out the new era's goal of "comprehensively promoting rural revitalization and prospering cultural undertakings and industries." As a pioneer and leader in China's rural revitalization practice, Zhejiang Province has taken the lead in carrying out the cultural development of villages by creating an environment that preserves the nostalgia and aiming to build a hundred Zhe-style vernacular dwelling characteristic villages within three years. Zhe-style dwelling houses have a long history and rich culture, especially in the northwestern and central regions of Zhejiang Province, where numerous traditional ancient villages are located. Incorporating them into the construction of livable, business friendly and beautiful rural areas. On a macro level, it is not only conducive to promoting traditional Zhe-style culture but also aligns with the strategy of rural revitalization. At the micro level, by drawing on and inheriting the architectural wisdom of Zhe-style residential buildings, sustainable development ideas, and practical experience can be provided for contemporary architectural design, which promotes the development of residential design towards a more environmentally friendly and sustainable direction (Fang et al., 2022; Zhang et al., 2023). Therefore, exploring the cultural inheritance methods of traditional Zhe-style residential buildings has become a new way to protect local architectural culture and save local resources.

G. Zhang · J. Lv · Q. Lin (✉) · M. Fang
School of Civil Engineering and Architecture, Zhejiang University of Science and Technology, Hangzhou 310023, China
e-mail: linqiaoyuan@foxmail.com

G. Zhang
e-mail: Zjhz_zgc@sina.com

G. Zhang
Zhejiang Southeast Architectural Design Group CO. LTD, Hangzhou 310023, China

Zhejiang-Singapore Joint Laboratory for Urban Renewal and Future City, Hangzhou 310023, China

B. He et al. (eds.), *Urban Climate Change Adaptation*, Advances in Science, Technology & Innovation,
https://doi.org/10.1007/978-3-031-65088-8_20

1.1 Analysis of the Concept of New Traditional Zhe-Style Residential Buildings

In 2015, the Zhejiang Provincial Government of China proposed the construction of Zhe-style residential buildings. "Zhe-style residential buildings" refer to traditional residential buildings with Zhejiang regional cultural characteristics. They are an important genre of traditional Han Chinese residential buildings formed by the combination of the migration of Central Plains culture to the south and the gathering of ethnic groups in Vietnam (You et al., 2023). Decorating with exquisite wood and brick carvings, the roof is mostly in the form of horsehead walls or flying eaves, presenting an elegant and exquisite overall feature (Wang, 2022). Zhe-style residential buildings have unique styles in terms of architectural structure, decorative art, and garden design. Representative works include the first residential building in Jiangnan-Shen Yuan, and the ancient architectural complex—Xiling Seal Engraving Society (Sun & Zhang, 2019). Zhe-style residential buildings not only showcase the rich historical and cultural heritage of the Zhejiang region but also represent a precious heritage of traditional Chinese architecture.

On the basis of traditional Zhe-style residential architecture, combined with new traditional theory and modern design concepts and technologies, the concept of new traditional Zhe-style residential architecture is proposed, which is innovative and improves the architecture level via retaining traditional cultural characteristics to meet the needs of modern life. Traditional decorative elements such as wood carving and brick carving are usually retained, while the structure, materials, and functions are updated and improved to meet the needs of modern life. Showcasing the charm of traditional culture provides a modern and comfortable living environment.

1.2 Concept Analysis of Landscape Genes

Just as genes are the basic units of biological inheritance, landscape genes, as the basic units of regional cultural "inheritance," are unique cultural factors shaped by historical heritage that serve as indicators for identifying the local cultural landscape (Maselli et al., 2024). Landscape genes specifically refer to the inherent cultural factors unique to a landscape that distinguish it from other landscapes. They are the basic unit of landscape "inheritance", that is, certain factors that have been passed down from generation to generation and are different from other landscapes (Jiang et al., 2023). They are important reference parameters for traditional settlement feature recognition. It emphasizes that landscape is a product of the interaction between human activities and the natural environment, carrying rich cultural connotations and historical memories.The formation, standardization, and evolution of traditional rural dwelling styles in Zhejiang Province also follow the genetic laws of landscape genes. They are historical products resulting from the continuous integration of external and local cultures and constitute an important part of the metabolic process of traditional village landscapes. In the exploration of the cultural landscape gene variation process within architectural metabolism, the modern art movement, architectural theorist Geddes, first proposed the concept of "New Tradition," aiming to address the inheritance contradiction between the "innovative" and the "eternal" aspects of genes (Fang et al., 2022). Since then, architects have been breaking free from traditional inherent forms, emphasizing unity and other techniques, and integrating traditional elements or symbols into modern architecture, thereby endowing buildings with a spirit of the new era (Ganobjak et al., 2023). However, simply imitating traditional architectural styles or symbolically incorporating traditional decorations does not truly engage with tradition. Therefore, it is necessary to have a deep understanding of the innovation, development, and application of new traditional theories in contemporary social conditions, that is, to advocate creative transformation and redesign on the basis of traditional culture, and to explore from the perspective of new traditions how to achieve modern functional needs and sustainable development goals while retaining traditional architectural features. Meanwhile, from the perspective of landscape genes, this paper explores the cultural genes and historical origins of the architectural layout, courtyard landscape, and architectural details of Zhe-style residential buildings, and explores how to achieve sustainable inheritance and development of architectural culture by emphasizing cultural traditions and landscape genes. This article aims to (1) Figure out the new method for landscape graph of Zhe-style dwelling houses; (2) Explore distinctive spatial morphology for Zhe-style dwelling houses; (3) Identify the cultural elements of Zhe-style dwelling houses.

2 Methodology

2.1 Genetic Identification Models for Zhe-Style Dwelling Houses

There are four basic models for identifying the landscape genes of traditional settlements: the two-dimensional form recognition model: analyze the horizontal and vertical arrangement characteristics of landscape elements based on the local planar morphology, facade features, and enclosure structure of the settlement. the three-dimensional form recognition model: Analyze the basic characteristics of individual or iconic buildings in the settlement through layout form and spatial imagery characteristics. the structural feature analysis model: analyze the overall landscape characteristics of traditional settlements by analyzing their individual buildings, architectural groups, spatial layout characteristics, and landscape construction techniques. and the visual and perceptual analysis model: Analyze the characteristics of traditional settlements in terms of architectural technology, shaping techniques, decorative arts, cultural beliefs, etc. through various methods such as observation, on-site visits, investigations, access to historical and cultural data, and investigation of natural environmental characteristics (Zhang & Long, 2023). However, modern architectural structures are primarily composed of reinforced concrete, so the third model is updated to material feature analysis to better adapt to the current needs of gene identification and construction. Through the four identification models, the genetic composition of Zhe-style dwelling houses can be broken down into nine distinctive cultural elements: house layout, facade composition, gable form, street and alley spaces, wind and rain corridors, roof design, building materials, overall color scheme, and decorative carvings as shown in Fig. 1.

2.2 Identification of Zhe-Style Architecture Culture

In terms of house layout, traditional dwellings exhibit a symmetrical arrangement influenced by feudal etiquette (Yu et al., 2016). Among them, the most distinctive feature is the central courtyard, around which the layout takes the form of "Zhe-style " character shape, "Sun" character shape, or "H" shape. In terms of facade composition, the facades emphasize left–right symmetry, with narrowing high walls on both sides to create "well-mouth windows." The juxtaposition of high and low horsehead walls is a typical feature of Zhe-style dwelling houses. In terms of street and alley spaces, the width of traditional commercial streets in southern regions during the Ming and Qing dynasties is generally between 3.0 and 6.0 m, with a D/H ratio between 1.0 and 2.0 (D representing street width and H representing building height) (Yang et al., 2022). In terms of wind and rain corridors, eaves corridors are widely used in Zhe-style architecture, usually with overhanging eaves above and connected to the side of the house, serving as a common feature that connects indoor and outdoor spaces and enhances spatial hierarchy. In terms of roof design, traditional sloping roofs come in various forms such as gable roofs, suspended roofs, and hardtop roofs, with hardtop

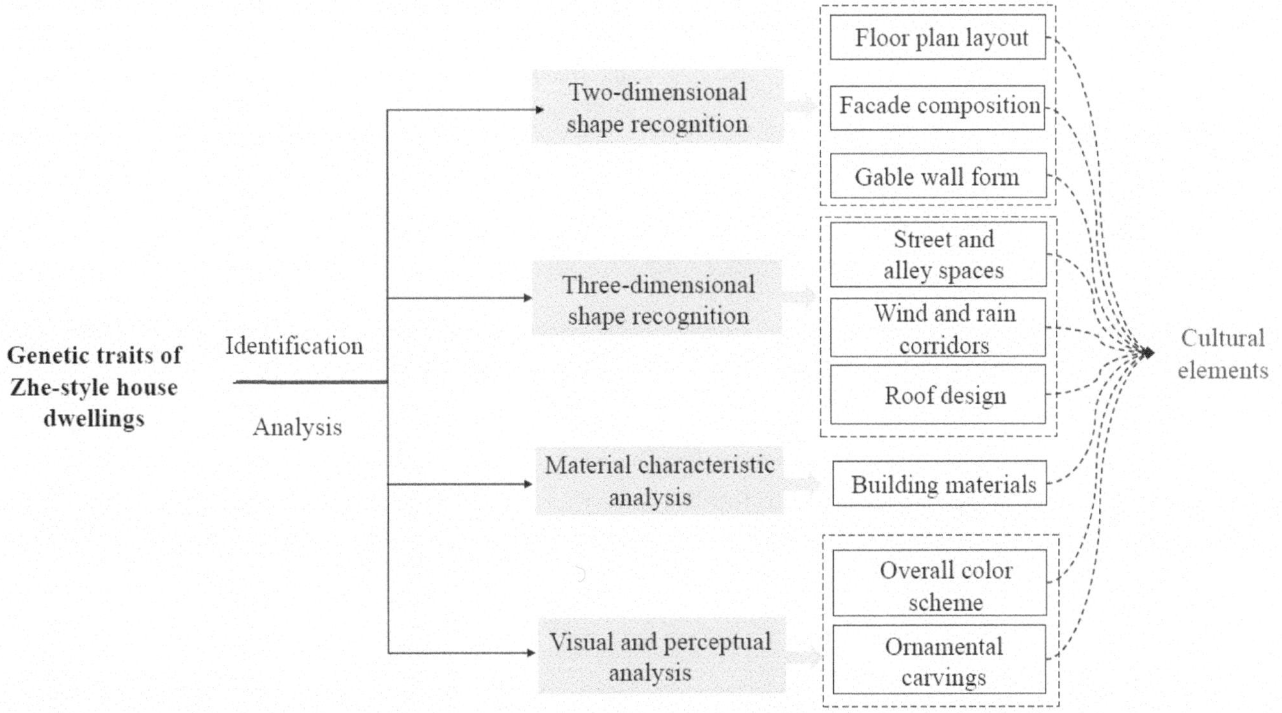

Fig. 1 Establishment and classification of genetic elements of Zhe-style—style folk dwellings. *Source* Drawn by authors

roofs being common for ordinary residences depending on the building's status. In terms of building materials, bricks, wood, and stones are the basic materials used for construction. Wood is often used for structural columns, railings, doors, windows, and other facade decorations, while bricks and stones are primarily used for enclosure structures. In terms of overall color scheme, traditional Zhe-style dwellings are generally characterized by black, white, and gray tones, with pink walls and dark-colored tiles, blending harmoniously with the natural surroundings of green mountains and white clouds. In terms of decorative carvings, the three types of carvings including the intricate carvings on green brick gate covers, the stone carvings on latticed windows, and the wood carvings on lintels and pillars are renowned in Zhe-style architecture (Wang et al., 2022), making the buildings exquisite and exceptional. The intricate carved windows not only serve as decorative elements, reducing the oppressive feeling of high walls, but also provide lighting, ventilation, and define spatial boundaries. The nine distinctive cultural genetic elements, through metaphorical expressions, convey people's wishes for a better life, as shown in Table 1. These distinctive cultural genes are deeply rooted in the history and traditional architectural practices of the central Zhejiang region, reflecting the values, beliefs, and lifestyles of the local community. By retaining these formal elements and integrating them into contemporary design, Zhejiang's architectural culture can be continued and promoted, which is conducive to connecting with the rich historical heritage of the region and cultivating a sense of pride and continuity in local cultural heritage.

3 Construction of Zhe-Style Dwelling Houses

Wengsen Street is located in the southeastern part of Shuangmiao Township, Xianju County, Zhejiang Province. From a human geography perspective, it belongs to the culturally diverse region where "Wu-Yue Culture" centered in northern Zhejiang merges, and it possesses a rich natural and cultural heritage. The plot is nestled among mountains and water, and the nearby Gaoqian ancient village preserves the authentic charm of numerous Ming and Qing dynasties' Zhe-style dwellings. It is considered the "first village" of traditional culture in central Zhejiang, providing a solid foundation for the construction of Zhe-style humanistic landscapes and inheriting the local architectural trends as shown in Fig. 2.

3.1 Expression of Two-Dimensional Morphology

3.1.1 Cultural Significance of Floor Planning

Starting with the "Zhe-style" shape layout in the traditional Zhe-style courtyard as the basic plan, gradually evolve the design as shown in Fig. 3. The houses are arranged in a basic symmetrical layout from north to south and east to west, to some extent echoing the symmetric structure of traditional rituals. Additionally, the central passage replaces the traditional courtyard, connecting the four rooms in the north and south as well as the auxiliary spaces on the left

Table 1 Extraction and analysis of cultural elements of Zhe-style dwelling houses

Recognition content	Cultural elements	Distinctive structure	Element interpretation	Symbolism
Zhe-style dwelling houses	Mountain-shaped walls	Horse head walls	Fire and wind resistance	Comfortable and pleasant living
	Layout design	L-shaped, H-shaped, etc.	Traditional etiquette	Four waters converge at the hall
	Facade composition	Modular system	Rhythm and cadence	Harmony and unity
	Street and alley spaces	Street and lane proportions	Pleasant scale and proportions	Peaceful coexistence
	Roofing style	Gabled and hip roofs, etc.	Architectural hierarchy	Harmony between heaven and humanity
	Covered walkways	Single eave corridors	Transitional spaces	Hospitality and etiquette towards guests
	Building materials	Wood, brick, and stone materials	Adaptation to local conditions	Natural simplicity
	Overall color palette	Black, white, gray, and wood colors	Pink walls and dark tile roofs	Confucianism and taoism aesthetics
	Carved decorations	Doors, windows, and railings	Exquisite decorative patterns	Auspicious and fortunate

Source Drawn by Authors

Fig. 2 Current situation of traditional residential buildings in Shuangmiao Township, Xianju County. *Source* Photographed by authors

and right. The layout with "multiple bedrooms" provides suitable guest rooms for rural tourism, reflecting the flexible application of traditional house types in the context of the new era. In addition, the layout structure of shops on the ground floor and residences on the upper floors, following the traditional street-market layout, provides entrepreneurial space for rural talent returnees, which contributes to the revitalization of the modern rural economy.

3.1.2 Cultural Significance of Facade Composition

Lothar Ledderose summarized Chinese characters into a module system, consisting of five hierarchical levels as shown in Fig. 4: element, module, unit, sequence, and ensemble (Liu & Li, 2020). The entire composition of the facade can be referenced based on this system, where Zhe-style elements undergo basic arrangement and combination to form modules with specific properties. Different modules are combined to form units in different ways, and the units are arranged in sequences with dividing walls on both sides. Ultimately, multiple sequences combine to form a unified ensemble system. In detail, cultural elements were show in the first row as Fig. 4a-1, e-1. Cultural elements were extracted as shown in the second row as Fig. 4a-2, e-2. Cultural recombination was constructed as units which could be shown in third row as Fig. 4a-3, e-3. Then, multiple units are combined into a series which could be shown in forth row as Fig. 4a-4, e-4. Finally, Each series forms a comprehensive and exquisite architectural morphology.

3.1.3 Cultural Significance of Gable Wall

The gable wall form is an important architectural cultural element in Zhe-style houses, with the horsehead wall being its typical symbol. The horsehead wall is simplified at the

Fig. 3 Plan deduction from traditional dwelling houses. *Source* Drawn by authors

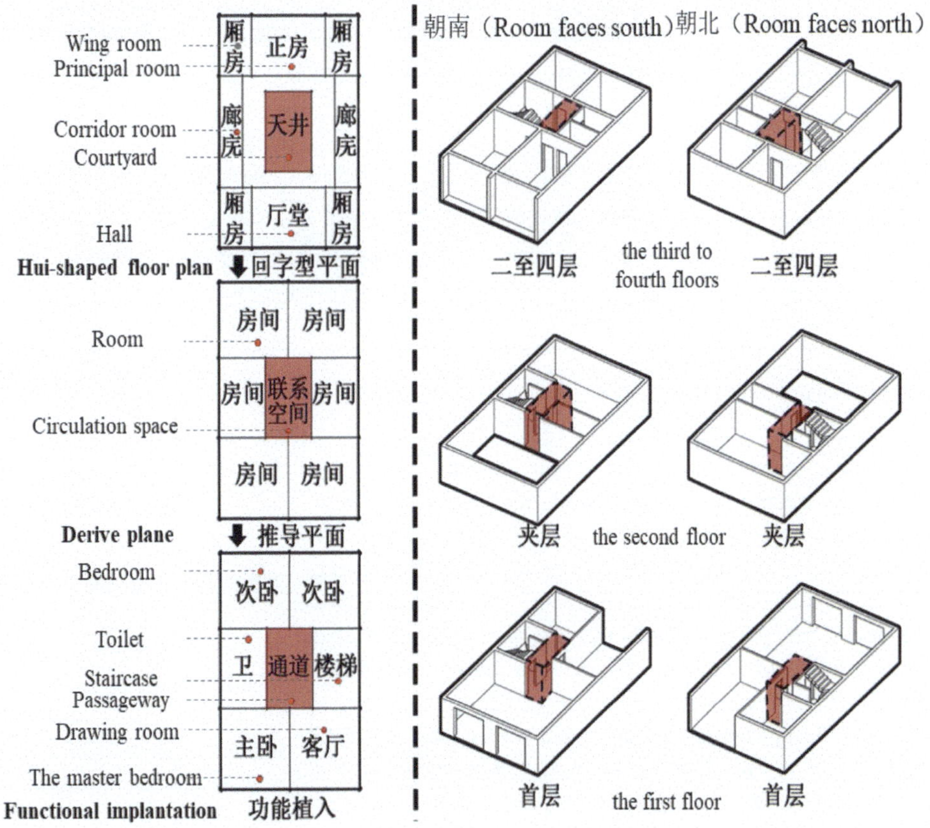

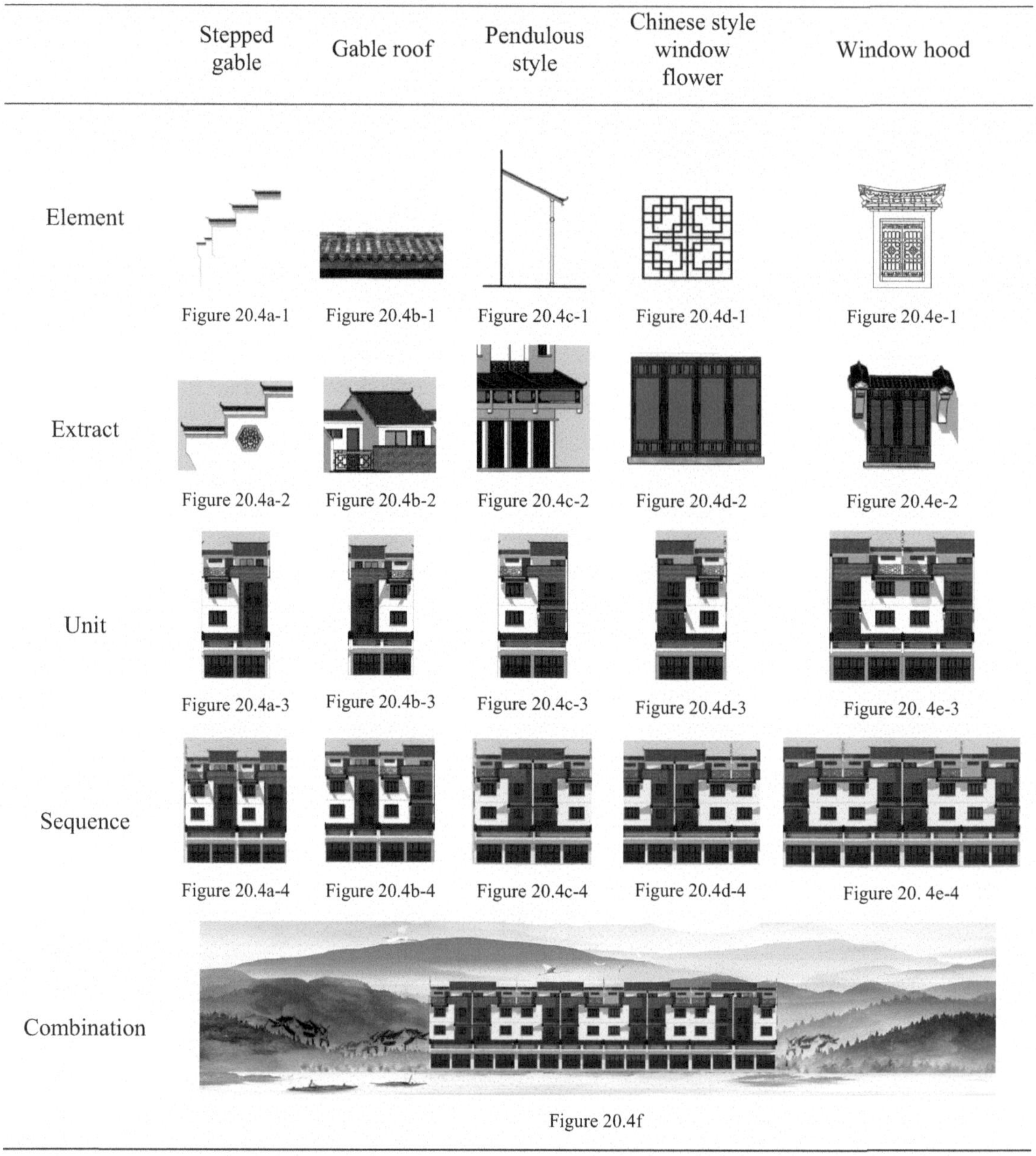

Fig. 4 Composition of Zhe-style façade. *Source* Drawn by authors

eave and adorned with carved window patterns, enhancing its cultural decorative function while also improving ventilation and air circulation. The exposed white walls at the corner recess, dividing walls in the lower-level veranda, and projecting walls on both sides of the window eaves not only correspond to the horsehead wall above but also further divide the building facade in a vertical and horizontal manner, adding to the charm of Zhe-style architecture. The design process of the gable wall is derived from the Element in Fig. 4, which extracts architectural cultural genes. Meanwhile, the zone in the red dotted markings strengthens the decorated effect on the white wall as shown in Fig. 5.

3.2 Continuation and Expression of Three-Dimensional Morphology

3.2.1 The Culture of Street Aspect Ratio

Wengsen Street continues the traditional Zhe-style pattern of blending and integration, opening up towards the surrounding rural landscapes, integrating nature into them, and achieving an organic fusion of distinctive spaces, cultural landscapes, and architectural functions. As a commercial street with distinctive characteristics, the architectural landscapes extend inward along the street while connecting the scattered

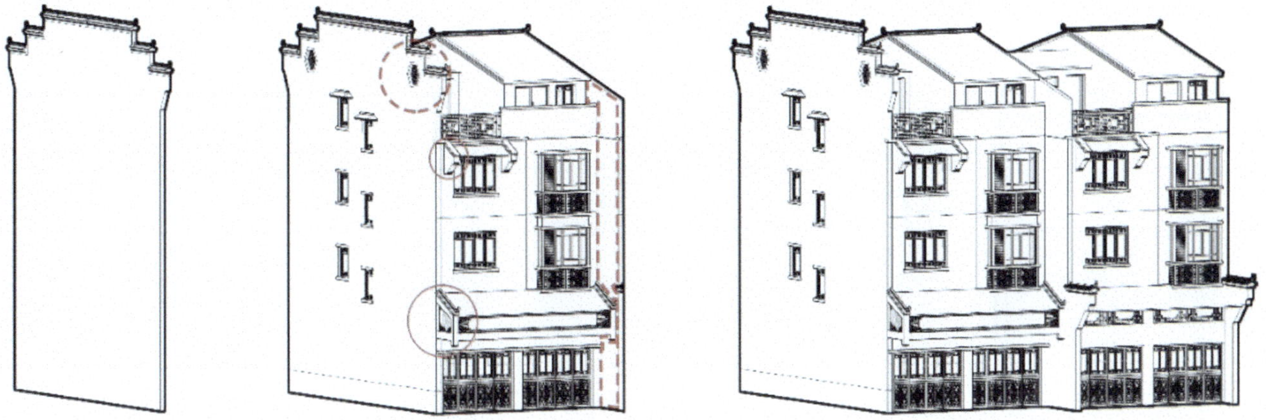

Fig. 5 The process of cultural expression of gable elements. *Source* Drawn by authors

residential buildings in the surrounding area, enhancing the culture of a settlement centered around the street. By emphasizing public spaces, interconnected courtyards, and harmonious integration with the natural environment, it is conducive to promoting social interaction, community life, and a sense of belonging within the community. The comfortable scale of street interactions provides a strong sense of definition and cohesion, allowing people to have a relatively unobstructed view while moving within, thereby giving them a sense of belonging and stability (Fan et al., 2022). W means the width of Wengsen Street, and H means the width of Wengsen Street. The W/H ratio is approximately 1.37, in line with the traditional value of Jiangnan streets about 1–2 (Fang et al., 2022) as shown in Fig. 6.

3.2.2 The Culture of Wind and Rain Corridor

The Wind and Rain Corridor draws inspiration from the traditional single-sloped corridor of southern architecture, providing a transitional transportation space. The eave pillars are abandoned, replaced by "hanging flower pillars" that divide the corridor, creating a more open ground floor "gray space" and providing a wider field of vision, which helps rural visitors locate shops and enhances the atmosphere of the market. The sides of the eaves are adorned with delicate Chinese window flowers, enriching the originally plain white walls with cultural heritage decorations and reflecting the charm of traditional dwellings as shown in Fig. 7.

3.2.3 The Culture of Roof Morphology

The roof design follows the construction hierarchy of traditional Zhe-style dwellings, featuring a traditional hard-sloped roof. The silhouette of the sloped roof is more in line with the gentle and meandering character of people in the southern region compared to a flat roof. Utilizing these elements helps to create architectural sensory designs with

Fig. 6 The aspect ratio of thewengsen street. *Source* photographed by authors

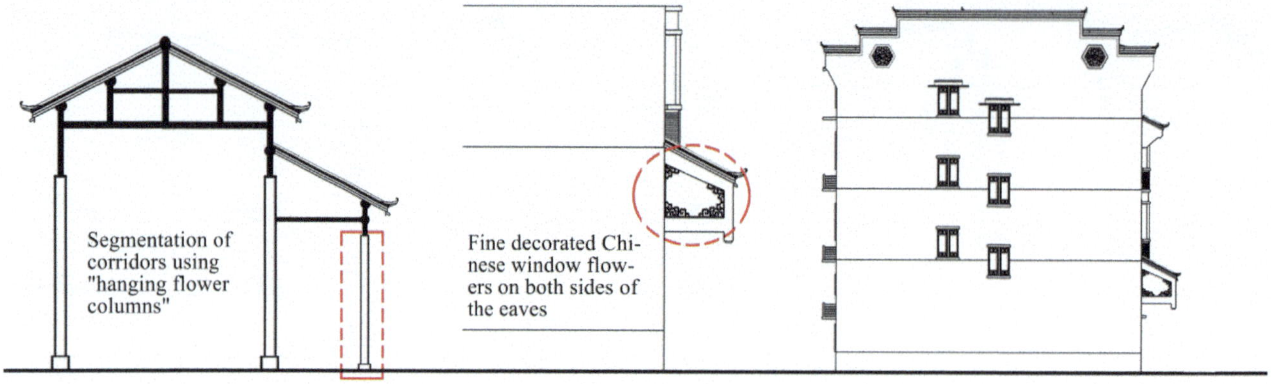

Fig. 7 Cultural expression of wind and rain corridor. *Source* Drawn by authors

visual impact and cultural significance, the layered combination of roofs creates variations in height between building masses, interweaving and resembling the undulating effect of mountains, blending with nature and engaging in a dialogue with the surrounding mountains as shown in Fig. 8. And by integrating these elements into the new traditional Zhe-style houses, the culture of Zhe-style houses can be continued, enriching the architectural environment with symbolic and aesthetic value. From a distance, the roofs of Wengsen Street resemble an ink-wash landscape painting that evokes a sense of nostalgia.

3.3 Continuity and Expression of Material Characteristics

Wood, blue tiles, green bricks, and white lime are important construction materials used by traditional Zhe-style residents. With the development of the economy and modernization, traditional building materials have been updated, with new brick materials replacing traditional green bricks, significantly improving insulation and energy efficiency. White coatings have replaced white lime, enhancing surface strength, improving waterproofing and stain resistance. Especially in mountainous areas with large temperature differences between day and night, the new exterior materials are resistant to rapid temperature changes without cracking. This not only continues the appearance of traditional architecture but also better protects the building structure, reducing the impact of external environmental changes on human comfort indoors.

3.4 Continuity and Expression of Sensory Elements

3.4.1 Overall Color Culture

"Ink and ink smoke charm" is the main color tone of "Zhe-style residential buildings", The overall color scheme of the new traditional Zhe-style residential buildings continues the black, white, and gray tones of traditional Zhe-style dwellings, by using simple colors and combining pink walls and black tiles, a light Chinese landscape painting is outlined, rendering the characteristics of the water town to the extreme (Li et al., 2022). At the same time, enhancing the sense of hierarchy within the "unit-sequence-ensemble" system and expressing the aesthetic pursuit of simplicity and naturalness. On the other hand, deep red, a commonly used traditional Chinese primary color, possesses festive characteristics and has always been loved by the public (Zou et al., 2022). Applying it to the bay windows adds a lively element to the unified black, white, and gray color scheme, enhancing both the style and the traditional visual experience as shown in Fig. 9.

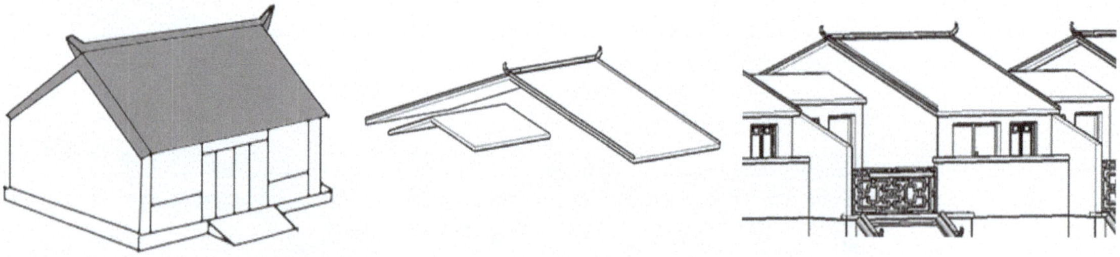

Fig. 8 Cultural expression of roof morphology. *Source* Drawn by authors

Fig. 9 Intentional image of color culture in new traditional Zhe- style residential buildings. *Source* Drawn by authors

3.4.2 Culture of Carved Decorations

The carving techniques for wooden doors and windows in central and western Zhejiang are extremely advanced, with a compact and complete composition, mainly featuring abstract grid flowers, dramatic stories, flowers and plants, and landscape landscapes (Zhang & Long, 2023). Railings, doors and windows, as important components in contact with people, are further optimized through the integration of Zhe-style three-dimensional carving art. Simplified carved window sills, combined with aprons, form Chinese-style lattice doors and windows, which not only preserve the form of traditional Chinese window frames with a modern minimalist approach but also effectively block street noise. The design of the railings adopts a simple and clear traditional floral lattice pattern, serving as a facade decoration that conceals the air conditioning units, enhances the aesthetics of the structure, adds surface depth, and imbues more cultural symbolism as shown in Fig. 10.

4 Conclusions

Zhe-style dwellings possess distinctive architectural and cultural characteristics, and their cultural heritage not only greatly contributes to the development and formation of rural culture but also becomes an indispensable and integral part of rural cultural revitalization. This study proposes that (1) Through four identification models and nine distinctive cultural elements constitute the genetic makeup of traditional Zhe-style dwelling houses; (2) it analyzes the inheritance modes for nine cultural elements, and showcases the process of heritage-driven design in the construction of Zhe-style dwellings in the new era; (3) The construction of Zhe-style dwellings in Wengsen Street goes beyond superficiality, encompassing modern material and spiritual-cultural needs, emphasizing architectural aesthetics and humanistic care, promoting the excellent traditions of architectural culture. Thus, it was discovered that the creation of new traditional Zhe-style residential buildings from the perspective of landscape genes can emphasize regional characteristics, and achieve harmonious integration between architecture and environment through understanding the local natural environment, historical culture, and social customs. At the same time, emphasizing the inheritance of landscape genes also helps to promote the sustainable development of buildings, and to promote cultural identity and social cohesion in communities. Through careful design of architectural aesthetics, spatial layout, and material utilization, we aim to shape the architectural form with Zhe-style residential characteristics and provide a theoretical basis for the protection, inheritance, and revitalization of traditional villages and buildings in Zhe-style rural revitalization projects, it can al so provide rich cultural connotations and sustainable development paths for contemporary architectural design, promoting innovation and inheritance of traditional Chinese architectural culture.

Acknowledgements This research was supported by Zhejiang University of Science & Technology (NO. F701104K02; 0,101,104,502).

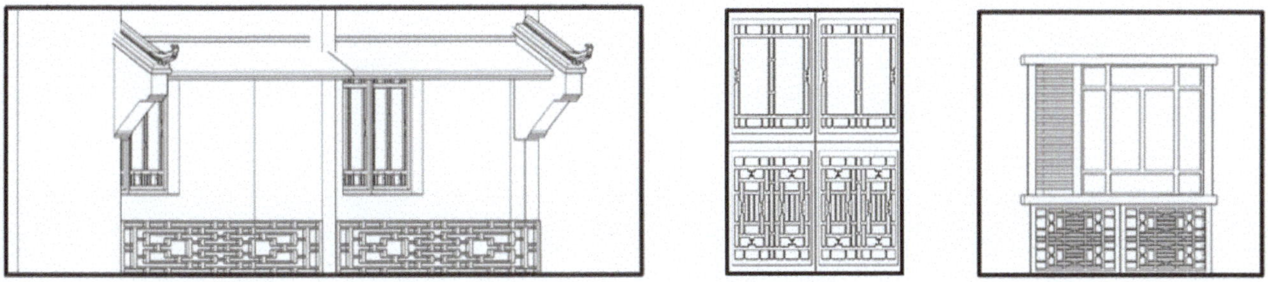

Fig. 10 Expression of decorative cultural elements. *Source* Drawn by authors

References

Fan, Z., Chen, C., & Huang, H. (2022). Immersive cultural heritage digital documentation and information service for historical Figure metaverse: a case of Zhu Xi, Song Dynasty, China. *Heritage Science, 10*(1), 148. https://doi.org/10.1186/s40494-022-00749-8

Fang, J., Zhang, Y., Zhang, Y., Guo, H., & Sun, Z. (2022, Apr 28). High-definition survey of architectural heritage fusing multisensors—The case of Beamless Hall at Linggu Temple in Nanjing, China. *Sensors (Basel), 22*(9). https://doi.org/10.3390/s22093369

Ganobjak, M., Brunner, S., Hofmann, J., Klar, V., Ledermann, M., Herzog, V., Kampfen, B., Kilian, R., Wehdorn, M., & Wernery, J. (2023, Oct 13). Current trends in aerogel use in heritage buildings: Case studies from the aerogel architecture award 2021. *Gels, 9*(10). https://doi.org/10.3390/gels9100814

Jiang, W., Yao, K., Lin, Q., Zhao, Y., & Lu, D. (2023, Nov 22). Optical fiber sensing-aided 3D-printed replacement parts for enhancing the sensing ability of architectural heritage. *Micromachines (Basel), 14*(12). https://doi.org/10.3390/mi14122135

Li, Y., Liu, Y., & Wen, Y. (2022). New algorithm of traditional Chinese medicine and protection of intangible cultural heritage based on big data deep learning. *BioMed Research International, 2022*, 1645204. https://doi.org/10.1155/2022/1645204

Liu, X., & Li, M. (2020 May). Safeguarding intangible cultural heritage to promote mental healthcare in China: Challenges to maintaining the sustainability of safeguarding efforts. *International Journal of Social Psychiatry, 66*(3), 311–313. https://doi.org/10.1177/0020764020904752

Maselli, G., Cucco, P., Nestico, A., & Ribera, F. (2024, Jun). Historical heritage-MultiCriteria Decision Method (H-MCDM) to prioritize intervention strategies for the adaptive reuse of valuable architectural assets. *MethodsX, 12*, 102487. https://doi.org/10.1016/j.mex.2023.102487

Sun, Z., & Zhang, Y. (2019, Jan 25). Accuracy evaluation of videogrammetry using a low-cost spherical camera for narrow architectural heritage: An observational study with variable baselines and blur filters. *Sensors (Basel), 19*(3). https://doi.org/10.3390/s19030496

Wang, L. (2022). Evaluation of the practical effects of environmental measures in the conservation of architectural heritage in Yan'an based on recurrent neural networks. *International Journal of Environmental Research and Public Health, 2022*, 3749482. https://doi.org/10.1155/2022/3749482

Wang, X., Li, H., Wang, Y., & Zhao, X. (2022, Aug 20). Assessing climate risk related to precipitation on cultural heritage at the provincial level in China. *Science of the Total Environment, 835*, 155489. https://doi.org/10.1016/j.scitotenv.2022.155489

Yang, J., Wang, L., & Wei, S. (2022, Dec 30). Spatial variation and its local influencing factors of intangible cultural heritage development along the grand canal in China. *International Journal of Environmental Research and Public Health, 20*(1). https://doi.org/10.3390/ijerph20010662

You, X., Zhang, Y., Tu, Z., Xu, L., Li, L., Lin, R., Chen, K., Chen, S., & Ren, W. (2023, Feb 28). Research on the sustainable renewal of architectural heritage sites from the perspective of extenics-using the example of tulou renovations in LantianVillage, Longyan City. *International Journal of Environmental Research and Public Health, 20*(5). https://doi.org/10.3390/ijerph20054378

Yu, H., Verburg, P. H., Liu, L., & Eitelberg, D. A. (2016, Jun). Spatial analysis of cultural heritage landscapes in Rural China: Land use change and its risks for conservation. *Environmental Management, 57*(6), 1304–1318. https://doi.org/10.1007/s00267-016-0683-5

Zhang, H., & Long, S. (2023). Evaluation of attraction and spatial pattern analysis of world cultural and natural heritage tourism resources in China. *Plos One, 18*(8), e0289093. https://doi.org/10.1371/journal.pone.0289093

Zhang, Z., Zou, Y., & Xiao, W. (2023). Exploration of a virtual restoration practice route for architectural heritage based on evidence-based design: A case study of the Bagong House. *Heritage Science, 11*(1), 35. https://doi.org/10.1186/s40494-023-00878-8

Zou, H., Liu, Y., Li, B., & Luo, W. (2022, Oct 11). Sustainable development efficiency of cultural landscape heritage in urban fringe based on GIS-DEA-MI, a case study of Wuhan, China. *International Journal of Environmental Research and Public Health, 19*(20). https://doi.org/10.3390/ijerph192013061

Vegetation as Adaptive Project: Regenerating Porta Romana Shaping a New Character

Kevin Santus

Abstract

Nature-based solutions working on adapting urban spaces could inform the imperative transition of the contemporary project. From a design perspective, this means reshaping our cities' relations and physical experience, overcoming the idea of just a technical answer to the climate crisis. Thus, an urgent question should address how architecture could translate technical/design tools into spatial, experiential, and compelling projects. The contribution focuses on the Porta Romana (Milan, Italy) transformation, in which the author collaborated in contact with the developer (COIMA REM), attended design briefs, and interviewed the design teams (among others with MDP, Diller Scofidio + Renfro, SOM). Porta Romana regeneration focuses on a neglected railyard to host the Olympic Village for the Winter Olympic games in 2026 and an urban park, structuring a new adaptive space. The project is driven by the scope of using nature as an adaptive tool, rethinking the urban experience, and tackling urban climate fragilities such as floods and heat island. The project presents nature-based solutions to structure a new design vision for the city, regaining proximity between nature and architecture. This contribution highlights architectural opportunities and solutions that the climate-related issues could arise, critically addressing the project of the grounds and crownings.

Keywords

Climate crisis · Nature-based solutions · Urban regeneration · Architectural project · Neighbourhood scale

K. Santus (✉)
Dipartimento di Architettura e Studi Urbani, Politecnico di Milano, Milano, Italy
e-mail: kevin.santus@polimi.it

1 Introduction

1.1 The Form of Naturality in Regeneration Processes to Face the Climate Crisis

The global climate breakdown directly impacts spatial construction (Turan, 2019). Often perceived as a technical issue, architects are now requested to implement strategies to cope with the climate crisis, considering the micro-climate condition and possible climate risks such as floods or urban heat islands, and to transform the physical space consequentially (Dawson, 2019).

However, this theme opens a new possible condition for the project beyond the simple application of technical solutions. Indeed, the transformations that are renovating parts of urban areas could engrave a new character on the space, rethinking, for example the balance between minerality and naturality, not only implementing vegetation, but radically rethinking the morphological relevance of vegetation, ground, and in general naturality, in the construction of space (Peleman et al., 2022). Hence, this new project condition needs to be understood with a double perspective. The first is the quantitative one, referring to the impacts that these transformations could have. So effectively increasing the urban water drainage or reducing the local temperature during heat waves. On the other hand, bringing back vegetation in the city project means rethinking the cultural relationship between the minerality of the city's architecture and the naturality, going over the duality that characterized the work of architects for centuries in urban areas (Silva, 2021).

To do so, the current discussion on the transformation of the project and the climate crisis needs to overlook current projects, trying to seize the design implication and those elements which can be impactful in addressing the risks prompted by the climate crisis and, at the same time, resonate in the construction of the form.

B. He et al. (eds.), *Urban Climate Change Adaptation*, Advances in Science, Technology & Innovation,
https://doi.org/10.1007/978-3-031-65088-8_21

2 Method

This contribution presents the results and reflection developed through an on-field observation conducted with the operators of COIMA REM (developers of the project), which took place in 2022 during the finalization of the Winter Olympic Village project phases. The process of regeneration of the Porta Romana railyard was carried out, among others, by SOM, MDP, Diller Scofidio+Renfro, and Arup. The project is used as a case study where a series of critical reflections presenting how nature-based solutions are deployed and become a climate-adaptive tool to face the new risks that are spreading and increasing worldwide in metropolitan areas.

The article reports the study conducted using project guidelines and design reports, explicitly focusing on the role of vegetation as a potential design tool to adapt the urban space and impact the architectural construction. Thus, the article gathers all the public materials related to the regeneration process and focuses on the parts where nature emerged as the project's main feature, finding a strong connection with the ecological and adaptive aim of the regeneration. Starting from this, to deepen the adaptive potential of using nature, the article frames nature as an architectural theme that could inform the current practice and shape a new character of the space, strongly related to the new climate and environmental condition.

More specifically, to spotlight those themes, the contribution used two types of study. The first concerned the study and gathering of public reports and documents related to the regeneration of Porta Romana. Secondly, the contribution presents a critical reflection and interpretation of a series of dialogues with the design teams, which has been vital to deepen the idea of using vegetation not as an ideological or technical answer to a need, but rather as proper architectural construction of space. These dialogues in the form of interviews, the documents, and the project's interpretation are then synthesized and presented in the following contribution.

3 Result and Discussion

3.1 Porta Romana: A Former Railyard Regeneration for a Future-Oriented Project

During the mid-1800s, railroad yards were developed in Milan, contributing significantly to developing and expanding national freight rail lines to and from the city. Focusing on the southern metropolitan area, we find one of the main former yards opened at the end of the nineteenth century: Porta Romana railyard. Once considered one of the beating hearts of the Milanese industry, it supplied valuable raw materials to the many factories and workshops in the neighborhood.

The sequence of urban plans, Beruto's one made in 1884, the Pavia-Masera in 1909, and the Albertini in 1934, defined the expansion of the city around Porta Romana, expanding into the surrounding agricultural plain between Cerchia dei Bastioni, Roggia Vettabbia, and Cavo Redefossi. In this area's evolution, the Beruto Plan was crucial in outlining the morphology of the city's expansions: rail networks were expanded, and plans for building growth in the suburbs, beyond the old Spanish walls, and redevelopment in the historic center were implemented (Oliva, 2008). The planned blocks present a character of introversion, as buildings were placed around their perimeter, while industrial areas were placed on the edges between the city and the countryside. In this context, the railyard had a prominent role in constituting a functional apparatus of the town and a morphological interruption for the fabric (Reggiori, 1947). Furthermore, it becomes evident that the railyard is not only a vast empty space separating the city center from the suburbs, but it also constitutes a division between a more compact and permeable urban fabric to the north, following the design of the Beruto Plan of 1884, and a more fragmented fabric to the south, characterized by various intervals between manufacturing and industrial settlements. Together with the built fabrics, the area developed a vast and varied system of green areas, including the nearby Alessandrina Ravizza Park, the Industrial Memories Park, the Marcello Candia Garden, and long tree-lined avenues, which extend to the Parco Agricolo Sud. This vast natural area dialogues with the surrounding countryside. Together with the vegetation system, the railyard sits on an extensive water system that includes numerous waterways buried over the years, which connect to the Roggia Vettabbia and the Cavo Redefossi. These two waterways were created to manage and distribute water within the city, but over time, they have been covered to facilitate the city center's viability.

For years, the site maintained its importance for the city's metabolism; however, over the past few decades, rail freight transport has declined due to structural changes and competition from road transport. This has led to a crisis in the rail sector, with many facilities in Milan that were once essential to the city's economy and infrastructure now abandoned and unused. The role of Porta Romana railyards transformed, and the tracks crossing the big urban void became used only as a passage for the urban and suburban trains. Inside the yard area, greenery grew, allowing vegetation to grow spontaneously and regain its space, creating a proper third landscape.

Starting from the beginning of the twenty-first century, a rising interest in urban void regeneration produced a fertile horizon of renovation, which also had something in common with the other city's railyards. In 2005, LAND with AIM promoted a study named "Raggi Verdi," which focuses on the ecological potential of the decommissioned railway network (LAND, 2014), and in recent years, become of particular interest, especially along the southern railway belt, such as that of Porta Romana and San Cristoforo. This study aimed to improve the ecological functionality of the municipal area by establishing new ecological corridors on abandoned railway lines and increasing the city's natural green spaces.

After ten years, in 2017, the municipality signed a strategic vision for the former railyards, indicating their renovation's priorities and aims. In this period, new themes emerged, globally and locally, and issues such as the climate crisis also took the front in the local Milanese agenda (Comune di Milano, 2017). Hence, in 2020, a partnership between Milano municipality and local developers, with COIMA SGR as the management company of the Fondo di Porta Romana, launched the competition for the definition of the master plan for the regeneration of Porta Romana (Fig. 1). Following, further competition was decided for the design of the Olympic village, which would occupy the area west of the yard, thus initiating the regeneration process. In this competition, the urban regeneration looked at the project as a potential driver for an ecological transformation of the area, where being able to address the climate risks through the design of architecture and open spaces, so tackling problems related to the heat island effect and floods, increasingly occurring in the metropolitan area. From that, the guidelines for the design spotlight the attention on green spaces, which, also in consideration of the municipal laws, should cover at least half of the plot, while the architecture should be able to implement nature-based solutions such as green roofs.

The challenge defined the possibility of transformation, where the former railyard could promote ecological sustainability through the creation of green spaces and ecological corridors. An idea of the city that embraces pluralism and promotes dialectical relations between different fragments of the city and rethinking the relationship between minerality and nature, could reveal a potential attitude of the project as "an ecological action that pushes the existing into the future, transforming waste into prominent figures" (Cozza & Valente, 2014, pp. 72–73). In this regeneration resides the possibility to think at architecture with a new alliance with nature, in which the latter is not considered as a punctual solution, as a feature that could impact the urban character of a place, which gives new meaning to the complexity of the contemporary project, entangling environment, sustainability, and project.

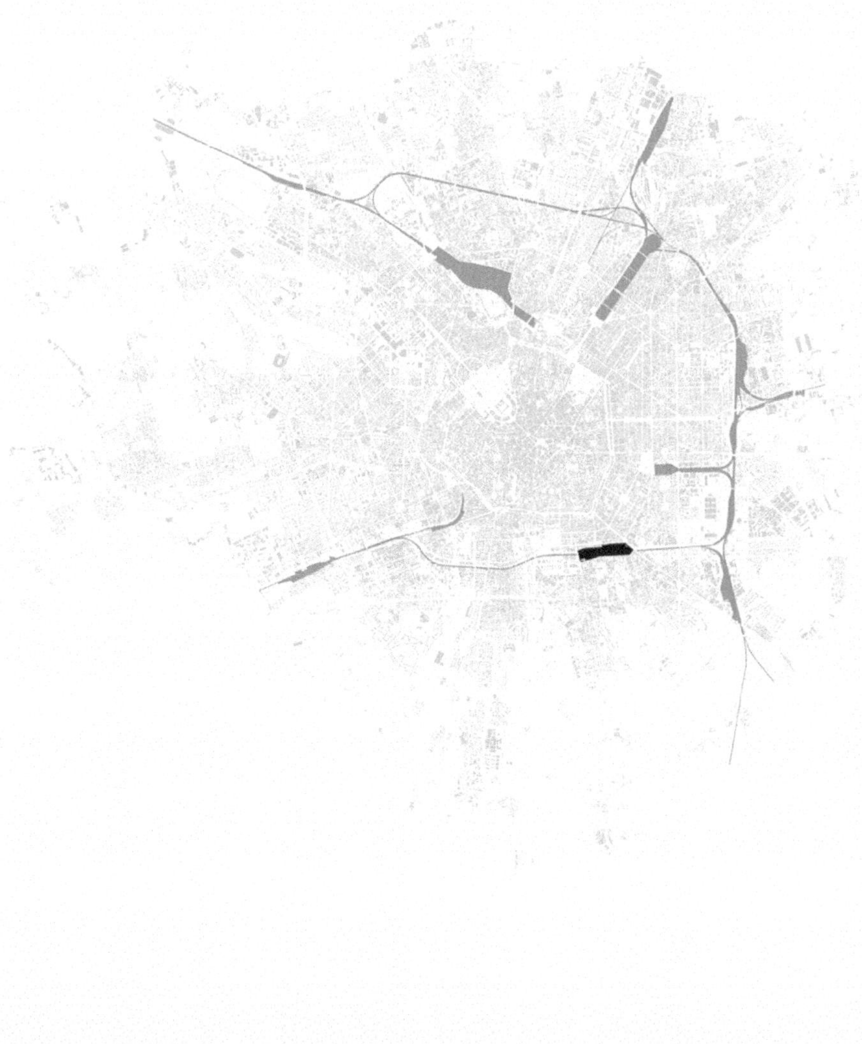

Fig. 1 Urban identification of Porta Romana railyard within the urban railway system of Milan. Drawing by the author

3.2 Porta Romana: Transforming a Neglected Void into an Adaptive System

Due to its location and geographical and spatial characteristics, Milan is a city with high environmental vulnerability to the impacts of climate change. In fact, there is a steady increase in temperatures and the frequency of precipitation. In the face of these climate risks, cities like Milan are increasingly turning to nature as an effective design tool to enhance resilience and address environmental challenges, as demonstrated by the recent approval of the PAC (Comune di Milano, 2019), which has in the climate adaptation a pillar for the ecological transition to be riched in the next three decades. By incorporating natural elements and spatial devices into urban design and architecture, Milan, can mitigate the impact of climate change and create a more sustainable and livable city. More specifically, by introducing nature-based solutions in the design of new buildings and public spaces, the aim is to contrast the heat island effect and increase the permeable areas to implement the drainage capacity to mitigate the disruptive impact of floods.

The renewal of the Porta Romana railyard (Fig. 2) started framing the regeneration as a possibility of implementing an ecological reservoir for the city, where buildings and public space could work in synergy to set a new adaptive system. This works at the neighborhood level, highlighting the intermediate scale as a blending point between architecture and the city, worked in a direction that sees adaptation as a process to rebalance the presence of nature in the city.

Nature, specifically vegetation, was framed as ecological and social infrastructure, presenting a unique opportunity to introduce new green fragments in the city, effectively addressing the ongoing environmental emergency. Indeed, a key aspect was the expansion of green spaces and biodiversity, as outlined in the metropolitan area's forestation plan. The redevelopment of disused railway zones, as readable in the guidelines for the competition, could play a vital role in this vision, allowing for the creation of substantial green areas that seamlessly integrate with the existing public spaces and urban fabric. Moreover, in Porta Romana, the competition also asked to use vegetation and ground as project tools to rethink the urban orography, bridging the gap created by the rail tracks (Fig. 3). The Program Agreement outlines a comprehensive approach that goes beyond merely creating new parks in former railway yard areas with both environmental and social purposes. Indeed, the guidelines for the masterplan asked to use vegetation as possible revitalization of the green spaces along the railway belt, aiming to establish a cohesive ecological network within the urban territory. The forthcoming public green spaces in the railway yard will act as physical connections, like stepping stones, linking to the linear green layout of the railway buffer strips and the prominent environmental systems in this urban section. In this, also the building should be part of this systematic ecological rethinking of the project, actively participating as adaptive elements.

Hence, the transformation project was based on a solid Science-based approach with a scientific assessment of climate change scenarios and potential impacts as an ex-ante analysis of the project's configuration. In this way, the design team was able to develop a practical usage of adaptive devices, but also understand which climate risks should face. It was clear that it was necessary to implement green and blue infrastructure to reduce the future impacts of flooding and heat waves, using regeneration as a possibility to reset a microclimate. Thus, the morphological understanding of the area went toward building massing and creating a large portion of vegetated space, which draws green strips in the plot.

Looking at the masterplan emphasizes the presence of a central park, which takes the shape of a hill reconnecting the southern and northern parts of the city, while also allowing the presence of a suspended forest, implementing the urban biodiversity. Close to these, a sequence of eco-zones creates a green corridor which, together with raingardens, implement the permeability of the ground. Interestingly, these spaces are not thought to host human presence, having abundant vegetation and different species of local animals.

Moreover, through the use of green courtyards and green roofs, the building transforms the railyard into an adaptive system, where nature plays a prominent role in defining the character of regeneration. Finally, looking at the Olympic Village architecture, it is possible to notice a metal grid on which vegetation could grow, implementing the shading and working in a bioclimatic improvement of the microclimate through nature.

3.3 Spatial Elements to Design Climate Risks Adaptation

Considering the regeneration of Porta Romana, it is possible to identify three elements that could be seen as design strategies that are able to face specific climate risks. This doesn't mean that they create an abacus of solutions. Instead, they represent devices from which it is possible to reflect on the role of nature as an adaptive element (van Eekelen, 2020).

Moreover, it is crucial to notice how the three elements are merged in a system in which their effectiveness is derived from the mutual synergy they create. This is clearer when looking at the water management system, in which all vegetation systems work as infrastructure (Dover, 2015)

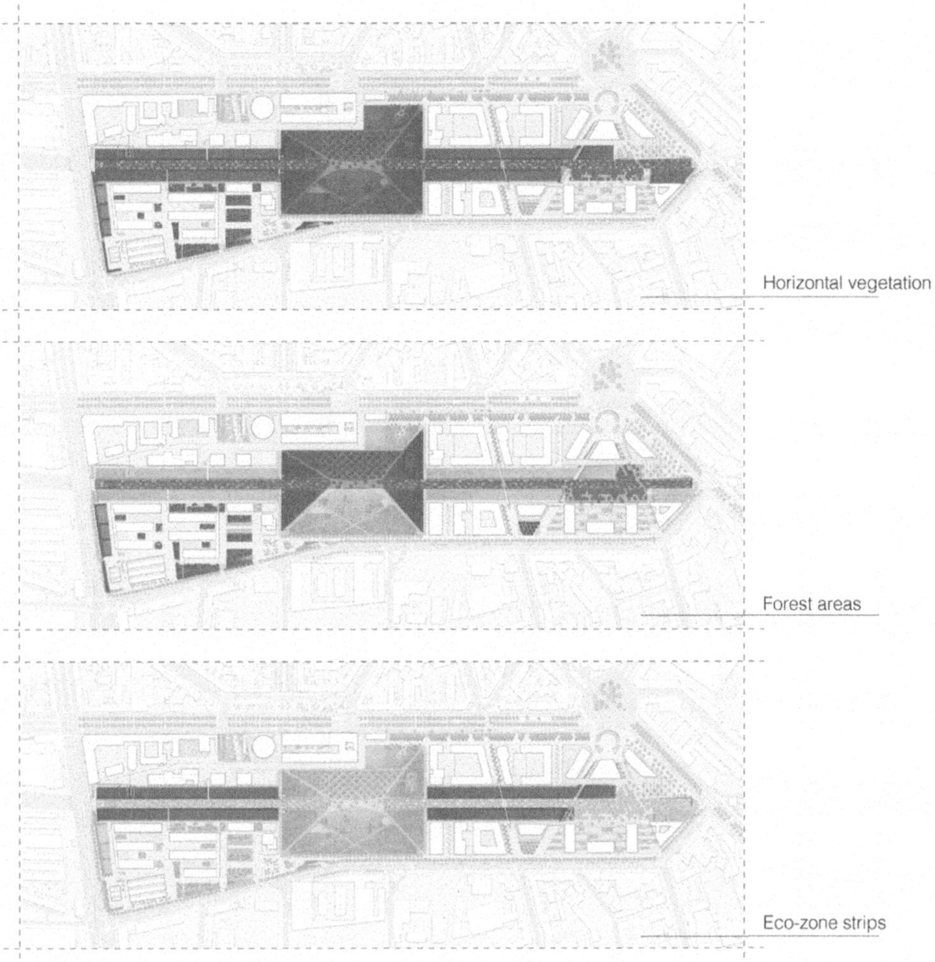

Fig. 2 The vegetation composition reveals the various layers which structure the new masterplan. Elaboration by the author

to manage the water infiltration. Similarly, the heat island effect mitigation results from a spread green system, where the whole is more than the sum of the single parts.

The three elements (Figs. 4 and 5) synthesized are:

Eco-zones: As visible at the border of the rail track, they establish green corridors and wildlife habitats throughout the plot, which allows for better biodiversity and encourages species migration and adaptation to changing environmental conditions. The creation of such areas was conducted through a selection of species that are part of the Po Valley and Milanese area, to allow better integration with the local climate conditions, but also taking into account the climate forecast for the coming years. This allows the project of green vegetated areas to structure spaces capable of adapting to the urban environment and being adaptive to changes in the years to come (Kabisch et al., 2017).

Horizontal vegetation and grounds: Integrating green infrastructure, such as urban parks, green roofs, and porous surfaces, helps absorb and manage excess rainwater during heavy storms. These green spaces act as sponges, reducing the risk of flooding and easing the pressure on traditional drainage systems. Working on the horizontal level allows the implementation of nature-based solutions, such as rain gardens, and green infrastructure, which can help manage stormwater, mitigate flooding, and support overall climate resilience. Moreover, these elements of adaptation hint at the typological potential of reimagining the building crowning as artificial grounds (Gissen, 2019) while highlighting the potential of the ground works.

Urban Forests: Not only as an action of generic forestation, planting trees in dense areas to create urban forests can mitigate the urban heat island effect, and construct thick and solid volumes of nature amid the urban minerality. Trees provide shade, evaporative cooling, and improve air quality, making the city more comfortable and resilient during heat waves, so tackling the regulation of the microclimate and rethinking the experience of public green space (Treib, 2021). An example is the courtyards within the Olympic Village buildings, which are designed to

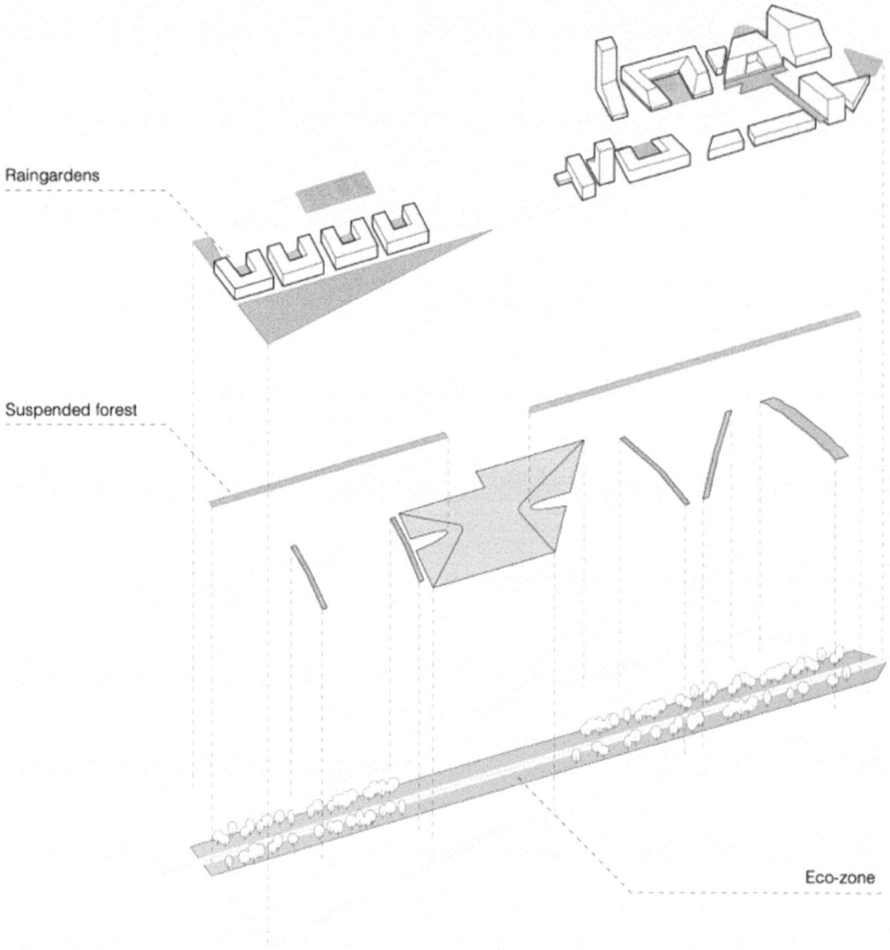

Fig. 3 The drawing synthesizes the elements composing the overall layout using vegetation as a feature. Drawing by the author

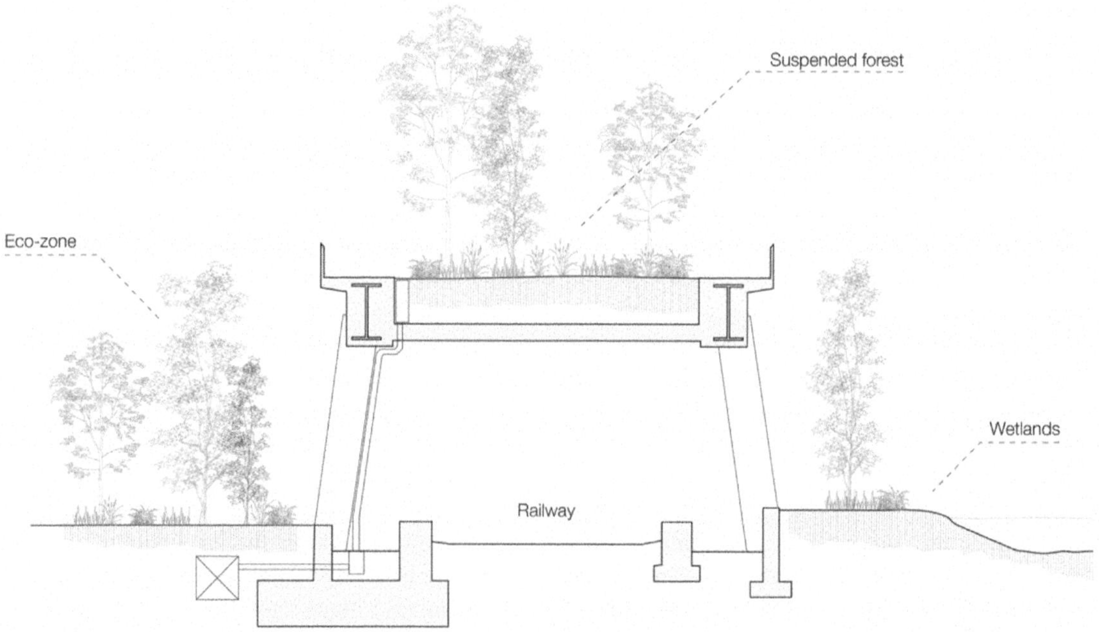

Fig. 4 The three vegetation components used merge the eco-zone, the suspended forest, and wetlands. Drawing by the author

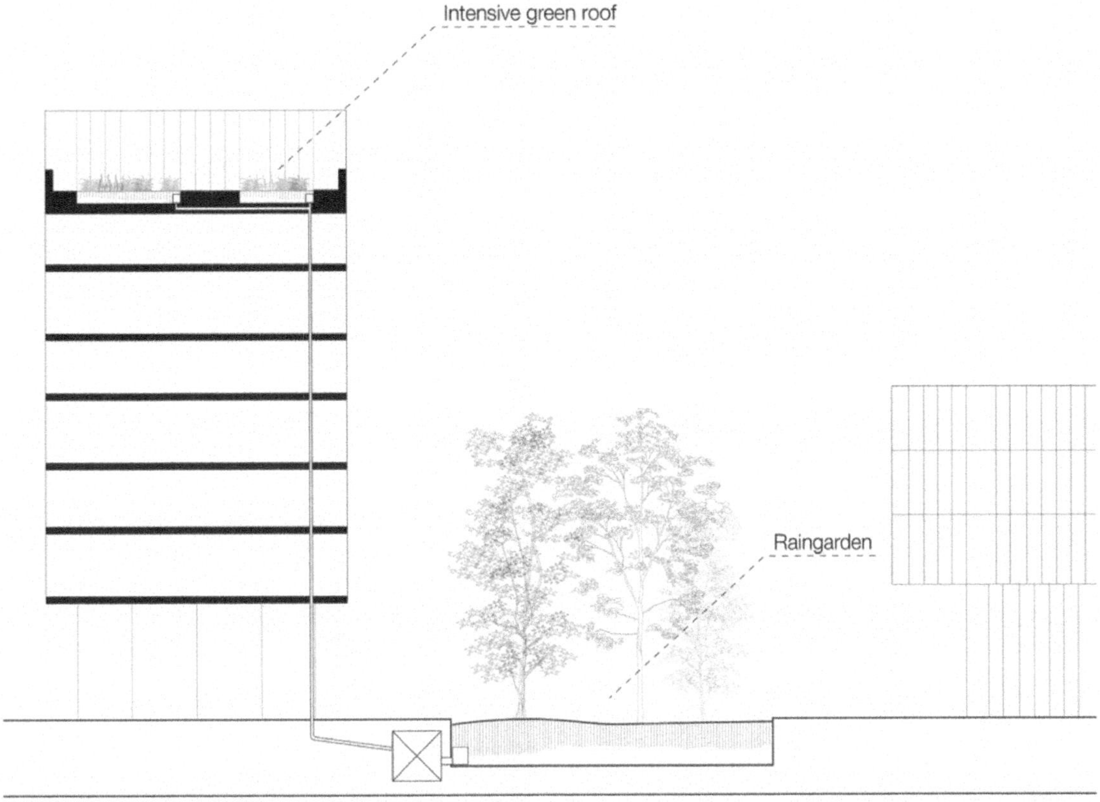

Fig. 5 Raingardens and intensive green roofs are designed with the built environment. Drawing by the author

accommodate green areas with very lush forest-like vegetation. This is to generate an internal character to the courts capable of combining minerality and nature and mitigating the heat island effect.

Together, these three elements display a space design method that overcomes the separation between architecture and nature (Latour, 2020), and draws ways to face the climate risk through space design. Here, the technical knowledge is related to the form and materials of space, in which nature becomes a living element and the connotative feature.

4 Conclusion

4.1 Vegetation as an Architectural Theme

Looking at the project, it's possible to synthesize two methodological approaches to nature to increase the adaptiveness of the project and the different scales. Primarily, looking at the work of MDP and Elizabeth Diller with LAND, there is a clear landscape orientation, which could help in understanding how the presence of greenery could impact the definition of the contemporary ecological character of our cities. Using vegetation to adapt the urban space means rethinking the minerality of spaces, their usage, the biodiversity present in the town, and lastly, reconsidering the control of the space in time. There is a shift in looking at nature, often pervasive, but still controlled, which, as presented in Porta Romana, frames nature as a tool of control. Raingardens, as much as the eco-zone, are used to configure a sustainable urban drainage system, but at the same time, they impact the character of the space. Secondly, using vegetation as an adaptive tool spotlights the relationship between architecture and nature. Green roofs and green façades connote relevant typological parts of architecture, in which, for example, the crowning element of the building becomes an artificial ground, while growing vegetation on a metal structure creates a scenario and interface between the architecture and public space.

Green spaces could be considered not only for their vital ecological role; indeed, the project shows how they serve as the backbone of new settlements. More than just a figure/background relationship, vegetation is considered a design tool to build the experience of the space. Moreover, beyond its environmental significance, Porta Romana Park holds great potential as an inclusive public space, fostering community connections. At the same time, the new arrangement of open spaces and public areas will facilitate a connection between the new settlements and the existing peripherical neighborhoods. In this term, the usage of nature also becomes a morphological theme of the project, building the space in terms of masses, giving directions, etc., which is part of the architectural construction of the plot.

If, on the one hand, the presence of nature in the project for the Olympic Village, and in general for the regeneration of Porta Romana, is connected to necessities raised from the competition guidelines and following the municipal regulations; on the other hand, it is possible to reflect on the potentialities that using nature could be produced. In other words, the presented project shows how nature-based solutions could become an experiential feature in the city and how they can constitute architectural devices to work in an adaptation perspective toward climate risks. In other words, nature become an architectural theme where "The sustainability pursuit results in the need for a new approach to the relationship between architecture and nature. The elements of the natural world, so far desired mainly due to their aesthetics, have now become useful elements, necessary for gaining energy, saving water, and improving air quality. [...] Natural elements have gained a new role of shaping development and public spaces and addressing the new, unified perception of architectural and urban space." (Haupt, 2014, p. 2).

Because of that, a needed reflection concerns the new character of the urban space, where the project uses nature-based solutions as tools to define a climate infrastructure for the architectural project. Here, the architectural project using vegetation could "de-emphasizes the physical boundary between residence and surrounding nature and establishes instead a spiritual threshold. [...] Today, unfortunately, nature has lost much of its former abundance, just as we have enfeebled our ability to perceive nature. Contemporary architecture, thus, has a role to play in providing people with architectural places that make them feel the presence of nature. When it does this, architecture transforms nature through abstraction, changing its meaning. When water, wind, light, rain, and other elements of nature are abstracted within architecture, architecture becomes a place where people and nature confront each other under a sustained sense of tension. I believe it is this feeling of tension that will awaken the spiritual sensibilities latent in contemporary humanity." (Ando, 1991, p. 75). In this way, adaptation won't be only a technical answer to the climate risks. Still, it will be an opportunity to question the contemporary role of architecture, typologies, and morphological definition, constructing a built environment more capable of interpreting climate and environmental features.

For this article, the author used only public information concerning the regeneration project of the Porta Romana railyard. The author thanks the team of COIMA REM and the design team involved in the Olympic Village Project, in which the author took part during the Ph.D. traineeship.

This article is part of an on-field analysis presented in the Ph.D. thesis by Kevin Santus, supervised by Prof. Ilaria Valente (DAStU) and Monica Lavagna (DABC) at the Politecnico di Milano, within the Ph.D. in Architecture, Urban, and Interior Design, coordinated by Prof. Alessandro Rocca.

Acknowledgements For this article, the author used only public information concerning the regeneration project of the Porta Romana railyard. The author thanks the team of COIMA REM and the design team involved in the Olympic Village Project, in which the author took part during the Ph.D. traineeship.

This article is part of an on-field analysis presented in the Ph.D. thesis by Kevin Santus, supervised by Prof. Ilaria Valente (DAStU) and Monica Lavagna (DABC) at the Politecnico di Milano, within the Ph.D. in Architecture, Urban, and Interior Design, coordinated by Prof. Alessandro Rocca.

References

Ando, T. (1991). Beyond horizons in architecture. In K. Frampton (ed.), Tadao Ando. The Museum of Modern Art.

Comune di Milano. (2019). *Piano Aria e Clima (PAC).*

Comune di Milano. (2017). *Documento di visione strategica, Scali ferroviari, Allegato AdP.*

Cozza C., & Valente I. (Eds.) (2014). La freccia del tempo. In *Ricerche di architettura delle infrastrutture*. Pearson.

Dawson, A. (2019). Extreme Cities. In *The peril and promise of urban life in the climate change*. Verso.

Dover, J. W. (2015). *Green infrastructure: Incorporating plants and enhancing biodiversity in buildings and urban environments.* Taylor & Francis Ltd.

Gissen, D. (2019). Nature. *AA Files, 76*, 126–129.

Haupt, P. (2014). Architecture vs. nature: a reinvented relationship. *Eco-Architecture, 142*. https://doi.org/10.2495/ARC140041

Kabisch, N., Korn, H., Jutta, S., & Bonn, A. (2017). *Nature-based solutions to climate change adaptation in Urban Areas: Linkages between science, policy and practice*. Springer Nature.

LAND. (2014). Raggi Verdi. In *Green Vision for Milano*. Aades.
Latour, B. (2020). La crise sanitaire incite à se préparer à la mutation climatique. *Le Monde, 2020*(04), 01.
Oliva, F. (2008). *L'urbanistica di Milano*. Hoepli.
Peleman, D., Ronner, E., Barcelloni, M., & Viganò, P. (2022). Exploring the Soil—Not a Sentimental Journey. *OASE | Journal of Architecture—A project of the soil, 110*, 4–15.
Reggiori, F. (1947). Milano. 1800–1943. Edizioni del Milione.
Silva, C. (2021). Developing climate-responsive cities: Exploring the environmental role of interstitial spaces of Santiago de Chile. *Journal of the British Academy, 9*, 107–142.
Treib, M. (2021). *The aesthetics of contemporary planting design*. Oro Editions.
Turan, N. (2019). *Architecture as measure*. Actar Publisher.
Van Eekelen, B. (Ed). (2020). Building with nature. In *Creating implementing and upscaling nature-based solutions*. NAI/010 Publsher.

The Carbon Pricing Policy in Indonesia Towards Paris Agreement Implementation

Joni Jupesta, Alue Dohong, and Bandung Sahari

Abstract

Carbon pricing is a policy instrument aimed to accelerate Paris Agreement goal. Due to its vast natural resources, Indonesia intends to promote Nature-Based Solutions (NBS) as part of its national and global climate mitigation plan, as it also yields other co-benefits in terms of climate adaptation and biodiversity protection. The new carbon exchange has been launched to expedite carbon trading in Indonesia. This study examines the policy impacts of carbon pricing in Indonesia through emissions trading schemes, crediting mechanisms, result-based payments, and carbon taxes in the FOLU, energy, and waste sectors.

Keywords

Paris Agreement · Climate change mitigation · NBS · Carbon pricing · Indonesia

1 Introduction

The latest IPCC 6th Assessment Report revealed total anthropogenic greenhouse gas (GHG) emissions during the period from 2010 to 2019 as cumulative net CO_2 emissions since 1850 (Shukla et al., 2022). In 2019, approximately 34% (20 $GtCO_2eq.$) of total anthropogenic GHG emissions originated from the energy supply sector, 24% (14 $GtCO_2eq.$) from industry, 22% (15 $GtCO_2eq.$) from agriculture, forestry, and land use (AFOLU), 15% (8 $GtCO_2eq.$) from transport, and 6% (3.3 GtCO2eq.) from buildings. The primary pathways to achieve the goals of the Paris Agreement involve reducing GHG emissions (ER). Achieving net-zero GHG emissions requires further carbon dioxide removal (CDR) through negative emissions technologies. This net-zero target has been adopted by numerous countries, cities, and businesses. Within climate mitigation strategies, the role of CDR is acknowledged as having risks for successful mitigation, prompting proposals to segregate targets for ER and CDR. For instance, the industry sector, contributing 24% of all GHG emissions, requires significant decarbonization in the medium (by 2030) and long term (by 2050).

All countries that ratified the Paris Agreement have announced their targets and measures for GHG emissions reduction by 2030, referred to as Nationally Determined Contributions (NDCs). The first Global Stock Take (GST), scheduled to commence by the end of 2023, will measure progress toward these NDC targets (UNFCCC, 2023). Market mechanisms serve as one of the instruments to achieve NDCs, outlined in Article 6.2, 6.4, and 6.8 of the Paris Agreement. Article 6.2 provides guidance on cooperative approaches between countries, while Article 6.4 focuses on trading mechanisms in the public and private sectors. Article 6.8 establishes a working program, including financing to support climate mitigation/adaptation, technology transfer, and capacity building. Article 6.4 succeeds the Carbon Development Mechanisms, which were market mechanisms under the Kyoto Protocol and were supervised by the UN body known as the Article 6.4 Supervisor Body. Article 6.4 serves both non-market and market purposes, as illustrated in Fig. 1.

J. Jupesta (✉)
United Nations University-Institute for the Advanced Study of Sustainability (UNU-IAS), 5–53–70 Jingumae, Shibuya-ku, Tokyo 150-8925, Japan
e-mail: jjupesta@yahoo.com

A. Dohong
Ministry of Environment and Forestry (MOEF), Jl. Jenderal Gatot Subroto, Gedung Manggala Wanabakti Blok I 2nd Floor, 10270 Jakarta, Republic of Indonesia

B. Sahari
Center for Transdisciplinary and Sustainability Science (CTSS), IPB University, Campus IPB Baranangsiang, Jl. Raya Pajajaran No.27, Bogor 16127, Indonesia

B. He et al. (eds.), *Urban Climate Change Adaptation*, Advances in Science, Technology & Innovation,
https://doi.org/10.1007/978-3-031-65088-8_22

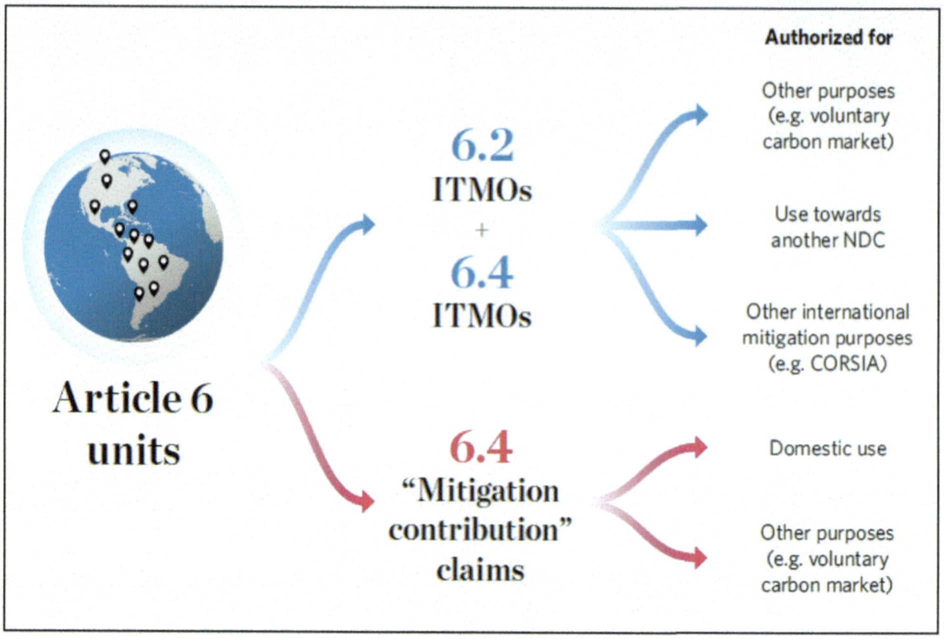

Fig. 1 Article 6 of Paris Agreement (TNC, 2023)

In 2022, a novel unit termed "mitigation contribution" was delineated under Article 6.4 for the first time. These units do not necessitate a corresponding adjustment, are non-authorized, and can be utilized for various purposes, including domestic mitigation pricing schemes, results-based climate finance, or domestic price-based measures, aiming to reduce emission levels in the host party. Additionally, these units can be employed in other markets, such as domestic or voluntary carbon markets. The social cost of carbon signifies the optimal carbon price, serving as an indicator of the cost margin associated with emitting carbon into the atmosphere. According to Tol's study, this cost has escalated from $9 per tCO_2 to $40 per tCO_2 for a high discount rate and from $112 per tCO_2 to $525 per tCO_2 for a low discount rate (Toll, 2023).

By the end of September 2023, the new carbon exchange in Jakarta was launched to expedite carbon trading in Indonesia. This study aims to evaluate the policy impacts of the carbon pricing policy in Indonesia. The Indonesian government aims to achieve Net Zero GHG emissions in the FOLU sector by 2030 and overall, by 2060 (Nurbaya et al., 2023). Due to its extensive natural resources, Indonesia intends to promote Nature-Based Solutions (NBS) as part of its national and global climate mitigation plan, as these solutions offer additional benefits in terms of climate adaptation and biodiversity protection. Incorporating NBS into economic valuation is crucial (World Bank, 2023). The subsequent section will delve into carbon pricing in Indonesia, followed by an analysis of its policy impact. The final part of this study entails an overarching evaluation of the carbon pricing policy concerning the implementation of the Paris Agreement.

2 Carbon Pricing in Indonesia

With a population of 278.1 million people in 2022, Indonesia ranks as the fourth most populous nation globally. The government recognizes the significance of addressing climate change within the national development plan, evident through active participation in the United Nations Framework Convention on Climate Change (UNFCCC) negotiations. Figure 2 illustrates the latest updates on National Determined Contributions (NDCs) from various sectors in Indonesia. The most substantial reduction in greenhouse gas (GHG) emissions is observed in the forestry and other land uses (FOLU) sector, accounting for a 17.4% decrease compared to Business As Usual (BAU), followed by a 12.5% reduction from the energy sector and a 1.4% decrease from the waste sector. Sectors like agriculture and IPPU contribute less than 1% to the total GHG emissions reduction compared to BAU.

The Kyoto Protocol (KP), adopted by the UNFCCC in 1997 and enforced in 2005 [8], was ratified by the Government of Indonesia (GoI) under Law No. 17/2004. Before ratification, Indonesia had prior experience with the Clean Development Mechanism (CDM) market mechanisms of the Kyoto Protocol. The Ministry of Environment served as the focal point for UNFCCC and functioned as the Designated National Authority (DNA) for the CDM. Alongside the CDM, Indonesia initiated a national market mechanism through the National Council on Climate Change (NCCC) in 2008. This voluntary market mechanism, known as the Nusantara Carbon Scheme (NCS), resembles the CDM but primarily operates as a voluntary domestic carbon offset mechanism within

Sector	GHG Emission Level 2010* (MTon CO2-eq)	GHG Emission Level 2030 MTon CO_2-eq			GHG Emission Reduction				Annual Average Growth BAU (2010-2030)	Average Growth 2000-2012
					MTon CO_2eq		% of Total BaU			
		BaU	CM1	CM2	CM1	CM2	CM1	CM2		
1. Energy*	453.2	1,669	1,311	1,223	358	446	12.5%	15.5%	6.7%	4.50%
2. Waste	88	296	256	253	40	43.5	1.4%	1.5%	6.3%	4.00%
3. IPPU	36	69.6	63	61	7	9	0.2%	0.3%	3.4%	0.10%
4. Agriculture	110.5	119.66	110	108	10	12	0.3%	0.4%	0.4%	1.30%
5. Forestry and Other Land Uses (FOLU)**	647	714	214	-15	500	729	17.4%	25.4%	0.5%	2.70%
TOTAL	**1,334**	**2,869**	**1,953**	**1,632**	**915**	**1,240**	**31.89%**	**43.20%**	**3.9%**	**3.20%**

Notes: CM1= Counter Measure 1 (unconditional mitigation scenario)
CM2= Counter Measure 2 (conditional mitigation scenario)

*) Including fugitive.
**) Including emission from estate and timber plantations.

Fig. 2 The National Determined Contributions (NDC) targets for Indonesia (MOEF, 2022)

Indonesia. It employs the Nusantara Carbon Unit (NCU) to denote one verified tCO_2 reduction. Several large state-owned enterprises, such as the National Bank of Indonesia and Garuda Indonesia, have participated in these domestic carbon offsets. Additionally, in 2014, the Ministry of Environment and Ministry of Forestry were merged (Maulidia & Jupesta, 2015).

The Paris Agreement, a legally binding international treaty on climate change succeeding the Kyoto Protocol which ended in 2015, came into force in 2016 (UNFCCC, 2023). The Government of Indonesia ratified the Paris Agreement in 2016 under Law No. 16 of 2016. Article 6 governs market mechanisms, while other financing mechanisms from the Forestry sector, such as Reducing Emissions from Deforestation and Land Degradation in Developing Countries (REDD+), are regulated under Article 5.2 of the Paris Agreement. In Indonesia, there exist four schemes for carbon pricing, comprising both trade and non-trade instruments. The carbon trading mechanisms encompass emissions trading schemes and crediting mechanisms, while non-trade instruments include carbon tax and result-based payments (RBP). As of 2023, the Indonesian government has announced a limited basis carbon tax specifically targeting coal power plants at a rate of 30,000 rupiah for one tCO_2eq. (exchange rate: 1 USD = 15,580 Rupiah). Result-Based Payment (RBP) constitutes payment for successful efforts to reduce carbon emissions in the REDD+ program. Indonesia has received over $100 million for REDD+ results for 2014–2016, equating to a total volume of 27 million tCO_2eq. submitted to the Green Climate Fund (GCF) for results-based payments (RBP) (GCF, 2023).

3 Policy Impacts

The carbon pricing policy has three key impacts. Firstly, the FOLU sector exhibits substantial potential for trade, either through Emissions Trading Schemes (ETS) or Crediting Mechanisms, owing to Indonesia's extensive forest area, estimated at 92.3 million hectares in 2006 based on the FREL submitted to UNFCCC in 2022 (Boer et al., 2022). The IPCC further emphasizes the significance of carbon dioxide removal, which encompasses natural-based solutions in climate change mitigation. Enhancing nature-based solutions for Carbon Dioxide Removal (CDR) to integrate them into the carbon market involves several pathways, including afforestation/reforestation, biochar, soil carbon sequestration, peatland rehabilitation, and mangrove preservation. Nature-based solutions offer multiple benefits not only in climate change mitigation but also in biodiversity preservation and ecosystem services. These solutions contribute to climate mitigation by regulating temperatures and serving as barriers against coastal erosion, as seen with mangroves in coastal areas. Forest ecosystems also sustain local and indigenous communities by providing livelihoods, biodiversity support, and ecosystem services. Successful implementation of Nature-Based Solutions requires enabling conditions and international support.

With careful implementation to ensure value and respect for systems, nature-based solutions present an opportunity to hasten this transition while mitigating global warming, bolstering climate resilience, and safeguarding biodiversity (Seddon, 2022). The discussion on carbon dioxide removal, including NBS, is underway in Article 6.4 by the

Supervisory Board of UNFCCC. Apart from NBS, the continuation of result-based payments for the REDD+ program will be deliberated at the 28th COP of UNFCCC scheduled for December 2023.

Another significant policy impact of the carbon pricing policy lies in promoting the deployment of renewable and clean energy, such as hydrogen. Given Indonesia's vast potential in renewable energy sources—namely solar, hydro, bioenergy, wind, geothermal, and ocean energy—the carbon pricing framework in the energy sector is envisioned to serve as a catalyst for accelerating the transition toward clean energy. Moreover, the Just Energy Transition Partnership (JETP), a multinational initiative spearheaded by the United States and Japan, pledges to mobilize both public and private finances to support Indonesia in phasing out coal-powered plants, adopting greener power sources like renewable-based hydrogen, carbon capture, utilization, and storage (CCUS), and advancing the sector's emissions reduction target by seven years to 2030 (Munthe & Nangoy, 2023). While the current carbon pricing doesn't explicitly outline clean energy technologies such as CCUS and renewable-based hydrogen, it's anticipated that these may be considered under the Paris Agreement's implementation, specifically within the realms of Article 6.2 and 6.4.

The third significant impact lies in waste management, particularly addressing plastic waste, which has become an urgent priority in accelerating climate mitigation efforts. Indonesia faces increasing global concerns regarding plastic pollution and ranks as the second-largest ocean plastic polluter. Estimates suggest between 0.27 to 0.59 million tonnes of plastic waste are discharged into Indonesian oceans. Currently, Indonesia's waste recycling rate, heavily reliant on the informal sector, captures less than 5% of the generated waste, with a meager plastic recycling rate of only 7% (MOEF, 2020). There exists a significant opportunity to fortify the recycling industry in Indonesia to effectively reduce and eliminate plastic pollution on land and in neighboring waters. Implementing a carbon tax for plastic producers could incentivize the adoption of clean technologies for plastic recycling, such as chemical recycling and carbon capture and utilization (CCU).

Moreover, air pollution presents a serious concern in major cities like Jakarta, surrounded by several coal-fired power plants (CFPPs). Various sources of pollution, notably CFPPs in proximity to Jakarta, significantly contribute to heightened pollution levels in the city. The Air Pollutant Index (API) has reached 121, indicating PM2.5 concentrations 8.7 times higher than the WHO's annual air quality guideline (IQAir, 2023). Emissions from coal-fired power plants (CFPPs) in Indonesia contribute significantly to health-related costs, estimated at USD 7.4 billion (Hasjanah, 2023). While the current carbon tax rate for CFPPs isn't effectively curbing air pollution, non-pricing measures such as environmental permits could encourage electricity producers to transition to cleaner energy technologies like CCUS or shift partially to renewable energy-based power sources.

4 Overall Evaluation

Indonesia's carbon pricing framework rests on two foundational pillars: mitigating climate change to alleviate the impact of global warming and expediting the transition toward sustainable development goals. The recent launch of the carbon exchange is anticipated to grow with broader participation from market actors. Carbon Dioxide Removal (CDR) emerges as a key element in global mitigation policies, offering more comprehensive benefits compared to current greenhouse gas (GHG) reduction and avoidance strategies. Indonesia can navigate the carbon markets by leveraging Nature-Based Solutions (NBS) and clean energy initiatives, including CCUS and renewable energy sources. However, potential trade-offs, such as those between food and energy security in biofuel production, need resolution. Addressing biofuel production challenges should prioritize technological innovation over land expansion, which risks deforestation and impedes achieving net-zero GHG emissions (Jupesta et al., 2012).

Considering the escalating GHG emissions from the waste sector, imposing a carbon tax can accelerate the adoption of cleaner production systems in waste management, such as chemical recycling and CCU. While the current carbon tax rate for coal-fired power plants (CFPPs) might not effectively curb air pollution, non-pricing measures like environmental permits could incentivize electricity producers to adopt cleaner energy technologies or transition partially to renewable energy-based power sources. Voluntary Carbon Market (VCM) mechanisms via crediting in both FOLU and the energy sector show significant potential due to increasing awareness of carbon trading's importance in climate mitigation. Strengthening credibility, transparency, and extensive capacity building are critical factors for improving carbon pricing mechanisms, aligning with Article 6.4, and ensuring effective Paris Agreement implementation toward NDC targets by 2030.

Acknowledgements This study does not reflect the view of the Ministry of Environment and Forestry Republic Indonesia.

References

Shukla, R., Skea, J., Reisinger, A., Slade, R., Fradera, R., Pathak, M., Al Khourdajie, A., Belkacemi, M., van Diemen, R., Hasija, A., Lisboa, G., Luz, S., Malley, J., McCollum, D., Some, S., & Vyas, P. (2022). Summary for Policymakers. In *Climate Change 2022:*

Mitigation of Climate Change. Contribution of Working Group III to the Sixth Assessment Report of the Intergovernmental Panel on Climate Change, Cambridge University Press, Cambridge, UK and New York, NY, USA.

UNFCCC. (2023). Technical dialogue of the first global stock take. Synthesis report by the co-facilitators on the technical dialogue. Bonn: UNFCCCC. https://unfccc.int/documents/631600. Accessed on November 15th 2023.

The Nature Conservancy (TNC). (2023). Article 6 Explainer. https://nature4climate.org/natures-solutions/latest-scientific-papers/article-6-explainer/. Accessed on November 25th 2023.

Toll, R. S. J. (2023). Social cost of carbon estimates have increased over time. *Nature Climate Change, 13*, 532–536.

Nurbaya, S., Justianto, A., Murniningtyas, S., Soegiri, E.W., Muttaqin, M.Z., Budiman, S., Salminah, M. (2023) FOLU Net Zink 2030: Indonesia Climate Actions Towards 2030, Jakarta: Ministry of Environment and Forestry of the Republic of Indonesia

World Bank. (2023). State and Trends of Carbon Pricing 2023, Washington DC: World Bank. https://openknowledge.worldbank.org/entities/publication/58f2a409-9bb7-4ee6-899d-be47835c838f. Accessed on October 25th 2023.

Ministry of Environment and Forestry (MOEF). (2022). Enhanced NDC: Submission to UNFCCC, Ministry of Environment and Forestry Republic Indonesia.

Maulidia, M., & Jupesta, J. (2015). Economic instruments to combat climate change in Asian Countries. In *Economic Instruments to Combat Climate Change in Asian Countries*, K. Matsumoto and A. M. Gao (eds.) (pp. 127–152). Alphen aan den Rijn: Wolters Kluwer.

Green Climate Fund (GCF). (2023). https://www.greenclimate.fund/project/fp130. Accessed on October 2 2023.

Boer, R., Anwar, S., & Margono, B. (eds). (2022). *Forest Reference Emissions Level (FREL), submission to UNFCCC*. Ministry of Environment and Forestry. https://redd.unfccc.int/media/2nd_frl_indonesia_final_submit.pdf. Accessed on November 15 2023.

Seddon, N. (2022). Harnessing the potential of nature-based solutions for mitigating and adapting to climate change. *Science, 376*, 1410–1416.

Munthe, B. C., & Nangoy, F. (2023, October 6). Indonesia delays plan to invest $20 bln in energy transition funds. *Reuters*. Retrieved from https://www.reuters.com/sustainability/climate-energy/indonesia-delays-plan-invest-20-bln-energy-transition-funds-2023-08-16/

Ministry of Environment and Forestry (MOEF). (2020). *National Plastic Waste Reduction Strategic Actions for Indonesia*. Ministry of Environment and forestry Republic of Indonesia.

IQAir. (2023, October 8). Jakarta air quality map. *IQAir website*. Retrieved from https://www.iqair.com/air-quality-map/indonesia/jakarta

Hasjanah, K (2023, August 21). Handling Air Pollution in Jakarta: The Role of Coal Power Plants Intervention Towards Early Retirement. *IESR website*. Retrieved from https://iesr.or.id/en/handling-air-pollution-in-jakarta-the-role-of-coal-power-plants-intervention-towards-early-retirement

Jupesta, J., Boer, R., Parayil, G., Harayama, Y., Yarime, M., de Oliveira, J. A. P., & Subramanian, S. M. (2012). Managing the transition to sustainability in an emerging economy: Evaluating green growth policies in Indonesia. *Environmental Innovation and Societal Transitions, 1*, 187–191.

GPSR Compliance

The European Union's (EU) General Product Safety Regulation (GPSR) is a set of rules that requires consumer products to be safe and our obligations to ensure this.

If you have any concerns about our products, you can contact us on ProductSafety@springernature.com

In case Publisher is established outside the EU, the EU authorized representative is:

Springer Nature Customer Service Center GmbH
Europaplatz 3
69115 Heidelberg, Germany

The manufacturer's authorised representative in the EU is Springer Nature Customer Service Centre GmbH, Europaplatz 3, 69115 Heidelberg, Germany. If you have any concerns regarding our products, please contact ProductSafety@springernature.com

Printed and bound by CPI Group (UK) Ltd, Croydon, CR0 4YY
01/10/2026
02225993-0002